Eric D. Wills

Biochemical Basis of Medicine

Eric D. Wills

Professor of Biochemistry in the University of London
at The Medical College of St Bartholomew's Hospital

Biochemical Basis of Medicine

1985 **WRIGHT** Bristol

Published by   John Wright & Sons Ltd, 823–825 Bath Road,
Bristol BS4 5NU, England.

British Library Cataloguing in Publication Data

Wills, E. D.
    Biochemical basis of medicine.
    1. Biological chemistry
    I. Title
    574.19′2′02461   QP514.2

ISBN 0 7236 0722 2

Typeset by   BC Typesetting
51 School Road, Bristol BS15 6PJ

Printed in   The Pitman Press
Great Britain by   Bath

Preface

During the past 50 years, biochemical principles, concepts and technology have become increasingly important for the understanding of disease, in clinical treatment and in medical research. It is therefore essential that the medical course should include an extensive groundwork of academic biochemistry and demonstrations of the application of biochemistry to medicine.

Although many excellent textbooks of basic academic biochemistry have been published during the past two decades, the author, after many years of teaching biochemistry in the basic medical sciences course, has come to support the view that a radical new approach to the subject is required in the medical course. This is because, as taught in conventional medical courses, biochemistry has tended to become an essentially academic discipline, divorced from clinical medicine. It is thus difficult for students to appreciate the great importance and significance of the applications of the subject in medicine.

Furthermore, during their basic medical and science courses, it is common experience that students often find difficulty in relating biochemistry to other disciplines such as anatomy, physiology and pharmacology—for example, in applying the biochemical concepts to the micro- and subcellular structures of the cell and, more important, to the functioning of the body as a whole, and thus the application of these concepts to disease processes.

This book is written with the object of overcoming these problems and presents the subject, in five Parts, from a new viewpoint. In the first Part, the biochemistry of the subcellular organelles is described in detail and the second Part deals with the biochemistry of the body as a whole. The third Part includes descriptions of specialized metabolism which occurs in many of the tissues of the body, whilst the fourth and fifth Parts are devoted to applications of biochemistry—the fourth to environmental hazards and the fifth to some biochemical aspects of diagnosis and treatment.

I have adopted an approach to the subject that involves the discussion of structural detail and metabolic processes in the context in which they occur in the cell or in the tissues. This has necessitated omitting detailed systematic descriptions of much basic material such as carbohydrate, lipid and protein structure, enzymology, basic metabolic processes such as glycolysis, citrate cycle, fatty acid oxidation, amino acid metabolism and protein synthesis, which subjects form the major proportion of current biochemical teaching and several biochemical texts. This decision is, I believe, quite justified because all these topics have been described repeatedly and so elegantly in many other text books that their inclusion is considered unnecessary in this book. A good knowledge of this basic material is, however, really essential to a full understanding of this book and students and their teachers are advised to consider how this may best be achieved before beginning the study of this book. However, in order to avoid the necessity of a second text being always available, basic biochemical information is summarized in a series of appendices for refreshment of memory of, for example, chemical structures or a metabolic pathway relevant to any particular chapter.

In addition to providing guidance for students, it is hoped that the approach adopted in the book may also be useful for teachers who wish to give serious thought to the type of basic biochemical syllabus that is desirable for a full understanding of the book and thus, also, for an understanding of the role of biochemical principles which are valuable in medicine. Although the book is written primarily with medical students in mind, it is hoped that many in the medical profession who wish to refresh their knowledge could find it useful and that it will be valuable for science students who wish to become aquainted with the applications of biochemistry to medical science.

Several have helped in the production of this book and the author would like to acknowledge the great help he has received from many associates and colleagues: to John Gillman who encouraged me to write the book and for much helpful advice; to the colleagues in my Department and especially Dr J. D. Hawkins, Dr D. M. G. Armstrong and Dr K. Brocklehurst who helped me by providing valuable information and for critical comments; to the many who have helped with the typing including my wife, Caroline Reddick, Alison Dowler and Sara Taylor; to Ray French for the preparation of many of the illustrations and especially to Jane Sugarman for her invaluable assistance as a subeditor.

Eric D. Wills

Acknowledgements

I would like to acknowledge the source of the following illustrations and tables:

Fig. 1.2: From C. Clarke and E. D. Wills (1980) *International Journal of Radiation Biology*, Vol. 38, pp. 21–30, by kind permission of the Publishers.

Fig. 1.3: From F. P. Altman (1979) *Progress in Histochemistry and Cytochemistry*, Vol. 9, No. 3, p. 40, by kind permission of the Author and Publishers, Gustav Springer-Verlag, Stuttgart.

Fig. 1.4: From M. T. Smith, N. Loveridge, E. D. Wills and J. Chayem (1979) *Biochemical Journal*, Vol. 182, p. 103, by kind permission of the Publishers.

Fig. 1.5: From C. Clarke and E. D. Wills (1978) *Journal of Steroid Biochemistry*, Vol. 9, pp. 135–139, by kind permission of the Publishers.

Figs. 1.6, 1,10, 6.2, 6.4, 31.1, 31.2: From K. E. Carr and P. G. Toner (1982) *Cell Structure*, by kind permission of the Authors and Publishers, Churchill Livingstone, Edinburgh.

Figs. 1.7, 1.8, 1.9, 1.11, 9.2: From R. V. Krstic (1979) *Ultrastructure of the Mammalian Cell*, by kind permission of the Author and Publishers, Springer-Verlag, Heidelberg.

Figs. 2.2, 2.3, 2.4, 6.6: Redrawn from W. C. McMurray (1973) in *Form and Function of Phospholipids*, eds G. B. Ansell, J. N. Hawthorne and R. M. C. Dawson, London, Elsevier.

Figs. 2.6, 2.7, 2.12: Redrawn from G. Weissman and R. Claiborne (1979) *Cell Membrane*, New York, HP Publishing.

Fig. 2.11: From G. Weissman and R. Claiborne (1979) *Cell Membrane*, by kind permission of the Artist and the Publishers, HP Publishing, New York.

Fig. 2.13: From Meyer-Overton (1902), quoted by R. Hober (1946) *Physical Chemistry of Cells and Tissues*, Edinburgh, Churchill Livingstone.

Table 2.1: From J. A. Lucy (1974) in *Cell in Medical Science*, eds F. Beck and J. B. Lloyd, Vol. I, London, Academic Press.

Fig. 3.2: Redrawn from W. E. Bowers, T. J. Finkenstaedt and C. De Duve (1966) *Journal of Cell Biology*, Vol. 32, pp. 325–337.

Fig. 3.4: From A. A. Aikman and E. D. Wills (1974) *Radiation Research*, Vol. 57, pp. 403–415, by kind permission of the Publishers.

Fig. 6.3: From A. H. Lehninger (1971) *The Mitochondrion*, by kind permission of the Author and the Publishers, The Benjamin/Cummings Publishing Company, New York.

Fig. 6.5: Reproduced from H. Baum (1974) in *Cell in Medical Science*, eds F. Beck and J. B. Lloyd, London, Academic Press.

Figs. 8.1, 8.2: Reproduced from F. Beck and J. B. Lloyd (eds) (1974) in *Cell in Medical Science*, Vol. I, London, Academic Press.

Figs. 9.1, 9.3: From T. S. Leeson and C. R. Leeson (1981) *Histology*, by kind permission of the Authors and Publishers, Holt-Saunders, London.

Figs. 9.7, 36.18: From S. C. Davison and E. D. Wills (1974) *Biochemical Journal*, Vol. 140, pp. 461–468.

Fig. 9.8: From E. Grzelinska, G. Bartosz, K. Gwozdzinski and W. Leyko (1979) *International Journal of Radiation Biology*, Vol. 36, pp. 325–334.

Fig. 9.9: From E. D. Wills (1971) *Biochemical Journal*, Vol. 123, pp. 983–991.

Fig. 9.11: From E. F. Gale, E. Cuncliffe, P. E. Reynolds, M. H. Richmond and M. J. Warring (1981) *The Molecular Basis of Antibiotic Action*, by kind permission of the Authors and Publishers, John Wiley, London.

Fig. 10.5: From J. J. Groen (1958) in *Essential Fatty Acids*, ed. R. G. Sinclair, London, Butterworths.

Fig. 11.3: From Dorothy Hollingsworth (1974) *British Nutrition Foundation Bulletin*, No. 12, by kind permission.

Tables 11.2, 11.3, 11.4, 11.5: From R. W. Swift and K. H. Fisher (1964) in *Nutrition*, ed. G. H. Beaton, Vol. 1, New York, Academic Press.

Table 11.6: From J. D. Marnack (1943) *Food and Food Planning*, London, Victor Gollancz.

Fig. 12.5: From R. G.Whitehead and P. G. Lund (1979) *Proceedings of the Nutrition Society*, Vol. 38, p. 69.

Fig. 13.9: From R. Reisert (1978) *American Journal of Clinical Nutrition*, Vol. 31, pp. 865–875.

Fig. 13.10: From A. Key (1970) *American Heart Association Monograph* No. 29.

Fig. 13.12: From P. N. Darrington, C. H. Bolton, M. Hartog et al. (1977) *Atherosclerosis*, Vol. 27, pp. 415–475.

Fig. 14.1: From F. G. Hopkins (1912), quoted in L. J. Harris (1938) *Vitamins and Vitamin Deficiencies*. Vol. 1. Blackiston, Churchill.

Fig. 14.2: From L. J. Harris (1938) *Vitamins and Vitamin Deficiencies*. Vol. 1. Blackiston, Churchill.

Fig. 14.12: From J. E. Dowling and G. Wald (1960) *Vitamins and Hormones*, Vol. 18, pp. 515–541, by kind permission.

Table 16.5: From S. E. Nixon and G. E. Mawer (1970) *British Journal of Nutrition*, Vol. 24, p. 241.

Fig. 16.11: From R. K. Crane (1965) *Federation Proceedings*, Vol. 24, pp. 1000–1006.

Fig. 20.8: From A. E. Harper, N. J. Benevengen and C. M. Wohlheuter (1970) *Physiological Reviews*, Vol. 50, p. 428.

Figs. 22.3, 22.4: From D. H. Marshall (1976) *Proceedings of the Nutrition Society*, Vol. 35, pp. 163–173.

Fig. 22.5: From G. W. Dolphin and I. S. Eue (1963) *Physics in Medicine and Biology*, Vol. 8(2), pp. 197–207.

Fig. 22.7: From F. Bicknell and F. Prescott (1946) *Vitamins and Medicine*, 2nd edn, by kind permission of the Publishers, William Heinemann Medical Books, London.

Fig. 22.20: From R. Spencer, M. Charman, P. M. Wilson et al. (1978) *Biochemical Journal*, Vol. 170, p. 93.

Figs. 23.4, 23.5: Adapted from G. F. Cahill (1970) *New England Journal of Medicine*, Vol. 282, pp. 668–675.

Fig. 24.1: From L. E. H. Whitby and C. J. C. Britton (1957) *Disorders of the Blood*, by kind permission of the Authors and Publishers, Churchill Livingstone, London.

Fig. 26.1: From N. Crawford and D. G. Taylor (1977) *British Medical Bulletin*, Vol. 33, pp. 199–206.

Figs. 26.20, 26.21, 26.22: From M. Levitan and A. Montague (1971) *Textbook of Human Genetics*, London, Oxford University Press.

Table 28.1: From J. D. Cook, C. A. Finch and N. J. Smith (1976) *Blood*, Vol. 48, pp. 449–455.

Fig. 29.1: (*a*) From Ch. Rouiller (1963) *The Liver*, New York, Academic Press; (*b*) from H. Elias and J. E. Pauly (1960) *Human Microanatomy*, by kind permission of the Authors and Publishers, DaVinci, Chicago.

Figs. 30.2, 30.4: Adapted from R. F. Pitts (1974) *Physiology of the Kidney and Body Fluids*, 3rd edn, Chicago, Year Book Medical Publishers.

Fig. 30.13: Adapted from A. Meister (1973) *Science*, Vol. 150, p. 3.

Figs. 31.3, 31.6, 31.7, 31.9, 31.11, 31.14: From J. M. Murray and A. Weber (1974) *Scientific American*, Vol. 230, pp. 58–71.

Fig. 31.10: From D. Mornet, R. Bertrand, P. Pantel et al. (1981) *Nature*, Vol. 292, pp. 301–306.

Fig. 31.15: From J. Squire (1981) *Nature*, Vol. 291, pp. 614–615.

Figs. 32.1, 32.2, 32.4, 32.13, 32.14: Reproduced from F. G. E. Pautard (1978) in *New Trends in Bio-inorganic Chemistry,* eds R. J. P. Williams and J. R. R. F. Da Silva, by kind permission of the Author and Publishers, Academic Press, London.

Figs. 32.3, 32.5, 32.15, 32.16: Reproduced from K. Simkiss (1975) *Bone and Biomineralisation*, by kind permission of the Authors and Publishers, Edward Arnold, London.

Figs. 32.10, 32.11: From D. J. Prockop, K. I. Kivirikko, L. Tuderman and N. A. Guzman (1979) *New England Journal of Medicine*, Vol. 301.

Fig. 33.4: From N. F. Anden, A. Pahlstrom, K. Fuye and K. Larsson (1966) *Acta Physiologica Scandinavica*, Vol. 67, p. 313, by kind permission of the Authors and Publishers.

Figs. 33.7, 33.10, 33.11, 33.12: *From Basic Neurochemistry,* 3rd edn (1981) eds G. J. Siegel, R. W. Albers, B. W. Agranoff and R. Kalzman, by kind permission of the Editors and Publishers, Little, Brown and Co., Boston.

Fig. 36.16: Redrawn from J. C. A. Knott and E. D. Wills (1974) *Biochemical Pharmacology*, Vol. 23, pp. 793–800.

Figs. 37.1, 37.2, 37.4: Redrawn from L. L. Madison (1968) *Advances in Metabolic Disorders*, Vol. 3, pp. 85–109.

Fig. 38.15: From J. F. Waterfall and P. Sims (1972) *Biochemical Journal*, Vol. 128, p. 265; P. Sims and P. L. Grover (1974) *Advances in Cancer Research*, Vol. 120, p. 165; J. Kapitulnik, W. Levin, A. H. Corney et al. (1977) *Nature*, Vol. 266, p. 378.

Figs. 39.1–39.11, 39.13, 39.14: From E. Schmidt and F. W. Schmidt (1967) *Guide to Practical Enzyme Diagnosis*, Mannheim, C. F. Boehringer and Soehne, with permission.

Fig. 39.12: From E. Schmidt and F. W. Schmidt (1976) *Brief Guide to Practical Enzyme Diagnosis*, Mannheim, C. F. Boehringer and Soehne, with permission.

Fig. 40.2: From D. W. A. Donald (1973) in *Anorexia and Obesity*, ed. R. S. S. Robertson. Edinburgh, RCP, pp. 63–70.

Fig. 40.15: From W. P. J. James and P. Trayhurn (1976) *Lancet*, Vol. ii, p. 770.

Figs. 42.7, 42.10, 42.13, 42.14, 42.21, 42.28: From I. Roitt (1977) *Essential Immunology*, by kind permission of the Author and Publishers, Blackwell Scientific, Oxford.

Fig. 42.20: From L. Stryer (1981) *Biochemistry,* San Francisco, W. H. Freeman.

Figs. 42.15, 42.16, 42.17, 42.18, 42.19, 42.24, 42.27: From A. T. Cunningham (1978) *Understanding Immunology*, by kind permission of the Author and Publishers, Academic Press, New York.

Contents

Part 2
Whole body metabolism

Part 3
Specialized metabolism of tissues

Part 4
Environmental hazards—detoxication

Part 5
Biochemical basis of diagnosis—disease and its treatment

Appendices

Part 1

Biochemistry of the cell and its metabolism

1

Biochemistry of the cell
and its metabolism

Chapter 1   Ultrastructure of the mammalian cell

1.1 Introduction

One of the fundamental and most important advances in biology was the discovery by Schleiden and Schwann, in 1838, that all animals and plants were composed of small units, the cells. Many of these appeared identical and tissues of animals appeared to be constructed of very large numbers of these small unit cells in a fashion similar to the utilization of bricks in construction of a building.

Since that time, these views have been modified in several ways. As fixed sections were originally observed, it was concluded that all cells were essentially rigid structures, but modern photographic techniques have shown that many cells are in a constantly dynamic state, flexible and often changing their shape. It is also clear that all tissues contain different types of cells, which may be distinguished by different structures within the cells and sometimes even by different metabolic activities.

Cells also possess properties which enable them to recognize cells of identical type and to distinguish foreign cells or those of another tissue. This property, which is probably dependent on the membrane glycoproteins (cf. Chapter 2), is clearly of great importance during embryonic development and during tissue growth and repair.

1.2 Methods of studying cell structure and function

Histology
This technique has been used from the very early days for cell investigation and it was observed that certain cells or components of cells, such as the nuclei, reacted strongly with some dyes whereas other dyes were less specific and stained the cytoplasm relatively uniformly.

Many hundreds of different dyes of varying degrees of usefulness have been discovered, through the years, for the study of cells or tissue sections and several of these dyes have proved extremely valuable for the study of cell structure. Unfortunately many of these staining reactions are only partially understood in precise chemical terms and thus the subject has tended to remain somewhat empirical although, nevertheless, a great deal of useful information has been obtained by the method. It has also been extremely valuable for the study of abnormal cells and is the basis of the science of 'histopathology'. Many diseases are diagnosed by this method, but for full details the reader should consult specialized texts.

Electron microscopy
The development of electron microscopy, started in the 1930s, enabled the very fine detail of cells to be studied. Specimens for electron microscopy have to be fixed, dehydrated and stained with osmium tetroxide for examination *in vacuo*. It must, therefore, be appreciated that electron microscopy images are to some extent artefactual, but nevertheless they have provided invaluable information about the ultrastructures of cells, enabling very fine detail of organelles such as the mitochondria to be studied.

Special electron microscopy techniques enable inorganic constituents of

3

cells to be measured but, in general, electron microscopy provides qualitative, rather than quantitative, information about cell structure and function.

Ultracentrifugation

Biochemical studies of the metabolic activities of the organelles demonstrated by electron microscopy could not begin until methods of separation of these organelles had been developed. This was first achieved by Schneider, Hogeboom and Pallade in the USA in the late 1940s. They showed that the organelles of a homogenate of a tissue such as rat liver could be separated by a procedure known as 'differential centrifugation'. The tissue is suspended in 0·25 M sucrose, the cells disrupted by means of a Potter homogenizer and then subjected to high-speed centrifugation. Separation is achieved as a result of differences in size and density and, to some extent, shape of the subcellular organelles. Centrifugal forces of sufficient magnitude and duration are used to produce almost complete separation of the organelles.

Density gradient centrifugation may also be employed. In this procedure the homogenate is layered on top of a continuous or discontinuous density gradient of sucrose solution and centrifugation is continued until the subcellular particles are in density equilibrium with the surrounding medium. The procedure for differential centrifugation, the more common method, is shown in *Fig*. 1.1.

Nuclei, being the heaviest particles, sediment readily, but they are not generally pure and are contaminated with whole cells and cell debris, so special methods have to be used to purify the nuclei (cf. Chapter 8).

The second fraction spun down after 10–20 min at approximately $10\,000 \times g$ contains most of the mitochondria, but this fraction is usually contaminated with the lysosomes. Special methods using density gradient centrifugation, must be used to separate the mitochondria and lysosomes although this combined fraction is often used for mitochondrial or lysosomal studies. This is possible because the two subcellular compartments contain quite different enzymes

and are involved in different metabolic processes. If the supernatant from the mitochondrial centrifugation is spun for 1 hour at $100\,000 \times g$, a fraction is deposited which is described as 'the microsomes' and contains the fragmented membranes of the endoplasmic reticulum and the ribosomes.

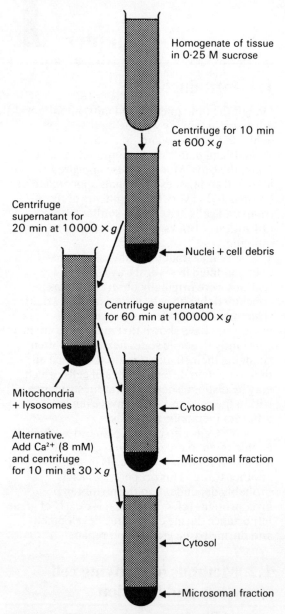

Homogenate of tissue in 0·25 M sucrose

Centrifuge for 10 min at $600 \times g$

Centrifuge supernatant for 20 min at $10000 \times g$

Nuclei + cell debris

Centrifuge supernatant for 60 min at $100\,000 \times g$

Mitochondria + lysosomes

Alternative. Add Ca^{2+} (8 mM) and centrifuge for 10 min at $30 \times g$

Cytosol

Microsomal fraction

Cytosol

Microsomal fraction

Fig. 1.1.   **Separation of subcellular organelles by centrifugation.**

The supernatant remaining after sedimentation of the microsomes is the cytoplasm or 'cell sap' and this contains most of the soluble protein and other molecules of the cell. These subcellular fractions have been the subject of intensive study for many years and they have provided detailed knowledge of the metabolic processes occurring in each subcellular organelle.

Problems arise, however, since without further purification, each fraction is contaminated with one of the others, e.g. mitochondria with lysosomes. Furthermore, measurements of enzyme activities of homogenates give a mean value of the enzymic activity of the tissue organelle, such as the mitochondria or lysosome, throughout the whole tissue. Due to the rather crude process of homogenization, distinction between different types of cells in any one tissue is impossible. Distinction is only possible using the more recent sophisticated technique of quantitative cytochemistry described in the next section.

Histochemistry and quantitative cytochemistry

About 30 years ago, as an offshoot of traditional histology, several investigators attempted to demonstrate metabolic activities in tissue sections by incubation with selected substrates. These substrates were synthesized or selected to produce a coloured or opaque product visible microscopically. The process often takes place in two stages, e.g.

$$A{-}B \xrightarrow[\textcircled{E}]{\text{Hydrolysis}} A{-}H + B{-}OH$$

Substrate

$$\downarrow C$$

$$B{-}OH$$

$$\backslash C$$

Complex

where C = coupling agent.

In the example shown, the tissue section containing the enzyme $\textcircled{E}$ is incubated for a specific time with the substrate A—B. Hydrolysis to A—H and B—OH is followed by the addition of a coupling agent C, which couples with B—OH in the example to form a coloured complex B(C)—OH which is shown clearly on microscopal examination. Using this technique, it is possible to localize special cell types and even special subcellular components which possess the enzyme $\textcircled{E}$ activity. Examples are shown in *Figs*. 1.2 and 1.3. For some enzymes it is possible to examine cells by electron microscopy in order to localize subcellular enzyme activity.

The extension of histochemistry to the single cell level is called 'cytochemistry' and as the measurement of specific substances, reactive groups, and enzyme-catalysed activities in single cells, usually located within a specific region of a tissue section, is called 'quantitative cytochemistry'. Basically, the procedures used in quantitative cytochemistry are very similar to those used in biochemistry, the most important difference being that, in quantitative cytochemistry, the coloured reaction product is precipitated within active cells, whilst in biochemical assays it is produced as a soluble product in a spectrophotometer cuvette. Quantitative cytochemistry has two major advantages over conventional biochemistry. Firstly, the measurements can be directly related to tissue histology. Thus, with tissues containing a variety of cell types, the biochemical functions of a single cell type can be studied with the cells maintained in a normal structural, and presumably functional, relationship with the other cell populations. Alternatively, cytochemistry permits the detection of biochemical differences between cells such as hepatocytes which might otherwise appear to be a single cell population. The second advantage of cytochemistry is that it is a non-disruptive technique. Thus the biochemical function of subcellular organelles and their membranes can be studied without isolating the organelles into a foreign medium, a procedure which, in itself, will almost certainly affect the reactions under study.

There are, however, three fundamental problems in quantitative cytochemistry. The first involves the preparation of sections free of measurable artefacts. In order to cut thin tissue sections of less than 20 μm thickness, so that

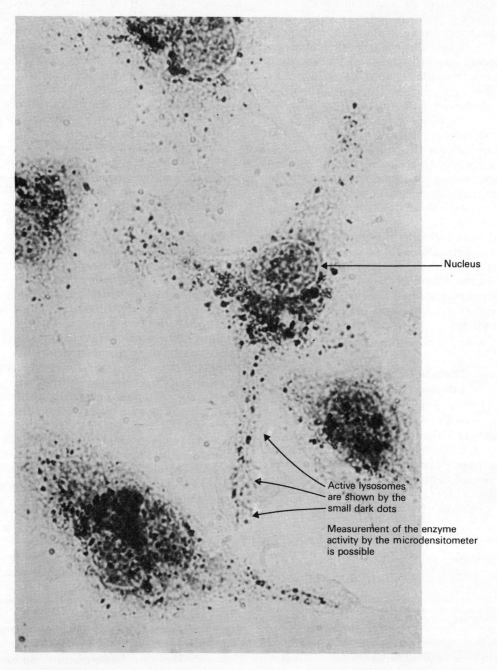

Fig. 1.2.   **The action of lysosomal acid phosphatase demonstrated by cytochemical methods in a cultured macrophage cell.**

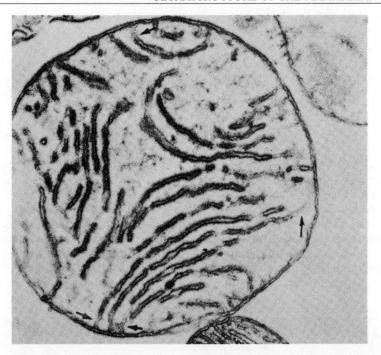

Fig. 1.3.   **Succinate dehydrogenase activity demonstrated cytochemically in heart mitochondria.**
The enzyme reduces a tetrazolium salt to an insoluble coloured formazan which can be seen and measured by light microscopy and using the microdensitometer. For electron microscopy, the formazan must be treated with osmium tetroxide. Magnification ×50 000.

biochemical activity can be directly related to detailed histology, the tissue must first be hardened. This can either be achieved by fixing the tissue and embedding it in paraffin wax or by freezing the tissue. It is well known that fixation causes an inhibition of enzymes, damage to the membranes of the subcellular organelles, and may also cause a shrinkage in the tissue volume. On the other hand, fresh frozen tissue sections, theoretically ideally suited for quantitative cytochemical studies, can be damaged due to the formation of ice during the freezing procedure, which causes damage both by its mechanical and by its attendant physicochemical effects. Frozen tissue sections must therefore be prepared which are free of ice artefacts. The second fundamental problem concerned the retention of 'soluble' enzymes and metabolites in a frozen section during the cytochemical

reaction. Frozen sections immersed in a buffer solution at pH 7–8 lose up to 60 per cent of their total nitrogen content in a matter of a few minutes, including all their 'soluble' enzymes and metabolites. The third major problem is that of measurement. Even when the enzyme remains active and is not released into the medium, the only form of measurement originally available was a semiquantitative visual assessment of activity based on the +, ++, and +++ form of assessment. In order to cut thin tissue sections of less than 20 μm thickness, the tissue must first be hardened. This is most conveniently achieved by supercooling the tissue to $-70\,°C$. It can then be sectioned, under certain controlled conditions without producing any apparent signs of ice artefact.

Two approaches have been developed in an attempt to overcome the problem of tissue disruption during incubation of the sections. These are (*a*) chemical fixation prior to incubation and (*b*) the use of inert stabilizing materials. Chemical fixation of the sections prior to incubation prevents much of the disintegration that occurs when unfixed frozen sections are incubated in an aqueous medium, but most fixation procedures are unacceptable because they cause enzyme inhibition and may also damage the membranes of subcellular organelles. The second approach is to use chemically inert cellular stabilizers which do not damage enzymes or membranes and is, therefore, of much greater practical use. The most widely accepted procedure for measuring precisely the total amount of a chromophore precipitated in an optically inhomogeneous manner is by means of a scanning and integrating microdensitometer. The basis of microdensitometric measurements, just as for spectrophotometry, is the Beer–Lambert law which relates concentration (*c*) of the chromophore to the measured absorbance or extinction (ϵ).

However, in microdensitometry the function to be measured is not normally the concentration of the coloured reaction product, but its mass per cell or per unit area of a section. In scanning and integrating microdensitometry, the area (*A*) to be measured is broken up optically (by means of a scanning device) into a large number of regions, each of area *a* and the absorption of each such region is measured separately. The results from all these regions (of area *a*) are

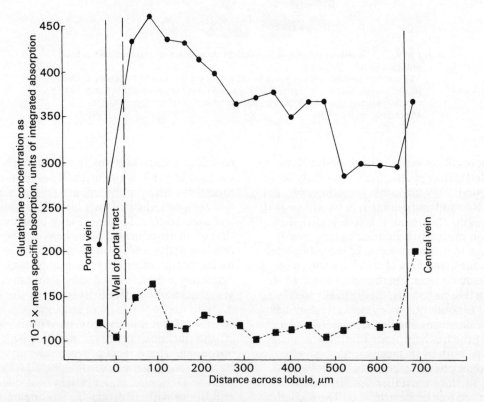

Fig. 1.4. **Measurement of glutathione in liver by quantitative cytochemistry.**
The concentration of glutathione varies considerably from the portal vein region to the central vein region. Conventional biochemical techniques cannot show these differences. (-- ■ --) Animal treated with diethylmaleate to deplete liver of glutathione; (–●–) control.

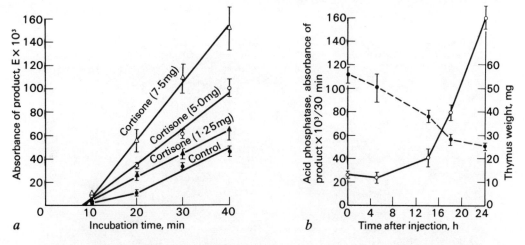

Fig. 1.5. **Measurement of lysosomal acid phosphatase activity in thymus section and the activating effect of cortisone.** Measurements were made by quantitative cytochemistry. In (*b*) (–) acid phosphatase; (- - -) thymus weight.

summed to give the integrated absorption of the specimen (area *A*). By suitable calibration, the mean integrated absorbance of the whole specimen can be determined. These methods have been widely applied to the measurement of molecular concentrations, and enzyme activities in subcellular organelles such as the lysosomes and for the study of many dehydrogenases. Examples are shown in *Figs*. 1.4 and 1.5.

1.3 Ultrastructure of typical cells

The structure of a typical mammalian cell viewed under the electron microscope shows several prominent features. A human small intestine cell is illustrated in *Fig*. 1.6. The following structural features may be recognized: the nucleus (N), the mitochondria (M), the Golgi apparatus (G), the granular or rough endoplasmic reticulum (GER) and free ribosomes (R).

All cells contain similar organelles, but the numbers and proportion of each can vary extensively from one cell type to another. Nevertheless, one of the most important discoveries of modern biochemistry is the demonstration that the biochemical function of each organelle, e.g. a ribosome or mitochondria, remains essentially constant despite varieties in number or form.

In addition to the main organelles illustrated in *Fig*. 1.6, many cells contain a system of tubules, illustrated in *Fig*. 1.7 and filaments illustrated in *Fig*. 1.8. The understanding of the structure and function of these organelles is at present very limited, but the analogy between the filaments and structures in muscle indicates that they may possess a contractile function and be able to alter the shape of the cell, enabling it to perform functions such as pinocytosis.

Microtubules are slender cylinders measuring about 20–30 nm in diameter. These are believed to be composed of repeating protein units and to be involved in intracellular transport of important molecules.

The great variability of cell structure, closely related to function, can be illustrated by the following examples. The chief cells of the gastric gland (*Fig*. 1.9) possess a large quantity of rough endoplasmic reticulum which is involved in protein synthesis and large numbers of distinctive vacuoles containing the precursor enzymes. A completely different cell type, the macrophage (*Fig*. 1.10), is primarily involved in the defence of the body against invading organisms. It therefore possesses very active pinocytotic activity and a very large

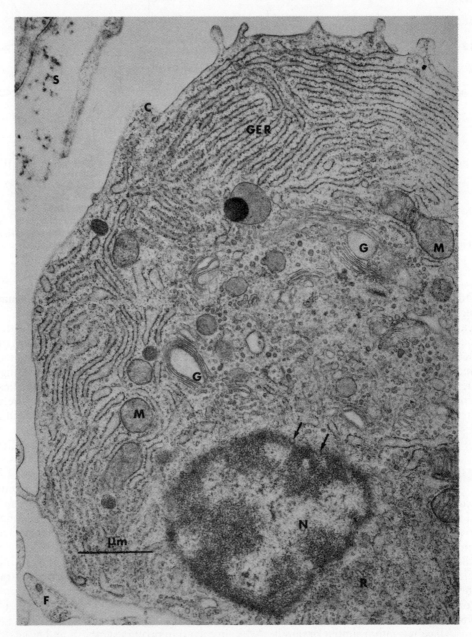

Fig. 1.6.   **General characteristics of cell structure as seen under the electron microscope.**
Human small intestine cell. Magnification ×26 000.

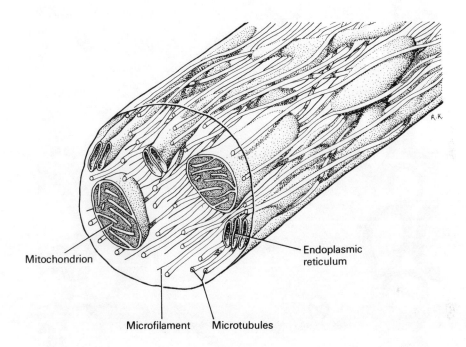

Fig. 1.7.   **Diagrammatic representation of microtubules.**
Nerve cell of cerebellum. Magnification ×26 000.

Microfilaments

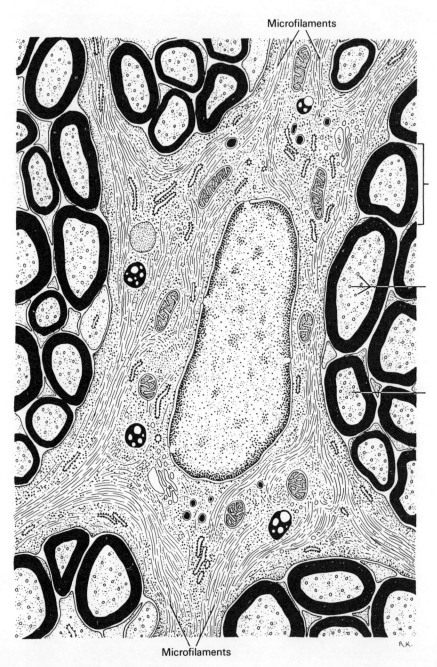

Microfilaments

Fig. 1.8.   **Diagrammatic representation of microfilaments.**
Astrocytes of cerebral cortex. Magnification ×20 000.

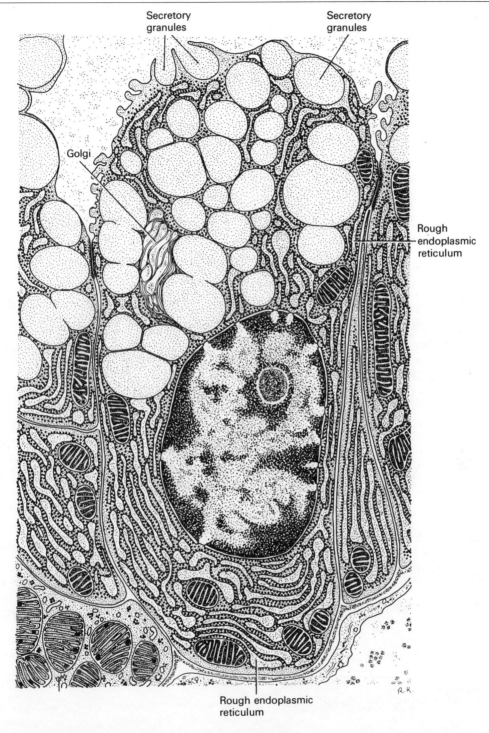

Fig. 1.9. **Cells of the gastric gland showing extensive rough endoplasmic reticulum and storage vesicles.** Magnification ×15 000.

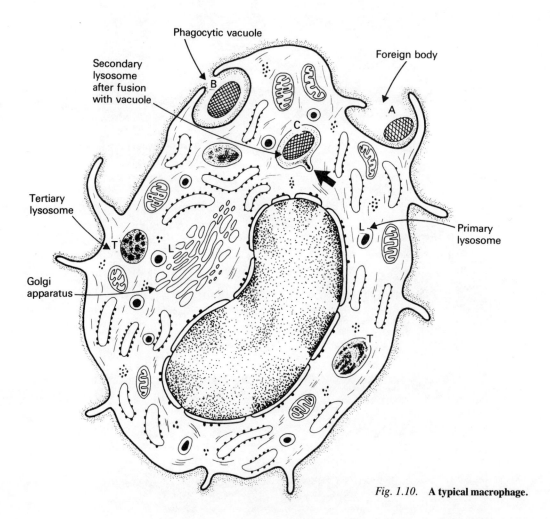

Phagocytic vacuole

Secondary
lysosome
after fusion
with vacuole

Foreign body

Tertiary
lysosome

Primary
lysosome

Golgi
apparatus

Fig. 1.10.   **A typical macrophage.**

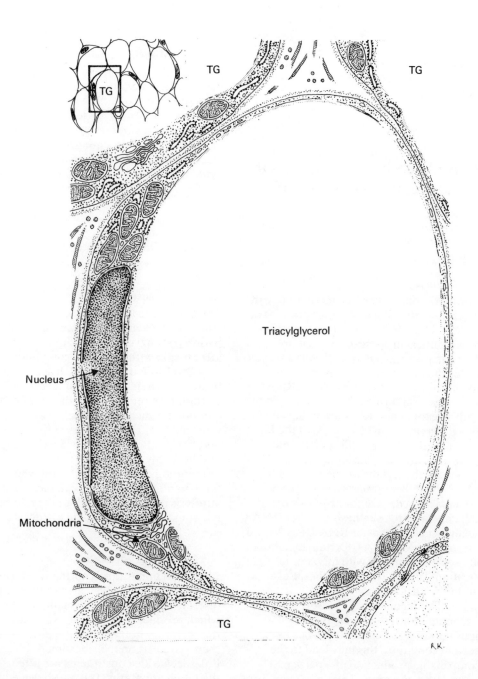

Fig. 1.11.   **An adipose tissue cell.**
TG = triacylglycerol. Magnification ×7000.

active complement of active lysosomes which digest the ingested particles (cf. Chapter 3).

Cells of the adipose tissue (*Fig*. 1.11) contain a very large quantity of stored tri-acylglycerol as a result of which the nucleus, mitochondria and all other cellular organelles are pressed into a very small volume around the periphery of the cell. This cell is almost entirely devoted to fat storage and its structure is thus adapted to this purpose.

1.4. Biochemical functions of the main subcellular components

These are described in detail in the following chapters of this book, but it would first be useful to summarize their essential functions.

Nucleus

The primary role of the nucleus is to carry the coding for synthesis of all cellular proteins on the DNA. The DNA in all cells of one human body is identical in constitution, but only part of the code is expressed in each cell type. DNA provides the information for the synthesis of many different messenger RNAs (mRNAs), each of which, in turn, codes for a protein of the cell. Histones and RNA are believed to be involved in repression of sections of the DNA code not used. DNA is replicated at cell division and thus each nucleus possesses DNA polymerases that synthesize new strands of DNA; RNA polymerases are also present. These enzymes catalyse the synthesis of mRNA and also the synthesis of other RNA molecules such as transfer RNA (tRNA) and ribosomal RNA (rRNA). Certain metabolic processes unconnected with DNA or its replication, such as NAD^+ synthesis, also occur in the nucleus.

The regulation of DNA repression and function is often severely disturbed in serious pathological conditions, such as cancer. In certain conditions this may lead to the uncontrollable production of a particular hormone from the normal gland or even from a tissue not normally producing it. Very large concentrations of hormones, such as parathyroid hormone, which can be released into the blood can cause severe consequences and even death (cf. Chapters 17 and 22).

Mitochondria

The mitochondria produce the energy for the cellular function, in the form of ATP from the oxidation of NADH or reduced flavoprotein formed by dehydrogenations occurring at several different stages in the citric acid cycle. The mitochondria contain an electron transport chain composed of several cytochromes and flavoproteins which oxidize NADH, the complete citric acid cycle and enzymes involved in fatty acid oxidation. The mitochondria also possess several elaborate systems for transporting molecules to and from the cytoplasm (cf. Chapter 7).

Lysosomes

The lysosomes are small sac-like organelles containing a very large complement of hydrolytic enzymes. They are normally active after fusion with a vacuole containing an ingested foreign particle such as a dead bacterium (heterophagy) or a vacuole containing a redundant or damaged cellular component such as a mitochondrion (autophagy). They are very active in the cells involved in the body's defence (macrophages) and probably also play an important role in immunology, processing the antigenic material (cf. Chapter 42). When a cell dies, extensive autolysis occurs as a result of the release of lysosomal enzymes into the cytoplasm.

Golgi apparatus

The Golgi apparatus is normally seen as a distinct membrane system containing many discrete vesicles. It is especially active in cells which produce proteins for export and forms the secretory granules for the proteins after their synthesis on the ribosomes. Polysaccharide constituents are often added in the Golgi apparatus. It is also believed to play an important role in the packaging of enzymes into primary lysosomes.

Endoplasmic reticulum: the ribosomes

All cells contain a dense interlacing network of membranes called the 'endoplasmic reticulum'. When prepared by ultracentrifugation, the fractions disperse into small fragments described as the 'microsomes'. The endoplasmic reticulum is of two types:

i.   Smooth endoplasmic reticulum, consisting only of membranes.

ii. Rough or granular endoplasmic reticulum to which are attached many ribosomes.

It is a controversial issue as to whether the structures of the membranes of the rough and smooth membranes are identical, but this is generally believed to be the case. The ribosomes attached to the endoplasmic reticulum are normally involved in synthesis of proteins for export from the cell, whereas ribosomes in the cytoplasm synthesize proteins for use within the cell. High density of rough endoplasmic reticulum is, therefore, characteristic of cells synthesizing secreted proteins (cf. *Fig*. 1.9).

The smooth membranes of the endoplasmic reticulum are very important in liver cells where they are primarily concerned with the oxidative metabolism and detoxication of many drugs and toxic organic molecules. Many of these substances are lipophilic and they readily associate with the lipid membranes of the smooth endoplasmic reticulum. One interesting aspect of the smooth membranes, especially in the liver, is that the quantity in the cell is not constant, but that it proliferates readily after treatment of animals with drugs which act as inducers (cf. Chapters 9 and 36).

Microtubules and microfilaments

Most cells contain demonstrable cytoplasmic microfilaments and many of these have been shown to be very similar to actin and a small number to myosin. The functions of these filaments is not certain, but they are believed to possess the power to modify the cell shape and to preserve morphological integrity in the face of mechanical stress.

Microtubules have been recognized in nearly all cells. Most show a periodic structure and are composed of polymerized macro-molecular subunits called tubulins. The tubules are often in a dynamic state, rapidly dissociating and reassociating. The true function of the microtubules, and whether it varies from one cell type to another, has not yet been established, but it has been suggested that they are concerned with the maintenance of cell shape, link subcellular organelles and control their movement or establish pathways for diffusion of small molecules.

2 Roles of extracellular and intracellular membranes: membrane structure and membrane transport

2.1 Introduction

The fact that all living cells are surrounded by a membrane, often called the 'plasma membrane', was apparent many years ago when cells were first studied by light microscopy. Its function appeared to be to keep the cell contents from dispersing, thus playing an important role in the cell, and in organism structure.

As microscopy, and particularly electron microscopy, developed to give improved magnification and definition, it gradually became apparent that membranes surrounded all the subcellular structures, for example the nucleus, mitrochondria and lysosomes. Furthermore, all cells were shown to possess a complex internal network of membranes, the endoplasmic reticulum.

These membranes within cells permitted many specialized subcellular metabolic processes to continue independently of metabolic processes in other cell components. Furthermore, all membranes exert a pronounced degree of specificity, allowing some molecules to pass through, whilst blocking the transport of others. These transport processes are clearly very important in the overall regulation of metabolism, in that they permit certain molecules to be metabolized in specific organelles. Many enzymes are also located with the membranes, for example on the endoplasmic reticulum, where they are available to metabolize molecules associated with this membrane. Thus, many lipophilic drugs entering the body are rapidly taken up by the liver, where they become associated with the endoplasmic reticulum membranes and are metabolized by the enzymes bound to the membranes.

Permanent integrity is of vital importance to all cells. Lytic agents, e.g. detergents, will solubilize membranes and release vital cellular components, e.g. haemoglobins from red blood cells, a process that can result from a snake bite with clear-cut serious consequences. Damage to membranes within cells can also cause serious consequences. For example, damage to lysosomal membranes results in leakage of hydrolytic enzymes, such as proteases and nucleases, that digest vital protein and nucleic acid components (cf. Chapter 3).

2.2 Membrane composition

All membranes are composed of lipids, proteins and, often, smaller proportions of carbohydrates.

Lipids

Lipids form 40–80 per cent of the total membrane constituents and, of the lipids, phospholipids are essential components of all membranes forming a major proportion of the lipid component of the membrane. The structures of these lipids are shown in *Fig*. 2.1. Thus 85 per cent of the lipid of the endoplasmic reticulum of liver is phospholipid and this type of lipid forms 62 per cent of the plasma membrane of a liver cell. Phosphatidylcholine (lecithin) is usually the most abundant phospholipid. Cholesterol, cholesteryl esters, triacylglycerols and free fatty acids occur in the neutral lipid fraction (*Table* 2.1).

GLYCEROPHOSPHOLIPIDS

Major importance in membranes

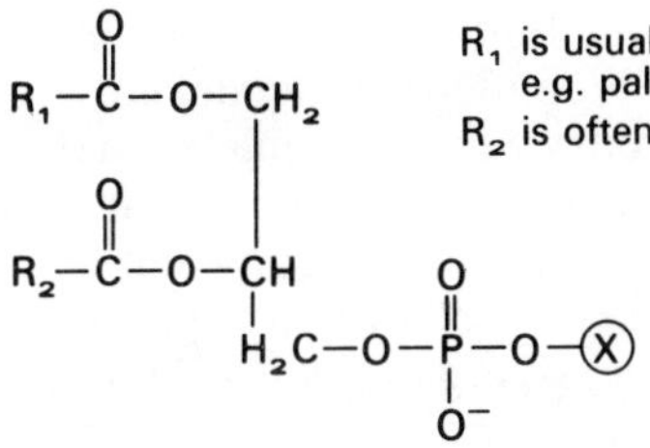

R_1 is usually a saturated or mono-unsaturated fatty acid, e.g. palmitic acid or oleic acid.

R_2 is often a polyunsaturated fatty acid, e.g. linoleic acid.

X = Choline

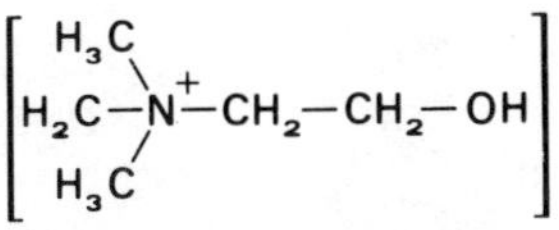

Phosphatidylcholine (PC)

X = Ethanolamine

$$[H_3{}^+N—CH_2—CH_2—OH]$$

Phosphatidylethanolamine (PE)

X = Inositol

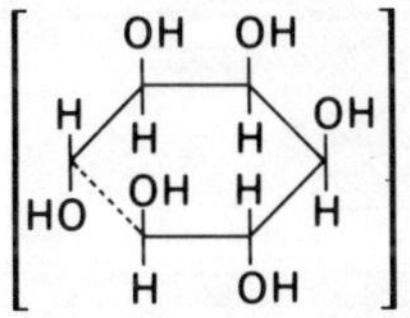

Phosphatidylinositol (PI)

X = Serine

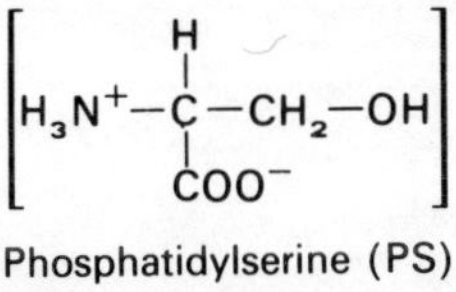

Phosphatidylserine (PS)

SPHINGOPHOSPHOLIPIDS

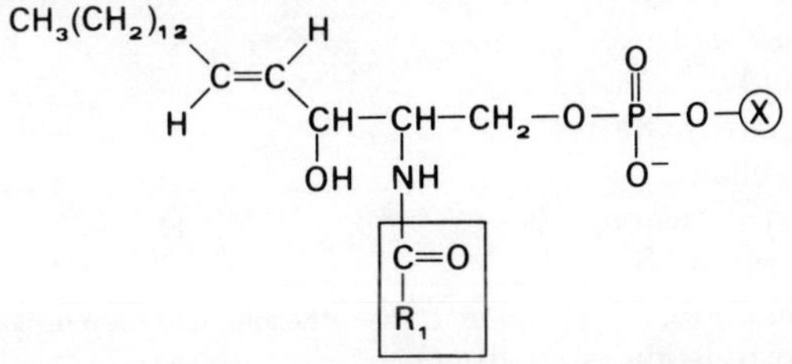

R_1 is usually a long-chain saturated or mono-unsaturated fatty acid, e.g. $C_{22:0}$ or $C_{24:1}$.

Fig. 2.1. **Phospholipids of major importance in membranes.**

Table 2.1 **Lipid composition of membranes**

Compound	Endoplasmic reticulum, % total	Plasma membrane, % total
Total phospholipids	84·9	61·9
Sphingomyelin	3·7 ± 1·1	18·9 ± 2·3
Phosphatidylcholine	60·9 ± 2·2	39·9 ± 2·8
Phosphatidylserine	3·3 ± 2·2	3·5 ± 1·8
Phosphatidylinositol	8·9 ± 2·3	7·5 ± 1·3
Phosphatidylethanolamine	18·6 ± 1·1	17·8 ± 1·5
Lyso-phosphatidylcholine	4·7 ± 3·4	6·7 ± 0·7
Lyso-phosphatidylethanolamine		5·7 ± 2·7
Total neutral lipids*	15·1	38·1
Cholesterol	24·6	34·5
Free fatty acids	40·6	35·1
Triacylglycerols	24·7	22·4
Cholesterol esters	10·1	8·0

* Each component of the total neutral lipids is given as a percentage of that total.

There is some variation between the lipid compositions of plasma membranes prepared from different tissues (*Fig.* 2.2), but the same tissue, e.g. red-blood-cell membranes, shows only minor variations when different species are compared (*Fig.* 2.3). The composition of the lipids of subcellular membranes are shown in *Fig.* 2.4; several important differences will be noted. The Golgi apparatus contains a high proportion of neutral lipid, particularly cholesterol; the mitochondrial membranes contain virtually no sphingosine, whereas the endoplasmic reticulum of kidney contains significant quantities.

It is important to note that many different (possibly 150–200) phospholipids exist in membranes. This is because each phospholipid contains two fatty acid molecules, the most commonly occurring fatty acids being shown in *Fig.* 2.5. The chain lengths of these fatty acids can vary between 12 and 22 and many can be unsaturated containing between 1 and 6 double bonds. In view of this, very large numbers of isomers of each phospholipid exist and many can be incorporated into membranes. The precise significance of the different types of fatty acids which, as phospholipid components, are found

in membranes is, at present, unclear but several lines of investigation indicate that the exact composition is important. Firstly, the fluidity or flexibility of membranes is dependent on the degree of unsaturation of the fatty acids forming the membrane and, as the degree of unsaturation increases, the membranes become more flexible and fluid. Secondly, because some polyunsaturated fatty acids, e.g. linoleic acid ($C_{18:2}$, ω-6), cannot be

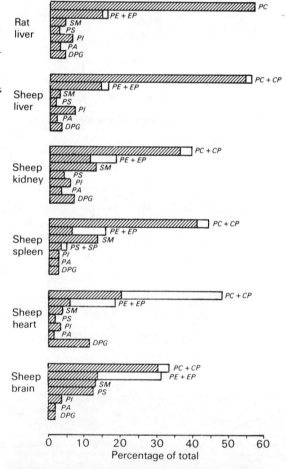

Fig. 2.2. **Phospholipid compositions of rat and sheep tissues.**
PC = phosphatidylcholine;
PE = phosphatidylethanolamine; PS = phosphatidylserine;
PI = phosphatidylinositol; SM = sphingomyelin;
DPG = diphosphatidylglycerol; CP, EP and
SP = plasmalogens.

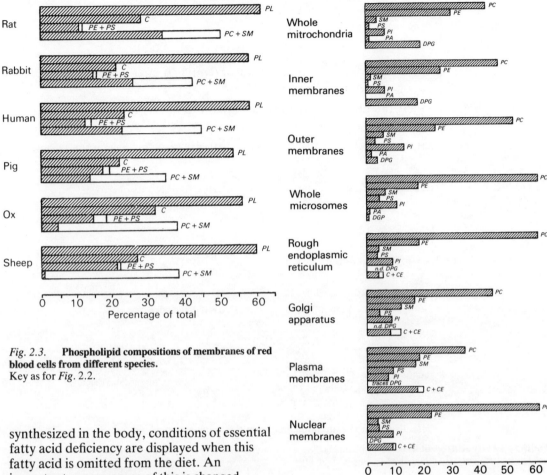

Fig. 2.3.   Phospholipid compositions of membranes of red blood cells from different species. Key as for *Fig. 2.2.*

Fig. 2.4.   Phospholipid compositions of subcellular organelles prepared from rat tissues. Key as for *Fig. 2.2.*

synthesized in the body, conditions of essential fatty acid deficiency are displayed when this fatty acid is omitted from the diet. An important consequence of this is changed membrane functions (cf. Chapter 9). Thirdly, if polyunsaturated fatty acids undergo peroxidation as a result of oxidative attack on the double bonds, polyunsaturated fatty acids can be destroyed with consequent loss of membrane structure and functions.

Studies on artificial membranes composed of lecithin show that cholesterol has a 'condensing effect' on the lecithin. As a consequence of the incorporation of cholesterol into these artificial membranes, the lecithin occurs in a smaller area at the lipid–water interface. It is uncertain whether these experiments can be extrapolated directly to membranes in cells, but it is clear that the proportion of cholesterol in the membrane will have an important effect on the membrane structure. The fatty acid composition of the membrane phospholipids is not rigidly fixed and can vary with dietary changes. For example, the proportion of linoleic acid in the phospholipids increases as the dietary intake increases which is clearly shown in the membranes of the endoplasmic reticulum. This can lead to changes in function (cf. Chapter 9), although the range of phospholipid structure changes that can be tolerated in the membrane composition without causing serious loss of function is not, at present, understood.

a Saturated and mono-unsaturated

$$\begin{cases} \text{Palmitic (16:0) } CH_3 - (CH_2)_{14} - COOH \\ \text{Stearic (18:0) } CH_3 (CH_2)_{16} - COOH \\ \text{Oleic acid (18:1) } (\omega\text{-9}) : CH_3 (CH_2)_7 - \overset{9}{C}H = \overset{10}{C}H - (CH_2)_7 - COOH \end{cases}$$

Polyunsaturated

$$\begin{cases} \text{Linoleic acid (18:2) } (\omega\text{-6, 9}) : CH_3 (CH_2)_4 - \overset{6}{C}H = \overset{7}{C}H - CH_2 \overset{9}{C}H = \overset{10}{C}H - (CH_2)_7 - COOH \\ \text{Linolenic acid (18:3) } (\omega\text{-3, 6, 9}) \\ CH_3 CH_2 \overset{3}{C}H = \overset{4}{C}H - CH_2 - \overset{6}{C}H = \overset{7}{C}H - CH_2 - \overset{9}{C}H = \overset{10}{C}H - (CH_2)_7 - COOH \\ \text{Arachidonic acid (20:4) } (\omega\text{-6, 9, 12, 15}) \\ CH_3 - (CH_2)_4 - \overset{6}{C}H = \overset{7}{C}H - CH_2 \overset{9}{C}H = \overset{10}{C}H - CH_2 \overset{12}{C}H = \overset{13}{C}H - CH_2 - \overset{15}{C}H = \overset{16}{C}H - (CH_2)_3 - COOH \end{cases}$$

b

Desaturation →

Chain elongation ↓

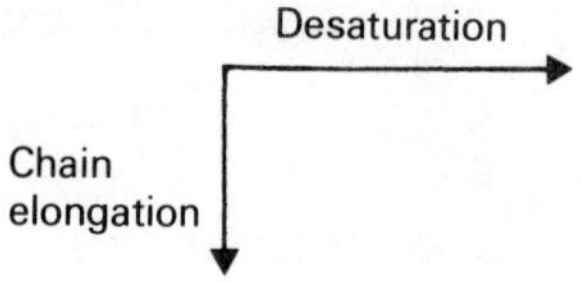

Linoleic acid (18:2) (ω-6, 9) → 18:3 (ω-6, 9, 12)

↓   ↓

20:2 → 20:3 (ω-6, 9, 12) → 20:4 (ω-6, 9, 12, 15)

(ω-6, 9)   ↓   Arachidonic acid

22:4 → 22:5

Linolenic acid (18:3) (ω-3, 6, 9) → 18:4 (ω-3, 6, 9, 12)

↓

20:4 (ω-3, 6, 9, 12)

↓

22:5 (ω-3, 6, 9, 12, 15) → 22:6 (ω-3, 6, 9, 12, 15, 18)

Note: Carbon atoms are numbered from the terminal methyl group

Fig. 2.5.   Fatty acid constituents of phospholipids.
(*a*) The constituents; (*b*) metabolism to longer-chain fatty acids.

Proteins

Proteins normally form a major proportion of most membranes, usually 50–70 per cent by weight, with the exception of myelin in the membrane of nerve which contains 90 per cent lipid. The relation of lipid:protein composition is shown in *Fig. 2.6*. Study of the membrane proteins presents many problems because of the difficulties arising from the process of dissociating them from membrane lipids, in order to obtain the pure proteins.

It is clear, however, that proteins are essential in membranes to play several roles: structural, transport, as enzymes and as recognition sites for like and unlike cells. A typical membrane, such as that of the red blood cell, will contain many different proteins, varying in molecular weight from 31 000 to 160 000. Many of these proteins are glycoproteins, i.e. they possess an oligosaccharide which is composed of 4 to 15 carbohydrate residues. Some of these proteins are composed of three parts: one containing the carbohydrate residue is on the outside of the membrane, a second part is within the membrane and a third within the cell contents as described above. The red-blood-cell membrane protein, glycophorin, is of this type. In this protein the oligosaccharides are attached to arginyl, threonyl and seryl residues (*Fig. 2.7*).

The membrane glycoprotein obligosaccharides are composed of nine sugars: the hexoses, glucose, galactose, maltose, and

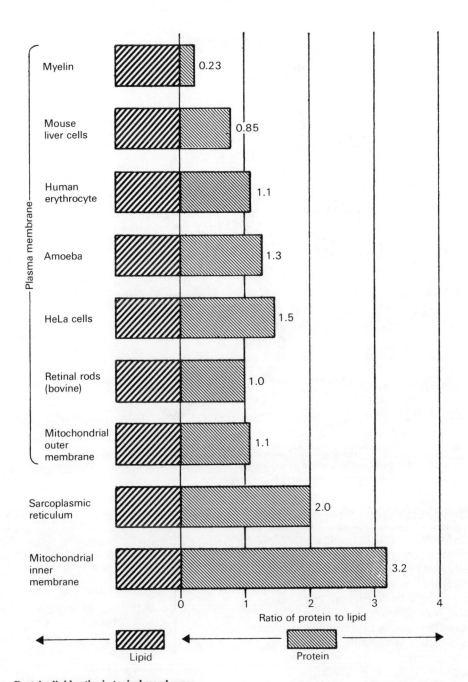

Fig. 2.6.   **Protein–lipid ratios in typical membranes.**

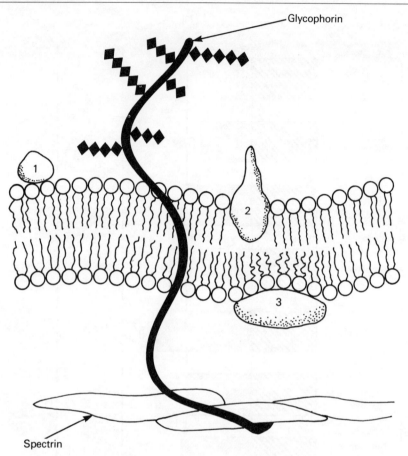

Glycophorin

1

2

3

Spectrin

Fig. 2.7. **The structure of glycophorin, a protein of the red-blood-cell membrane.**
1, 2 and 3 are membrane proteins lying on the surfaces of the membrane (1) or
partially immersed within the phospholipid molecules.

fucose; the pentoses, arabinose and xylose;
acetylglucosamine, acetylgalactosamine and
N-acetylneuraminic acid (sialic acid) (*Fig*. 2.8).
Although the component units of oligosaccharides are much less in number than those
of proteins, great variations of polymers can be
formed because the sugars can link at many
different sites. Branch points can be formed
and very large numbers of complex structures
be developed. For example, five different
sugars in a 13-residue oligosaccharide chain
can form 10^{24} different polymers.

The cell-membrane glycoproteins are
extremely important in cell recognition and
adhesion and this may be achieved by
enzymes, such as glycosyl tranferases, on the
surface of cells recognizing the carbohydrate
chain of another cell, and catalysing a linkage
with a protein on the membrane and the sugar
chain of another cell.

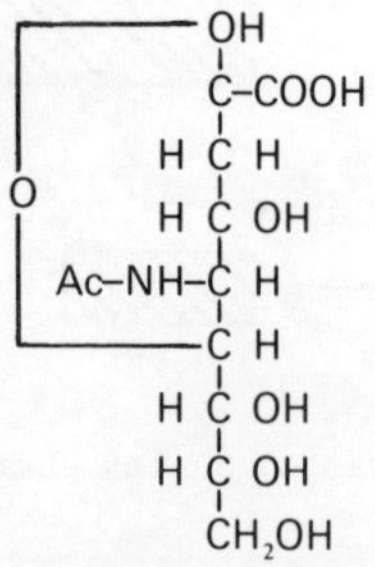

$$
\begin{array}{l}
\quad\ \ \text{OH} \\
\quad\ \ |\\
\quad\ \ \text{C–COOH} \\
\quad\ \ |\\
\ \ \text{H C H} \\
\quad\ \ |\\
\ \ \text{H C OH} \\
\quad\ \ |\\
\text{Ac–NH–C H} \\
\quad\ \ |\\
\quad\ \ \text{C H} \\
\quad\ \ |\\
\ \ \text{H C OH} \\
\quad\ \ |\\
\ \ \text{H C OH} \\
\quad\ \ |\\
\quad\ \ \text{CH}_2\text{OH}
\end{array}
$$

$$\text{Ac} = \text{CH}_3\text{CO–}$$

Fig. 2.8. **The structure of sialic acid
(*N*-acetylneuraminic acid).**

2.3 Membrane structure

All membranes are known to be composed of bilayers of phospholipids. A typical molecule of a phospholipid, such as phosphatidylcholine (lecithin), contains a strongly lipophilic component, the two fatty acid chains, and a strongly hydrophilic component, the choline and phosphate residues. This part of the molecule is also a zwitterion possessing a positive charge on the nitrogen of the choline and negative charges on the phosphate (*Figs*. 2.1 and 2.9).

In solution, phospholipids tend to form micelles; the lipophilic parts of several phospholipids associate by van der Waals attraction forces and the hydrophilic groups form a surface layer with the water interface (*Fig*. 2.9). A type of membrane monolayer can clearly be formed on imagining a very large micelle, but this would require an internal lipid environment for stability. Such structures do exist, for example fat droplets in the gut prior to absorption are in this form, and several lipoproteins in the plasma possess this type of structure (*Fig*. 2.9).

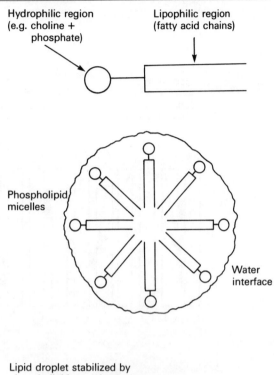

Hydrophilic region (e.g. choline + phosphate)

Lipophilic region (fatty acid chains)

Phospholipid micelles

Water interface

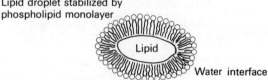

Lipid droplet stabilized by phospholipid monolayer

Lipid

Water interface

Fig. 2.9.   **Phospholipid micelles.**

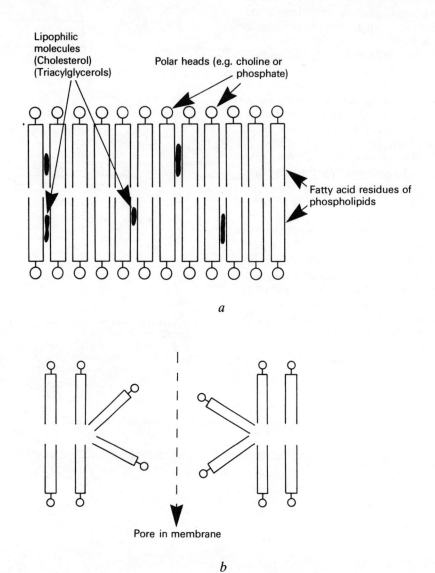

Fig. 2.10.   (a) Association of phospholipids to form membrane bilayer; (b) pore formation.

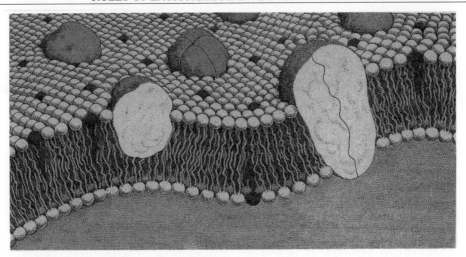

Fig. 2.11. **Diagrammatic concept of a typical membrane.**
Proteins lie on the surface or extend right through the phospholipid structure.

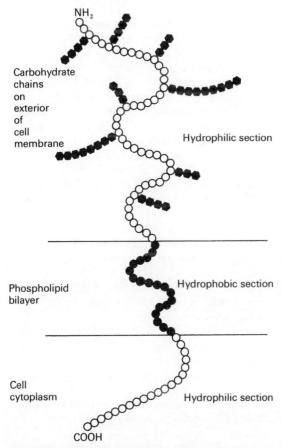

Fig. 2.12. **Glycoprotein molecule extending right through the phospholipid layer.**

Most cellular environments are, however, primarily not lipid, but aqueous. Stability of an aqueous environment is achieved by means of a bilayer. Pairs of phospholipid molecules associate so that lipophilic components are in close proximity and hydrophilic groups are presented both to the external and internal environments (*Fig.* 2.10). Lipophilic molecules such as cholesterol are incorporated deep within the fatty acid residues of the phospholipids.

Many protein molecules are associated with the membranes (*Fig.* 2.11) and these can lie on the outer surface, on the inner surface or they can extend right through the membrane (*Fig.* 2.12). The sugar chains are normally located on the outer surface of the membrane and they form important recognition sites, both for other cells and for important molecules such as hormones.

2.4 Membrane functions

Functions of membranes of cells and subcellular components may be divided into three categories:

  a. Forming a boundary and transport function
  b. Forming a matrix for the localization of important cellular functions, especially enzyme action
  c. Cell membrane proteins, including glycoproteins play important roles as hormone receptors and recognition sites.

In (a) it is clear that the external cellular membrane plays a very important function in defining the boundary of the cell and regulating the transport of many molecules, such as amino acids, carbohydrates, vitamins, oxygen and ions, that must be transported into the cell and waste products, such as urea and carbon dioxide, that must be transported out. Frequently, a two-way process is involved; thus a muscle cell can incorporate amino acids from the plasma when good supplies are available, but it can also release amino acids into the plasma in conditions of starvation. The two-way process must clearly be regulated and this is normally achieved by the action of hormones such as insulin. Similar considerations apply to the subcellular organelles such as the nucleus and mitochondria. Within the cell, the transport is often more complex and many metabolic intermediates are transported. This is often related to the location of specific enzymes and to whether these are inside or outside the organelle. Transport mechanisms of this type are well illustrated by the mitochondrial–cytoplasm relationships and are described in Chapter 7.

In (b) membranes also play a very important role within cells by forming matrices for the localization of enzymes. This function is shown by the mitochondrial matrix and the endoplasmic reticulum. In the mitochondria it is generally considered that the inner membranes are essential for localization and the correct orientation of enzyme components for maximum efficiency (as described in Chapter 6). This consideration may also apply

to the endoplasmic reticulum but, in addition, the membranes of the endoplasmic reticulum play a very important role in localizing lipophilic substrates, such as drugs and toxic molecules, in the correct orientation for metabolic attack by the oxidation enzymes of the membrane. This role is described in detail in Chapters 9 and 36.

In (c) the proteins of external membranes perform several important functions. They act as recognition sites for other similar cells and foreign cells are readily attacked by the body's immune system.

The 'blood group' substances of red blood cells A and B which identify different groups of cells are typical membrane glycoproteins. They are both composed of 11 or 13 monosaccharide units in the form of branched chains. In type A, two of the branched termination groups are N-acetyl-galactosamine units whilst in type B they are replaced by galactose as the terminal unit. Neither is present in type O.

Many cells of the body possess hormone receptor proteins as membrane constituents. These receptors are specific for peptide hormones such as glucagon or insulin or for steroids. An important consequence of the binding of many hormones to the membrane receptor is the activation of the enzyme adenylate cyclase which catalyses the formation of cyclic AMP from ATP. Cyclic AMP is then used to activate a protein kinase and thus activate a cascade of enzymes resulting in, for example, the catabolism of glycogen to liberate blood glucose (cf. Chapter 17). Steroid hormones, such as oestrogens, are however transported within the cell, bound to proteins, after uptake by a membrane receptor, to act eventually on the nuclear DNA.

The B-lymphocyte cell also possesses special immunocompetent receptors which are capable of recognizing and responding to foreign antigens. After stimulation with an appropriate antigen and with the aid of a T cell, the B lymphocyte matures into a large plasma cell which performs its defensive function (cf. Chapter 42).

It is also important to note that malignant tumour cell membranes show properties different from those of normal cells. They are less adhesive to other cells and can, therefore, break away more easily and form metastases. This phenomenon, of clear importance, has not yet been explained satisfactorily. It may be due to a loss of receptors for agglutination or due to the receptors being covered by other molecules rendering them ineffective.

2.5 Membrane transport

Molecules and ions which move into and out of cells can be categorized into five main classes:

  a. Water
  b. Metabolites, such as glucose and amino acids
  c. Gases: oxygen and carbon dioxide
  d. Ions: Na^+, K^+, Ca^{2+}
  e. Other organic molecules: drugs, toxic substances

Studies of membranes of all types have shown that transport of molecules across membranes can be divided into two main categories:

  i.   Simple diffusion
  ii. Active transport

Simple diffusion implies that the molecules move in response to a concentration gradient whereas, for active transport, a supply of energy, usually in the form of ATP, is required and the process is highly selective and specific. This is achieved by the action of carrier molecules.

A third form of transport, called 'facilitated diffusion' is also described. In this form of transport, the molecules diffuse across the membrane in response to the concentration gradient but a degree of selectivity is shown. The rates with which various similar molecules can be transported across the membrane can also be regulated by specific carriers but energy is not required.

Diffusion

Gases, such as oxygen and carbon dioxide, normally diffuse readily across biological membranes, the rate depending on the concentration gradient. Special proteins, such as myoglobin, which possess a high affinity for oxygen are often important in enhancing the rate of movement of oxygen from the haemoglobin on which it is transported. Most cells are freely permeable to water and water normally moves to balance the osmotic pressure changes caused by movements of electrolytes and, less commonly, of proteins. The movement of water across cell membranes is discussed in detail in Chapter 21.

Small organic molecules diffuse through membranes at a rate which is often proportional to their size. Small molecules diffuse more rapidly, but molecular size is not the only criterion. Over 80 years ago in a series of classical experiments, Overton showed that lipid solubility was an important criterion in regulating the movement of small molecules across membranes of cells. Generally the more lipid-soluble molecules penetrate membranes more rapidly and effectively than do hydrophilic molecules (*Table* 2.2, *Fig.* 2.13).

Table 2.2 **Permeability of cells to organic molecules**

Degree of penetration	Organic molecule
+ + + + +	Monohydric alcohols, aldehydes, ketones hydrocarbons, esters, weak organic acids and bases
+ + + +	Dihydric alcohols, amides
+ + +	Glycerol, urea
+ +	Tetrahydric alcohols
+	Hexahydric alcohols, carbohydrates, amino acid, salts of organic acids

From Meyer-Overton (1902) quoted by Hober, R. *Physical Chemistry of Cells and Tissues*, 1946, Edinburgh: Churchill.

This is a very important concept that has stood the test of time and is generally found to be true. It has very far-reaching medical implications. For example, we can predict that the more lipophilic the molecule, the more likely it is to be taken up by tissue cells: lipophilic toxic molecules can gain access to the cells and lipophilic drugs and anaesthetics can enter the brain cells. Some of the early observations of Overton on the

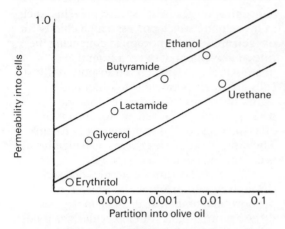

Fig. 2.13. **Relation between cell permeability and partition between an olive oil: water dispersion.**

Table 2.3 **Relation between narcotic effects and olive oil: water partition**

Narcotic	Narcotic concn, μm	Olive oil: water partition
Ethanol	600	0·03
Propanol	110	0·13
Valeramide	50	0·07
Ether	24	2·4
Salicylamide	2	14
Thymol	0·055	600

Narcotic effects were shown on tadpoles.
From Meyer-Overton (1902), quoted by Hober, R.
Physical Chemistry of Cells and Tissues, 1946, Edinburgh: Churchill.

narcotic effects and lipid solubility are summarized in *Table* 2.3. The whole problem of narcosis clearly cannot be explained on such a simple basis, but lipophilicity must be an important factor.

Recently, much progress in the understanding of membrane permeability has been obtained by the use of artificial membranes called 'liposomes': these are small globules surrounded by a phospholipid membrane. Different types of phospholipids can be used to construct the artificial membrane and many different molecules can

be incorporated within the globule or added to the surrounding medium.

Experiments with liposomes and natural membranes, such as the red-blood-cell membrane, have shown that membranes readily discriminate between anions and cations. Fixed positive changes in the membranes, presumably associated with the nitrogen of the phospholipids, impose a severe restriction on the movement of cations. In support of this concept, anions move out of liposomes much more rapidly than do cations. The size of the ion is important because the

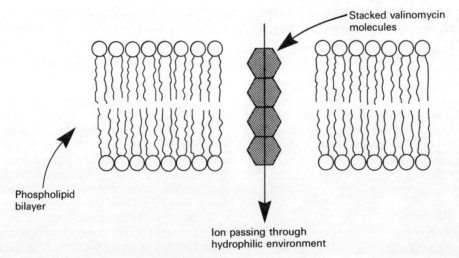

Fig. 2.14. **Formation of membrane pores by valinomycin.**

rate of Cl^- exchange across the membrane is much greater than that of NO_3^- or SO_4^{2-}.

Understanding of membrane permeability has been aided by the use of several cyclic peptides with antibiotic properties such as valinomycin (cf. Chapter 43). This peptide greatly increases the permeability of synthetic membranes to ions and, furthermore, shows a degree of specificity so that the permeability of ions is in the order $H^+ > Rb^+ > K^+ > Cs^+ > Na^+ = Li^+$.

Molecules of the valinomycin type have hydrophobic residues on the outside and polar residues within the ring. They may, therefore, form a membrane pore within a bilayer which enables polar molecules to pass through (*Fig.* 2.14).

Facilitated diffusion

This diffusion is mediated by carrier molecules and is selective, but it only occurs down a concentration gradient as shown:

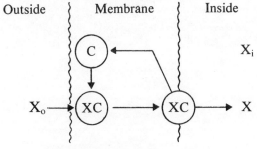

Concentration outside $= X_o$
Concentration inside $= X_i$
$$X_o > X_i$$
C = Carrier molecule

The entry of glucose into the red blood cell occurs by facilitated diffusion and shows properties typical of the process.

The carrier is specific for D-glucose because L-glucose enters the cell very slowly. The carrier shows saturation characteristics indicating that the carrier molecules act like enzymes, i.e. they possess active sites that can be saturated with the substrate, in this case glucose. The transport is not inhibited by inhibition of glycolysis, indicating that a supply of energy is not required, but it can

be inhibited both competitively and non-competitively. The drug phloretin causes changes in the glucose carrier which, in turn, causes inhibition of the transport.

The identity of the glucose carriers is unknown, but it is likely to be a protein with a high affinity for glucose. It has been calculated, however, that each red blood cell possesses 800 000 binding sites and that each site can transport 180 glucose molecules per second.

Active transport

Active transport is energy dependent and can occur against a concentration gradient:

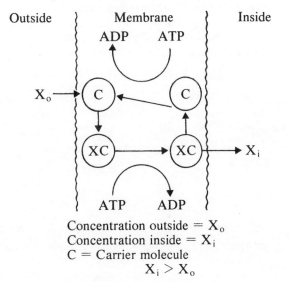

Concentration outside $= X_o$
Concentration inside $= X_i$
C = Carrier molecule
$$X_i > X_o$$

Active transport is a very important process enabling cells to accumulate molecules or ions from the environment against the concentration gradient. Conversely, contents of cells heavily loaded with electrolytes or metabolic products can be excreted against the concentration gradient. This is very important in the kidney which is frequently involved in the excretion of ions and unwanted molecules into the urine. Many cells also maintain their correct ion balance by active transport.

The mechanism of active transport is, except in a few cases, very poorly understood. It is generally believed that it must involve a carrier protein (C) whose conformation must be altered so that it has a high affinity for the molecule to be transported (X), becoming ($\overset{o}{C}$). This could involve a supply of energy from ATP through conversion to ADP. Alternatively, energy may be involved in the transport of XC across the membrane, or, both processes may require energy. The active transport system most intensively studied is the membrane ATPase of the red blood cell and many other membranes, and is described in detail in Chapter 25. In this system, the enzyme ATPase is believed to be the actual carrier, and the process involves an exchange of Na^+ and K^+, rather than a simple transport of a single ion or molecule, e.g. X across the membrane. A change in conformation of ATPase is essential for transport to occur.

Several other transport systems are known to be coupled with Na^+/K^+ exchange catalysed by membrane ATPase and this process is described in detail in Chapter 16.

The mechanism by which the interaction of ion transport and glucose transport occurs is not clear. It is possible that Na^+ could increase the affinity of the carrier for glucose or, alternatively, the Na^+ could stimulate the velocity of the glucose–carrier complex across the membrane.

Many exchange transport processes have also been described for the mitochondrial membrane and these studies have shown that many interlinked systems are possible; they are described in detail in Chapter 6. It appears, therefore, that the role of a single carrier as originally visualized for one metabolite is unlikely to describe an accurate picture of membrane transport which will normally involve several complex interlinked systems.

Chapter 3

Role of subcellular organelles: lysosomes

The lysosomes are small sac-like organelles found in most cells but much more abundant in some cell types and tissues than in others. For example, they are abundant in macrophages but lymphocytes have relatively small numbers of lysosomes. They contain within their membrane a large variety of hydrolytic enzymes.

Lysosomes are not static organelles but move through a life cycle within the cell and take on states described as 'primary', 'secondary' and 'tertiary'. The significance of these terms will be apparent later in the chapter, but it should be appreciated that the term 'lysosomes' applies to a heterogeneous collection of cellular organelles of related but slightly different composition and activity.

The lysosomes are not found in bacteria but they appeared very early in evolution and play an important role in protozoa such as amoeba where they form the digestive system of the cell and thus play a vital role in the nutrition of the cell. During the course of evolution this role for lysosomes has been partially retained but often modified to suit other requirements. For example, lysosomes of macrophages play an important part in the body's defence against disease.

3.1 Origin of lysosomal enzymes

Lysosomal enzymes are proteins and, like many other proteins, are synthesized in the rough endoplasmic reticulum of the cells. It is possible that a specific area of the reticulum is involved in the synthesis. The newly synthesized proteins are then transported to the smooth endoplasmic reticulum, where they pinch out a small area of membrane to form vesicles, in which they are transported to the Golgi apparatus. The precise function of this organelle is uncertain, but it is believed that glycoproteins are incorporated here and each enzyme molecule may become attached to a carbohydrate or glycoprotein molecule. The enzyme–glycoprotein complex surrounded by a membrane then leaves the Golgi apparatus as a 'primary' lysosome (*Fig.* 3.1).

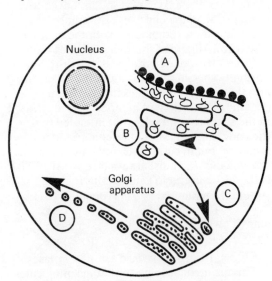

Fig. 3.1. **Schematic representation of biosynthesis of primary lysosomes.**
(*A*) Lysosomal enzymes are synthesized on the rough endoplasmic reticulum. (*B*) The complete lysosomal enzyme complement is then transported to the smooth endoplasmic reticulum where enclosure in a portion of membrane occurs. (*C*) The enzymes, now contained in a membrane sac, are transported to the Golgi apparatus. It is uncertain exactly what occurs in the organelle, but glycoproteins are incorporated into the membrane or into the enzyme complex or both (*D*).

3.2 The nature of the lysosomal enzymes

All the enzymes of the lysosomes catalyse the hydrolytic cleavage of covalent bonds, chiefly C—O, C—N or C—C. They are therefore termed *hydrolytic enzymes* or *hydrolases*. Expressed in simplest terms the reactions catalysed are of the following type:

$$A–B + H_2O \xrightarrow{\text{(E)}} A–H + B–OH$$

The enzyme (E) catalysing the reaction is usually named after the substrate (A–B) hydrolysed and in this example would be A−B−ase.

The precise mechanism of the hydrolysis is known for a few hydrolytic enzymes and an example is shown in Appendix 13.

As shown in Appendix 12, the International Union of Biochemistry (the Enzyme Commission) has placed hydrolases in class 3 of the enzyme list so that the code numbers of all the enzymes (EC numbers) in the lysosomes commence with the figure 3. The second figure of the code indicates the nature of the bond hydrolysed, the third figure normally specifies the nature of the substrate, and the final figure defines the particular enzyme. The system of numbering is illustrated in the scheme below. It is not intended that the code numbers should be committed to memory, but in current scientific literature the enzyme number must always be quoted and it is therefore useful to gain familiarity with the system. It is summarized in Appendix 12.

Examples of hydrolytic enzymes many of which occur in the lysosomes are:

EC 3.1 Enzymes hydrolysing ester bonds

EC 3.1.1 Carboxylic acid ester hydrolases
This group of enzymes will attack molecules such as simple esters, fats (triacylglycerols or triglycerides) and phospholipids.

EC 3.1.1.1 Esterases (non-specific),
e.g. methyl butyrate + $H_2O \rightarrow$ methanol + butyric acid
EC 3.1.1.3 Triacylglycerol lipase,
e.g. triacylglycerol + $H_2O \rightarrow$ diacylglycerol + fatty acids
EC 3.1.1.4 Phospholipase A_2,
e.g. phosphatidylcholine + $H_2O \rightarrow$ lyso-lecithin + fatty acids
EC 3.1.3 Enzymes hydrolysing phosphate esters
This group of enzymes hydrolyses the phosphate group from many different phosphate esters, such as glucose 6-phosphate or nucleotides.

EC 3.1.3.2 Acid phosphatase,
e.g. glycerol 3-phosphate + $H_2O \rightarrow$ glycerol + phosphate
This is a very important marker enzyme used for the study of lysosomal enzyme activity. Its activity can easily be determined by conventional methods and a simple histochemical method shows the location of this enzyme in the cells.
EC 3.1.3.4 Phosphatidate phosphatase,
phosphatidate + $H_2O \rightarrow$ diacylglycerol + phosphate
EC 3.1.3.31 Nucleotidase,
nucleotide + $H_2O \rightarrow$ nucleoside + phosphate
e.g. AMP + $H_2O \rightarrow$ adenosine + phosphate
EC 3.1.4 Enzymes hydrolysing phosphoric diesters
This is a very important group containing enzymes which hydrolyse the large nucleic acid molecules, DNA and RNA, into small fragments.

EC 3.1.4.5(6) Deoxyribonuclease I(II),
DNA + $(n-1)H_2O \rightarrow n$ oligodeoxyribo-nucleotides
EC 3.1.4.22(23) Ribonuclease I(II),
RNA + $(n-1)H_2O \rightarrow n$ oligoribo-nucleotides
EC 3.1.6 Enzymes hydrolysing sulphate esters
EC 3.1.6.1 Arylsulphatase,
e.g. phenol sulphate + $H_2O \rightarrow$ phenol + sulphate

EC 3.2 Enzymes hydrolysing glycosyl compounds
These enzymes hydrolyse many glycosides, and di-, tri- or polysaccharides into monosaccharides or short-chain saccharides

EC 3.2.1 Enzymes hydrolysing *O*-glycosyl compounds
EC 3.2.1.20 α-Glucosidase
Hydrolyses glucose linked by an α bond,

e.g. maltose + $H_2O \rightarrow 2\alpha$-glucose
EC 3.2.1.21 β-Glucosidase
X-β-glucose + $H_2O \rightarrow XH + \beta$-glucose
EC 3.2.1.24 α-Mannosidase
EC 3.2.1.35 Hyaluronoglucosaminidase
(hyaluronidase),
e.g. chondroitin sulphate + $H_2O \rightarrow$
N-acetylglucosamine + N-acetyl-
galactosamine
EC 3.2.1.50 N-Acetylglucosaminidase

EC 3.4 *Enzymes hydrolysing peptide bonds*

These enzymes hydrolyse large protein
molecules into small peptides or peptides into
amino acids. The lysosomal enzymes show very
close similarities to the digestive enzymes, such
as pepsin, which are also included in this
subclass.

EC 3.4.14.1 Dipeptidyl peptidase
(cathepsin C)
dipeptide–polypeptide + $H_2O \rightarrow$ dipeptide
+ polypeptide
EC 3.4.22.1 Cathepsin B
polypeptides + $H_2O \rightarrow$ peptides + amino
acids
EC 3.4.23.5 Cathepsin D
polypeptides + $H_2O \rightarrow$ peptides + amino
acids
Cathepsins B and D attack peptide chains
at different and specific amino acid
residues (cf. digestive enzymes,
Chapter 16)
EC 3.4.24.3 Collagenase
Degrades native collagen at the helical
region of the chain (cf. Chapter 32 for
discussion of collagen structure).

3.3 Investigation methods and properties of lysosomal enzymes

a. Methods of investigation

There are two very different approaches
available for studying lysosomal enzymes.
The lysosomes may be separated from a cell
homogenate by differential centrifugation as
described in Chapter 1. It is, however, very
difficult to obtain a pure preparation for the
lysosomes normally sediment with the
mitochondria. Fortunately, for many

experiments, the mitochondria do not
interfere, and many studies on 'lysosomes'
have used the mixed suspension. In these
preparations, the lysosomal enzymes
demonstrate 'latency', that is, when suspended
in an isotonic medium, the enzyme activity is
very low unless the lysosomal membrane is
disrupted, for example, by the addition of a
detergent.

Alternatively, lysosomes can be
studied histochemically. In these methods
described in Chapter 1 very thin frozen
sections are cut in a special cryostat. These
tissue sections are then immersed in a solution
of the substrate of the enzyme under
investigation (e.g. glycerol 3-phosphate for
acid phosphatase) and kept at the optimum pH
for a specified time. The product of the enzyme
reaction, which is formed at the site of
lysosomal enzyme activity, is visualized by
a specific staining method (*see Fig.* 3.4).
Recently it has become possible to make
quantitative measurements using this method
(cf. Chapter 1).

b. Specificity

Many of the lysosomal enzymes exhibit a very
low degree of specificity, for example esterase
will hydrolyse a very large number of different
esters and triacylglycerol lipase has little
specificity for particular acid residues, although
the rates of hydrolysis of different triacylglycerols vary. The enzymes hydrolysing
glucose-containing carbohydrates have a high
degree of specificity for the glucose residue and
for the α- or β-linkage, but virtually none for
the residue forming the glycoside.

Cathepsins, like the proteolytic
enzymes of the gastrointestinal tract, are
usually specific for certain amino acids in the
peptide chain (cf. Chapter 16).

c. Optimum pH

All the enzymes of the lysosome have their
optimum pH well into the acid range, generally
pH 4·0–5·0 (*Fig.* 3.2). This is unusual because
the optimum for most enzymes of the cell is
much closer to neutrality, frequently lying
between pH 6·5 and 7·5. Though it is very

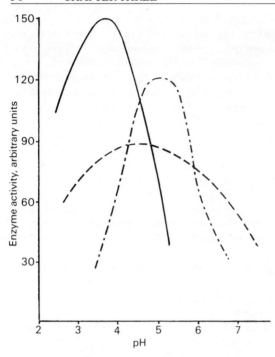

Fig. 3.2. **Optimum pH values for spleen macrophage lysosomal enzymes.**
(——) Cathepsin D; (–·–) ribonuclease; (– –) acid phosphatase.

difficult to measure the pH within the cytoplasm, it is generally assumed to be approximately neutral (pH 7·0). Within the lysosome, however, the pH may be much more acid.

The reason for the acid pH optima of lysosomal enzymes is not at all clear; it may serve as a protective mechanism, rendering an enzyme relatively inactive should it leak out into a normally functioning cell.

d. Sulphydryl enzymes

Some of the lysosomal enzymes, especially the cathepsins, must have at least one, and sometimes several, free —SH groups of cysteine residues for full activity. Such enzymes are termed 'sulphydryl' or —SH enzymes. If the —SH groups are oxidized to disulphides, —S—S—, or combined with an alkylating agent such as iodoacetamide,

$$ICH_2CONH_2 + E—SH \rightarrow$$

$$E—SCH_2CONH_2 + HI$$

the enzyme activity is destroyed. These enzymes are also very sensitive to metal ions, such as Cu^{2+} and Hg^{2+}, that combine with —SH groups.

3.4 The life cycle of the lysosome

Although this is not identical in all cells, the typical process is shown in *Fig.* 3.3, and it may take one of two pathways termed 'heterophagy' and 'autophagy'.

Heterophagy

The process is typified in the nutrition of a simple eukaryotic cell, such as an amoeba. Particles of food (X) are taken into the cell surrounded by a membrane which is sometimes called a *heterophagic vacuole*, and within the cell they fuse with a *primary lysosome*. Special recognition sites, containing the carbohydrate moieties of the membrane glycoproteins, must be present on the membranes of the vacuole and primary lysosome for recognition to take place. Fusion causes a *secondary lysosome* to be formed and active digestion of the ingested particle occurs. The food material, whether protein, carbohydrate or lipid, is very effectively hydrolysed by the many enzymes in the lysosome so that the fragments, usually amino acids or monosaccharides, can be

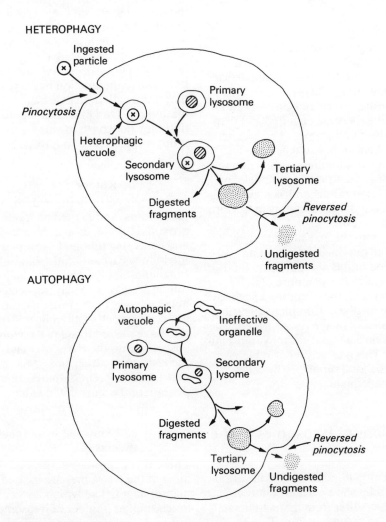

Fig. 3.3.   **Heterophagy and autophagy.**

released into the cytoplasm for further metabolism. Undigested particles that cannot be attacked by the enzymes sometimes remain; organelles containing such particles are termed *tertiary lysosomes*. The particles may be expelled by reversing the process of pinocytosis, but occasionally the particles may be retained within the tertiary lysosomes. Often this retention gives rise to pathological conditions, some of which are discussed below (*see* pp. 41–42). Heterophagy is a common occurrence in the cells of many mammalian tissues and its importance is discussed below and on p. 40.

Autophagy

The process is essentially similar to that of *heterophagy*, but the particle trapped in a vacuole is usually another subcellular particle or organelle (*Fig.* 3.3).

There are several reasons why certain cellular organelles become engulfed and then digested by lysosomal enzymes. The organelle, such as a mitochondrion, could be deteriorating and not fulfilling its function or, under conditions of restricted food intake or starvation, far fewer mitochondria may be needed to produce the cell's energy in the form of ATP (cf. Chapter 6).

There must be an elaborate signalling system in the cell by which deteriorating organelles or conditions of inadequate food supply are recognized. The nature of these stimuli is not yet clear but hormones, such as insulin, may be involved. Complete digestion of cellular organelles by the lysosomal enzymes will release amino acids and carbohydrates into the cytoplasm. This is clearly a desirable process when the food supply to the cell is depressed or inadequate.

3.5 Functions of lysosomes in the tissues

Lysosomes perform important roles in various tissues. These roles show certain common features, but they differ from tissue to tissue and are related to tissue function. These functions are summarized below.

a. General

In all tissues, lysosomes play an important 'autolytic' role. Dying or dead cells rapidly become more acid because impairment of the oxygen supply prevents the oxidation of the pyruvic acid produced by glycolysis (cf. Chapter 5 and Appendices 15 and 16) which is rapidly converted to lactic acid and, consequently, there is a rapid drop of pH in the dying cells. A fall in cell pH always causes activation of lysosomal enzymes that leak out into the cytoplasm and, in this new acid environment close to their optimum pH, they are very active. The enzymes rapidly digest all the cell contents, digested fragments entering the bloodstream to join the general metabolic pool. This is an important mechanism for disposing of degenerate or dead cells.

This process was originally believed to be the main function of the lysosomes and they were called 'suicide bags' by their discoverer, Du Duve. But it is now realized that they take part, in various tissues, in many more complex functions discussed below.

b. Kidney

Lysosomes occur in the endothelial cells of the glomerulus, and in the lining cells of the proximal tubule, loop of Henle, distal tubule and collecting tubule. Lysosomes of all parts of the kidney carry out autophagy of cellular organelles but lysosomes of the proximal tubule are likely to play a special role in heterophagy. In the proximal tubule protein molecules, such as albumin and haemoglobin, that have passed through the glomerulus are absorbed into the lysosomes and degraded by the lysosomal cathepsins. It has been estimated that 10–15 per cent of total serum albumin degradation occurs in the kidney.

c. Lymphoid tissue: spleen and thymus

These tissues contain several cell types but most of the lysosomes are located in the macrophages. The lymphocytes contain a much smaller number of lysosomes, as shown in *Fig.* 3.4, where slices of spleen and thymus tissue have been specifically stained to localize

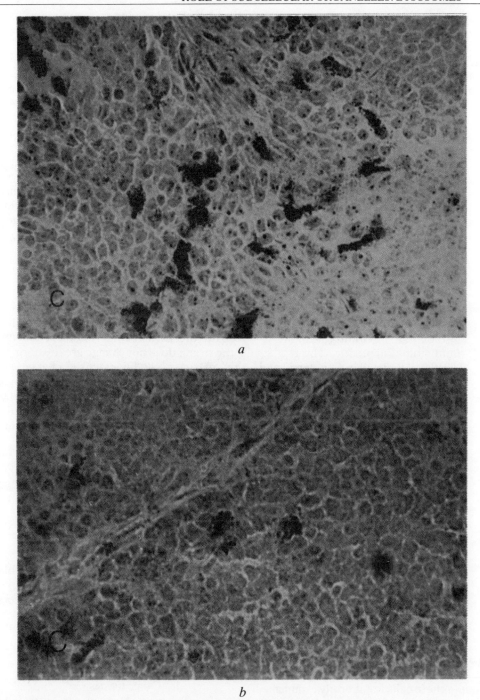

Fig. 3.4. **Histochemical staining of lysosomes in (*a*) spleen and (*b*) thymus.**
Sections of tissue (8-μm thick) are incubated in glycerol 3-phosphate medium as a substrate for acid phosphatase. Enzyme activity is shown by intense dark stain. Note the relatively small numbers of cells (macrophages) that stain intensely. The other cells in the tissues have very small numbers of weakly active lysosomes. Magnification $\times 200$.

one enzyme, acid phosphatase. It will be observed that only a relatively small proportion of the cells stain for this lysosomal enzyme; these are the macrophages. The functions of the lysosomes in lymphoid tissue are primarily concerned with two processes:

i.   The macrophages, by a process of endocytosis, rapidly take up foreign proteins or other foreign debris entering the bloodstream, and the spleen macrophages take up and engulf red cells which have become damaged or aged. Cells of the spleen or thymus which become damaged, for example by irradiation, are also taken up. Once inside the macrophages, the waste material is broken down by lysosomal heterophagy and subsequent digestion, as described above on pp. 37–38.

ii. The lysosomes of the macrophage also play an important role in the immunological process. This is discussed in detail in Chapter 42 and at this stage the process will simply be outlined. The macrophage is believed to take up antigens by endocytosis into its lysosomes, a step that appears to be essential before the lymphocytes can produce antibodies. However, the exact role played by the lysosomes in the macrophages and whether complete or partial digestion of the antigen occurs, is uncertain. It would appear that the lysosome is playing a distinct and different role that cannot easily be equated with the normal heterophagic and autophagic processes.

d. Nervous system

Lysosomes have been demonstrated in neurones and Schwann cells. Their precise function in these cells is not firmly established, but they are believed to play an autophagic role by removing damaged proteins and other organelles. Neuronal lysosomes can accumulate particles of lipids or

polysaccharides in disease conditions, as described below (*see* p. 41).

e. Bone

Lysosomes play an essential part in resorbing and remodelling the cells of the bone (*see* Chapter 32), a process which involves the osteoclast cells (rich in lysosomes) and possibly also the osteocytes. Enzymes, such as hyaluronidase, hydrolyse proteoglycans (e.g. hyaluronic acid), and those that hydrolyse the collagen fibres (peptidases and collagenase) are the most active during bone resorption, but not through the normal processes of autophagy or heterophagy. They are secreted from the cells to act on the bone matrix.

f. Uterus

Lysosomes occur in the phagocytic cells, macrophages, stromal cells, and epithelial cells of the uterus. The total lysosomal activity in the organ does not change greatly during the menstrual cycle, but extensive changes occur in the distribution of active enzymes amongst different cell types and within single cell types. The significance of these changes is unclear but autophagic activity is increased at the end of the cycle.

Post-partum involution of the uterus occurs rapidly and is entirely dependent on very active lysosomal activity of the phagocytic cells. Both heterophagy and autophagy are involved.

g. Mammary gland

Increased lysosomal activity is responsible for involution of the mammary gland at the cessation of lactation. Autophagic processes reduce the cytoplasmic contents and heterophagic processes remove dead cells and milk.

h. Other tissues

Lysosomes have also been demonstrated in many other tissues, such as the anterior pituitary gland, the testes, prostate and spermatozoa. In spermatozoa the lysosomal hyaluronidase, situated near the surface membrane, is essential for penetration of the ovum.

The lysosomes can sometimes play a special role in the release of stored material such as hormones, and a good example of this is seen in the adrenal medulla. This gland (cf. Chapter 17) stores and secretes catecholamine hormones, and particularly adrenaline, in the specialized 'chromaffin' cells which contain unique cellular organelles—'chromaffin granules'. It has been postulated that, after fusion of these granules with primary lysosomes to form secondary lysosomes, the hydrolytic enzymes bring about the secretion of adrenaline into the circulation possibly by releasing the bound hormone from a protein carrier. Any triggering mechanism for the release of the hormone would either stimulate the fusion of chromaffin granules with the hormone or activate the lysosomal enzymes.

3.6 Lysosomes in pathological conditions

During the past few years many studies have been made of the possible roles played by lysosomes in several pathological conditions. Many speculative hypotheses have been advanced and much further research is required to assess their validity.

We have noted already that lysosomal enzymes become very active during cell death and that it is almost inevitable that lysosomes will play some role in the development of nearly all pathological conditions of the cell. It is, however, very difficult to establish whether lysosomal changes are a primary event or secondary to other damage.

How may the activity of the lysosomes be altered? Several possibilities have been considered and investigated:

i.   Modification of the enzyme content of the lysosome or modification of the activity of the enzymes.

ii. Changes in the permeability of the lysosomal membrane. Leakage of enzymes could eventually lead to serious cellular damage.

iii. Retention of undigested particles in the 'tertiary lysosomal' state for prolonged periods.

Examples of pathological processes in which lysosomes have been implicated are listed below.

a. Excess of vitamin A
(cf. Chapter 14)

It has been known for many years that vitamin A in excess is very toxic and even fatal. High doses of the vitamin cause the cessation of growth of young animals, congenital malformations, and abnormalities and fractures of bone. Studies on cultured bone cells *in vitro* have shown that vitamin A (retinol) is very effective in releasing hydrolytic enzymes, e.g. cathepsins, from the cells into the media. Lysosomal enzymes from many other tissues are released by vitamin A and there is thus good evidence that excesses of vitamin A may act in this manner in the living body.

b. Arthritis

Arthritis is an inflammation of the joints and lysosomes play an important role in the disease. Although the whole process is complex and not completely understood, it is likely that joint damage in acute arthritis is caused mainly by lysosomal enzymes.

These enzymes may be released from the cells by bacteria or the toxins they produce, but it is possible that the condition may be caused by failure of the cells to synthesize a balanced complement of lysosomal enzymes. Surplus enzymes would then be secreted and if these are of the type which cause extensive bone degradation, e.g. cathepsins or hyaluronidase, then arthritic conditions would develop. Recent research has shown that attack on unsaturated fatty acids of lysosomal membrane phospholipids by free radicals could cause damage to the membrane by peroxidation (cf. Chapter 14) and consequent leakage of enzymes.

c. Genetic abnormalities

Several conditions are known in which one or more of the important lysosomal hydrolases are not synthesized. A metabolic block can then develop and 'polysaccharidoses' or 'lipidoses' can occur, as shown by the following examples:

Type II glycogenolysis or glycogen storage disease. These patients have very large deposits of glycogen in their tissues and suffer from serious muscular weakness due to the glycogen deposits in the muscle fibres. This has been traced to a deficiency of the enzyme acid α-glycosidase (maltase), which degrades glycogen to glucose. Glycogen particles are taken up by autophagy but, in the absence of the enzyme, cannot be digested and accumulate. These patients have a normal complement of enzymes for degradation of glycogen by phosphorolysis and are not hypoglycaemic.

Hurler syndrome. The Hurler syndrome is characterized by dwarfism, hepato-splenomegaly, skeletal deformities, and mental retardation. In this condition the lysosomes contain large deposits of mucopolysaccharides (proteoglycans) and glycolipids. Although other theories have been advanced, there is good evidence that the defect is caused by the absence of enzymes in the lysosomes to degrade proteoglycans and glycolipids.

d. Teratogenesis and cancer

Teratology, the study of congenital defects, has led to the study of teratogenic factors or chemicals which can cause abnormalities. The importance of the subject was brought to public notice by the dramatic effects of the drug thalidomide, but many compounds probably cause more subtle changes.

The dye trypan blue can cause teratogenic effects in rats. It is taken up into the lysosomes and can remain there for long periods. Furthermore, it has been demonstrated to inhibit the activity of many lysosomal enzymes. Many other drugs and chemicals may also behave similarly.

Although the role of lysosomes in causing teratogenic effects is speculative, it is known that lysosomes play an important part in remodelling processes during embryonic development. It is, therefore, possible that interference with their action could cause, ultimately, serious congenital defects.

There is experimental evidence that many chemical carcinogens, such as the polycyclic hydrocarbons (cf. Chapter 38), are concentrated in the lysosomes after uptake into cells. Subsequent mechanisms by which the lysosomes may be involved in carcinogenic processes are unclear. They may simply act as a store for the carcinogen, leading to long-term effects on other organelles, e.g. the nucleus, or they may cause specific release of lysosomal enzymes which cause serious cell damage but do not kill the cell. One possibility is that DNAase may be released from the lysosome by the carcinogen under certain conditions. This may enter the nucleus and cause damage to the DNA molecules, which could be transmitted to subsequent generations of cells.

Chapter 4   Role of subcellular organelles: peroxisomes

4.1 Historical and background

During their classic studies on the separation of the lysosomal fraction by centrifugation of rat liver homogenates, De Duve and his coworkers observed that certain oxidation enzymes, in particular uric acid oxidase, D-amino acid oxidase and the enzyme attacking hydrogen peroxide (catalase), were always associated together in a discrete fraction which could be separated from the lysosomes and mitochondria.

These enzymes were believed to be associated in special organelles called 'peroxisomes'. Subsequently, other tissues, e.g. the kidney, protozoa and plant seedling cells were also shown to contain similar organelles. Further research enabled the list of enzymes associated with the organelle to be extended to include a group catalysing certain special oxidation reactions.

4.2 Structure of the peroxisome

Most research has supported the view that, in liver and certainly in kidney, the 'peroxisomes' are identical with the organelles previously described by morphologists as 'microbodies'. Microbodies are characterized by the possession of a single limiting membrane and a fine granular matrix. They also have a denser inner core or nucleoid with a polytubular structure.

It is believed that the peroxisome proteins are formed on the endoplasmic reticulum and that the peroxisomes are then produced by a budding process.

4.3 Enzymic complement of the peroxisomes

Enzymes so far demonstrated to occur in mammalian tissues are listed in *Table* 4.1. In addition to these enzymes, several enzymes such as citrate synthase, isocitrate lyase and malate synthase have been demonstrated to occur in plant peroxisomes and may also occur in some mammalian peroxisomes.

Table 4.1 **Enzymes which occur in liver or kidney peroxisomes**

Enzyme	Reaction catalysed
Calatase	$H_2O \rightarrow H_2O + O_2$
Urate oxidase	Urate $\rightarrow$ Allantoin + H_2O_2
D-Amino acid oxidase	D-Amino acid $\rightarrow \alpha$-Ketoacid + H_2O_2
L-Amino acid oxidase	L-Amino acid $\rightarrow \alpha$-Ketoacid + H_2O_2
Glycolate oxidase	Glycolate $\rightarrow$ Glyoxylate + H_2O_2
α-Hydroxyacid oxidase	α-Hydroxyacid $\rightarrow \alpha$-Ketoacid
Glyoxylate oxidase	Glyoxylate $\rightarrow$ Formate
Isocitrate dehydrogenase	Isocitrate $\rightarrow \alpha$-Ketoglutarate

It will be noted that several of these enzymes produce hydrogen peroxide. This is destroyed by catalase which either catalyses the conversion of hydrogen peroxide to water and oxygen, or uses it in its own peroxidase action.

4.4 Biological functions of peroxisomes

A possible scheme that links the enzyme

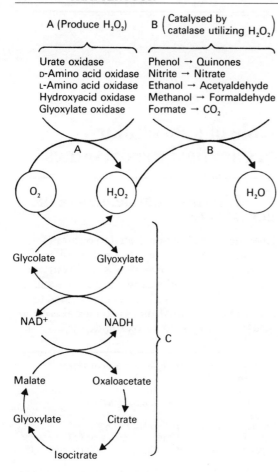

Fig. 4.1.   **Metabolic functions of the peroxisomes.**

It will be noted that, in the proposed scheme, the glycolate–glyoxylate system (C) can be maintained as a cyclic system by a series of reactions involving isocitrate dehydrogenase and malate dehydrogenase. The enzymes catalysing this part of the system have not yet been demonstrated in mammalian cells, so that this part of the scheme, although established in plant cells, is of doubtful significance in animal cells. However, the link of the A group of enzymes with the B group of reactions through H_2O_2 can occur without the cyclic glyoxylate system.

The (A–B) system is not coupled to energy conservation and represents a wasteful form of respiration. Furthermore, the rate of oxygen utilization is directly proportional to the oxygen tension. This fact suggests that the peroxisomes might be part of a protective mechanism against oxygen toxicity. In cells where the glycolate–glyoxylate system functions, the peroxisomes may also play a role in regulation of the balance between the $NAD^+/NADH$ ratios in the cell.

4.5 Evolutionary history of peroxisomes and its significance

It has been postulated by De Duve that the peroxisomes represent a very primitive cellular organelle evolved when cellular organisms first adapted to aerobic conditions. The establishment of oxygen in the atmosphere frequently gives rise, in spontaneous reactions, to the formation of hydrogen peroxide, necessitating the development of mechanisms for the destruction of this toxic product. The peroxisome may thus have developed as a protective organelle whilst energy was derived from anaerobic metabolism. In the course of evolution, mitochondria, believed to have arisen originally from parasitic bacteria, then gradually took over the utilization of oxygen in the cell providing an effective mechanism of energy conservation.

During the course of evolution, many key enzymes of the peroxisomes, such as those of the glyoxylate cycle, may have been lost; more recently, in man and apes, urate oxidase

systems discovered in peroxisomes is illustrated in *Fig.* 4.1. It must be appreciated that not all these enzymes have been detected in mammalian peroxisomes, some having, so far, been detected only in plants or protozoa.

The system proposed assumes that coupling takes place between a group of oxidases producing H_2O_2 (A) and a group of reactions (B) catalysed by catalase acting, in the presence of H_2O_2, as a peroxidase. This system is believed to occur in most peroxisomes including mammalian peroxisomes. The oxidation of ethanol and formate are likely to be of the greatest metabolic importance in mammals.

has also disappeared. In higher animals, it could, therefore, be the case that the peroxisomes have a very limited role in metabolism and merely act as a protection against increases of oxygen tension; they could represent a type of biochemical anachronism.

Chapter 5 Role of subcellular organelles: metabolism in the cytosol

5.1 Introduction

The cytosol, or cell sap, is the fluid which permeates the whole internal environment of the cell and is thus in contact with all cellular organelles. Consequently, it is clearly an important vehicle for the transport of metabolites from one organelle to another.

It is interesting that, for histologists, cytologists and electron microscopists, the cytosol has traditionally been of little or no interest since it has no structural form that can be recognized. For example, electron microscopy shows most of the cytosol as a clear background and in conventional histology the cytosol usually appears as a uniformly stained background. For these reasons, many biologists have tended to ignore the importance of the cytosol, but biochemical studies have shown that it is of great importance and that a large proportion of cellular metabolic activity occurs in this cellular compartment.

5.2 Preparation of the cytosol

For biochemical study, a dilute solution of the cytosol is normally prepared by ultra-centrifugation. After homogenization of the tissue in sucrose solutions as described in Chapter 1, the dilute homogenate is subjected to 60–90 min centrifugation at $100\ 000 \times g$ in the ultracentrifuge. The supernatant is described as the cytosol, although it is normally diluted by a factor dependent on the volume of tissue used for the preparation and on the volume of suspending medium used to prepare the homogenate. Preparation is relatively simple but the cytosol, as prepared, may be contaminated with proteins or small molecules that have leaked from partially damaged organelles. It is, for example, extremely difficult to prepare a cytosolic fraction free of enzymes that have leaked from lysosomes damaged in the homogenization.

5.3 Composition of the cytosol

The cytosol is a dilute aqueous solution containing a very large number of different water-soluble proteins, organic and inorganic molecules as listed in *Table* 5.1.

Table 5.1 **Constituents of the cytosol**

Proteins	
Enzymes	Enzymes involved in glycolysis, the pentose phosphate pathway, glycogen synthesis and degradation, amino acid metabolism and fatty acid synthesis
Other proteins	Many protein molecules involved in transport within the cell, e.g. carriers of metals, steroid hormones
Metabolites	All water-soluble metabolites are present such as those produced by glycolysis, the pentose phosphate pathway, amino acids and ketoacids
Inorganic constituents	All normally occurring cations such as Na^+, K^+, Ca^{2+}, Mg^{2+} and anions such as HCO_3^-, Cl^- and phosphate are present

5.4 Functions of the cytosol

Carbohydrate metabolism

The cytosol plays a very important role in the metabolism of glucose by the glycolytic pathways or glycolysis (*Fig.* 5.1). Glucose entering the cell is first phosphorylated by the enzyme hexokinase or glucokinase to form glucose 6-phosphate. The hexose phosphate, fructose 1,6-diphosphate, is split into two triose phosphate molecules which are then converted to pyruvate. If anaerobic conditions

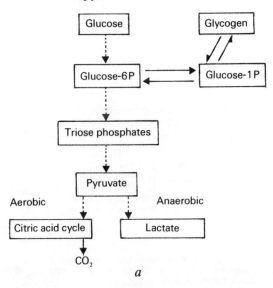

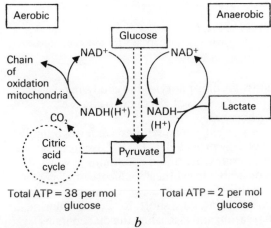

Fig. 5.1. **Glycolysis.**
(*a*) Major steps of glycolysis. (*b*) Aerobic and anaerobic glycolysis.

prevail, then pyruvate is reduced to lactic acid, a reaction which regenerates the NAD^+ from NADH and which is essential for the continued oxidation of the triose phosphate. In aerobic conditions, however, NADH may be reoxidized by the mitochondria and pyruvate enters the mitochondria where it is completely oxidized to carbon dioxide and water.

The process of glycolysis, shown in outline in *Fig.* 5.1, produces two molecules of ATP even under anaerobic conditions and thus plays an important role in energy production. Its second role is the production of pyruvate which is a very important substrate for oxidation by the citric acid cycle in the mitochondria. Stored glycogen can also be used as a substrate for glycolysis and this is first converted to glucose 1-phosphate.

The process of glycolysis is controlled at the stages shown in *Fig.* 5.2. This is essential because it may be necessary to store glucose as glycogen or to utilize it for energy production. An important site of control is the regulation of the enzyme phosphofructokinase which converts fructose 6-phosphate to fructose 1,6-diphosphate. This enzyme is allosterically inhibited by ATP or citrate and activated by ADP or AMP. If the supply of energy is adequate, as indicated by the excess availability of ATP or by the high concentrations of citric acid cycle intermediates, then glycolysis is switched off or dampened down. Conversely, it is stimulated by ADP or AMP. The enzyme forming pyruvate, pyruvate kinase, is similarly controlled.

The enzymes involved in glycogen synthesis and degradation occur mainly in the cytosol and the metabolic processes are summarized in *Fig.* 5.3. Glucose 1-phosphate plays a key role in glycogen degradation, being formed by the action of phosphorylase and converted to UDP-glucose for synthesis into glycogen (*Fig.* 5.3).

It is important to note that two entirely different enzyme systems are used for the synthesis of glucose 6-phosphate from glucose and for the hydrolysis of glucose 6-phosphate to glucose and phosphate. Glucose 6-phosphate is synthesized from glucose and ATP using the enzyme hexokinase

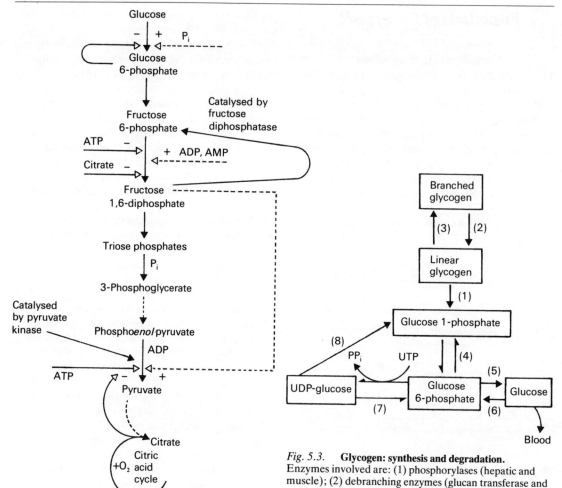

Fig. 5.2.   **Regulation of glycolysis.**
Sites of control indicated by open arrowheads.

Fig. 5.3.   **Glycogen: synthesis and degradation.**
Enzymes involved are: (1) phosphorylases (hepatic and muscle); (2) debranching enzymes (glucan transferase and amylo-1,6-glucosidase); (3) glycosyl (4,6) transferase; (4) phosphoglucomutase; (5) glucose 6-phosphate; (6) glucokinase (liver) or hexokinase (all tissues); (7) UDP-glucose phosphorylase; (8) glycogen synthase.

(or glucokinase), whereas glucose 6-phosphate is hydrolysed by glucose 6-phosphatase. This enzyme is not free in the cytosol, but is bound to the membranes of the endoplasmic reticulum where glycogen granules are also bound.

Synthesis and degradation of glycogen is controlled by a series of elaborate inter-related mechanisms (*Fig.* 5.4). The degradation of glycogen is initially activated by stimulation of the action of adenylate cyclase by adrenaline or glucagon. This causes activation of a protein kinase which in turn

activates phosphorylase so catalysing phosphorolysis of glycogen to glucose 1-phosphate.

Glycogen synthesis requires a supply of energy provided by UTP. Glycolysis can be put into reverse under conditions in which it is desirable to synthesize glucose or glycogen and this will usually occur when supplies of carbohydrates are short, for example during starvation and also when the diet contains large quantities of protein and little carbohydrate (*Fig.* 5.5). Starting materials are lactate, often formed as a result of muscle activity, or

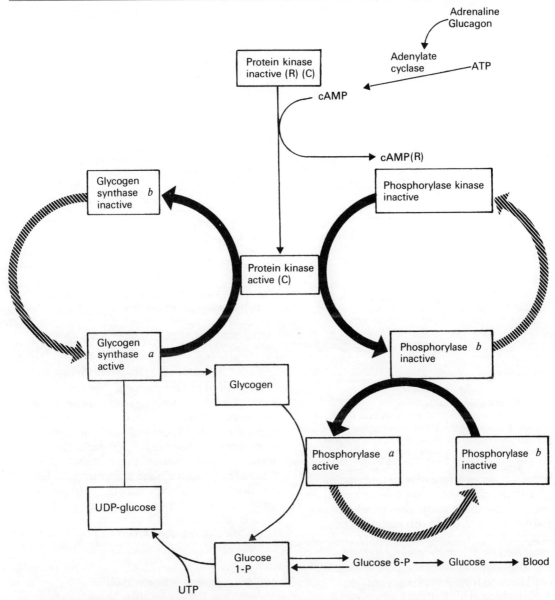

Fig. 5.4. **Regulation of metabolism of glycogen.**
Glycogen synthase *b* is glucose 6-phosphate dependent being activated by glucose 6-phosphate and inhibited by phosphate ions; glycogen synthase *a* is glucose 6-phosphate independent. Phosphorylase *a* is active and phosphorylase *b* is inactive but activated by AMP. ➡ Kinase reactions; ⇏ phosphatase reactions.

oxaloacetate which can be formed from any constituent of the citric acid cycle in the mitochondria and then transported into the cytoplasm, often as malate (cf. Chapter 7). As the action of pyruvate kinase cannot be reversed, phospho*enol*pyruvate is synthesized using phospho*enol*pyruvate carboxykinase. The metabolic reactions which follow proceed in the reverse direction to normal glycolysis, but fructose diphosphate is converted to fructose 6-phosphate by the diphosphatase enzyme.

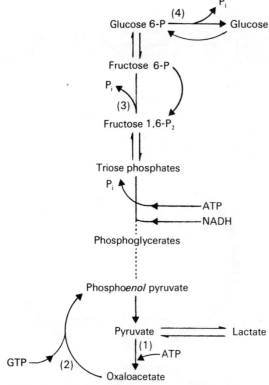

Fig. 5.5.   **Gluconeogenesis.**
Enzymes involved which are not part of the pathways of
glucose degradation are: (1) pyruvate carboxylase (biotin);
(2) phospho*enol*pyruvate carboxykinase; (3) fructose
1,6-diphosphatase; (4) glucose 6-phosphatase.

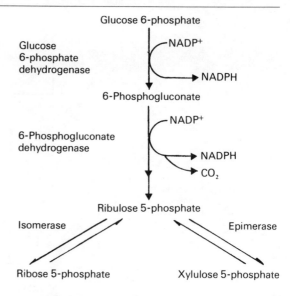

Fig. 5.6.   **Essential metabolic stages of the pentose
phosphate pathway (pentose shunt).**

As an alternative to undergoing
glycolysis, glucose 6-phosphate can enter
the pentose shunt in the pentose phosphate
pathway (*Fig.* 5.6). In this pathway, glucose
6-phosphate undergoes immediate oxidation
catalysed by a dehydrogenase to form
6-phosphogluconate and then a second
oxidation to form a pentose, ribulose
5-phosphate. This pathway serves two main
purposes: it is the main route of synthesis of
pentoses required for synthesis of nucleotides
and nucleic acids, and it also provides a supply
of NADPH.

The reduced form of NADP+
(NADPH) is essential for fatty acid synthesis
and also for the oxidation of enzyme systems
incorporating cytochrome P450 which are
used in the metabolism of xenobiotics on the
endoplasmic reticulum (cf. Chapter 36).

Citric acid cycle intermediates

Nearly all of the reactions of the citric
acid cycle (Appendices 17–19) can take place
within or outside the mitochondria, but only
certain metabolites readily pass through the
mitochondrial membrane. Of particular
importance are malate and citrate (cf.
Chapter 7). Both malate and citrate can be
oxidized by their respective dehydrogenase
enzymes and, alternatively, citrate can be split
by citrate lyase to reform oxaloacetate and
acetyl-SCoA (cf. *Fig.* 7.11, Chapter 7).

Amino acid metabolism

Very extensive metabolism of amino acids
takes place in the cytosol, and the reactions
include oxidative deamination,
decarboxylation and transamination.
Oxidative decarboxylation of α-ketoacids
formed from the branched-chain amino acids,
also takes place. These reactions, summarized
in *Fig.* 5.7, are not, however, always confined
to the cytosol because some of them, such as
transamination, can also occur within the
mitochondria.

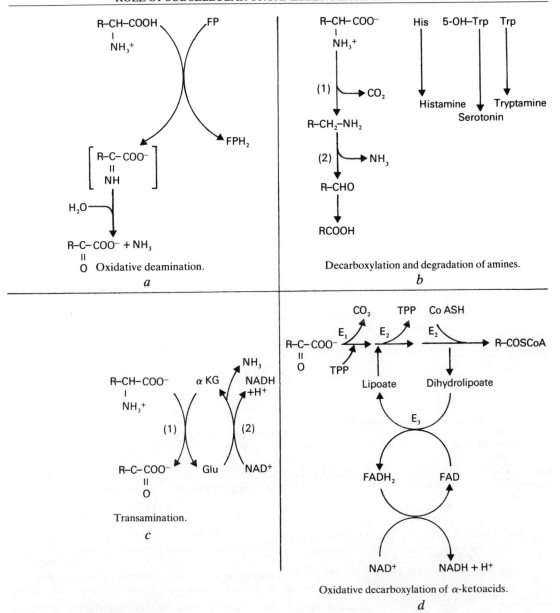

Fig. 5.7.   **Metabolism of amino acids.**

(*a*) FP is a flavoprotein for the majority of amino acids with the exception of the L-glutamate dehydrogenase enzyme where NAD^+ or $NADP^+$ is the acceptor.

(*b*) The enzymes are (1) decarboxylase and (2) monoamine oxidase.

(*c*) The enzymes are (1) transaminases, e.g. glutamate–oxaloacetate transaminase which catalyses aspartate $\leftrightarrows$ oxaloacetate and glutamate–pyruvate transaminase which catalyses alanine $\leftrightarrows$ pyruvate; (2) glutamate dehydrogenase; $\alpha KG = \alpha$-ketoglutarate; Glu = glutamate; note that reactions (1) and (2) are reversible.

(*d*) TPP = thiamin pyrophosphate.

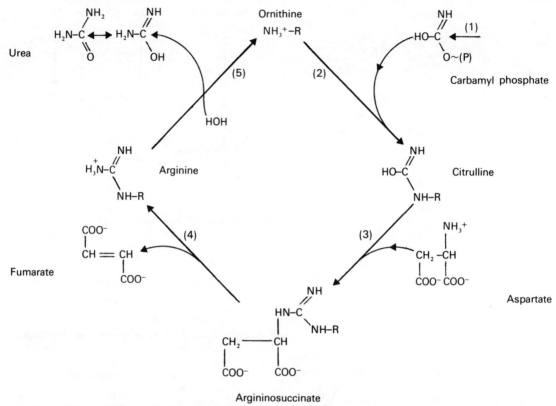

Fig. 5.8.   Major stages in urea formation (the ornithine cycle).
Enzymes involved are: (1) carbamylphosphate synthetase, ammonia dependent and activated by acetyl glutamate;
(2) ornithine carbamyl transferase (OCT); (3) argininosuccinate synthetase; (4) argininosuccinate lyase; (5) arginase.

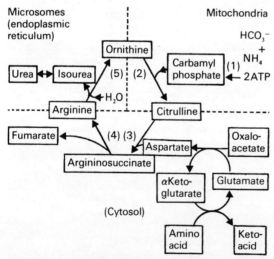

Fig. 5.9.   Subcellular locations of the ornithine cycle in the liver.
Enzymes are identified in *Fig.* 5.8.

For some metabolic pathways there is a complex inter-relationship between the reactions occurring in the cytosol and in the mitochondria. A good example is that of the ornithine cycle which is the major pathway for the synthesis of urea (*Fig.* 5.8), part of which occurs in the cytosol, part in the mitochondria and part in the endoplasmic reticulum (*Fig.* 5.9). In this system, the ornithine carrier plays an important role by transporting ornithine into the mitochondria and this is coupled with the transport out into the cytosol of citrulline (cf. Appendix 25).

Fatty acid oxidation

Fatty acids are transported in the plasma on albumin complexes and they are oxidized within the mitochondria. Before they can be transported into the mitochondria from the cytoplasm, they must first be converted to carnityl derivatives. This occurs in two stages: the fatty acids are initially converted to coenzyme A derivatives which then take up carnitine and are transported across the mitochondrial membrane as carnitine complexes (cf. Chapter 7).

Purine, pyrimidine and nucleic acid metabolism

In the cytosol many of the stages of purine, pyrimidine and nucleotide synthesis take place but some stages of nucleotide synthesis, for example that of NAD^+, may require that one of the stages occurs in the nucleus.

All cellular RNA and DNA is synthesized in the nucleus, but the attachment of amino acids to tRNAs for protein synthesis, catalysed by aminoacyl-tRNA synthetases, occurs in the cytosol.

In cells in which protein synthesis is occurring at a rapid rate, the concentration of the transferase enzymes in the cytosol is very high and they can form a high proportion of the total cytosolic protein.

Chapter 6 Role of subcellular organelles: mitochondria and energy conservation

6.1 Introduction

By the end of the last century, it was known from microscopical studies that most cells contained small elongated bodies to which the name 'mitochondria' was given. Although several researchers had indicated that these organelles could play a major role in cellular oxidation, their true function remained obscure for nearly 50 years. A very important development, however, came in 1948 when it was demonstrated that the subcellular organelles of homogenized tissues could be separated by differential centrifugation. Homogenates prepared in sucrose solution (0·25 M), and then subjected to centrifugation at varied speeds (measured as multiples of the force of gravity, g) and varied periods of time, yield purified suspensions of organelles such as nuclei, mitochondria or microsomes. Mitochondria sediment, after removal of the nuclei, by centrifugation for 15–20 minutes when subjected to a force of $(10\ 000–12\ 000) \times g$ in the ultracentrifuge, a typical scheme of separation being shown in *Fig.* 6.1.

It had been known for over two hundred years that most cells and tissues consumed oxygen, but it was generally assumed that the oxygen consumption was relatively uniform throughout the cell, and that all components of the cell utilized oxygen to an equal extent. Studies of separated mitochondria showed, however, that this was not so and that a large percentage of the oxygen consumption by cells occurred in the mitochondria. Further investigations demonstrated that the oxidation of metabolites conserved the energy produced in the chemical

form of ATP, and that mitochondria were the main site of formation of ATP from ADP in living cells.

Mitochondria are complex organelles, possessing inner and outer membranes, and even their own DNA that

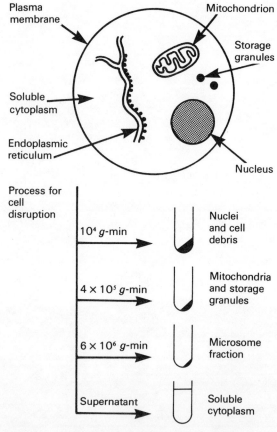

Fig. 6.1. **Separation of subcellular components by centrifugation.**

54

can code for some, but not all, of their own proteins. This apparent degree of independence shown by the mitochondria, has initiated the proposition that mitochondria were originally bacteria which, long ago, in the course of evolution, became assimilated into eukaryotic cells, to carry out their oxidative metabolism in a type of symbiosis.

A typical cell, such as a rat liver cell, normally contains about 800 mitochondria, but renal tubule cells contain about 300. Sperm cells may contain only 24 mitochondria but the egg cell of the sea urchin contains about 14 000.

6.2 Mitochondria in typical cells

When viewed under the electron microscope mitochondria usually appear as sausage-shaped bodies which can be compressed into sphere-like bodies. In some tissues such as liver or stomach parietal cells the mitochondria are distributed rather haphazardly throughout the cytoplasm (*Fig.* 6.2) but in others, such as muscle, they are packed systematically around the myofibrils (*Fig.* 6.3). In adipocytes the mitochondria surround the large globules of triacylglycerol.

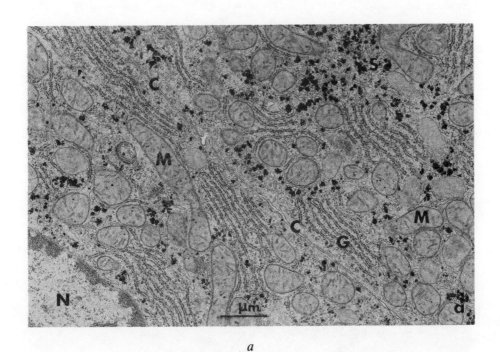

a

Fig. 6.2.   **Mitochondria in typical cells.**
Mitochondria in liver cells (*a*) and in gastric parietal cells of the stomach lining (*b*).
M = mitochondria; N = nucleus. Magnification: (*a*) × 16 000, (*b*) × 15 000.

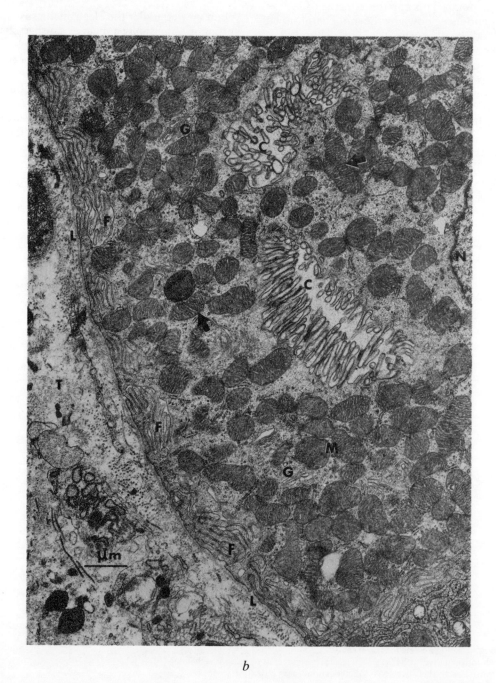

b

Fig. 6.2.   continued

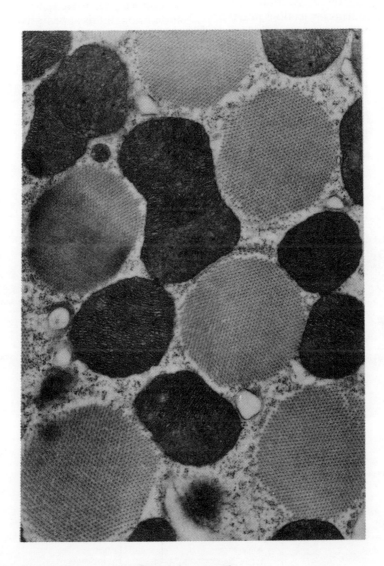

Fig. 6.3.   **Packing of mitochondria around myofibrils in insect muscle.**
Magnification × 40 000.

6.3 Mitochondrial structure

Typical mitochondria are about 3 μm long and 0·5–1 μm in diameter. The whole organelle is surrounded by an 'outer membrane'. An inner membrane forms projections into the mitochondrial contents that are called 'cristae', effectively dividing the mitochondria into a series of compartments (*see Fig.* 6.4). The space between the membranes is termed the 'peripheral space' and that between the invaginations the 'intracristae' space. It is not certain whether these spaces are continuous and although this appears likely, it is possible that the connections are only by pores.

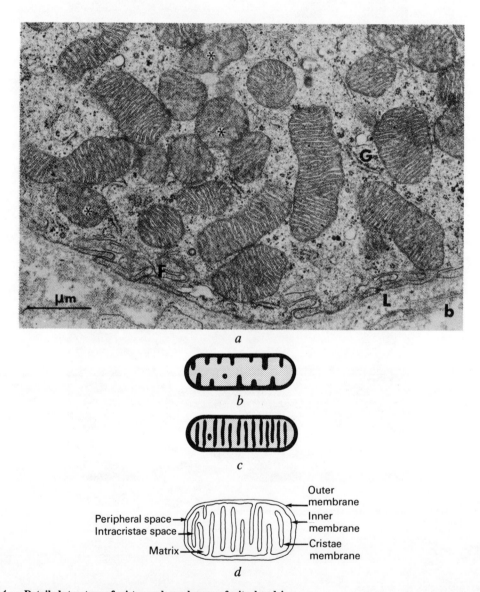

Fig. 6.4.   **Detailed structure of cristae and membranes of mitochondria.**
(*a*) Electron micrograph of kidney mitochondria showing cristae. (*b*) Cristae in liver mitochondria; (*c*) cristae in kidney mitochondria; (*d*) mitochondrial membranes.

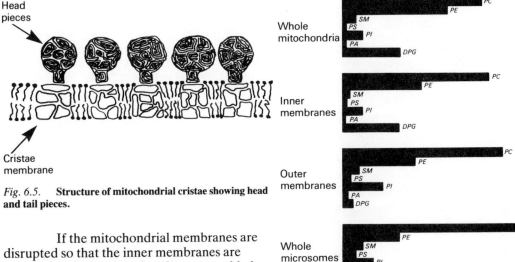

Fig. 6.5.   **Structure of mitochondrial cristae showing head and tail pieces.**

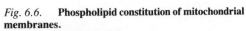

Fig. 6.6.   **Phospholipid constitution of mitochondrial membranes.**
PC = phosphatidylcholine;
PE = phosphatidylethanolamine; PS = phosphatidylserine;
PI = phosphatidylinositol; PA = phosphatidic acid;
SM = sphingomyelin; DPG = diphosphatidylglycerol.

If the mitochondrial membranes are disrupted so that the inner membranes are exposed, it can be seen that they are studded with a large number of spherical particles about 8–9 nm in diameter. The particles or 'head pieces' appear to be attached to the matrix facing surface of the membrane by 'stalks' about 4 nm long and 3 nm wide. Treatment of the inner membrane with detergents will cause the 'head pieces' to become detached. These experiments indicate that the detailed structure of the cristae may be represented as shown in *Fig.* 6.5.

6.4 Composition of the mitochondria

Mitochondria contain about 200–300 mg lipid/g protein and the majority of the lipid, over 90 per cent, is phospholipid. Phosphatidylcholine and phosphatidylethanolamine are the major phospholipids, smaller quantities of sphingomyelin, phosphatidylserine and phosphatidylinositol also occurring. Of special importance in the mitochondrial membranes is the large proportion of diphosphatidylglycerol and this is particularly abundant in membranes of cardiac mitochondria. A relatively large proportion of the phospholipids are in the plasmologen form (*Fig.* 6.6). The inner membrane is unusual in containing very little cholesterol. Most of the proteins in solution in the aqueous compartment of the mitochondria

have an enzymic function, but only about half of the proteins of the inner membrane are believed to play a catalytic role and others may thus play 'structural' or 'core' roles.

6.5 Functions of the mitochondria

The major functions of the mitochondria, summarized in *Table* 6.1, are involved mainly with 'energy trapping'. Essentially the process involves the transformation of energy made available by the oxidation of many molecules normally produced by ingested or stored foods into a form which can be utilized by the cellular processes. This energy must be in a specific

Table 6.1 **Summary of major functions of the mitochondria**

1. Operation of the citric acid cycle and production of NADH, i.e. the generation of reducing equivalents
2. β-Oxidation of fatty acids producing acetyl-SCoA, reduced flavoproteins and NADH
3. Oxidation of NADH by means of electron transport coupled with the synthesis of ATP
4. Accumulation of divalent actions such as Ca^{2+}
5. Transport of H^+ out of the mitochondria coupled to ATP synthesis

form, mainly adenosine triphosphate, 'ATP'. The discovery of the importance of ATP as a source of energy for muscle contraction arose from the pioneering work of Meyerhof and Lohman in Germany, but the fundamental significance of ATP was described originally by Lipman who proposed that, in all cells, ATP was involved in a cyclic process essential for the provision of cellular energy:

molecular changes of ATP to ADP and inorganic phosphate must be considered to explain the reason for the energy which is released and made available.

Utilization of ATP, reactions in group II, occurs in all subcellular fractions, the plasma and intracellular membranes, the cytosol, endoplasmic reticulum and nucleus, as well as in the mitochondria, but production of ATP from ADP and inorganic phosphate occurs primarily in the mitochondria within all cells. It must not be overlooked, however, that ATP can be synthesized in the cytosol by a process termed 'substrate level phosphorylation'. This occurs in two separate reactions of glycolysis catalysed by the enzyme glyceraldehyde phosphate dehydrogenase and enolase (*Fig.* 5.2, Chapter 5 and Appendix 14).

The mitochondria, however, perform the very important function of carrying out

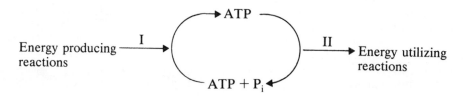

Energy utilizing reactions, include for example muscular work, all synthetic processes in the cell, and ion transport across membranes. The importance of ATP depends on the fact that it is a 'high energy phosphate' compound, as can be shown by the fact that it releases a large quantity of energy on hydrolysis. Other phosphate compounds such as glucose 6-phosphate are described as 'low energy phosphate' compounds, releasing much less energy on hydrolysis. Originally, ATP was described as possessing 'energy-rich phosphate bonds' indicated: AMP~Ⓟ~Ⓟ or ADP~Ⓟ.

This was a useful concept, and may still be useful when considering metabolism from a basic standpoint, but the term 'energy-rich bond' has been strongly criticized on the grounds that the bond linking the phosphate is not actually 'energy-rich', and the total

oxidations and trapping the release of energy in the form of ATP, the process known as 'oxidative phosphorylation', because the oxidation is coupled with the phosphorylation of ADP to form ATP.

The importance of the utilization of oxygen by all animals was established in the eighteenth century by Priestley and it was assumed by early biochemists that molecules in the cell would become oxidized by the addition of oxygen. This view turned out to be incorrect and addition of oxygen proved rare. The first clear understanding of cellular oxidation was provided by Weiland and Thunberg in Germany who demonstrated the importance of dehydrogenation reactions, catalysed by 'dehydrogenases', which remove hydrogen atoms in pairs from selected metabolites.

Dehydrogenation is also a form of oxidation because oxidation is now known to involve transfer of electrons, the loss of electrons being described as oxidation and the gain of electrons as reduction. The interconversion of the two forms of iron, Fe^{2+} (ferrous) and Fe^{3+} (ferric) which is important in biological oxidation and occurs within the cytochrome components, is one of the simplest forms of an oxidation and reduction, or redox, system:

$$Fe^{3+} + e \xrightarrow[\text{Reduction}]{} Fe^{2+}$$
$$\xleftarrow[\text{Oxidation}]{}$$

Dehydrogenation involves loss of hydrogen atoms, normally in pairs and is also an oxidation because a hydrogen atom can be regarded as composed of a proton (H^+) + an electron (e).

Electrons can also be transferred as a negatively charged hydrogen H^-, called a 'hydride ion' which is important when NAD^+ is reduced to NADH.

The main function of the dehydrogenase enzymes, many of which oxidize intermediates of the citric acid cycle (Appendix 17), is to produce NADH from NAD^+. These reactions do not produce available energy but, by forming NADH, provide the mitochondria with a very important source of energy because the mitochondria is designed to oxidize NADH and from this oxidation to produce available energy. NADH is oxidized by an electron transport system. Electrons flow in a clearly defined sequence, through a flavoprotein, iron–sulphur proteins, coenzyme Q and thence through a series of cytochromes b, c, c_1, a and a_3. Cytochrome a_3 is the terminal oxidase transferring 4 electrons to an oxygen molecule to form, with 4 protons, 2 molecules of water. Exactly how this final stage occurs is not clear.

Uptake of 2 electrons by an oxygen molecule will form hydrogen peroxide (H_2O_2) and uptake of 1 electron by oxygen will form the superoxide radical O_2^- which is a very reactive and toxic species. Enzymes are present which can destroy hydrogen peroxide (catalase) and the superoxide radical (superoxide dismutase). It is clear that the normal process is for an uptake of 4 electrons to occur but how this is reliably achieved is

still not resolved. The main components of the electron transport chain are shown in Appendix 21. The sequence of the components is dependent on the redox potentials (E_0) of their oxidized and reduced forms, the system becoming gradually more positively charged as water is eventually formed. Thus the E_0 of the $NAD^+/NADH$ system is -0.32 V, for oxidized/reduced cytochrome it is $+0.25$ V and for the reduction of oxygen to form water it is $+0.82$ V.

The energy released during the passage of electrons along the electron transport chain is 'trapped' or 'conserved' and used for the synthesis of ATP from ADP and inorganic phosphate. Exactly how this occurs is uncertain and this is despite the intense investigations of the process in many laboratories for nearly 40 years. It is generally considered that 3 molecules of ATP are formed for each molecule of NADH oxidized and that 1 ATP is formed at each of the 3 sites indicated. However, even this stoichiometry has recently been questioned.

Numerous theories have been put forward to explain the mechanism of this very important metabolic process, the coupling of oxidation and phosphorylation. Originally it was suggested that a chemical coupling mechanism was involved, a theory based on the mechanism of ATP formation during substrate level phosphorylation, for example during glycolysis.

In this theory, the oxidation of XH_2 where X is one of the components

$$XH_2 + A \rightarrow X + AH_2$$

of the electron chain is assumed to take place in the presence of a molecule B with which it could form a high-energy derivative, thereby trapping the released energy:

$$B + XH_2 + A \qquad\qquad X{\sim}B + AH_2$$

High-energy compound

$$X{\sim}B + P_i \longrightarrow X{\sim}\boxed{P_i} + B$$

$$X{\sim}P + ADP \longrightarrow ATP + X$$

B and X would then return to the system and be used in a cyclic manner where X undergoes further reduction. Several variations of this mechanism have been put forward. For example, the energy available in the intermediate $X \sim B$ could be used directly in the synthesis of ATP from $ADP + P_i$ without the formation of $X \sim \textcircled{P}$ as an intermediate.

The hypothesis initiated many years of intensive research for intermediates of the type illustrated by $X \sim B$ and $X \sim P$, but these could not be extracted or purified. On account of the failure to identify high-energy intermediates, the chemical coupling theory has tended to be abandoned.

A second theory, based on the discovery that an ATP synthetase enzyme was dependent on two proteins F_1 and F_0 located on the stalks which protruded into the cristae of the mitochondria (*Fig.* 6.5), has postulated that the energy released by electron transport causes a conformational change directly or indirectly in the ATP synthetase $F_0 – F_1$ complex. This energy is believed to be used to energize the complex and thus synthesize ATP. This theory is attractive and may be a true explanation, but it is very difficult to prove experimentally.

The theory which has received most attention and discussion is one put forward by a British biochemist, Mitchell, called the 'Chemiosmotic Theory'. Mitchell, who was awarded the Nobel Prize in 1982, proposed that the energy released by the electron transport chain is used to pump protons from inside the mitochondria to the outside. This would set up a potential difference and thus a source of energy which could be utilized as the protons pass back into the mitochondria. This energy enables the ATP synthetase enzyme to synthesize ATP from ADP and P_i (*Fig.* 6.7a). The theory supposes that protons are continuously pumped from the mitochondria and then flow back again. It does not require the intervention of high-energy intermediates.

Experimental investigations have supported the existence of proton transport, but difficulties arise when attempts are made to explain the close links between electron transport and proton ejection. In an attempt to overcome these difficulties Mitchell has

proposed a loop hypothesis illustrated in *Fig.* 6.7b. The transport components are visualized as being alternate H and electron acceptors; the hydrogen acceptors, e.g. C, eject protons as they become oxidized. Thus, for example, C becomes reduced by accepting 2 electrons from B^{2-} and 2 protons from the matrix. The protons are ejected from CH_2

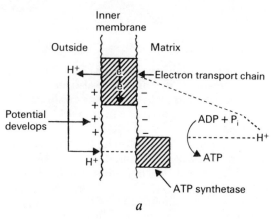

a

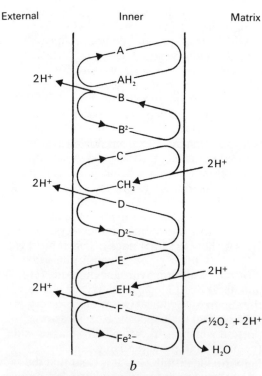

b

Fig. 6.7. **Mitchell's chemiosmotic theory.**
(*a*) Basic concept; (*b*) loop hypothesis.

leaving a pair of electrons which are used to reduce D to D^{2-}.

As set out, the theory is simple and attractive but difficulties arise when it is attempted to fit known carriers into the scheme. The $NAD^+/NADH$ system would appear to fit as A/AH_2 and the transfer of electrons to flavoprotein (B/B^{2-}) is likely. Furthermore, the final stage (F/F^{2-}) which transfers 2e to $\frac{1}{2}O_2$ would appear to fit with the action of cytochrome oxidase.

One problem is, however, to link in a hydrogen carrier, for example that described as E/EH_2 in the part of the chain normally presumed to contain only cytochromes, which are electron carriers. Mitchell has suggested that coenzyme Q which can act as a hydrogen carrier may intervene between electron carriers at certain stages, but no good experimental evidence exists to support this concept.

In summary, the chemiosmotic theory provides a valuable working hypothesis to explain the coupling of oxidation to ATP synthesis but many details of its operation remain to be resolved. Furthermore, as discussed in Chapter 7, the stoichiometry of the relation of ATP synthesis to proton movement is complex and cannot yet be equated with a simple plan such as Mitchell has proposed.

Uncoupling and inhibition of electron transport

The electron transport chain clearly plays a vital role in the provision of energy for all cell processes and, as a consequence, any molecules which interfere with its function are generally extremely toxic. One of the best known of these is cyanide which avidly binds to the free ligand of iron in all cytochromes, and cytochrome a_3 or cytochrome oxidase is especially vulnerable. Cyanide will cause the cessation of oxygen uptake and inhibition of ATP synthesis in almost all cells of the body and is thus extremely toxic. In small doses, it is likely to cause death by its effect on the respiratory centre.

Other molecules such as hydrogen sulphide (H_2S) and carbon monoxide will also bind to iron in the cytochromes and inhibit electron transport. Other naturally occurring toxins such as rotenone, a toxic extract of plants used by South American natives, blocks electron transport from NADH to coenzyme Q and the antibiotic, antimycin A, blocks transfer of electrons from coenzyme Q to cytochrome c. The net effect of all these inhibitors is to block both oxygen uptake and ATP synthesis. This is because ATP synthesis and oxygen uptake are normally tightly 'coupled'.

A very interesting group of drugs has been discovered which are 'uncoupling' agents. These do not inhibit oxygen uptake but inhibit the synthesis of ATP and, of this group, dinitrophenol is best known and has been tried clinically as a drug to treat obesity. One drug, the antibiotic oligomycin, specifically inhibits ATP synthesis, and drugs of this type cause the energy made available by oxidation of NADH to be released as heat and to thus be unavailable for chemical synthesis.

Uncoupling of mitochondrial ATP synthesis may, under certain situations where an extra supply of heat is required, be a valuable type of physiological regulation and it is possible that the process could be naturally hormonally controlled. Certain mitochondria and especially those contained in 'brown fat' may be naturally uncoupled and produce heat during hibernation and their function may be related to the onset of obesity (cf. Chapter 40).

Reduced to its simplest concept, the operation of the mitochondria in the conservation of energy is represented schematically in *Fig.* 6.8. Of the substrates available to the mitochondrion, pyruvate produced from glucose by glycolysis is an important substrate, and other substrates, such as α-ketoglutarate produced by deamination of glutamic acid, can also be used in the citric acid cycle for the production of NADH. Fatty acids entering the cell from the plasma or released from stored triacylglycerols cannot enter the mitochondria unless they are first bound to carnitine, being transported through the membrane as carnitine complexes (*see* Chapter 7). Inside the mitochondrion, the carnitine is released and the fatty acids are oxidized by β-oxidation (*see* Appendix 20) to acetyl-SCoA, with the production of reducing

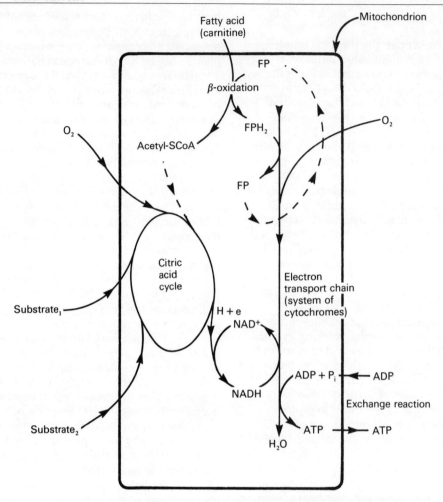

Fig. 6.8.   **Schematic representation of the role of the mitochondria in energy conservation.**
FP = flavoprotein.

equivalents in the form of reduced flavoprotein ($FADH_2$) and NADH; additionally, acetyl-SCoA becomes available for oxidation to a substrate for the citric acid cycle (*see* Appendix 17).

   Mitochondrial membranes are impermeable to NAD^+ or NADH, the latter being effectively transported across the mitochondrial membrane in the form of

a substrate, such as malate, that is then dehydrogenated (to oxaloacetate) with reduction of NAD^+ to NADH. Both ATP formed within the mitochondrion and ADP in the cytosol are able to exchange with each other across the membrane (*see* Chapter 7). The accumulation by mitochondria of divalent cations and the transport of proteins across the mitochondrial membrane are discussed in Chapter 7.

6.6 Relation of mitochondrial structure to enzyme activity and function

The outer membrane, intermembrane space, inner membrane and matrix are all known to be associated with specific and localized functions of the mitochondria.

Outer membrane

This membrane is freely permeable to most small molecules and is not generally believed to play a very important role in mitochondrial metabolism, although several molecules involved in lipid metabolism are associated with this membrane. These include the fatty acid elongating enzyme, glycerol phosphate acyltransferase, choline phosphotransferase, phosphatidate phosphatase and phospholipase A.

Intermembrane space

Several enzymes are located in this compartment of the mitochondria and especially those involved in nucleotide metabolism, including adenylate kinase, nucleoside diphosphate kinase, nucleoside monophosphate kinase and creatine kinase.

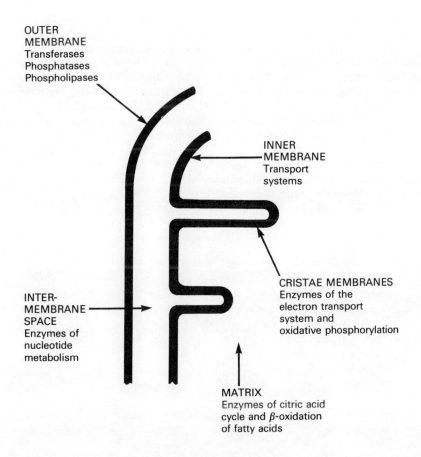

OUTER
MEMBRANE
Transferases
Phosphatases
Phospholipases

INNER
MEMBRANE
Transport
systems

INTER-
MEMBRANE
SPACE
Enzymes of
nucleotide
metabolism

CRISTAE MEMBRANES
Enzymes of the
electron transport
system and
oxidative phosphorylation

MATRIX
Enzymes of citric acid
cycle and β-oxidation
of fatty acids

Fig. 6.9. **Localization of important mitochondrial enzymes.**

Inner membrane

This membrane, as discussed in detail in Chapter 7, contains all the transport systems regulating the entry and exit of metabolites into and out of the mitochondria and the very important exchange reactions.

The extensions of the inner membrane that form the cristae are functionally very important since all the components of the electron transport chain, the flavoproteins and the cytochromes, are located on these membranes. Cytochrome c is localized on the outward facing surface of the membrane and can be removed relatively easily. Most of the other components are, however, firmly bound.

Matrix

The matrix consists of a fluid mitochondrial 'cytoplasm', containing all the enzymes of the citric acid cycle, for the β-oxidation of fatty acids and other dehydrogenases, such as glutamate dehydrogenase. It is, therefore, a very important component of the mitochondria and responsible for the production of the reducing equivalents in the form of NADH which will be used by the electron transport chain.

The matrix also contains the enzymes involved in mitochondrial replication. Enzyme localization is shown diagrammatically in *Fig*. 6.9.

6.7 Replication of mitochondria

It is very doubtful if mitochondria ever appear *de novo* in a cell. They are believed to arise from the maternal zygote and during cell division become partitioned between the two daughter cells.

Mitochondria contain a form of DNA that is quite unlike that found in the eukaryotic nucleus. It is double stranded, circular in form and similar to that found in bacteria. A full complement of enzymes exist in the mitochondria for protein synthesis, including RNA polymerase and amino acid activating enzyme, plus transfer RNA and ribosomes, so that mitochondria are able to synthesize proteins coded for by their own DNA. The information available in the DNA is, however, only sufficient to code for a few proteins and most of the proteins found in mitochondria are coded for by nuclear DNA and synthesized in the cellular cytoplasm.

The reason for the strange dual system for synthesizing mitochondrial proteins is unclear. The proteins coded by mitochondrial DNA may not be available in the cytoplasm, or the system that codes and synthesizes proteins within the mitochondria may be a vestige of the original bacteria from which the mitochondria were derived.

Chapter 7

Role of subcellular organelles: inter-relationships of the mitochondria and cytosol

To reduce discussion of the roles of the mitochondria and cytosol to the simplest terms, the main functions of the cytosol can be described as preparation of metabolites for oxidation in the mitochondria and synthesis of important food stores such as fat and glycogen, and the main function of the mitochondria as provision of an energy supply for synthesis of molecules required in the cell and other processes. Energy generated in the mitochondria is normally in the form of chemical energy, ATP, but the energy derived from oxidation processes can also be diverted to produce heat or accumulate Ca^{2+} (cf. Chapters 6, 40).

Many enzymes that utilize ATP, including those involved in synthetic processes, are located outside the mitochondria; consequently a large traffic in ATP from mitochondria to cytosol must be essential for efficient functioning of the cell. Alternatively, ATP-requiring metabolic processes that originate in the cytosol, may be diverted into the mitochondria so that ATP is utilized there and the products returned to the cytosol. The ornithine cycle which produces urea is a good example of this type of metabolism (*see* p. 70).

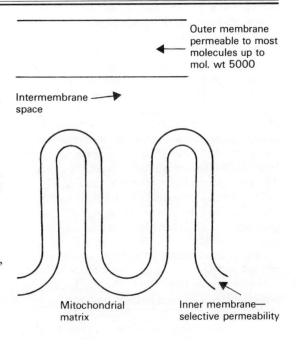

Outer membrane permeable to most molecules up to mol. wt 5000

Intermembrane space

Mitochondrial matrix

Inner membrane— selective permeability

Fig. 7.1. **Mitochondrial membranes.**

7.1 Membranes of the mitochondria

Mitochondria are surrounded by two membrane systems that have very different biochemical properties. The outer membrane has a high lipid:protein ratio and is permeable to most molecules with a molecular weight of 5000 or less. The inner membrane has a low lipid:protein ratio, and is highly invaginated with a large surface:volume ratio (*see* Fig. 7.1).

The permeability of the mitochondrial membrane is highly selective and most hydrophilic substances cannot penetrate the lipid bilayer. Specialized carrier proteins within the membrane catalyse the transport of metabolites through the bilayer.

67

7.2 Mitochondrial transport systems

The activity of different mitochondrial transport systems varies both from one tissue to another and according to the main functions of the mitochondria. Many processes are, however, common to mitochondria in all tissues. Synthesis of ATP, which occurs in all mitochondria, requires the entry of oxygen, ADP, phosphate and electron-rich substrates, such as pyruvate and fatty acids. The products of the oxidation of these reactions are water, carbon dioxide and ATP which must then be transported out. Oxygen, water and carbon dioxide are freely permeable and require no specific transport system. Under some conditions, HCO_3^- can also penetrate the membrane even though it is a negatively charged ion. Specialized carrier systems are, however, available for transport of phosphate, pyruvate, and fatty acids across the membrane. In addition, several exchange systems exist. The most important of these are the ADP–ATP exchange, the malate–α-ketoglutarate exchange and the glutamate–aspartate exchange. One of the main reasons for the existence of these shuttle systems is the transport of reducing equivalent across the mitochondrial membrane, since the membrane is impermeable to NADH. The mechanism by which this is achieved is described on p. 73.

A tricarboxylate carrier system will catalyse the exchange of citrate, isocitrate, phospho*enol*pyruvate and malate. The activity of this system, which is very low in heart but high in liver, catalyses the exchange of any of these metabolites with any other. During the operation of these transport systems, an effective regulation of H^+ balance across the membrane must be maintained, and this is achieved by balancing the influx and efflux of protons.

It is generally believed that the transport of metabolites across the mitochondrial membrane occurs on protein carriers that behave very much like enzymes: they can be saturated with substrate, they exert a high degree of substrate specificity, and they are inhibited by substrate analogues and by sulphydryl reagents. Kinetic studies of transport indicate that the activation of energy involved in transporting most metabolites is high and usually within the range 20–30 kcal.

As will be discussed later in the chapter (pp. 70–74), several carriers transport protons in addition to the substrate and, in these systems, the proton is usually bound to the protein and not to the substrate.

7.3 Classification of carrier types

Carrier systems so far discovered can be classified into four categories (shown in *Fig.* 7.2):

Type (i). An anion, for example pyruvate or phosphate, is transported across the membrane in association with a proton, but dissociation occurs within the membrane so that ions are found inside the mitochondria.

Type (ii). Anions of equal charge are exchanged. Typically, two dicarboxylic acid ions may be involved.

Type (iii). Exchange of neutral metabolites is involved.

Type (iv). Anions are exchanged, resulting in an imbalanced electron potential. This gives rise to the description 'electrogenic'. The most important example of this type is the ATP–ADP exchange.

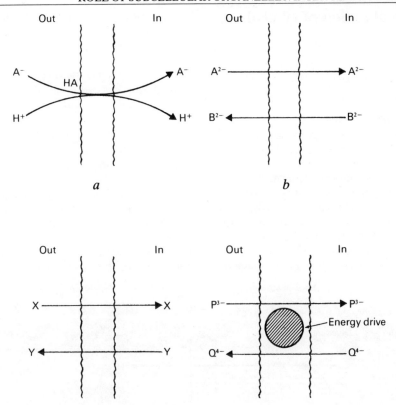

Fig. 7.2. **Types of carrier across the mitochondrial membrane.**
Type (*a*) proton compensated—electroneutral; type (*b*) electroneutral exchange of
anions; type (*c*) neutral metabolite transport not involving protons; type (*d*)
electrogenic exchange of anions.

7.4 Classification of carriers

This classification can be made as below:

a. *Proton compensated—electroneutral*
   Phosphate carrier
   Pyruvate carrier
   Glutamate carrier
   Ornithine carrier
   Tricarboxylate carrier

b. *Electroneutral exchange*
   Dicarboxylate carrier
   α-Ketoglutarate carrier

c. *Neutral metabolite carrier*
   Glutamine carriers
   Carnitine–acylcarnitine carrier
   Neutral amino acid carrier

d. *Electrogenic carrier*
   Glutamate–aspartate exchange
   Adenine nucleotide carrier

7.5 Mode of action of typical carriers

Below examples of the action of specific carriers are considered.

a. Pyruvate carrier

Pyruvate ions produced by glycolysis are protonated to form pyruvic acid which is transported across the membrane. In the mitochondria, dissociation occurs and pyruvate is metabolized by the citric acid cycle to CO_2 and H_2O which then diffuse freely through the membrane. Outside, in the cytosol, the H_2O and CO_2 reform H_2CO_3 which ionizes to H^+ and HCO_3^-, the H^+ being used to reform pyruvic acid. The net result is an exchange of

HCO_3^- for pyruvate in the cytosol. Details of the pyruvate carrier are given in *Fig.* 7.3.

b. Ornithine carrier

This carrier is essential for the operation of the ornithine cycle and the synthesis of urea (*see* Appendix 25). The synthesis of citrulline from ornithine, NH_3 and CO_2 requires ATP, so the precursors are taken up into the mitochondria where they utilize the ATP which is synthesized there.

It is likely that ornithine is transported into the mitochondria on the ornithine carrier in exchange for H^+ which is, in turn, provided by the transport of a molecule of citrulline out of the mitochondria. For details of the ornithine carrier *see Fig.* 7.4.

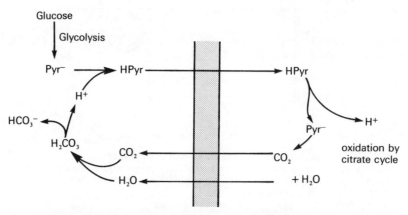

Fig. 7.3.   **Action of the pyruvate carrier showing how proton balance across the mitochondrial membrane is maintained.**
Pyr^-, pyruvate; HPyr, pyruvic acid.

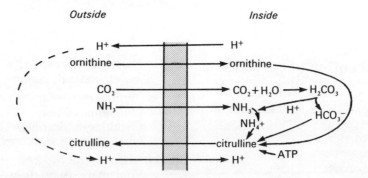

Fig. 7.4.   **The mode of action of the ornithine carrier.**

c. The tricarboxylate carrier

This carrier system can transport citrate, isocitrate, *cis*-aconitate and malate, all of which exchange with each other. Proton compensation is also necessary when a tricarboxylic acid, such as citrate, exchanges with a dicarboxylic acid, such as malate. The carrier system is important for the transport of citrate out of the mitochondria in order to provide a substrate for synthesis of fats (*see* p. 74). For details of the tricarboxylate carrier *see Fig.* 7.5.

Outside   *Inside*

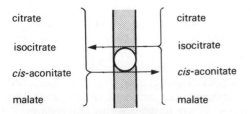

Fig. 7.5.   **The tricarboxylate carrier.**
Proton compensation is necessary when citrate or other tricarboxylic acids exchange with malate.

d. The dicarboxylate carrier and the α-ketoglutarate carrier

These carriers catalyse the electroneutral exchange of divalent anions and the specificities of the two carriers are shown in *Fig.* 7.6. The two carriers have similar functions, but the use of inhibitors and kinetic studies clearly demonstrate that they are distinct. It should be noted that not all the molecules transported on carriers are valuable metabolites. Malonate,

for example, is a powerful inhibitor of succinate dehydrogenase.

e. The glutamine carrier

During metabolic acidosis, homeostasis and pH regulation is maintained by several mechanisms and, of these, excretion of NH_4^+ by the kidney is very important. The conversion of NH_3 to NH_4^+ provides an important means by which protons can be excreted, and the excretion of NH_4^+ also helps to conserve Na^+ which would otherwise be lost in large quantities in acidosis.

Outside   *Inside*

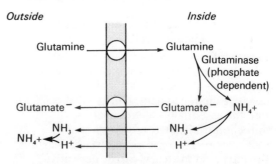

Fig. 7.7.   **The glutamine carrier.**

NH_4^+ is formed by the hydrolysis of glutamine stored in the kidney, but as this hydrolysis takes place in the mitochondria a glutamine carrier system is required. Glutamine is exchanged for glutamate and a proposed mechanism of operation is shown in *Fig.* 7.7; it must be appreciated, however, that some details are still controversial.

Glutamine passes through the membrane on the carrier and is hydrolysed to glutamate and NH_4^+ by glutaminase.

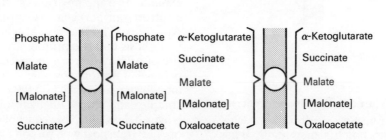

Fig. 7.6.   **The dicarboxylate carrier and the α-ketoglutarate carrier.**

Glutamate is then transported back to the outside of the membrane carrying one negative charge, but this is balanced by H^+ transport. The net result of the operation of this system is a replacement of glutamine by glutamate and NH_4^+ in the cytosol.

In view of the fact that the glutamine carrier is clearly of great importance in the supply of NH_4^+ during acidosis, it has been suggested that the regulation of the activity of the carrier plays a major role in the control of acidosis. The exact mechanism of this control has not yet been elucidated.

f. The carnitine–acylcarnitine exchange carrier

Fatty acids are important substrates for mitochondrial respiration in many tissues, particularly the heart. Fatty acids carry a negative charge at neutral pH so that transport into the essentially electronegative mitochondrial matrix requires a special mechanism. The process involves converting the fatty acid to an acyl-SCoA derivative, and the transfer of acyl groups across the membrane to a carnitine complex (*see Fig*. 7.8).

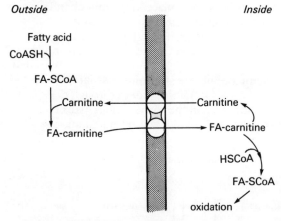

Fig. 7.8.   **The carnitine carrier system.**
FA, fatty acid.

g. The neutral amino acid carrier

This carrier transports neutral amino acids across the mitochondrial membrane. It has no proton, cation or energy requirement.

h. Glutamate–aspartate exchange

Entry of glutamate into the mitochondria in exchange for the efflux of aspartate is facilitated by a supply of energy and the carrier system is, therefore, described as 'electrogenic'. The glutamate enters together with a proton. This fact demonstrates that the electrical nature of the transporters is not dependent on the ionic state of the transported substrate, but on the nature of the carrier and its capacity to bind and transport protons.

i. The adenine nucleotide carrier

This very important carrier system exchanges ATP with ADP. ATP is associated with four negative charges whereas ADP has three negative charges. The system is believed to be electrically balanced, since synthesis of each molecule of ATP involves the liberation and transport of four protons, three of which are transported back during ATP synthesis. The carrier system is electrogenic and requires energy (*see Fig*. 7.9).

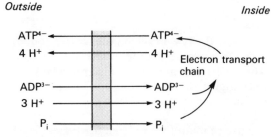

Fig. 7.9.   **The ATP–ADP carrier system.**

The transport of protons across the mitochondrial membrane forms an important part of Mitchell's chemiosmotic theory to explain the trapping of energy from the electron transport chain (Chapter 6 and Appendix 21). However, Mitchell's original hypothesis that two protons are formed at each site of ATP synthesis is incorrect and the balance of proton movements across the membrane during ATP–ADP exchange is complex and not yet finally resolved. Nevertheless, the basic concept of coupling of ATP transport with proton transport is clearly valid.

7.6 Roles of carrier systems in metabolism

The carriers in the mitochondrial membrane play important roles in regulation of metabolism and the interdependence of the mitochondria and cytosol. We shall give some examples of their involvement in metabolic processes.

a. Effective transport of NADH into the mitochondria

The mitochondrial membranes are not permeable to pyridine nucleotides, such as NAD^+ or NADH, but a supply of NADH is essential in the mitochondria since it is the main substrate for the electron transport chain and, therefore, for the energy supply in the form of ATP. The problem is solved by the use of the dicarboxylate transport system (*see Fig.* 7.10).

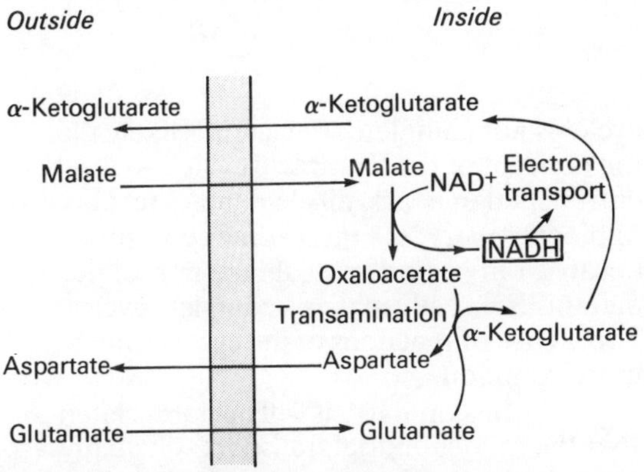

Fig. 7.10. **Effective transport of NADH into the mitochondria—the malate–aspartate shuttle.**

Malate is transported into the mitochondria in exchange for α-ketoglutarate. Inside the mitochondria, malate is oxidized by malic dehydrogenase to provide the NADH which is used as a substrate for the electron transport chain. Oxaloacetate is also formed and this undergoes transamination in the mitochondria to form aspartate that is then transported out in exchange for glutamate. During the transamination reaction with oxaloacetate, α-ketoglutarate is formed from glutamate that has exchanged with malate. The net result is the exchange, in the cytosol, of α-ketoglutarate for malate and aspartate for glutamate; in the mitochondria, NADH is provided for oxidation by the electron transport chain (*see* Chapter 6 and Appendix 21).

b. Transport of citrate from the mitochondria

In conditions of excess availability of substrates for oxidation and of limited energy requirement, there is clearly a need to store a large proportion of the available energy supply. Conversion of metabolites to fat is one of the most effective means of doing this.

The process involved is illustrated by the transport of citrate out of the mitochondria when the need for its oxidation is very low (*see Fig.* 7.11). Citrate is transported out of the mitochondria in exchange for malate and, in the cytosol, citrate is cleaved by the enzyme citrate lyase to form oxaloacetate and acetyl-SCoA. The oxaloacetate can be reduced by malate dehydrogenase and NADH to form malate which will exchange with citrate. The acetyl-SCoA, produced by citrate cleavage, can be synthesized into fatty acids but this also requires a supply of NADPH, which may be supplied by oxidation of a proportion of the citrate by isocitrate dehydrogenase. This oxidation produces α-ketoglutarate and NADPH. α-Ketoglutarate may also be metabolized to malate by the reactions of the citrate cycle to maintain the supply of malate for citrate exchange.

c. Citrate oxidation

A third example which is the reverse of the situation described above shows the role of the mitochondrial carriers in citrate oxidation. Citrate is transported into the mitochondria in exchange for malate, but since citrate in its fully ionized form possesses three negative charges and it exchanges with malate possessing two negative charges, the citrate ion must carry a proton (*see Fig.* 7.12). Within the mitochondria, the citrate is oxidized by the

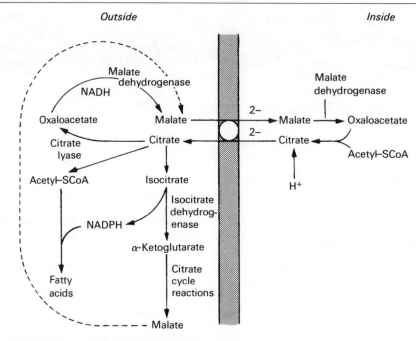

Fig. 7.11. **Transport of citrate out of the mitochondria.**

reactions of the citrate cycle to produce ATP, H_2O and CO_2. H_2O and CO_2 are transported across the membrane to form H_2CO_3 which then dissociates to form the H^+ required for citrate transport.

This metabolic pathway for the conversion of citrate to malate is a good illustration of the use of part of the citric acid 'cycle' reactions and it will be noted that 'the

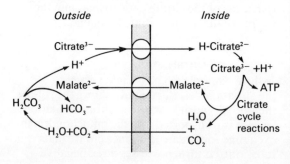

Fig. 7.12. **Citrate transport into the mitochondria for oxidation.**

cycle' is not complete. In fact, it is likely that the concept of the 'citric acid cycle', so firmly entrenched in nearly all elementary text books of biochemistry, is a misleading concept because only a part, probably minor, of the metabolism really makes a complete cycle. Partial use of reactions of the cycle is much more common.

In summary, it will be appreciated that the mitochondrial transport system must play very important roles in metabolism, moving some metabolites into the mito-chondria for oxidation and other metabolites out for synthesis into storage products. It appears likely that further research will demonstrate that carrier systems play very important roles in the regulation of metabolism. It is possible that they are hormonally controlled and thus they will eventually be shown to be as, or more, important as sites of regulation than the enzymes catalysing the metabolic processes themselves.

Chapter 8 Role of subcellular organelles: the nucleus

8.1 Introduction

The observation that the cells of most multicellular organisms possess a large prominent organelle described as the nucleus was made very soon after microscopical investigations began. The structure stained readily and clearly with basic dyes and was prominent in most histological preparations.

Early on it was thought that the nucleus played an important role in controlling cellular metabolism, but the mechanism of control was not understood until the role of its component DNA was discovered: the DNA carries the genetic information for the manufacture of all proteins in the cell. Codes on sections of DNA are 'switched on' or 'switched off' to allow, or terminate, the release of the code for the synthesis of special proteins within the cell and by this means DNA can control the cellular metabolism.

Using techniques of ultra-centrifugation (cf. *Fig.* 1.1, Chapter 1), it became possible to separate nuclei and to study the chemical composition and metabolism of the nucleus in great detail and, thus, biochemical observations began to complement those made with the light and electron microscope. Nuclei are normally the heaviest particles in the cells and they separate relatively easily after centrifugation at relatively low sedimentation values (cf. Chapter 2). These preparations are, however, usually impure and contaminated with cell debris and unbroken cells. Other methods have to be used to obtain pure nuclei which involve sedimentation through organic solvents or in very strong (e.g. 2 M) sucrose solution.

Despite its prominence and apparent importance, the nucleus is not essential for life since bacteria do not possess nuclei, the DNA being free in the bacterial cytoplasm.

8.2 Structure of the nucleus

Membrane

The nucleus is surrounded by a nuclear envelope which is seen in the electron microscope to be composed of a double layer. Each membrane is about 8 nm thick and the space between varies between 10 to 50 nm. The outer membrane appears to be continuous with the endoplasmic reticulum (*Fig.* 8.1).

The nuclear envelope is characterized by large numbers of circular pores often arranged in hexagonal arrays. The pores are usually 60–90 nm in diameter, but may be larger in some cells such as oocytes. Often the pores appear to contain a granule which does not completely fill the pore (*Fig.* 8.2).

The function of these pores is uncertain, but it is generally believed that they are present to allow large molecules such as RNA to pass through. Small molecules are normally transported through the membrane readily, but specific transport systems may be essential for some.

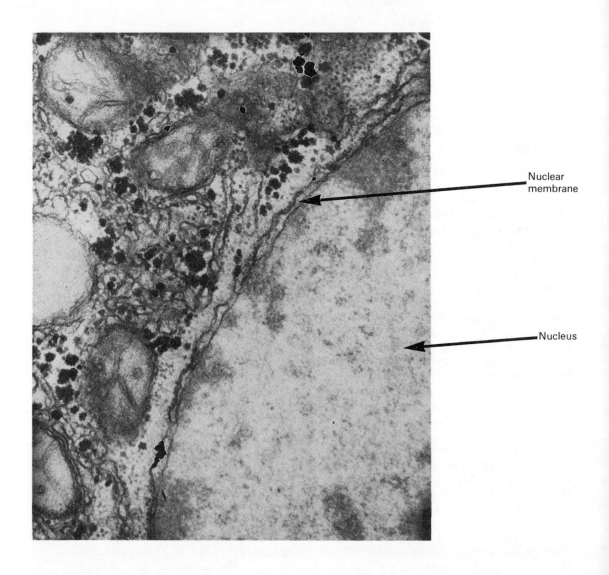

Nuclear
membrane

Nucleus

Fig. 8.1.   **Nuclear membrane.**
Electron micrograph of the nucleus showing the double layer of the nuclear membrane. Magnification × 75 240.

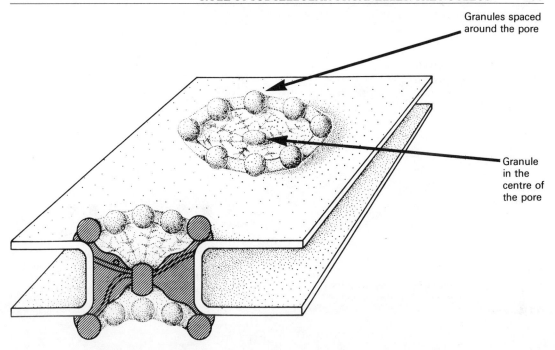

Granules spaced
around the pore

Granule
in the
centre of
the pore

Fig. 8.2.   **Diagrammatic representation of a pore in the nuclear membrane.**

Chromatin

A prominent feature of the nucleus is the
network of chromatin fibres that stain darkly
with basic stain and with Feulgens' stain
indicating that they contain DNA. DNA
accounts for 30 per cent of the dry weight of
the fibres at interphase.

The chromatin fibres may be
condensed into a form described as
heterochromatin which is usually concentrated
around the periphery of the nucleus; in the
remainder of the nucleus the fibres are
dispersed. The average diameter of the
chromatin threads is about 23 nm, but
scattered along the fibres are protuberances
(nu bodies) which are believed to be caused
by supercoiling of the DNA (shown
diagrammatically in *Fig.* 8.5 below).
Supercoiling is necessary to pack the DNA,
which in a human cell would stretch to
3 metres, into a nucleus of 5 μm in
diameter.

Chromatin proteins

Associated with the DNA in the chromatin are
a number of proteins, the most important of
which are the very basic proteins called
'histones'. There are five main groups of
histones as shown in *Table* 8.1.

Table 8.1 **Groups of histones**

Name		Type	Arg, %	Lys, %
F1	I	Lysine-rich	2·5	26·3
F2b	II	Moderately lysine-rich	6·7	16–17
F2a2	IIb	Moderately lysine-rich	11·5	10·5
F2a1	IV	Arginine-rich	14·1	9·7
F3	III	Arginine-rich	13·4	8·8

All the histones contain a large
percentage of lysine or arginine and they
were originally named, based on this criteria,
'lysine-rich' or 'arginine-rich', but more
recently other systems based on F numbers or

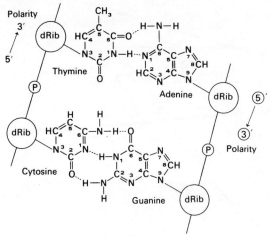

Fig. 8.3.   **Base-pairing in double strands of DNA.**
(– – –) Hydrogen bonding; dRib, deoxyribose;
Ⓟ phosphate.

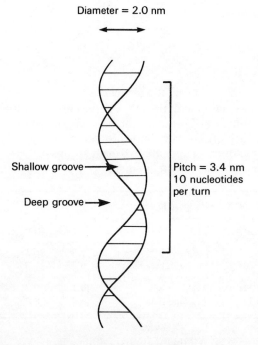

Fig. 8.4.   **Watson–Crick model for DNA.**

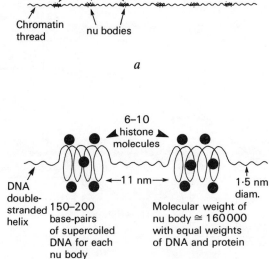

a

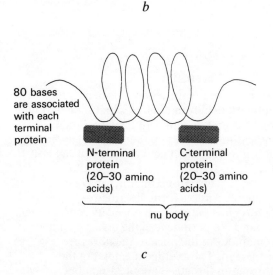

c

Fig. 8.5.   **Chromatin thread and nu body (nucleosome)
supercoiling of DNA.**
Three stages of magnification of the nu bodies
(nucleosomes) are shown diagrammatically in (*a*), (*b*)
and (*c*).

Roman numerals have been used (*Table* 8.1). The histones are relatively small protein molecules with a molecular weight within the range 15 000–20 000 and a section of DNA, of molecular weight 10^6, would have approximately 50 histone molecules attached to it. Normally, an approximately equal proportion of each group of histone proteins is bound to DNA. Some of the proteins, and especially F2A1, have remained remarkably constant throughout the course of evolution because the constitution of histone prepared from calf thymus and that from pea seedlings differs in only 2 amino acids out of 102! Changes occur in the histone composition during special cell division processes such as spermatogenesis, in which the lysine-rich histones are replaced by arginine-rich histones, and then by protamines which contain up to 80 per cent arginine.

Histones are modified by attachment of acetyl and phosphoryl groups which are often turned over at a very fast rate. Acetylation occurs on α-NH_2 groups and on the ϵ-NH_2 groups of lysine; phosphorylation of the —OH of serine is also frequent.

In addition to the histones, several less well characterized proteins are found in association with chromatin. About 30 components of the group have been described and they are much more acidic than histones, normally containing about twice the percentage of acidic to basic amino acids. Some of these proteins are important enzymes and DNA polymerase is likely to be a component of this group.

Not all the nuclear protein is associated with the chromatin and many other proteins are present, some associated with DNA. Schematic concepts of the structure of DNA, supercoiling and the association with histone proteins are shown in *Figs*. 8.3, 8.4 and 8.5.

The nucleolus

Usually, densely staining granules (the nucleoli) can be detected by light microscopy in the nucleus; selective stains give a strong reaction for RNA, but stain weakly for DNA. The number of nucleoli varies from one cell type to another and they disappear during cell division, at prophase, becoming associated with their satellite chromosomes, probably at the organizer region.

The nucleoli are built up from bead-like structures, 15 nm in diameter, containing dense particles of RNA embedded in a matrix. Nucleolar DNA codes for ribosomal RNA with special arrays of bases coding for 28-S and 18-S rRNA. RNA polymerases are found in the nucleoli, as are ribonucleases, which are enzymes essential for RNA synthesis and RNA degradation. The experimental evidence therefore indicates that nucleoli are primarily involved with the synthesis of rRNA for the ribosomes.

8.3 Metabolism in the nucleus

DNA synthesis

Replication of DNA is a very important function of the nucleus and must occur at every cell division. DNA synthesis has been very extensively studied in bacteria and most accounts of DNA replication describe the process taking place in bacteria which have no nucleus. In addition to DNA synthesis, taking place during division, some repair of DNA is probably a frequent occurrence.

Even in rapidly dividing cells, DNA synthesis does not occur continuously but during a limited period described as the 'S' phase. This is preceded by a G_1 period and succeeded by a G_2 period and the whole sequence of events is described as the 'cell cycle', as illustrated in *Fig*. 8.6. In the human body, under normal conditions, some cells such as those of the liver rarely divide unless the tissue is damaged, whereas those of the intestinal mucosa or bone marrow are dividing steadily.

Synthesis of DNA is initiated at many sites of replication, sometimes called replicons, and there may be up to 10^4 sites in a single human nucleus. Replicons may exist at the attachment sites of the chromatin threads to the nuclear membrane (*Fig*. 8.7).

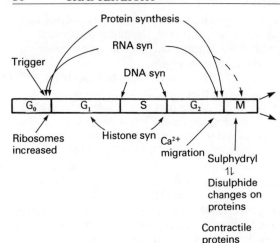

Fig. 8.6. **Summary of sequence of events in the cell cycle.**
G_0 is the resting phase for cells not undergoing rapid
division, e.g. most liver cells. In these cells, a 'trigger
event' is needed to initiate the cell division process. The
numbers of ribosomes, the rates of RNA and protein
synthesis are increased early in G_1 and these processes
continue until the end of G_2. Synthesis of histones
commences during G_0, and the formation of at least some
of these proteins is necessary before DNA synthesis
commences. DNA synthesis is confined to S phase. G_2 is
usually a much shorter period than G_1, and is a preparation
for the division process which occurs in M phase.

Several complex events occur during M phase,
and significant biochemical changes occur including the
development of contractile proteins regulated by a release
of Ca^{2+} and changes in ratio of —SH to —S—S— in some
of the proteins.

In a typical mammalian cell growing in culture the
cell cycle time is 20–24 h.

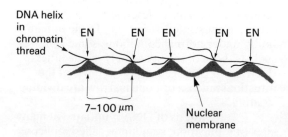

Fig. 8.7. **Role of nuclear membrane in the initiation of
DNA synthesis.**
EN—sites of initiation.

DNA replication

In simplest terms, the replication of DNA
involves the use of one strand as a template to
direct newly incorporated bases into formation
of A–T or G–C base-pairs. The synthesis in all
cells requires a primer DNA, bases in the form
of deoxynucleotides (deoxynucleoside
triphosphates), dATP, dGTP, dCTP and dTTP
and the enzyme DNA polymerase which exists
in several forms. Synthesis always occurs in the
$5' \rightarrow 3'$ direction, which relates to the
numbering of the deoxyribose ring. The
process is shown schematically in *Fig.* 8.8.

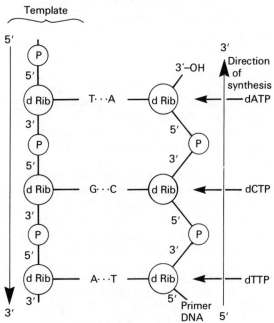

Fig. 8.8. **Basic concept of DNA replication (single strand).**

In all cells, both prokaryote and
eukaryote, DNA exists in a double-stranded
helix which is usually supercoiled. It is,
consequently, essential for the DNA to be
unwound prior to synthesis, so that relaxation
of the supercoiled DNA has to occur. This
is believed to be achieved by a number of
'unwinding proteins' which maintain the DNA,
or at least portions of it, in a state in which it
can act as a template.

Although the DNA polymerase enzymes from bacteria have been extensively studied and purified, much less is known about the polymerases of mammalian cells. Cultured mammalian cells, such as calf thymus and human KB-type cells, contain two DNA polymerases: one is found in the cytoplasm and has a high molecular weight, whilst the other is found in the nucleus and has a much lower molecular weight (40 000–50 000). The relationship between these enzymes is unclear, but it is possible that the cytoplasmic DNA polymerase migrates to and acts in the nucleus. Unlike the bacterial DNA polymerase, the eukaryotic enzymes have no exonuclease activity and no ability to catalyse deoxyribonucleotide pyrophosphate exchange.

Replication of DNA in mammalian cells is initiated at several different sites so that the double strand appears to have a series of beads attached which gradually grow bigger and eventually merge (*Figs*. 8.9 and 8.10). The multiple initiation of replication enables the entire DNA strand to replicate at a very fast rate and a chromatin thread, which may be 1 m long in a typical mammalian cell, is completely replicated in S phase, in about 8 hours. This rate is 100 times faster than that occurring in a typical bacterial cell.

Initiation may begin at sites where the DNA is attached to the membrane and is started by a nuclease splitting a section of a DNA chain. This end which is formed by the split can then act as a template for the start of the synthesis of a new DNA chain. Initiation occurs at multiple sites and both chains of DNA are replicated simultaneously. Ends of newly replicated DNA strands are bound by ligase enzymes and nucleases snip off the attachment to the original DNA strand so that two new strands of DNA are formed (*Fig*. 8.10). Isotopic labellng of precursors shows that histone synthesis always accompanies DNA synthesis in eukaryotic cells. These proteins are believed to be essential for maintaining the structure of DNA as a nucleoprotein complex.

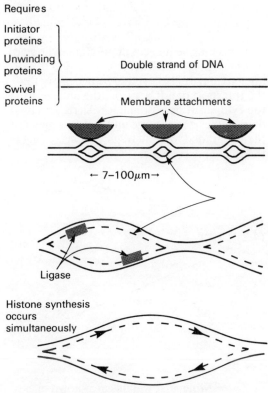

Fig. 8.10. **Summary of sequence of DNA synthesis in the nucleus.**

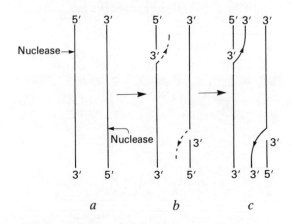

Fig. 8.9. **Sequence of major steps in the replication of double-stranded DNA in the nucleus.**
(*a*) Splitting of DNA chains by nuclease enzyme; (*b*) replication of DNA by DNA polymerase; (*c*) rejoining of original strands and newly synthesized DNA by ligases.

Synthesis of RNA

Apart from limited synthesis of RNA in mitochondria, all cellular RNA synthesis occurs in the nucleus. Many types of RNA are required in cellular function, as shown below.

a.   Low-molecular-weight RNAs.
    i.   Chromosomal RNA.
    ii.   4–8-S RNAs.
    iii.   Transfer RNA (tRNA).
b.   Ribosomal RNAs and its precursors including 5-S, 7-S, 18-S, 28-S, 32-S and 45-S RNAs.
c.   High-molecular-weight RNAs with sedimentation coefficients ranging from 50 S to 200 S. They are rich in A–U base-pairs and do not appear in the cytoplasm. This species of RNAs may be precursors for other types of RNA.
d.   Messenger (mRNA).

RNA synthesis is dependent on the enzyme RNA polymerase and at least six different polymerases have been identified.

Cytoplasmic mRNA frequently contains long sequences of poly(adenylic acid) residues [poly(A)]. These are found in the nucleus and are believed to be necessary for the transport of mRNA from the nucleus to the cytoplasm. Ribosomal RNA (rRNA) is first synthesized as large units, 45 S, which are then degraded to small fragments according to the following sequence:

ribosome, whilst the 28-S rRNA forms one of the rRNA molecules which constitute the large component of the ribosomes (cf. Chapter 9). The ribosomes contain between 80 and 100 different proteins and these are believed to be added in the nucleus during the formation of the rRNA.

Synthesis of nicotinamide adenine dinucleotide (NAD⁺)

The final reaction involved in the synthesis of NAD^+, i.e.

Nicotinamide mononucleotide (NMN) +

$$ATP \rightleftharpoons NAD^+ + Pyrophosphate$$

is catalysed by an ATP:NMN adenylyl-transferase which is exclusively located in the nucleolus to which it is tightly bound. The cell is therefore dependent on nuclear function for its entire supply of NAD^+.

Energy supply for the nucleus

Nuclei possess the entire glycolytic sequence of enzymes and normally obtain their energy supply from glycolysis. Some nuclei, such as those of lymphocytes which are very large, occupying about 60 per cent of the cell, can

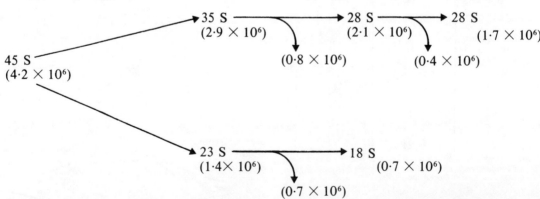

Molecular weights are shown in brackets.

The 18-S rRNA forms part of the rRNA of the smaller component of the

however carry out oxidative reactions of the citric acid cycle and of the electron transport chain and, therefore, obtain their energy by these processes.

Chapter 9

Role of subcellular organelles: the endoplasmic reticulum

9.1 Nature of the endoplasmic reticulum

By studying electron microscope photographs, it can be shown that most cells contain a complex branching network of membranes. This was originally described for chicken fibroblast cells and called the 'endoplasmic reticulum'. The membranes are extensively folded to form vesicles, tubules and large flattened sacs. This network of membranes divides the cytoplasm into two main regions, one region, consisting mainly of vesicles, being enclosed within the matrix and the other region forming the cytoplasmic matrix. When examined under the electron microscope, the membranes appear in parallel pairs (*Fig.* 9.1) and diagrammatic representations of the membranes are shown in *Fig.* 9.2.

An additional important feature of the membranes can be distinguished in most areas. Parts of the membranes are 'rough' or 'granular' on account of the attached ribosomes, whilst the remaining parts are 'smooth' or 'agranular' (*Figs.* 9.1 and 9.2). Intensive study, by Palade, of the cells of the endoplasmic reticulum demonstrated that there is a close correlation between the amount of rough membranes and the quantity of proteins exported from the cell. Examples of this are shown by cells of the liver which export plasma proteins, by plasma cells which synthesize antibodies and by fibroblasts which produce collagen. The proteins produced in the ribosomes penetrate the cavities of the endoplasmic reticulum where they are stored in vesicles formed by budding off from the membranes; these vesicles then migrate to the cell surface where they release their contents. Proteins that form components of the cells are usually synthesized on ribosomes not attached to the endoplasmic reticulum.

The division of the cytoplasm into two compartments, by means of the membranes of the endoplasmic reticulum, results in the possibility of the formation of ionic gradients across the membranes and thus of electrical potentials. It is known that Na^+–K^+-dependent ATPases, similar in nature to the enzyme present in the red-blood-cell membrane (Chapter 25), are present in these membranes and must be important in the regulation of ionic gradients. The significance of these ionic gradients across the endoplasmic reticulum in many cells is not understood, but they are clearly important in cells of the central nervous system and the muscle. In muscle, a specialized endoplasmic reticulum is present called the 'sacroplasmic reticulum' and it is important as an intracellular conducting system.

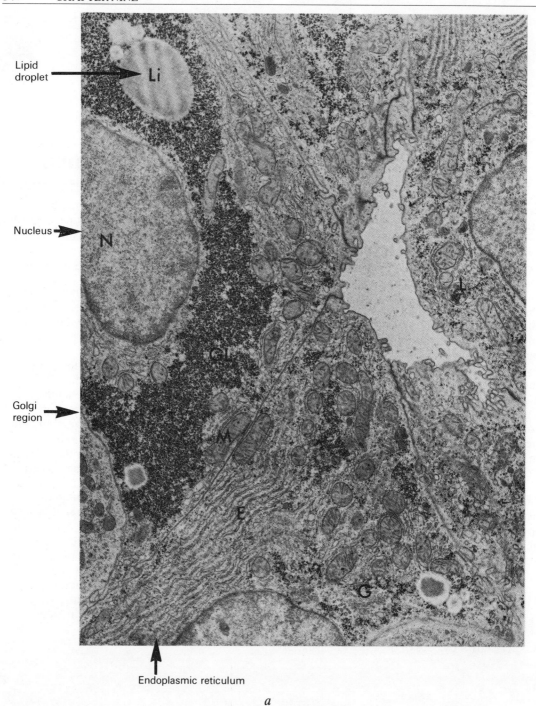

Lipid droplet

Nucleus

Golgi region

Endoplasmic reticulum

a

Fig. 9.1.   **The endoplasmic reticulum as seen under the electron microscope.**
(*a*) Rat liver cell, × 13 000; (*b*) pancreatic cells, × 8000; (*c*) pancreatic cells under high magnification showing the ribosomes, × 80 000.

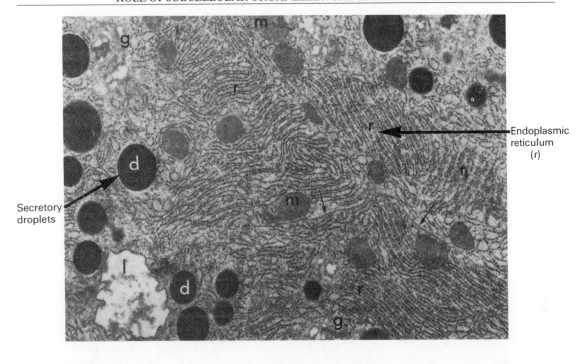

b

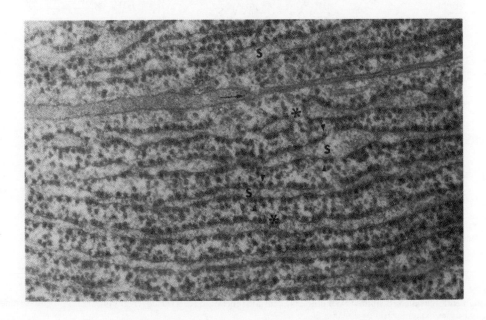

Fig. 9.1.   continued   *c*

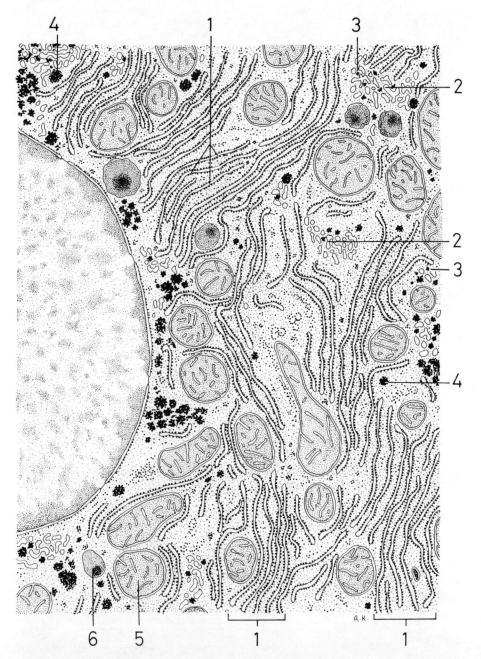

Fig. 9.2.   **Diagrammatic representation of the endoplasmic reticulum in a typical liver cell.**
(1) Rough endoplasmic reticulum; (2) smooth endoplasmic reticulum; (3, 4) glycogen granules; (5) mitochondrion;
(6) peroxisomes. Magnification × 25 000.

9.2 Relationship of the endoplasmic reticulum to other cellular organelles

a. Golgi region

The endoplasmic reticulum is known to be closely associated with the 'Golgi region', first identified in nerve cells but now known to be an important organelle in all mammalian cells. The Golgi region is an extension of the smooth membranes of the endoplasmic reticulum and consists of an intricate pattern of flattened double membranes, closely associated with the nucleus (*see Fig.* 9.3).

All the roles of the Golgi region are not yet established, but the organelle is known to be important in the processing of lysosomal enzymes. These are synthesized in the ribosomes of the rough endoplasmic reticulum and are transported to the Golgi apparatus where carbohydrate residues are introduced to convert them into glycoproteins that are then packaged into vesicles (cf. Chapter 3).

b. Mitochondria—nucleus

The endoplasmic reticulum appears to have an important relationship with the mitochondria and it is possible that the mitochondrial outer membrane may be continuous with the membrane of the endoplasmic reticulum. If so, this link would provide an important pathway for the passage of metabolites between these organelles.

A similar relationship is believed to exist between nuclear membranes and the endoplasmic reticulum. Many important molecules, such as mRNA and tRNA, must be transported out of the nucleus for use in protein synthesis on the ribosomes and a continuous membrane linkage is clearly of great importance to these processes.

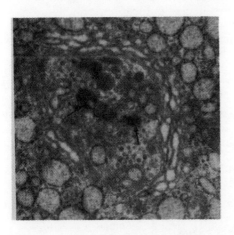

a

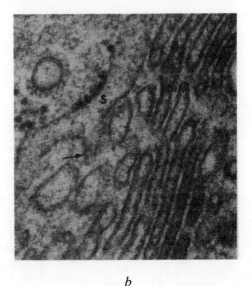

b

Fig. 9.3.   **The Golgi region.**
(*a*) The bowl-shaped Golgi region of an intestinal epithelial cell is shown, × 16 000; (*b*) the flattened Golgi region of a jejunal epithelial cell is shown, × 66 000.

9.3 Separation of the endoplasmic reticulum by ultracentrifugation —the 'microsomes' or 'microsomal fraction'

Although the electron microscope can provide important information about the overall structure and location of the endoplasmic reticulum in cells, it gives little or no information about the biochemical components of the membranes or of their metabolic roles. For biochemical study, a cell fraction, known as the 'microsomal fraction' may be prepared by ultracentrifugation.

The microsomal fraction is usually prepared by centrifuging the homogenate of tissue in 0·25 M sucrose at $100\,000 \times g$ for 60 min after removal of cell debris, nuclei, mitochondria and lysosomes according to the scheme shown in *Fig.* 9.4. The material prepared by this process, frequently referred to as 'microsomes' or 'microsomal fraction', is not structurally related to those organelles present in the cells, but consists mainly of small spherical vesicles formed from the disrupted endoplasmic reticulum. The fractions will normally contain ribosomes from the rough endoplasmic reticulum but, by using more sophisticated centrifugation methods, the ribosome-containing rough membranes and the smooth membranes can be separated (*see Fig.* 9.5).

In recent years, a method for the preparation of the microsomal fraction has been developed which requires only low-speed centrifugation. After removal of the cell debris, nuclei, mitochondria and lysosomes, the supernatant is treated with a diluted solution (8 mM) of calcium salt in 12·5 mM sucrose which causes precipitation of the 'microsomes'. These are readily separated by centrifugation at only $30 \times g$ for 10 min. Both the chemical analysis and the metabolic activities of the fraction prepared in this way are identical to those of the fraction prepared by high-speed centrifugation, a fact which has been verified by many experiments in the author's own laboratory.

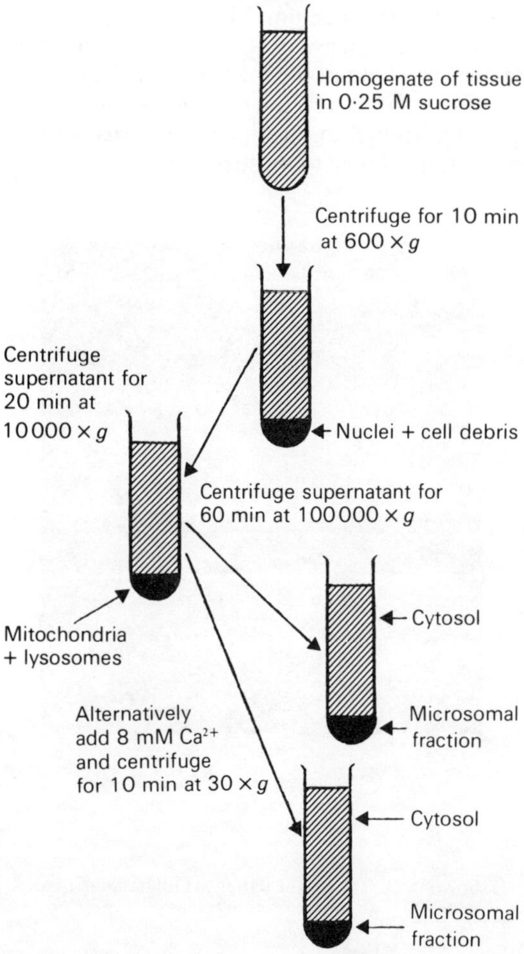

Homogenate of tissue in 0·25 M sucrose

Centrifuge for 10 min at $600 \times g$

Nuclei + cell debris

Centrifuge supernatant for 20 min at $10\,000 \times g$

Mitochondria + lysosomes

Centrifuge supernatant for 60 min at $100\,000 \times g$

Cytosol

Microsomal fraction

Alternatively add 8 mM Ca²⁺ and centrifuge for 10 min at $30 \times g$

Cytosol

Microsomal fraction

Fig. 9.4. **Preparation of the microsomal fraction.**

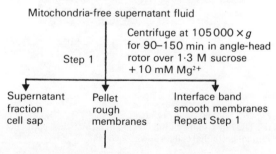

Mitochondria-free supernatant fluid

Step 1

Centrifuge at $105\,000 \times g$ for 90–150 min in angle-head rotor over 1·3 M sucrose + 10 mM Mg²⁺

Supernatant fraction cell sap

Pellet rough membranes

Interface band smooth membranes Repeat Step 1

Fig. 9.5. **Summary of procedure for preparing submicrosomal fractions from rat liver.**

9.4 Structure and composition of the endoplasmic reticulum

The main constituents of the microsomal fraction are protein, phospholipids, RNA and glycogen (*see Table* 9.1). In smooth membrane preparations, the quantities of RNA and glycogen are much less than in 'whole microsomes' and the protein:phospholipid ratio changes so that it is close to 2:1; preparations of rough membranes contain a larger proportion of RNA (*Table* 9.1).

The overall structure of the endoplasmic reticulum may be represented by a model similar to that used for membranes in general, but the protein content is greater than in the exoplasmic membranes. It will be noted that proteins may be aligned on the surface of the membrane lipid or they may traverse the phospholipid bilayer structure (*see Fig.* 9.6). At least 13 different proteins have been separated from preparations of endoplasmic reticulum by gel electrophoresis. Their molecular weights vary between 16 000 and 170 000 (*see Table* 9.2).

Phospholipids form the major proportion of the membrane lipids and the percentages of cholesterol, triacylglycerols and fatty acids are relatively small (*see Table* 9.3). Phosphatidylcholine is the major phospholipid in the membrane followed by phosphatidyl-ethanolamine; sphingomyelins occur in a relatively low concentration(*see Table* 9.4 and Appendix 6).

A large proportion of the fatty acids of the membrane phospholipids are polyunsaturated fatty acids such as linoleic, arachidonic acid and docosahexaenoic acid

Table 9.1 **Chemical composition of whole microsomes and submicrosomal fractions of rat liver**

| Fraction | Composition, mg per g equiv. liver | | | |
	Protein	Phospholipid	RNA	Glycogen
Whole microsomes	19.4 ± 2.2	7.5 ± 1.3	3.8 ± 0.2	43.8 ± 9.6
Smooth membranes	7.6 ± 1.5	3.5 ± 0.7	0.1 ± 0.06	2.0 ± 0.5
Rough membranes	4.5 ± 0.5	1.3 ± 0.3	2.1 ± 0.4	8.0 ± 2.0

Table 9.2 **Proteins of the endoplasmic reticulum membranes (separated by gel electrophoresis)**

Component	$10^{-3} \times$ Molecular weight
S1	171 (Glycoprotein)
	140
S2	104
S3	81
	72
	65
S4	60
S5	54
S6	49
S7	35
S8	30
	19
S9	16

From Blackburn G. R., Bornens M. and Kasper C. B., 1976, *Biochim. Biophys. Acta,* Vol. 436, pp. 387–398.

Table 9.3 **Lipid composition of the endoplasmic reticulum membranes**

| Lipid | Lipid composition, mg/g liver of* | |
	Whole membrane	Smooth membranes
Phospholipids	6.3	2.2
	(78)	(82)
Cholesterol	0.57	0.22
	(7)	(8)
Cholesterol esters	0.07	0.032
	(1)	(1.5)
Triacylglycerols	0.46	0.20
	(6)	(7)
Free fatty acids	0.7	0.032
	(8)	(1.5)

* Percentages are given in parentheses.
From Glaumann H. and Dallner G., 1968, *J. Lipid Res.,* Vol. 9, pp. 720–729.

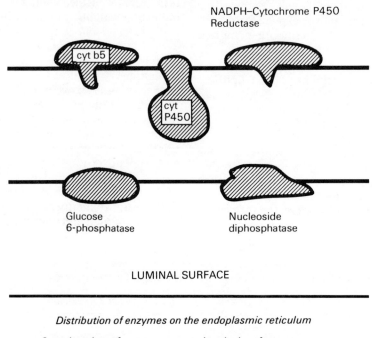

CYTOPLASMIC SURFACE

NADPH–Cytochrome P450
Reductase

cyt b5

cyt
P450

Glucose
6-phosphatase

Nucleoside
diphosphatase

LUMINAL SURFACE

Distribution of enzymes on the endoplasmic reticulum

Cytoplasmic surface	*Luminal surface*
Cytochrome b_5	Nucleoside diphosphatase
NADH–cytochrome b_5 reductase	Glucose 6-phosphatase
NADPH–cytochrome *c* reductase	Cytochrome P450
Cytochrome P450	β-Glucuronidase
ATPase	
5′-Nucleotidase	
Nucleoside pyrophosphatase	
GDP–mannosyl transferase	

Fig. 9.6.   **Location of enzymes in the membranes of the endoplasmic reticulum.**

($C_{22:6}$; Appendix 7). Furthermore, the fatty acid compositions of phosphatidylcholine and phosphatidylethanolamine are not identical (*see Fig.* 9.7). Many efforts have been made to establish whether the smooth membranes and membranes carrying ribosomes are of identical composition. Several investigations have been unable to detect any significant differences, but it has also been suggested that the phospholipid:cholesterol ratio is much greater in rough membranes (15:1) than in smooth membranes (4:1).

Fatty acid components of phospholipids are essential for the correct conformation of the membrane and for effective functioning of the enzymes located in the membrane. If the fatty acids are split from the membrane by phospholipases or the unsaturated bonds are destroyed by peroxidation, the structure is severely disrupted (cf. Chapter 14). This inevitably leads to impairment of function of, for example, the enzymes involved in oxidative drug metabolism. A diagrammatic

Table 9.4 **Phospholipid composition of the endoplasmic reticulum membranes**

Phospholipid	Total phospholipids, %, in		Smooth membranes*
	Whole membranes		
	a^*	$b\dagger$	
Phosphatidylcholine	47·5	53·0	46·4
Phosphatidylethanolamine	19·0	20·2	17·5
Phosphatidylserine	8·5	6·6	8·0
Sphingomyelin	5·8	3·7	6·6
Phosphatidylinositol	10·0	7·5	11·1

* From Glaumann H. and Dallner G., 1968, *J. Lipid Res.*, Vol. 9, pp. 720–729.
† From Davison S. C. and Wills E. D., 1974, *Biochem. J.*, Vol. 140, pp. 461–468.

representation of the disordered membrane is shown in *Fig.* 9.8 and the fall in activity of one of the membrane-bound enzymes in *Fig.* 9.9.

It was originally believed that the fatty acid composition of the membrane phospholipids remained constant, but recently it has been established that the composition, in fact, varies with the type of fat fed to the animal. A comparison of the effects of feeding purified diet containing different fats on the fatty acid composition of the endoplasmic reticulum phospholipids is shown in *Table* 9.5. This composition reflects directly the fatty acid composition of the dietary lipids. Thus the percentage of linoleic acid ($C_{18:2}$) is very high after feeding on corn oil which contains a larger percentage of a triacylglycerol

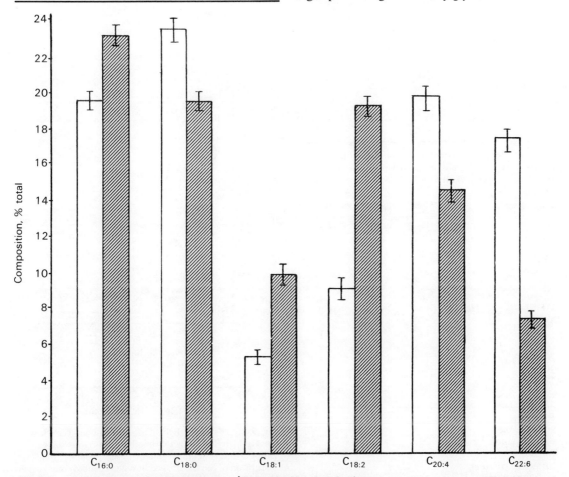

Fig. 9.7. **Composition of phosphatidylcholine and phosphatidylethanolamine.**
The fatty acid compositions of phosphatidylcholine (▨) and phosphatidylethanolamine (▢) of the liver endoplasmic reticulum.

Normal structure   Disordered structure

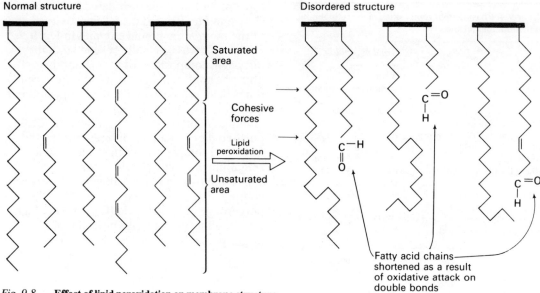

Fig. 9.8.   **Effect of lipid peroxidation on membrane structure.**

Table 9.5 **Percentage fatty acid composition of liver microsomal phospholipids of rats fed on purified diets containing 10 per cent fat for 10–25 days.**

| Fatty acid | Mean composition, %, on diet of | | | |
	Coconut oil	Lard	Corn oil	Herring oil
12:0	1·1	—	—	—
14:0	4·3	—	—	—
16:0	28·5	22·8	21·2	22·3
16:1	9·2	4·8	2·2	3·4
18:0	14·5	17·9	15·9	16·0
18:1	23·5	26·5	15·3	17·8
18:2	7·6	10·3	25·1	5·1
20:1	—	—	—	1·6
20:4	8·8	13·5	17·8	6·4
20:5	—	—	—	8·7
22:5	—	—	—	1·5
22:6	2·6	4·3	2·5	17·0
Rate of aminopyrine demethylation (nmol formaldehyde/min per mg protein)	3·26	3·5	5·03	6·53

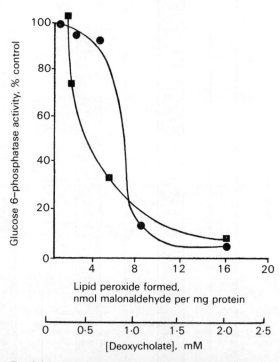

Fig. 9.9.   **Fall in activity of membrane-bound enzymes.** Loss of activity of a membrane-bound enzyme of the endoplasmic reticulum (glucose 6-phosphatase) after disruption of the membrane by a detergent (or bile salt) or by peroxidation.

(●) Effect of deoxycholate; (■) effect of peroxidation.

containing linoleic acid. Herring oil and coconut oil, on the other hand, contain a very low concentration of linoleic acid and thus the membrane phospholipid contains low concentrations of this fatty acid. The proportions of $C_{20:5}$ and $C_{22:6}$ are, however, very high after feeding on herring oil.

These changes in fatty acid composition lead to changes in activity of membrane-bound enzymes and particularly in the enzyme system involved in oxidative drug metabolism (cf. Chapter 36; *see also Table* 9.5).

9.5 Functions of the smooth membranes of the endoplasmic reticulum

The functions of the endoplasmic reticulum are well established in some tissues such as the liver, but not understood so well in others. Functions of the liver endoplasmic reticulum are summarized in *Table* 9.6.

Many of these functions will be described in detail in other chapters but it is useful at this stage to review the overall picture of the processes which are located in this component of the cell. Of special interest is the enzyme glucose 6-phosphatase which hydrolyses glucose 6-phosphate into glucose + inorganic phosphate. This enzyme is of vital importance for the regulation of blood sugar since all glucose leaving the liver, where it is stored as glycogen, must first be converted to glucose 6-phosphate prior to hydrolysis to glucose (cf. Appendices 14, 15).

Most of the enzymes involved in glucose and glycogen metabolism are located in the cytosol and it is, therefore, unusual that this enzyme is membrane bound. The membrane is important for enzyme function, because if the membrane is destroyed, for example by detergents, by phospholipase or by peroxidation of the fatty acids, the activity of the enzyme falls to very low values (*Fig.* 9.9). It is not possible to prepare a pure specimen of glucose 6-phosphatase free of membranes and the membranes must, therefore, be essential for the active conformation of the enzyme.

Table 9.6 **Functions of endoplasmic reticulum in liver**

Membranes	*Function*
Rough	Synthesis by ribosomes of proteins for export
Smooth	Conversion of glucose 6-phosphate by glucose 6-phosphatase to glucose
	Plasma lipoprotein synthesis
	Synthesis of triacylglycerols, phospholipids and part of the synthetic pathway in cholesterol synthesis
	Bile salt synthesis and secretion—7α- and 12α-hydroxylation of cholesterol and conjugation of cholic acid
	Inactivation of steroid hormones which frequently involves hydroxylation
	Detoxication of xenobiotics
	Oxidative metabolism by cytochrome P450 system and conjugation reactions, e.g. glucuronyl transferase
	Oxidative metabolism of carcinogens, e.g. polycyclic hydrocarbons to active epoxides and diols

The role of the endoplasmic reticulum in plasma lipoprotein synthesis is described in Chapter 19 and the synthesis of bile salts in Chapter 29.

Details of the metabolism of foreign compounds and the role of the endoplasmic reticulum in this metabolism are described in Chapter 36.

9.6 Functions of the rough membranes of the endoplasmic reticulum—the ribosomes

As mentioned earlier the rough endoplasmic reticulum contains large numbers of small bodies, called 'ribosomes' which are specifically concerned with protein synthesis.

The overall process involved in protein synthesis is summarized in Appendix 28. The ribosomes play several important roles in the process: they accept the mRNA with the correct base sequence which, as a series of triplet bases, codes for the particular protein being synthesized; they accept the tRNA carrying amino acids; they link the amino acids, by means of a peptidyl transferase, forming peptide bonds and then a peptide chain, and finally the ribosome moves along the mRNA one codon at a time to allow the next tRNA carrying its amino acid to slot into position (Appendix 32).

All ribosomes are known to be composed of two components, one nearly twice as large as the other and often represented diagrammatically as resembling a 'cottage loaf' (*Fig.* 9.10). In view of the fact that ribosomes and ribosomal fragments have been studied extensively by ultra-centrifugation, their properties are often described by sedimentation coefficients or 'S' values described originally by Svedberg (hence Svedberg S units). This sedimentation coefficient is dependent on the size, shape and density of the particles relative to the density of the solvent, and when centrifugations are

carried out under standard conditions, it is primarily dependent on the molecular weight of the particle.

Ribosomes of prokaryotes have been studied very extensively and been shown to consist of a whole particle of sedimentation coefficient 70 S made up of two subparticles of coefficients 30 S and 50 S. The 30-S component is mainly concerned with the binding of the mRNA and the 50-S component with the alignment of the tRNAs, peptide bond formation and translocation (*Fig.* 9.10; *see* Appendix 28). The ribosomes are composed of RNA molecules, in addition to mRNA and tRNA, and several proteins. Very elegant electron microscope studies have enabled a relatively clear picture of the bacterial ribosome to emerge and this is shown in *Fig.* 9.11.

The ribosomes found in mammalian cells are larger than those which occur in bacteria. They measure approximately 20–22 nm across, as opposed to the 18 nm of bacterial ribosomes, have a total sedimentation coefficient of 80 S and are composed of two units, 40 S and 60 S (*Fig.* 9.10). Detailed studies have been made of the ribosomes of rat liver and these are shown in *Table* 9.7.

Bacterial (prokaryotic)

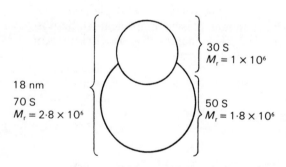

30 S
$M_r = 1 \times 10^6$

18 nm
70 S
$M_r = 2 \cdot 8 \times 10^6$

50 S
$M_r = 1 \cdot 8 \times 10^6$

Eukaryotic

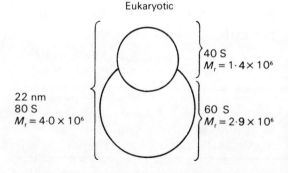

40 S
$M_r = 1 \cdot 4 \times 10^6$

22 nm
80 S
$M_r = 4 \cdot 0 \times 10^6$

60 S
$M_r = 2 \cdot 9 \times 10^6$

Fig. 9.10.   **Typical bacterial and eukaryotic ribosomes.**

Table 9.7 **Ribosomes of rat liver**

Subunit	$10^{-6} \times$ Molecular weight	Sedimentation coefficient, S
40 S	1·4	36·9
60 S	2·9	56·3

The smallest subunit of the liver ribosomes contains 1 molecule of 18-S RNA, and about 30 proteins of total molecular weight $0 \cdot 7 \times 10^6$; it is composed of approximately equal weights of protein and RNA (*see Fig.* 9.12).

The larger subunit of the ribosome contains 30 molecules of RNA, 5 S (mol. wt = 39 000), 5·8 S (mol. wt = 51 000), and 28 S (mol. wt = $1 \cdot 7 \times 10^6$) and 45–50 proteins. In the 40-S subunit, the total molecular weight of the proteins is 707 000; the molecular weights varying between 11 200 and

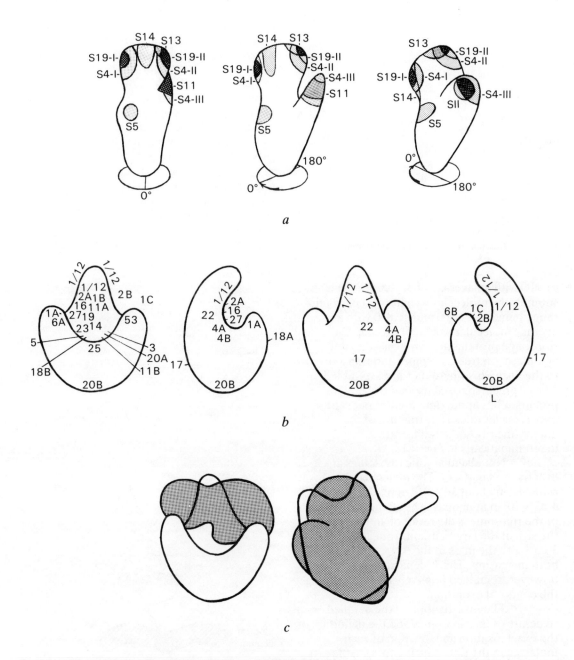

Fig. 9.11.   **Diagrammatic representation of the bacterial ribosome.**
(*a*) Views of the 30-S subunit; protein positions are indicated by S numbers. (*b*) Views of the 50-S subunit; likely protein positions are indicated by numbers.
(*c*) Possible modes of association of 30-S and 50-S subunits.

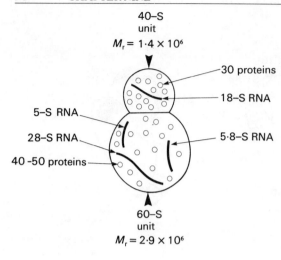

Fig. 9.12.   **Components of eukaryotic ribosomes.**

ribosomes can manage this with 53 proteins, whereas 80 proteins are required for eukaryotic cells. The functions of all these extra proteins is unclear, but some are known to be needed to help attach the ribosomes to the endoplasmic reticulum. These proteins have been termed 'ribophorins'. Other proteins may be necessary for the regulation of ribosomal function but others may simply have no function and be redundant.

In many eukaryotic cells, ribosomes synthesizing proteins may be active free in the cytoplasm or when attached to the endoplasmic reticulum. All proteins which are exported from cells, secretory proteins, are synthesized on the ribosomes attached to the endoplasmic reticulum and this system frequently plays an important role in excretion,

41 500 with an average of 21 400. In the 60-S subunit, the molecular weights of the protein range between 11 500 and 41 800 with an average molecular weight of 21 200. The ribosomal proteins are synthesized in the cytoplasm on free polysomes and transported to the nucleolus where they are assembled on a large precursor of ribosomal RNA. This precursor has approximately 100 per cent excess nucleotide. The structure of the eukaryotic ribosome is illustrated diagrammatically in *Fig.* 9.12.

Not all eukaryotic ribosomes are exactly the same size. The molecular weights range from about $3 \cdot 9 \times 10^6$ in plants to $4 \cdot 55 \times 10^6$ in mammals. The change in the mass of the ribosome is the result of an increase in the size of the large subunit from $2 \cdot 4 \times 10^6$ to $3 \cdot 05 \times 10^6$, the mass of the RNA and protein both increasing. The 40-S particle has, however, remained relatively constant during the course of evolution.

The investigation of the detailed structure of the ribosomes and the definition of the exact location and function of each molecule in the ribosome is a formidable task and only limited progress has been made so far in the study of prokaryotic ribosomes.

Although both prokaryotic and eukaryotic ribosomes carry out nearly identical processes of protein synthesis, bacterial

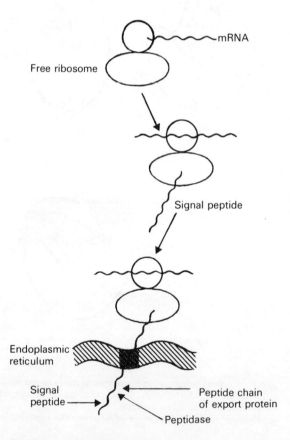

Fig. 9.13.   **Role of signal peptide.**

as it does, for example, for the lipoproteins from the liver (cf. Chapters 19 and 29).

All ribosomes in any one cell are believed to be identical, some functioning free and some whilst being attached to the endoplasmic reticulum. The attachment of the ribosomes to the endoplasmic reticulum is accomplished by means of a 'signal peptide', which is composed of sequences of about 15–30 amino acid residues, largely hydrophobic in nature. The mRNAs for secretory proteins are believed to first become attached to free ribosomes which begin to synthesize the peptide chain. The first 15–30 amino acids from the signal peptide locates the ribosome on to the endoplasmic reticulum and is subsequently split by a peptidase. The probable sequence of events is illustrated in *Fig.* 9.13.

Part 2

Whole body metabolism

Chapter 10 Nutrition: general aspects

10.1 Introduction

Although, throughout history, nutrition has been of fundamental importance to man's development, health and welfare, it is remarkable that nearly all scientific investigations and reports on nutrition are less than 150 years old.

One of the earliest texts was written by a physician to the London Hospital, J. Pereira in 1843. It was entitled *A Treatise on Food and Diet* and contained several accurate analyses of common food. About 30 years later a *Treatise on Food and Diet* by Pavy was published which contained fairly accurate descriptions of the role of protein, carbohydrates and fats in human nutrition. At that time vitamins had not yet been discovered. It is interesting to note that a description of a nutritional experiment appears in the Bible in the first book of Daniel 1–5. This describes the appearance of youths for training as courtiers in the court of Nebuchadnezza, King of Babylon in 607 BC. They were given a daily portion of the King's meat to eat and wine to drink but protested and asked to be given pulse to eat instead. They requested: 'Prove thy servants, I beseech thee, 10 days and let them give us pulse to eat and water to drink. Then let our countenances be looked upon before thee and the countenances of the youths that had eaten of the King's meat.'

This started the earliest recorded nutritional experiment, 'and after 10 days, their countenances appeared fairer and fatter in flesh than all the children which did eat the portion of the King's meat. So the steward took away their meat and gave them pulse.'

After reading the following chapters, it would be useful to reflect on the validity and implications of this experiment!

10.2 Components of an adequate diet

Typical dietary components which are needed for a mammal such as man are set out in *Table* 10.1 and *Fig.* 10.1.

Table 10.1 **Components of natural diets**

1. *Energy component*
   (Fat), carbohydrate, protein

2. *Essential component*
   Essential amino acids, fatty acids
   Vitamins, inorganic

3. *'Inert' components*
   (not digested and/or absorbed)
   i. Large molecules
   ii. Small molecules

4. *Toxic components*
   i. Large molecules
   ii. Small molecules

Components of group 1 supply, normally as a consequence of their oxidation to CO_2, H_2O and other end-products, energy to enable 'life', as we know it, to exist. Energy must be constantly supplied for muscular contraction, ion transport, maintenance of nerve potentials, and chemical synthesis. Carbohydrates, fat or protein may be oxidized to provide this energy, normally in the form of

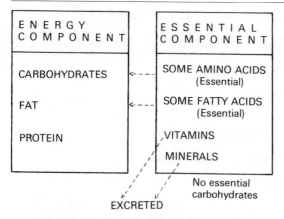

Fig. 10.1. **Constitution of an adequate diet.**

ATP. To a certain extent, these sources of energy are interchangeable but excess quantities of fat cannot be oxidized without the provision of minimal amounts of carbohydrate or protein.

Components of group 2 are classified as 'essential' because they cannot be synthesized by the human body. Clearly all metals such as Na$^+$, K$^+$, Ca^{2+}, Fe^{2+}, Mg^{2+} and others fall into this category and small quantities are required each day to replace losses in the urine, faeces or skin. Few anions are essential. Phosphate is necessary in the diet but occurs widely and iodine is essential for the formation of thyroxine in the thyroid. Fluoride is not essential but becomes incorporated into dental enamel and prevents tooth decay. In addition to essential inorganic components, several organic molecules must be incorporated in the diet because the human body cannot synthesize them. Included in these groups are 10 amino acids, described as the 'essential amino acids', some polyunsaturated fatty acids, described as 'essential fatty acids', and several special organic molecules most of which are required in very small quantities, the 'vitamins'.

Although, to a certain extent, essential amino acids, fatty acids, and vitamins fall into a single category, that of being essential in the diet, there is an important distinction. If large quantities of essential amino acids or fatty acids are consumed, they can be used as a source of energy, i.e. they become equivalent to any of the normal components of group 1. This is not true for the vitamins, because if large quantities of vitamins are consumed they are not utilized as energy but excreted unchanged or metabolized to inert products (*Fig.* 10.1).

Although consumption of the dietary components (listed in groups 1 and 2) in correct daily quantities would provide a healthy and adequate diet, most people do not eat purified proteins, fatty acids and vitamins. They eat an enormous variety of food: complex mixtures of lobster, cream doughnuts, chips, chocolate mousse, brown bread, cabbage, margarine, lettuce and both prepared and natural foods. As a consequence, many other components are taken into the diet and these can be classified as 'inert' in group 3 or 'toxic' in group 4.

A proportion, possibly large, of any meal can be described as 'inert', having no nutritive value, usually because it is not completely, or only very incompletely, absorbed. Many plant polysaccharides, such as cellulose, fall into this category because they are not attacked by the host's digestive enzymes, even though some digestion by bacterial enzymes occurs. The polysaccharides and other woody material (lignin) forming the 'dietary fibre' fall into this group, and, as discussed in Chapter 16, this is now thought to be of considerable value in the diet. Some small organic molecules and inorganic ions can

Table 10.2 **Toxic substances in diet**

A. *Intentional additives*
   1. Specific nutrients
   2. Flavours—colours
   3. Antioxidants
   4. Preservatives

B. *Incidental additives*
   1. Pesticides
   2. Plant growth regulators
   3. Animal growth promotors (antibiotics)
   4. Fertilizers or derivatives

C. *Plant toxins*

D. *Microbes: microbial toxins*

fall into this category; for example SO_4^{2-} is absorbed only to a limited extent, it cannot be utilized in the body and is excreted in the urine.

Nearly all food contains small quantities of toxic substances, as listed in *Table* 10.2. They can be added during processing, for example to preserve the food, or gain entry to the food incidentally, for example during crop spraying or treatment with fertilizers. It must also be appreciated, however, that many plants contain toxins,

several of which are extremely dangerous and have caused many thousands of human deaths and vast numbers of more serious illnesses. It is thus completely erroneous to believe that all 'natural' food is fine for health and that all processed food is harmful.

In addition, many forms of food can be contaminated with micro-organisms which produce dangerous toxins. A full discussion of the problem of food toxins is beyond the scope of this chapter but a summary is listed in *Tables* 10.3–10.8.

Table 10.3 **Food additives and toxicants**

Intentional additives: Added during processing of food.

Antioxidants: Added in a maximum concentration of 0·01 per cent to prevent peroxidation of polyunsaturated fatty acids.

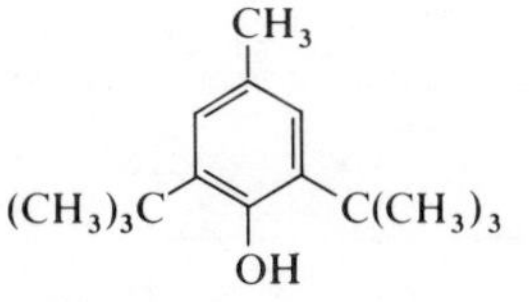

2,6-Di-*tert*-4-methylphenol (BHT) 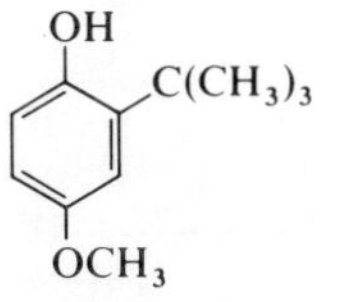   *β-tert*-Butyl-4-methoxyphenol (BHA)

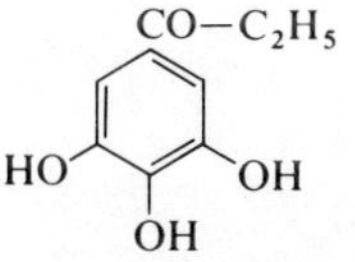

Propyl gallate 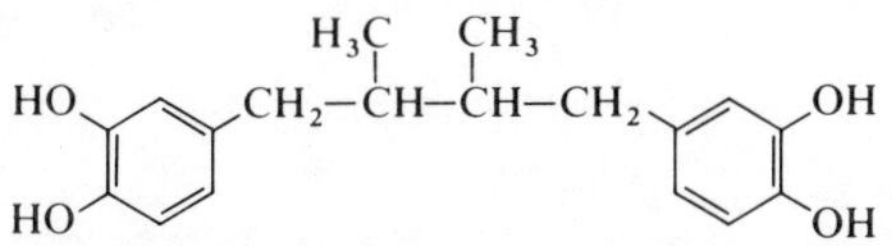   Nordihydroguaiaretic acid (NDGA)

Table 10.4 **Food additives and toxicants**

Accidental additives: Taken up by plants after treatment of crops; lipid-soluble compounds can be stored in animal fat depots

i.   *Insecticides*

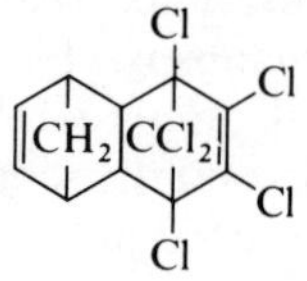

Aldrin 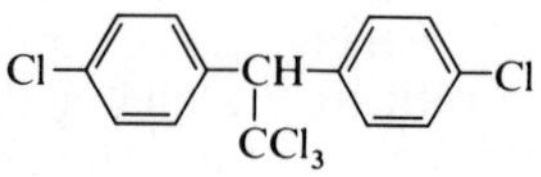   DDT

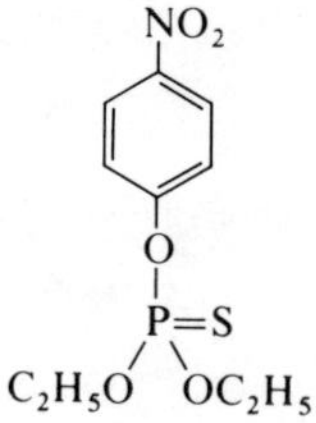

Parathion

ii. *Herbicides*

2,4-Dichlorophenoxyacetic acid

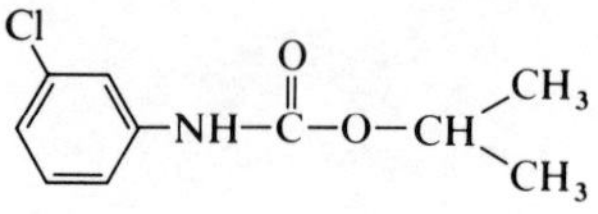

Carbamates

Trichlorobenzoic
acid

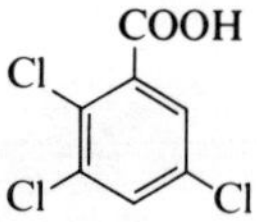

iii. *Plant growth regulators*

Auxins: indole-3-acetic acid

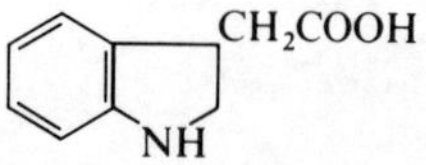

Table 10.5 **Plant toxins: natural constituents of plants**

i. *Nitrates—nitrites*

Nitrates are concentrated by some plants from soil or formed by bacterial oxidation of NH_4^+; NO_3^- and NO_2^- can oxidize haemoglobin to methaemoglobin

Nitrites may take part in formation of potentially carcinogenic nitrosamines in the animal body

ii. *Oxalates—phytates*

High concentration in spinach, rhubarb and cocoa, inhibit absorption of metals such as Ca^{2+} and Fe^{2+} by forming insoluble complexes

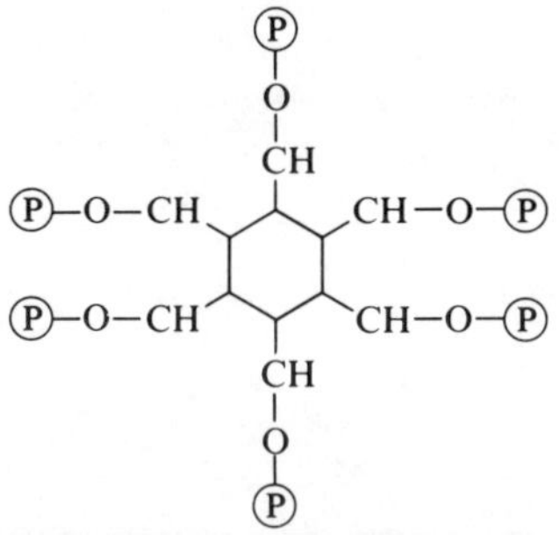

Phytic acid

iii. *Lathyrogens*

Lathyrism, characterized by degeneration of the spinal cord and paralysis, is one of the oldest recorded human diseases arising in populations eating seeds of *Lathyrus* (vetches); major problem in India and N. Africa

$\beta(\gamma, \alpha$-Glutamyl)amino-propionitrile
(osteo-lathyrogen)

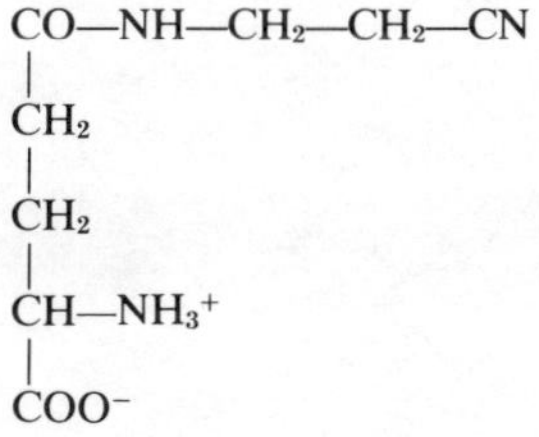

L-α, γ-diaminobutyric acid (+ derivatives)

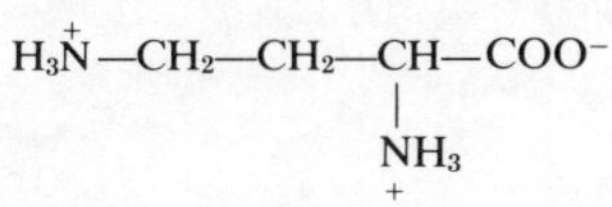

(neurolathyrogen)

Table 10.5 continued

---

iv. *Toxic lipids*
Many plant oils contain unusual fatty acids which are potentially toxic
Erucic acid is an important constituent of 'rape' or 'mustard seed' oils which may contain 20–40 per cent of the acid; causes impaired growth, energy utilization and mitochondrial function

Erucic acid: $CH_3(CH_2)_7$—$CH=CH$—$(CH_2)_{11}$—$COOH$

Branched-chain fatty acids
Branched-chain fatty acids (derived from phytol) may be incorporated into myelin lipids
e.g. 3,7,11,15-tetramethylhexadecanoic acid

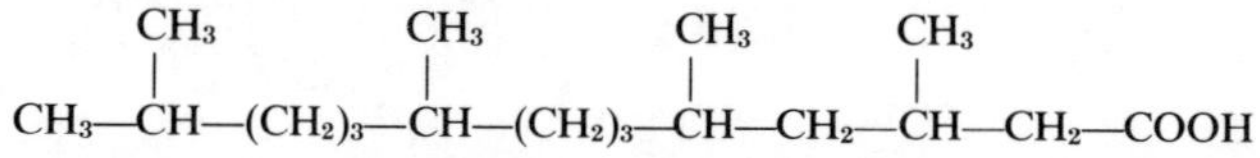

$$
\begin{array}{cccc}
CH_3 & CH_3 & CH_3 & CH_3 \\
| & | & | & | \\
\end{array}
$$

CH_3—CH—$(CH_2)_3$—CH—$(CH_2)_3$—CH—CH_2—CH—CH_2—$COOH$

Cycloprene fatty acids: constituents of several seed oils; suppress animal growth

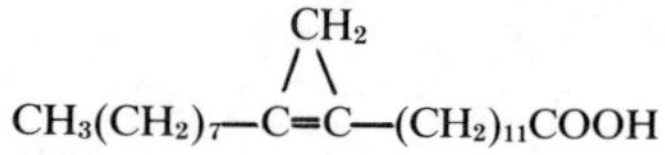

$$
\begin{array}{c}
CH_2 \\
/ \backslash \\
\end{array}
$$
$CH_3(CH_2)_7$—$C=C$—$(CH_2)_{11}COOH$

---

v. *Cyanogens*
Plant glycosides which produce cyanide as a result of enzymic hydrolysis; found in many fruits, yam, peas, maize and millet
Production of cyanide depends on methods of preparation of food

Limamarin

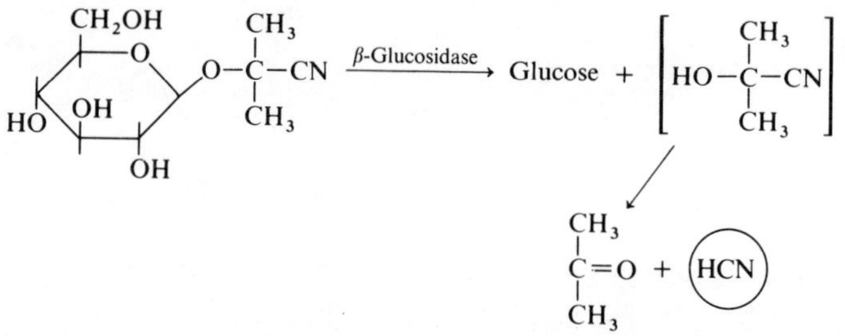

Laetrile

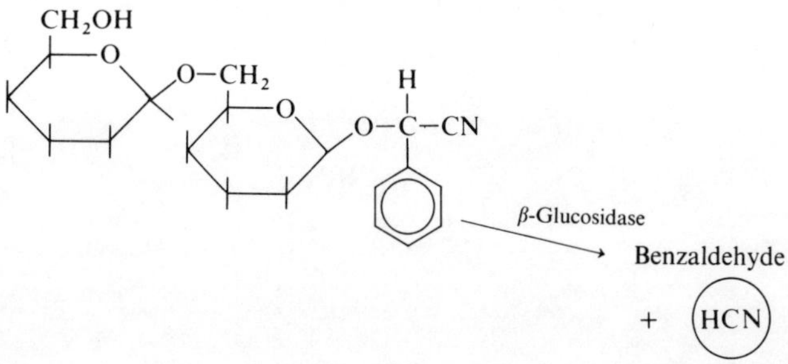

Table 10.5 continued

vi. *Goitrogens*
Thyroid antimetabolites occurring in many plants, especially the brassica family; apart from NaSCN, compounds are thioglucosides and converted to goitrogens by the action of 'thioglucosidase'
Thiocyanate blocks uptake of I^- but goitrin blocks coupling of iodotyrosine and iodination of monoiodotyrosine

Sodium thiocyanate

β-Thioglucosides

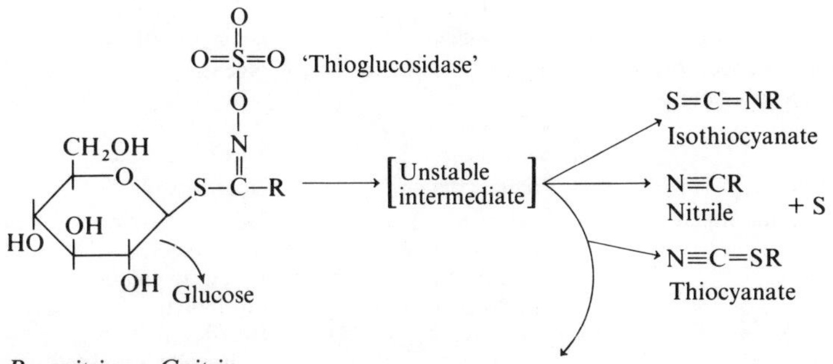

Progoitrin → Goitrin

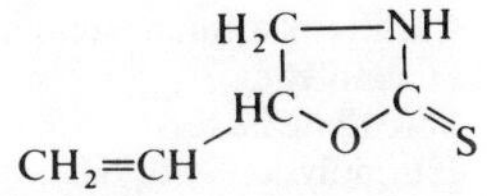

vii. *Phenolic compounds*

Phlorizin—glucoside of *phloretin*
Phlorizin occurs in leaf, root bark, fruit core and seeds of fruits such as apple (300–400 mg/kg); toxic in kidney producing glycosuria

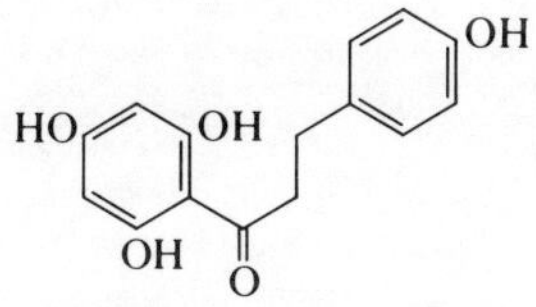

Tangeritin
Flavonoid found in skin of tangerine, orange and mandarin.
Cytotoxic effects on embryonic cells

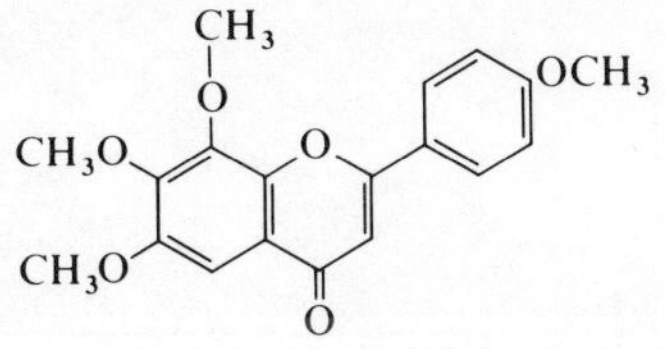

viii. *Lectins* (Phytoagglutinins)
Plant proteins, usually glycoproteins mol. wt 60 000–120 000 found in over 500 species, including many leguminosae; agglutinate red blood cells and may be specific for group A or B
Denatured by heat and lose toxicity

Table 10.6 **Some bacterial exotoxins causing human disease and found in foods**

Many species of bacteria produce a wide range of toxic compounds. These may be small molecules such as amines or toxic macromolecules, which are synthesized on the raw or cooked food or formed in the gut after ingestion of the food. Bacterial protein exotoxins are of great practical importance and often extremely dangerous. The neurotoxic exotoxin of *Clostridium botulinum* is 3×10^6 times as toxic as an equivalent dose of strychnine.

Species of bacteria	Human disease	Exotoxins	Major effects
Clostridium botulinum (the botulins formed depends on type)	Botulism	Botulinum— A, B, C, D, E, F	Neurotoxic
		Haemagglutinin	Blood coagulation
Cl. welchii	Enteritis necroticans and gas-gangrene	$\alpha, \beta, \gamma, \delta, \epsilon, \eta, \iota, \theta$, Toxin	Haemolytic and dermonecrotic and lethal
		κ-Toxin	Proteolytic
		μ-Toxin	Spreading factor
Shigella dysenteriae	Dysentery	Haemorrhagin	Neurotoxin
Staphylococcus aureus	Pyogenic infections	$\alpha, \beta, \gamma, \delta, \epsilon$, Toxin	Dermonecrotic and haemolytic
		Hyaluronidase	Spreading factor
		Coagulase	Blood coagulation
		Staphylokinase	Fibrinolytic
		Enterotoxin	Emetic
		Leukocidin	Kills white blood cells
Streptococcus pyogenes	Pyogenic infections Tonsillitis	Dick-toxin	Erythrogenic
		Streptolysin-O	Cardiotoxic and haemolytic
		Streptolysin-S	Haemolytic
		Hyaluronidase	Spreading factor
		Streptokinase	Fibrinolytic
		Streptodornase	DNAase
Vibrio cholerae	Cholera	Enterotoxin	Diarrhoeagenic
		Haemolysin	Haemolysis

Table 10.7 **Fungal toxins**

Two groups of nutritionally important fungal toxins can be described:
a. Toxins of fungi used for food, e.g. mushrooms
b. Toxins of fungi accidentally present in food, e.g. aflatoxins, yellow rice

Citreoviridin
Produced by *Penicillin toxicarium* on mouldy (yellow) rice; it causes death by cardiovascular and respiratory failure, and is possibly an antagonist of ubiquinone

Cyclochlorotidine
Produced by *Penicillin islandicum* on mouldy rice; hepatotoxic and causes cirrhosis in humans; it causes a rapid depletion of liver glycogen and severe hypoglycaemia; long-term treatments with small doses cause liver carcinogenesis

Aflatoxins
Produced by *Aspergillus flavus*, frequently on mouldy peanuts
Potent inducer of hepatic carcinogenesis

Compound I   *Compound II*

At least 8 different aflatoxins are known with different substituents on a, b, c and d: these are normally H, OH or OCH_3

Fagicladosporic acid
Produced by *C. epiphyllum*, ; causes oedema and haemorrhages

$$CH_3-(CH_2)_9-CH=CH-(CH_2)_9-CO-SH$$

Table 10.8 **Shell fish toxins**

Many shell fish are known to be extremely poisonous when eaten but this is likely to be caused, primarily, by bacterial contamination or by poisonous plankton trapped in the gills

Saxitoxin is produced by a dinoflagellate *Gonyaulux catanella* which becomes trapped in the gills of mussels and clams; common on Pacific coast of US, it is neurotoxic and causes paralysis. It is one of the most toxic small molecules known and it is not normally destroyed by cooking

10.3 Causes of malnutrition

Although deficiency of essential nutrients in the diet will give rise to symptoms of malnutrition it must not be assumed that this is the only cause. Inadequate intake is described as a 'primary deficiency', but secondary deficiencies, caused often by inadequate absorption from the digestive tract, can also occur. For example, deficiencies of vitamin B_{12} are entirely, and those of folic acid mainly, due to absorption failure. Whatever the primary cause, the end-results are similar (*Fig.* 10.2).

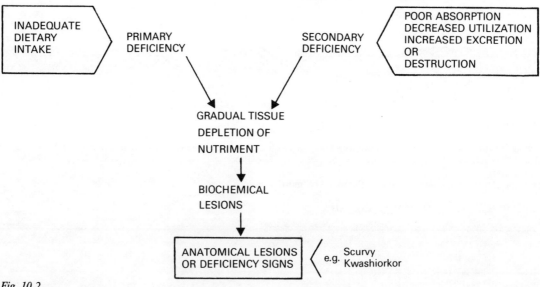

Fig. 10.2.
Causes of malnutrition.

10.4 Nutritional methodology

Faced with the vast range of foodstuffs which are available, it is clearly important to devise adequate methods for the evaluation of their nutritional value.

In general, the methods used fall into two general categories:

i. Deprivation and replacement feeding studies on groups of animals
ii. Epidemiology, which can be applied to man.

Deprivation and replacement

When experiments are carried out in this category, groups of laboratory animals such as rats are fed a diet deprived of, for example, protein and graded quantities of pure protein or foodstuff containing protein are then fed to ascertain the response. Although apparently simple to carry out, such experiments involve many decisions concerning, for example, the species to use, how many animals should be used and the length of time for which the experiment is carried out. When the design of the experiment has been perfected, many criteria can be used to ascertain whether the animals are benefitting or are being harmed by the diet and these criteria are listed in *Table* 10.9.

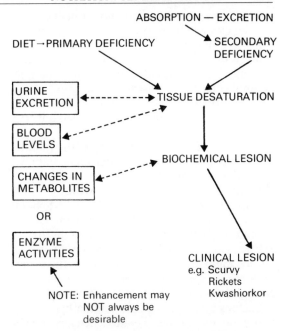

Fig. 10.3. **Biochemical determinations in nutritional evaluation.**
In the boxes: Biochemical evaluation.

Table 10.9 **Criteria of nutritional benefit**

1. Gain in live weight
2. Gain in body energy stores
3. Gain in body nitrogen
4. Gain in specific nutrients
5. Effects on general health
   (disease, resistance or *promotion*)
6. Effects on reproduction
7. Effects on longevity
8. Alterations in specific   } Biochemical
   metabolic pathways   } evaluation

It will be noted that biochemical measurements can often be used and blood or urine samples from humans can give a good indication of the dietary status of any individual; the principles are illustrated in *Fig.* 10.3. Many of the vitamins are converted

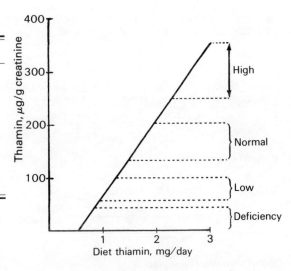

Fig. 10.4. **Diet–urinary thiamin.**

to coenzymes essential in metabolism and deficiencies will therefore show up as reduced enzyme activities. Alternatively, a urine analysis may indicate the vitamin status of the individual for a particular vitamin, for example thiamin (*Fig.* 10.4). In several instances, however, the deficiency will not become apparent until a 'loading test' is carried out. An example is shown for vitamin B_{12} in *Table* 10.10. From the pathology of the blood it is impossible to distinguish between vitamin B_{12} and folate deficiencies (cf. Chapter 24). Vitamin B_{12}, but not folate, is however required for the metabolism of branched-chain amino acids, such as valine or isoleucine and, if vitamin B_{12} is lacking, metabolism of these amino acids terminates in the formation of methyl malonate which is excreted in the urine. Loading patients with valine or isoleucine clearly shows the existence of a vitamin B_{12} deficiency after measurement of methyl malonate in the urine. Folate deficiency can be studied by loading the subject with histidine and analysing the urine for formiminoglutamic

acid which requires folate coenzyme for further metabolism.

Epidemiology

An entirely different approach to nutritional problems is that of epidemiology. This technique involves the assessment of the nutritional status of groups or communities and careful analysis of their dietary habits. The procedure can be used to study animals, for example sheep grazing on different pastures, but it is particularly valuable for human studies. The great advantage is that the whole population being studied exists under

	Benedictines	Trappists	
Number	181	168	
Calories per day	2523	2625	
Calories (%) from:			
Protein	13	14	
Animal protein	6	5	(s)
		1	(w)
Total fat	35	26	(s)
		7.5	(w)
'Animal' fat	20	10	(s)
		3	(w)

Table 10.10 **Methyl malonate excretion in vitamin B_{12} deficiency: loading tests**

Subjects	No treatment	Valine load	Isoleucine load
Normal	0·2 15·3	0·1 15·7	0·5 23·1
Vitamin B_{12} deficient	0·3 264	4 348	2·8 730
Folate deficient	0·7 6·0	0·8 21·6	1·0 16·1

Methyl malonate, mg/day, under

Ranges of values given

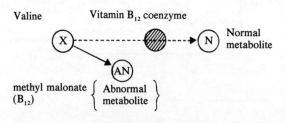

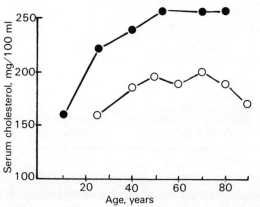

Fig. 10.5. **Epidemiological investigations: study of Trappist and Benedictine monks.**
–●– Benedictine; –○– Trappist. s = summer; w = winter.

completely natural conditions with no experimental restrictions and thus the conclusions are likely to be valid and applicable to human society. The disadvantage is that many individual variations of dietary intake must occur during the experimental period and little control is possible. Some closed communities, for example monks in monasteries, provide an excellent population for study. In the example shown in *Fig*. 10.5, the monks in Benedictine and Trappist monasteries were compared. Samples of their plasma demonstrated clearly that intake of animal fat was associated with increased concentration of plasma cholesterol, and thus also with the incidence of arterial disease. Many other epidemiological investigations have confirmed and extended these observations.

10.5 Nutritional problems in modern society

Until relatively recently, perhaps until the 1960s, nutrition was often considered entirely in terms of 'adequacy'. It was assumed that it was only necessary for humans to consume 'adequate' quantities of energy, protein, essential fatty acids, vitamins and minerals to attain good health. Recent developments in nutrition have, however, demonstrated that the subject is much more complex than had hitherto been supposed.

Consumption of natural foods

It is essential to appreciate that not all ingested foods readily become available to the body from the gastric intestinal tract for energy or other purposes and a proportion is always wasted. Complex interactions between foodstuffs may also occur during preparation and cooking. This may lead to the formation of unavailable products which are unusual and thus not digested, or even toxic products. An example of a toxic product is that of lysinoalanine which causes kidney damage in rats (*Fig*. 10.6). Toxic constituents may also occur in the food, either as natural constituents

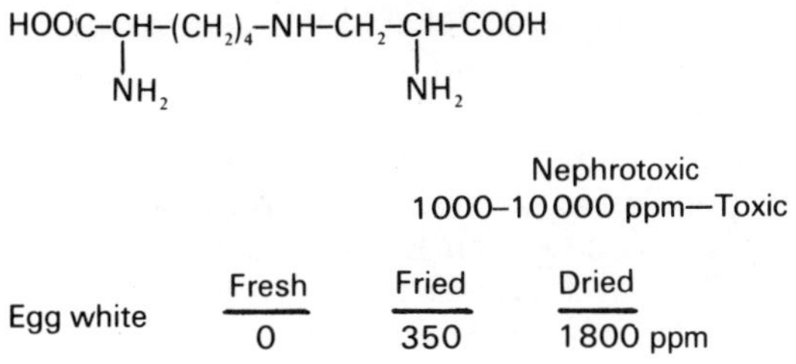

HOOC–CH–(CH2)4–NH–CH2–CH–COOH
NH2
NH2
Nephrotoxic
1000–10000 ppm—Toxic
Egg white
Fresh
0
Fried
350
Dried
1800 ppm

Fig. 10.6. **Lysinoalanine formed in protein after mild alkali treatment.**

or as additives which have been described in detail in *Tables* 10.3–10.8.

Excessive dietary intakes

Excessive food consumption is also a major problem. Although this is normally considered in relation to obesity, many other desirable foodstuffs may be toxic in excess, examples being inorganic iron and vitamin D. The effect of toxicity excess is shown in *Fig*. 10.7*a*. Here the effects of foodstuffs A and B which are toxic in excess only are compared and are shown superimposed on the distribution curve for foodstuff intake of the population. Owing to the shape of the curve A, only a small percentage of the population will be at risk from the consumption of excess quantity of A (Ⓧ on *Fig*. 10.7), but a large percentage will be at risk from the consumption of B which is presumed to have a linear response (Ⓨ on *Fig*. 10.7).

Frequency
B
A
Acceptable risk
Intake
X
Y
Health risk
a
Frequency
C
Intake
X
Y
Health risk
b

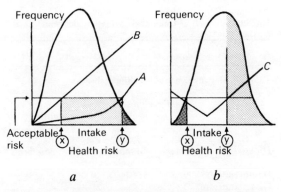

Fig. 10.7. **Deficiencies and excesses of dietary components in relation to acceptable risks.**
(*a*) Toxic in excess only; (*b*) toxic in excess and deficiency disease.

Foodstuff C has both a known deficiency disease and is toxic in excess, so two lines are plotted that intersect to give a desirable intake (*Fig.* 10.7*b*). Two fractions of the population are at risk, one from excess consumption Ⓧ and another from deficiency Ⓨ. Typical deficiencies and excesses are shown in *Table* 10.11 and these can often vary in different parts of the world (*Table* 10.12). The problem of excessive intake and deficiency disease leads to the need to evaluate the most desirable intake of any particular nutrient, or the optimum dose (*Fig.* 10.8). It is often extremely difficult to arrive at a correct decision about the optimum intake when taking a large body of data into account and the divergences of opinion which often exist. For example, it has been agreed for many years in the UK that an intake of 30 mg/day of ascorbic acid (vitamin C) is adequate, but, in the USA, nutritionists believe that 100 mg/day is essential. These divergent views were based on several experiments with human volunteers. If the vitamin C intake is reduced to 10 mg/day all symptoms of the deficiency disease, scurvy,

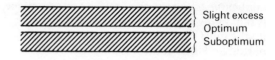

Fig. 10.8. **Optimum doses.**

can be suppressed but the plasma concentration of ascorbic acid remains low; however a maximum plasma concentration of vitamin C can only be maintained when the vitamin C intake is about 75 mg/day, a fact which has led to the belief in the USA that an intake of 75–100 mg/day is essential.

Problems of this type, many of which exist, lead to wide-ranging discussions concerning the value of slight excesses and the danger of a suboptimal dose (*Fig.* 10.8).

Micro-organisms in the gastrointestinal tract

Many foodstuffs are acted upon by micro-organisms in the digestive tract as described in detail in Chapter 16. Many of the products are toxic or can have important pharmacological actions, decarboxylation of amino acids being a typical example.

$$\text{Tyrosine} \xrightarrow[\substack{\text{Bacterial} \\ \text{decarboxylase}}]{} \text{Tyramine}$$
$$\text{Active in brain}$$

The extent of a reaction such as this will clearly depend on the rate of amino acid absorption and the nature and number of bacteria available to carry out the reaction in the intestine. It is thus important to appreciate that molecules not present in the ingested food may be absorbed into the blood.

Psychological problems related to food

Through long tradition, intake of food is a ritual closely woven into the fabric of man's

---

Table 10.11 **Deficiencies and excesses in diet**

Deficiency	Food	Excess toxicity
Marasmus	Energy	Obesity
—	Sucrose	Obesity—dental caries
Essential fatty acid deficiency	Fat	Obesity—heart disease
Wasting	Protein	[Gout]
Anaemia	Iron	Siderosis
Rickets	Calciferol	Hypercalcaemia

Table 10.12 **Prevalence of nutritional disease**

Disease	Developed countries Per-centage	Developed countries $10^{-6} \times$ Number	Developing countries Per-centage	Developing countries $10^{-6} \times$ Number
Dental caries	99	1060	10	175
Under nutrition	3	28	25	434
Anaemia	5	54	30	525
Obesity	25	267	3	52
Xerophthalmia	0	0	1	18

social life and thus many sociological and psychological problems relate to food. For example, most people find colourless food unattractive and the food manufacturers find it essential to colour and flavour food to make it attractive. Some aspects of psychological problems associated with food intake are discussed in Chapter 40.

Nutrition and disease

One of the most important nutritional discoveries of the past two decades is the appreciation that certain diets tend to lead to the development of some diseases, whereas others appear protective. These diets are otherwise nutritionally adequate.

  One of the first clear observations on this subject was the discovery that intake of saturated fat was related to a high level of circulating cholesterol and thus to a high incidence of coronary heart disease; in contrast a diet high in polyunsaturated fat caused the plasma cholesterol to be lowered and the incidence of coronary disease to be reduced. An example of this type of observation is shown for the monks in *Fig.* 10.5.

  More recently, high intakes of fibre have been shown to protect against colon cancer, diverticulosis and varicose veins (cf. Chapter 16).

  Of particular interest are the recent observations on the possibility that certain forms of cancer are related to diet, the possible

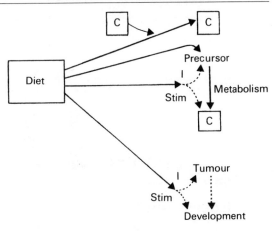

Fig. 10.9. **Diet and cancer.**
☐C = carcinogen.

effects of the diet being shown in *Fig.* 10.9. The carcinogen ☐C may be a dietary component or be acquired during food processing, for example during cooking. Alternatively, the diet may supply a carcinogen precursor or provide components which enhance or inhibit the metabolism of a procarcinogen from other sources to an active carcinogen. Dietary components could also stimulate or inhibit the development of the tumour. Many tumours and especially those of the gastrointestinal tract, such as the colon, are believed to be diet dependent, but carcinogens are likely to be transported on plasma lipids and thus gain ready access to many organs.

Chapter 11

Nutrition: energy requirements and the supply of energy by oxidation of foodstuffs

11.1 Energy units

The basic energy unit is the small calorie (c) which is the heat energy required to raise the temperature of 1 g of water from 14·5 to 15·5 °C. This is equivalent to 4·185 joules, the energy unit used to describe mechanical energy.

The 'small calorie' (c) is not convenient for use in nutritional studies and it is replaced by the 'large calorie' (C) or kilocalorie (kcal) which is equivalent to 1000 small calories and this unit has been used in nutrition for many years. Unfortunately, the term 'calorie' has been used in nutrition texts and in much of the popular scientific literature when the kilocalorie should have been used. The confusion is not, however, of serious practical impact because, in general, values are usually considered in relation to one another and absolute values of energy are not normally of major interest.

However, recently, the Commission on SI units proposed that the kilojoule (kJ) should replace the kilocalorie for nutritional studies and that all energy units expressed as calories should be multiplied by 4·185 and described as kilojoules. This decision is very unfortunate since the majority of the general public has become used to the concept of the calorie as an energy unit for nutrition and the proposal to use the term 'joule' is very confusing. In general, the energy units are expressed as 'calories' in this chapter although reference will be made to conversion to joules.

11.2 Energy supply and utilization

It is important to appreciate that the first law of thermodynamics, i.e. 'energy can neither be created nor destroyed' or 'all systems must be in energetic equilibrium with their environment', applies to all living organisms, and to all animals, including man. As a consequence, the energy for all living processes that require energy must be supplied from the environment in the form of chemical energy, i.e. as food. An animal body has a specific chemical energy requirement and cannot, for example, utilize heat energy from a hot stove or electrical energy to supply its energy needs.

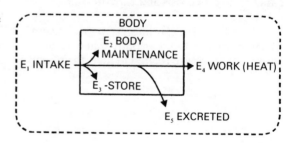

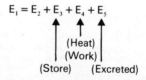

Fig. 11.1. **Energy balance in animal body.**
E_2 = 'basal metabolism' or 'basal metabolic rate' (BMR).

The basic principles are shown in *Fig.* 11.1. The energy intake in the form of oxidizable foodstuffs or E_1 is utilized for several purposes which may be subdivided into E_2 (body maintenance), E_3 (stores, normally glycogen or fats), E_4 (work that normally produces heat as a by-product) and E_5 (excreted energy). E_5 will include the heat energy associated with the excreta and the chemical energy potentially available from the oxidation of molecules not digested in the digestive tract and absorbed.

Several important deductions can be made from *Fig.* 11.2.

a. If the energy supply (E_1) is excessive, then the energy passing into store (E_3) increases substantially and obesity gradually develops (cf. Chapter 40).

b. If the energy supply (E_1) ceases, or is insufficient, then energy must be supplied by the store (E_3) because energy for body maintenance (E_2) must continue. The tissues encounter several problems when they utilize body energy stores and these are discussed in Chapter 23.

c. Work or muscular activity (E_4) always produces heat as a by-product. It is well known that walking briskly on a cold day is often a pleasurable experience, whereas standing or waiting on a similar day can be very unpleasant.

11.3 Basal metabolism

An important component of the energy requirement is E_2 which is required for body maintenance. This energy is utilized for many body processes essential for life: the activity of the heart, conduction of nerve impulses, ion transport across membranes, reabsorption in the kidney and many synthetic processes. This energy requirement is often referred to as the 'basal metabolism' or the 'basal metabolic rate' (BMR). This rate depends on many factors, as listed in *Table* 11.1. The surface

Table 11.1 **Basal metabolism (BMR)**

BMR depends on:

1. Body size (surface area)
2. Body composition (fat, water)
3. Age
4. Sex
5. Health
6. Climate
7. Hormones

Man:	38–40 cal·m^{-2}·h^{-1} ≡ 160 kJ
	1600 cal/day ≡ 6500 kJ
Woman:	36·6 cal·m^{-2}·h^{-1} ≡ 150 kJ
	1150 cal/day ≡ 4700 kJ

Table 11.2 **Effect of body weight on BMR**

Weight, kg		BMR, cal/day	
Men	*Women*	*Men*	*Women*
—	35	—	1654
—	40	—	1823
45	45	2447	1987
50	50	2643	2146
55	55	2833	2300
60	60	3019	2451
65	65	3201	2599
70	70	3379	2743
75	—	3553	—
80	—	3725	—

Note: Effects on the total energy requirements for reference man and woman are shown.

Table 11.3 **Effect of age on BMR**

Age, years	Percentage of reference, %	BMR, cal/day	
		Men	*Women*
20–30	100·0	3200	2300
30–40	97·0	3104	2231
40–50	94·0	3008	2162
50–60	86·5	2768	1990
60–70	79·0	2528	1817
70	69·0	2208	1587

Note: Effects on the total energy requirements for reference man and woman are shown.

Table 11.4 **Effects of environmental temperatures on BMR**

Mean external annual temperature, °C	Percentage reference, %	BMR, cal/day	
		Men	Women
−5	104·5	3344	2404
0	103·0	3296	2369
5	101·5	3248	2335
10	100·0	3200	2300
15	97·5	3120	2243
20	95·0	3040	2185
25	92·5	2960	2128
30	90·0	2880	2070

Note: Effects on the total energy requirements for reference man and woman are shown.

area is most important and is directly proportional to the basal metabolism. A normal adult man requires 38–40 $cal \cdot m^{-2} \cdot h^{-1}$ (160 $kJ \cdot m^{-2} \cdot h^{-1}$) or about 1600 cal/day (6500 J/day) and adult women require slightly less (*Table* 11.2). The requirement decreases with age, increases with body weight and decreases with increase in the environmental temperature (*Tables* 11.2–11.4).

Disease or malnutrition can have a significant effect on the basal metabolic rate, which decreases significantly during starvation or even semi-starvation and can increase during an infection. Several hormones cause

changes in the BMR, the most important of which is thyroxine (cf. Chapter 17). In hypothyroidism–myxoedema, the BMR may be reduced by 40 per cent and it increases significantly during overactivity of the thyroid.

11.4 Energy for work activity

This component, described as E_4 in *Fig.* 11.1, can vary extensively from one individual to another and from one day to another. The term 'work' may refer to the physical activity involved in the work process itself, but it can also depend on the energy involved in travelling to and from work, for example walking or cycling, or on the energy involved in playing games. For an individual lying at rest, this component (E_4) is nearly zero so that the calories required for the BMR are sufficient.

All human activities do, however, require energy and typical examples are summarized in *Table* 11.5. These range for a 70-kg man from 70 cal/h for dishwashing to 1140 cal/h for rowing in a race. The daily energy requirements for three typical adult individuals are shown in *Table* 11.6.

Comparing the clerk with the female typist, it will be noted that the former's basal metabolic energy requirement (1800 cal) is considerably greater than that of the female

Table 11.5 **Energy cost of activities exclusive of basal metabolism and influence of food**

Activity	Energy, $cal \cdot kg^{-1} \cdot h^{-1}$	Activity	Energy, $cal \cdot kg^{-1} \cdot h^{-1}$
Boxing	11·4	Sawing wood	5·7
Carpentry (heavy)	2·3	Sewing by hand	0·4
Cello playing	1·3	Singing in a loud voice	0·8
Dishwashing	1·0	Sitting quietly	0·4
Dressing and undressing	0·7	Skating	3·5
Driving a car	0·9	Standing at attention	0·6
Eating	0·4	Standing relaxed	0·5
Horse riding (trot)	4·3	Sweeping a bare floor—with a broom	1·4
Knitting a sweater	0·7	Sweeping with vacuum-cleaner	2·7
Laundery (light)	1·3	Swimming (2 miles/h)	7·9
Lying still, awake	0·1	Tailoring	0·9
Painting furniture	1·5	Typewriting rapidly	1·0
Paring potatoes	0·6	Violin playing	0·6
Piano playing (Mendelssohn's songs)	0·8	Walking rapidly	3·4
Rowing in a race	16·0	Washing floors	1·2
Running	7·0	Writing	0·4

Table 11.6 **Examples of energy expenditure per day**

Typist		Clerk		Metal worker	
Age, years	26	Age, years	30	Age, years	45
Height (cm)	165	Height (cm)	175	Height (cm)	165
Weight (kg)	60	Weight (kg)	75	Weight (kg)	65
Basal metabolism, 24 hours	1480	Basal metabolism, 24 hours	1800	Basal metabolism, 24 hours	1600
Dressing & undressing 1·5 hours	75	Dressing & undressing 1 hour	50	Dressing & undressing 1 hour	50
Meals 1·5 hours	40	Meals 1·5 hours	40	Meals 1·5 hours	40
Walking 1 hour	140	Walking 1 hour	240	Walking 1 hour	240
Typing 8 hours	320	Clerical work 8 hours	240	Work 8 hours	1440
Sitting 3 hours	85	Sitting 2·5 hours	75	Sitting 4·5 hours	125
Dancing 1 hour	250	Cycling 2 hours	400		
Totals	2390		2845		3495

typist (1480 cal). This is primarily because the male requires more energy than the female and also because his weight (75 kg) is substantially more than that of the typist (60 kg). However, as he is slightly older than the female typist, the difference will be slightly reduced on this account. Both the clerk and the typist use relatively small proportions of their energy (320 cal and 240 cal, respectively) on their office work which is not energy demanding. Cycling consumes a considerable fraction of the total energy required by the clerk during the day!

The metal worker has a smaller basal metabolic requirement than the clerk (1600 cal as compared with 1800 cal) and this is because he is older and of lower weight. His work requirement (1440 cal) is, however, very high and thus his total energy requirement for the day is substantially more than that of either the clerk or the typist.

The varied energy requirements of different individuals under different circumstances has led to the establishment of the concept of 'Reference Man' and 'Reference Woman' by the Food and Agricultural Organisation (FAO) of the United Nations and the World Health Organisation (WHO). The typical adult male has been calculated to require 3200 cal/day and the typical adult female 2300 cal/day (*Table* 11.7). These values form a useful yardstick when considering the energy requirements of various individuals in different situations.

Table 11.7 **Reference man and woman**

Man		Woman
25	Age, years	25
65	Weight, kg	55
10	Temperature, °C	10
8	Work, h	8
4	Sedentary, h	4
3	Recreation, h	3
3200	← Total cal → in 24 h	**2300**
13 376	←kJ→ in 24 h	**9614**

11.5 Energy supply from foodstuffs

To provide human daily energy requirements, carbohydrates, fat or protein may be oxidized. In order to determine the theoretical energy, any foodstuffs can be oxidized in the laboratory artificially by means of a bomb calorimeter. A measured weight of foodstuff is placed in a sealed container containing oxygen and burned by means of an electrical discharge. The heat so produced is measured by the calorimeter. This experiment shows that about 680 kcal are produced by the combustion of 1 g carbohydrate.

$$C_6H_{10}O_5 + 6O_2 \rightarrow 6CO_2 + 5H_2O + \text{ENERGY}$$

(162 g)   (680 kcal)

(glycogen—stored unit of carbohydrate)

$$\frac{680}{162} = \underline{4\cdot1 \text{ cal/g}} \equiv (17\cdot1 \text{ J/g})$$

For glucose (180 g) the figure is slightly less:

$$\frac{680}{180} = 3\cdot8 \text{ cal/g}$$

$$RQ = \frac{CO_2}{O_2} = 1$$

and 7657 kcal from the combustion of 1 g of a typical fat, such as tripalmitin.

$$C_{51}H_{98}O_6 + 72\cdot5 O_2$$
$$\rightarrow 51CO_2 + 49H_2O + \text{ENERGY}$$

(806 g)   (7657 kcal)

$$\frac{7657}{806} = \underline{9\cdot5 \text{ cal/g}} \equiv (39\cdot7 \text{ J/g})$$

$$RQ = \frac{CO_2}{O_2} = \frac{51}{72\cdot5} = 0\cdot707$$

It is not possible to obtain an accurate measure of energy provided by the oxidation of protein using the bomb calorimeter because this

instrument causes the oxidation of nitrogen to nitrogen oxides such as NO_2. In the animal body these are not formed and nitrogen is excreted as urea. Indirect methods have to, therefore, be used and, as a general rule, dietary protein produces about 4·1 cal/g, the same figure as for carbohydrate.

It is important to appreciate that the total energy theoretically provided by oxidation of each foodstuff is not available to the body. The subdivision of the total provided by a typical food is shown in *Fig.* 11.2. It will be noted that the energy actually available to the tissue, the metabolizable energy, is significantly less than the total energy which is measured by oxidation of the food in the bomb calorimeter. There are several reasons for this: firstly, some of the energy is wasted because it is resistant to digestive enzymes; secondly, after absorption of the digested product some of the absorbed food energy is wasted on account of losses in the urine, sweat and faeces. Thus, for example, the urine always contains small quantities of molecules such as lactic and amino acids which could provide energy after oxidation. In the example shown in *Fig.* 11.2 it will be noted that most of the energy is used for basal metabolism, but some is stored, about 25 per cent used for 'work', and some is utilized for the so called 'specific dynamic effect' or 'specific dynamic activity'.

This phenomenon of loss of available energy whenever food is ingested was, remarkably, first discovered 100 years ago by the German physiologist, Rubner. He observed that consumption of food always increased the heat production in man and animals and, furthermore, that this increase was always much greater after consumption of protein than after eating fat or carbohydrate. Typical losses are shown in *Table* 11.8. This is an important aspect of energy provision by different foods because it shows that, contrary to popular belief, protein is not a good source of energy.

The full explanation for the high value for the specific dynamic effect of protein has still not been resolved, but it is generally believed that the flood of amino acids into the plasma following a protein meal will stimulate a large number of synthetic processes, all of

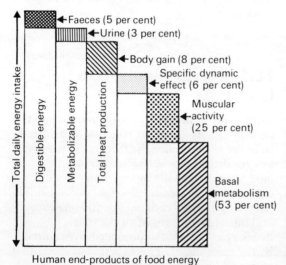

Human end-products of food energy

Fig. 11.2. **Typical disposal of food energy.**

Table 11.8 **Specific dynamic effect of foodstuffs: the 'heat increment' of food**

Foodstuff	Energy 'wasted'
Fat	$2\cdot5 \rightarrow 4\%$
Carbohydrate	$5 \rightarrow 6\%$
Protein	$20 \rightarrow 30\%$

Effect varies with:
1. Time after ingestion of meal
2. Nutritional state
3. Composition of diet

which require energy. These include the synthesis of stored triacylglycerols from the ketoacids formed from the amino acids and the synthesis of the large quantity of urea formed from the amino nitrogen. Increased intake of proteins always causes a much faster rate of protein turnover which is dependent on a much faster rate of protein synthesis. This process requires a great deal of energy and each peptide bond requires the provision of 5 moles of ATP + GTP and protein synthesis probably accounts for the major fraction of energy required for the specific dynamic effect.

The existence of the specific dynamic effect demonstrates that protein is not a good source of energy and, furthermore, that it is much more expensive than fat or carbohydrate. Fat is a good source of energy, producing 9·1 cal for each g oxidized, but fat cannot be utilized exclusively as a source of energy without oxidation of carbohydrate since ketosis will occur. The reasons for this are explained in Chapter 23. This concept that oxidation of fat requires co-oxidation of carbohydrates was neatly stated many years

ago: 'Fats burn in the flame of carbohydrate fire'.

The best energy source for humans is carbohydrate in the form of starch. Unfortunately, during recent years, starch has been highlighted as a potential cause of obesity whereas high intakes of dietary fat are probably much more likely to be an important factor.

The type of food used for energy production in the human body varies considerably in different parts of the world. In the developing world, cereal starch is a major energy source but, in Europe and the USA, fat and protein have become major sources of energy. Surveys in the UK have demonstrated that the proportion of energy supplied by fat has increased steadily during the past 25 years, whereas that supplied by carbohydrate has decreased. Affluence tends to increase fat consumption (Table 11.9 and Fig. 11.3). This is a very unfortunate trend, because it is now well established that high intakes of saturated dietary fat correlates with an increase in arterial disease and thrombosis (cf. Chapter 19).

Table 11.9 **Current food consumption in UK**

	Average values
Daily calorie intake per person	3050
Estimated needs	**2350**

Sources of energy	% total	Trend
Protein	11	↔
Fat	42	↑
Carbohydrates	42	↓
Alcohol	5	↑

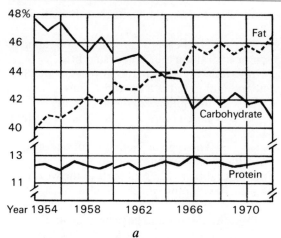

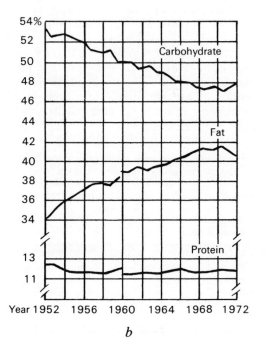

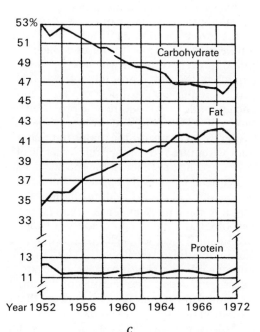

Fig. 11.3. **Percentages of energy derived from carbohydrate, fat and protein.**
(*a*) National Food Survey: percentage of energy derived from carbohydrate, fat and protein; income group A1 1954–1972. (*b*) National Food Survey: percentage of energy derived from carbohydrate, fat and protein; income group C 1952–1972. (*c*) National Food Survey: percentage of energy derived from carbohydrate, fat and protein. Old-age pensioner families 1952–1972.

Chapter 12 Nutrition: proteins in the diet

12.1 Introduction

The fact that it is essential for man to consume a regular supply of protein must have been apparent from very early times. Man became aware that, when protein is lacking in the diet, serious illness and eventual death result. From early history, man appears to have become obsessed with the importance of eating protein and the cult of hunting animals for food, begun many thousands of years ago, is almost certainly a result of this obsession. A strange concept developed, which many still believe today, that increase in dietary protein will lead to greater deposition of muscle and therefore to increased strength and vitality. It is now known that there is virtually no scientific basis for this concept, and an adequate protein intake is all that is necessary for health. In fact, excess intake of protein is likely to be unhealthy. This obsession with the eating of meat in earlier days can be seen from the stocks of food taken in by the Earl of Bedford in 1654 (*Table* 12.1).

The cult of eating large amounts of protein was scientifically endorsed by the famous German chemist Liebig in the early nineteenth century. He believed that muscle action actually consumed muscle protein which must be replaced by an adequate, or relatively large, intake of protein. This view is largely incorrect but, as we shall see in the following section, muscle protein is lost under certain conditions, although at a much slower rate than Liebig considered likely. Unfortunately the great eminence of Liebig caused many to endorse his views and the leading British nutritionalist of the mid-nineteenth century, Playfair, in Edinburgh supported the idea of the necessity for a large protein consumption, recommending intakes of 60–185 g of protein per day.

The fact that such a large intake of protein is unnecessary was first suggested by the American nutritionalist, Chittenden, who, in 1905 by experimenting on himself, proved that an intake of 40–50 g protein per day would maintain perfect health. Subsequent research has adequately vindicated Chittenden's views, but the cult of high protein intake, and especially eating animal protein foods, is strongly ingrained into our way of life, and is often involved with sociological and religious taboo.

Table 12.1 **The extent to which meat predominated in the diet of wealthy people can be judged from the record of a week's purchases of stores for the Woburn Household of the Earl of Bedford in April 1654**

Purchases	£	s	d
One bullock of 68 stone	–	–	–
2 sheep	–	–	–
1 calf	–	–	–
A quarter of mutton	–	4	6
A side of veal	–	7	6
10 stone 4 lb of pork	–	19	3
1 pig	–	2	10
2 calves' heads	–	1	10
4 capons	–	8	6
12 pigeons	–	5	6
20 lb of butter	–	10	0
Eggs	–	3	0
Crayfish	–	1	10
A peck and a half of apples	–	1	9
Bread	–	1	6
2 pecks of oatmeal	–	2	8
Yeast	–	1	8
Six bushels of fine flour	–	–	–

12.2 Protein turnover

Of basic importance in the understanding of the role of dietary protein is the concept of protein turnover. This concept arose primarily from the pioneering research of Schoenheimer in the USA in the 1940s who was, for the first time, able to study the metabolism of proteins and amino acids labelled with the stable isotope of nitrogen (^{15}N) in animal tissues. He observed that, if an amino acid such as leucine labelled with ^{15}N, was fed to an adult animal in a stable condition, a considerable proportion of the ^{15}N became incorporated into other amino acids and proteins, and was then quite slowly excreted. He concluded that all proteins in the body were in a constant state of flux, being steadily synthesized and degraded. He also suggested that the cells of the body possessed a type of 'amino acid pool' which was in dynamic equilibrium with the amino acids formed from the degradation of body protein, and with those taken into the diet. It was also apparent that a steady loss from the pool occurred even when no amino acids, as protein, were being ingested in the diet. This fact had been established about a hundred years earlier by the German nutritionalist Voitt, but was usually ignored. These observations may be summarized as the concept of 'protein turnover', shown in *Fig.* 12.1. The amino acid pool can be regarded as a bath equipped with a small drain which it is impossible to plug (X) and a large overflow (Y). As a result of this, it is impossible to

prevent loss of amino acids, but if the dietary protein intake is high then the excess amino acids are metabolized and the nitrogen excreted as urea (Y in *Fig.* 12.1).

In view of the fact that there is a constant loss of nitrogen as urea, it is clearly essential for there to be regular dietary intake of protein. Further investigations of protein turnover have shown that the subject is complex because the rate of synthesis and degradation of every protein in the body is probably different. For many proteins within cells, it is very difficult to measure turnover accurately, but, generally, the pattern of turnover tends to fall into one of three groups shown in *Fig.* 12.2.

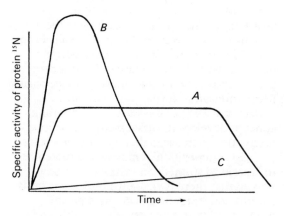

Fig. 12.2. Protein turnover.
Measurement of the rate of incorporation of ^{15}N of an amino acid, such as labelled leucine, into typical proteins and the loss of label during protein degradation.

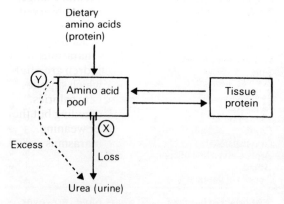

Fig. 12.1. Protein turnover.

Group A proteins are rapidly synthesized, have a finite life, and are degraded rapidly. Typical of this group is haemoglobin, which is normally stable for the life-span of the red blood cells: about 120 days in man. Group B proteins are rapidly synthesized, but are equally rapidly degraded. Their half-life is therefore very short. Typical of these proteins are the plasma proteins, such as plasma albumin. Proteins of group C turn over very slowly and have a very long half-life. In some tissues, such as the brain and bone, there appears to be two pools of protein, one relatively unstable and turning over rapidly,

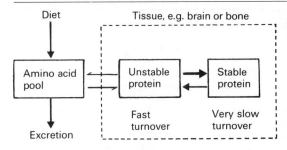

Fig. 12.3.   **Slow and rapid turnover rates of proteins.**

whilst the other is much more stable and turns over very slowly (*Fig.* 12.3).

The turnover of proteins of the body as a whole will clearly represent the summation of all those of the individual proteins of the body. Typical values are shown in *Table* 12.2 where it will be noted that, in man, approximately half the body protein is turned over every 5–6 months. In animal experiments, the rate of protein turnover is markedly enhanced by increasing the protein intake. It will be noted (*Table* 12.2) that the rate of turnover of serum proteins in rat can be increased almost six times by the inclusion of a high percentage of protein in the diet.

Table 12.2   **Rates of turnover of body protein**

	Half-life (days) of	
	Total body	Liver/serum proteins
Rat	17	6–7
Man	158	10–20

Effect of diet (rat)	
Diet	Half-life of plasma protein (days)
No protein	17
25% protein	5
65% protein	2·9

12.3 Consequences of removal of protein from the diet

If protein is removed from the diet of animals a series of complex pathological changes takes place. Young animals cease to grow but, in young and old, the bone becomes rarified, cell proliferation is depressed, collagen formation in connective tissues is much reduced, the skin atrophies, hair is lost, the muscles atrophy and there is decreased enzyme formation in the liver and decreased enzyme secretion from organs such as the pancreas. The condition is extremely serious and death will ultimately occur.

In human societies in many parts of Africa, South America and Asia, serious protein deficiency is found in young children, especially in the 1 to 3-year-old age group. The condition was first described in West Africa in 1933 and given the native name 'kwashiorkor'. True kwashiorkor patients suffer only from protein deficiency, but many young children also suffer from energy (or calorie) deficiency in addition. This condition is known as 'marasmus'. For many young victims the symptoms are not so clearly defined, and the condition observed tends to lie in between the two extremes.

Kwashiorkor $\rightleftharpoons$ Marasmus

$$\begin{bmatrix} \text{Calories} \\ \text{adequate} \\ \text{Protein} \\ \text{deficiency} \end{bmatrix} \quad \begin{bmatrix} \text{Calories} \\ \text{and} \\ \text{protein} \\ \text{deficiency} \end{bmatrix}$$

Characteristic symptoms of the two conditions are shown in *Table* 12.3. Symptoms of marasmus are similar to those of severe starvation, but kwashiorkor patients show characteristic signs of oedema, mental changes and fatty infiltration of the liver, not observed in marasmus.

It is generally believed that the development of kwashiorkor or marasmus occurs after weaning. Young children are often put on a poor inadequate diet, which supplies little protein, and often inadequate calories. Growth is normal during breast feeding but the growth rate rapidly declines after weaning. Early weaning tends to lead to marasmus whereas late weaning tends to lead to kwashiorkor (*Fig.* 12.4).

More intensive studies of the two conditions during recent years have, however, shown that inadequate energy intake may be

Table 12.3 **Characteristics of marasmus and kwashiorkor**

Features	Marasmus	Kwashiorkor
Essential features		
1. Oedema	None	Lower legs, sometimes face, or generalized
2. Wasting	Gross loss of subcutaneous fat, 'all skin and bone'	Less obvious; sometimes fat, blubbery
3. Muscle wasting	Severe	Sometimes
4. Growth retardation in terms of body weight	Severe	Less than in marasmus
5. Mental changes	Usually none	Usually present
Biochemistry/pathology		
1. Serum albumin	Normal or slightly decreased	Low
2. Urinary urea per g of creatine	Normal or decreased	Low
3. Urinary hydroxyproline index	Low	Low
4. Serum free amino acid ratio	Normal	Elevated
5. Anaemia	May be observed	Common; iron or folate deficiency may be associated
6. Liver biopsy	Normal or atrophic	Fatty infiltration

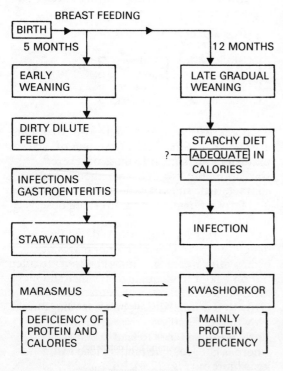

Fig. 12.4. **Protein–calorie or protein–energy malnutrition in infants.**

much more common than was originally supposed. For example, studies of 1–2-year-old children in Ghana, Guatemala, Jamaica, Polynesia, Thailand and Uganda showed that their energy intake varied within the range 218–360 kJ/kg body weight. It can be calculated that the basal metabolic requirements for children of this age is 330 kJ/kg and they are likely to need a total intake of about 430 kJ/kg, but protein intake tended, in fact, to be close to adequate. This therefore raises the question as to how, and why, the two separate conditions develop. Recent research has shown that this may be related to the hormonal response to the malnutrition condition. The normal response to starvation is for the secretion of cortisol to be substantially increased and, as a result, the tissue proteins are catabolized, whereas, in kwashiorkor, this is much less marked. Comparison of groups of children in Uganda and Gambia suffering from protein–energy malnutrition illustrates this (*see Fig.* 12.5). Children studied in Gambia had lost much more weight than those in Uganda and tended to suffer from marasmus, whilst those in Uganda suffered from kwashiorkor.

These conditions appear to be closely related to plasma hormone levels. Cortisol was

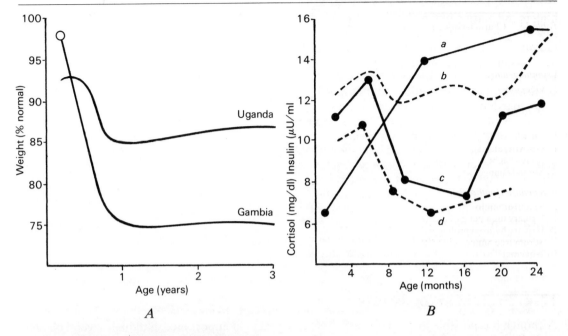

Fig. 12.5. **Endocrines in protein–energy malnutrition. Comparison of children in Gambia and Uganda.**
(*A*) Body weight changes.
(*B*) Plasma cortisol and insulin concentrations.
—— Cortisol, – – – – insulin,
(*a*) and (*d*) in Gambia and (*b*) and (*c*) in Uganda.

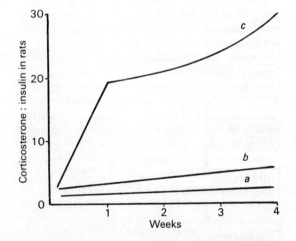

Fig. 12.6. **Plasma corticosterone : insulin ratios after feeding energy-deficient or protein-deficient diet to rats.**
(*a*) Control; (*b*) protein deficient; (*c*) energy deficient.

Table 12.4 **New concepts of marasmus and kwashiorkor**

	Marasmus	*Kwashiorkor*
Diet	Inadequate for all nutrients	'More' adequate in energy
Endocrine regulation	'Normal' own tissues consumed	'Poor'
Mobilization of fatty acids and amino acids	'Normal'	'Poor' Amino acid deficiency ↓ Plasma albumin falls ↓ Oedema

high in the plasma of Gambian children but low in Ugandan children, and thus the Gambian children showed much greater weight loss. Insulin showed the reverse pattern. Experiments in rats show that the corticosterone:insulin ratio responds much more to a deficiency of energy than to a deficiency of protein (*Fig.* 12.6). These studies have led to the concept that, for several reasons not clearly understood but possibly genetic, children of different groups react in different ways to the stress of protein–calorie malnutrition (*Table* 12.4). It should also be noted that infections, and especially those of the gastrointestinal tract, can play a very important role in malnutrition causing serious malabsorption of the inadequate food supply.

Table 12.5 **Protein contents of typical foods**

Protein		
Content	Percentage	Food
High	18–28	Meat, fish, cheese Peanuts (roasted) Peas (dry)
Moderate	10–18	Nuts (fresh) Egg Shell fish
Low	6–10	Bread Cereals Chocolate
Very low	1–5	Fruit Vegetables (fresh)
None	0	Sugar Margarine Coke

12.4 Protein in foodstuffs

Although it is possible to eat pure carbohydrate in the form of sucrose or nearly pure fat in the form of margarine or butter, no foods normally available in the human diet are composed entirely of protein. Foods are, therefore, described as protein rich or protein poor, depending on their protein content.

Many years ago it was noted that most proteins contain approximately 16 per cent nitrogen and that protein supplies nearly all the dietary nitrogen. If, therefore, the total nitrogen is determined in any foodstuff and the figure obtained multiplied by 6·25, a very close estimate of the protein content of the food is obtained. The method used for total nitrogen determination is that developed by Kjeldahl, many years ago, in which, by incineration with concentrated sulphuric acid, the nitrogen in the protein is converted quantitatively into ammonium sulphate which can be easily measured.

This calculation gives an accurate measure of protein in many foods, such as meat, but overestimates the protein content of some plant foods such as leaf and root vegetables. This is because in some of these plants, the amino acids and peptides form a relatively large proportion (20–50 per cent) of the total nitrogen. In view of the fact that all proteins are digested in the stomach and

intestine to amino acids, this may not be of practical importance, although several plants contain a very high percentage of certain amino acids used for nitrogen storage, particularly glutamine and asparagine and thus the amino acid spectrum will not be similar to that of a typical food protein. Percentages of protein in foodstuffs are shown in *Table* 12.5. Although it is meat, fish and eggs which are known as good sources of protein, it is less well appreciated that some vegetable food, for example dried peanuts, contains a similar percentage of protein to meat. Furthermore, the protein content of bread cannot be ignored, since if substantial quantities are eaten, bread can easily supply one-half or more of the daily protein requirements.

12.5 Daily protein requirements

As discussed at the beginning of the chapter, man, for many years, has been obsessed with the concept of consuming very large daily intakes of protein. Even the medical experts of the past supported this belief, and in 1865, the daily intake recommended ranged from 57 to 184 g and, in 1880, Voitt recommended 120 g per day. At the beginning of the century, American nutritionists questioned the need for large daily intakes of protein and they were eventually shown to be correct.

It is now possible to calculate accurately the requirements by measuring nitrogen excretion. If healthy adults are transferred from a normal diet containing protein to one which contains no protein but is adequate in all other respects, the nitrogen excretion in the urine steadily declines over a period of 5–6 days, but it does not approach zero and, for human adults, continues at a rate of approximately 2·69 g nitrogen per day (*Fig.* 12.7). In addition, much smaller amounts of nitrogen are lost in the faeces and from the skin.

Taking all the total losses into account it can be calculated that:

Total daily nitrogen loss = 54 mg N per kg body weight per day

Total daily protein loss = 0·34 g protein per kg body weight per day
For a 70-kg adult the daily loss = 24 g protein per day

This figure is the minimum protein intake and assumes complete utilization of all the protein ingested. In practice, the utilization of the amino acids of proteins is about 70 per cent efficient for most normal diets, so that the 24 g must be multiplied by 10/7 to give an estimate of minimum intake of approximately 34 g per day for an average adult.

It is, therefore, clear that original estimates of protein needs were greatly exaggerated and that it is possible to remain healthy on approximately 25 per cent of the daily intake originally proposed. The expert food committees of the United Nations and World Health Organisation have, however,

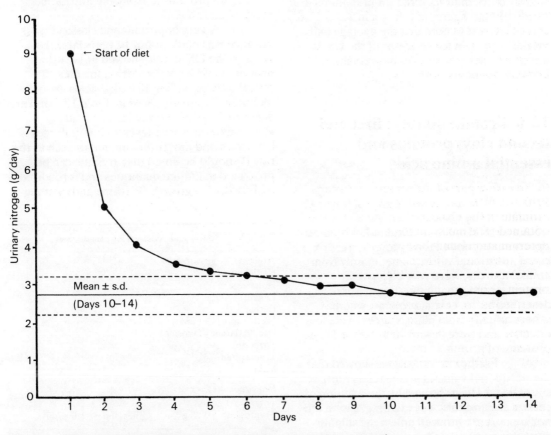

Fig. 12.7. **Effect of protein-free diet, adequate in energy, on urinary nitrogen excretion.** Eighty-three subjects are fed on a protein-deficient diet.

Table 12.6 **Recommended minimum protein requirements**

Age (years)	Protein* (g per day)
0–1	13–16
1–2	19
2–3	21
3–5	25
5–7	28
7–9	30
9–12	36
12–15	46
15–18 (girls)	40
15–18 (boys)	50
18–35 (men)	45
35–65 (men)	43
18–55 (women)	38

* Net protein utilization = 70%

decided to recommend a daily intake which is slightly more than required for minimum basic needs and the figures are shown in *Table* 12.6. It is of interest to note that the average daily intake of protein for residents of the UK is about 90 g per day and this has remained constant for many years.

12.6 Protein quality: first and second class proteins and essential amino acids

In the earlier part of the century, between 1910 and 1916, as a result of animal feeding experiments by Osborne and Mendel in the USA and by Hopkins in England, it became apparent that not all food proteins were of equal nutritional value. Some, mainly from animal sources, were very effective in supporting the growth of rats and were described as 'first class' proteins, whereas others, usually from plant sources, were less effective and were described as 'second class' proteins.

Further investigations showed that the nutritional value of second class proteins could be greatly improved by the addition of certain amino acids. For example, casein did not support growth well unless the sulphur-containing amino acid cysteine was added, whereas growth on the maize protein zein, was

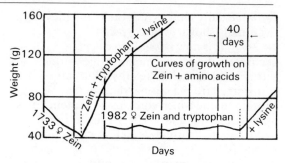

Fig. 12.8. **Growth of rats fed on maize protein (Zein) with additions of tryptophan and lysine.**

very poor unless two amino acids, lysine and tryptophan, were added (*Fig.* 12.8). Certain proteins therefore appeared to be of poor nutritional value because they lacked certain amino acids, and so research began to switch away from proteins to study the nutritional value of amino acids.

A very important and classical advance was made about 10 years later, by Rose in the USA, who showed in 1921 that protein could be omitted from the diet completely providing 10 amino acids were included. These are listed in *Table* 12.7 and are described as the 'essential amino acids'. Requirements are believed to be similar for both man and most other mammals such as the rat. It should be noted that arginine can be produced in limited quantities and is required mainly during growth. Tyrosine and cysteine

Table 12.7 **Amino acids essential for rat and man**

Arginine ← growth only
Histidine
Isoleucine
Leucine
Lysine
Methionine [Cysteine]
Phenylalanine [Tyrosine]
Threonine
Tryptophan
Valine

Requirements for man
Total ≃ 6·3 g/day
Tryptophan ≃ 0·25 g   Others 0·8 → 1·0 g /day

Amino acids in square brackets are semi-essential.

are sometimes described as semi-essential because they can be synthesized, but only one particular amino acid, phenylalanine, can be converted to tyrosine and methionine to cysteine.

Although experiments on essential amino acid requirements were originally carried out on animals, especially rats, several attempts have been made to assess requirements in human subjects. These have been generally attempted using nitrogen balance studies as described on p. 135 of this chapter. The subject will go into negative nitrogen balance if a diet is supplied deficient of an essential amino acid.

The early experiments on the ten essential amino acids required by rats showed that, in short-term requirements, young men or women did not require arginine or histidine for maintenance of nitrogen balance and that methionine could be 80–90 per cent spared by adequate cysteine and phenylalanine 70–75 per cent spared by tyrosine. Histidine was, however, essential for infants less than 7 months old. Longer-term feeding studies, however, showed that histidine was essential for adult men, especially for erythropoiesis.

Studies of fetal humans have indicated that cysteine may also be deficient during early development because liver cystathionase activity which can form cysteine is very low.

It is important not to confuse the minimum requirements for essential amino acids which is about 6·25 g/day and the total minimum daily protein requirement which is, theoretically, about 24 g/day. The essential amino acids must be provided, but additional amino acids, usually taken as protein, must also be in the diet to make up the minimal nitrogen intake required to compensate for the daily nitrogen losses. The nature of the extra amino acids is not of serious consequence and they can be either essential, non-essential or a mixture of both. Theoretically, an adequate quantity of a single amino acid, such as glutamate or alanine, would suffice.

The discovery of the essential amino acids clarified the basis for the division of proteins into first class and second class groups. The first class proteins, such as meat and eggs,

contain all the essential amino acids, whereas second class proteins are deficient in one or more of these amino acids. As a consequence second class proteins are inadequately utilized in the body.

12.7 Concept of 'biological value' and 'chemical score'

An important step forward in evaluating the nutritional value of proteins was made in the 1940s by two British nutritionalists, Block and Mitchell. They showed that it was possible to assess the nutritional value of any protein quantitatively by measuring the protein (or nitrogen) contents of the food taken by an animal over a specific period and relating this to the nitrogen excreted during this same period.

They proposed that this value should be called the 'biological value' of a protein. Sometimes, more recently, it is described as the net protein utilization (NPU) and is expressed:

'Biological value' or NPU of a protein:

$$= \frac{N_I - N_E}{N_I} \, (\%)$$

where

N_I = nitrogen intake

and

N_E = nitrogen excreted

The biological value can theoretically vary between 100 per cent when no nitrogen is excreted and zero per cent when none of the protein is utilized and all of it is excreted. In practice, biological values normally fall between 100 and 50 per cent. Most animal proteins, such as meat, fish and eggs, have very high biological values, close to 100 per cent whereas those of plant proteins are much lower (*Table* 12.8). The concept of the biological value can best be understood from the practical experiment shown in *Fig.* 12.9, which can be carried out on human subjects or on animals. If protein is removed from the diet, but other components are fed in adequate quantities, the nitrogen excreted falls for a few days and then

Table 12.8 **Typical biological values**

| Protein | Biological value in | |
	Man	Rat
Egg albumin	91	94
Beef muscle	67	69
Casein	56	51
Wheat gluten	42	65

levels out at about 3 g/day as shown in *Fig.* 12.9. Protein is now fed to the subject to give a quantity of nitrogen exactly equal to that which is being lost. When egg protein is fed, no change in the nitrogen excreted occurs, i.e. the excretion due to the ingested protein is zero and thus the biological value is close to 100 per cent. All the protein is being used by the tissues and no extra nitrogen is excreted. If, however, the same quantity of nitrogen is fed in the form of wheat protein, then a marked increase of nitrogen excretion occurs due to poor utilization of this protein. The value for N_E is therefore substantial and the biological value is close to 50 per cent.

Experiments of this type are slow, expensive and time consuming, but it was shown by Block and Mitchell that the biological value of a protein could be related to its amino acid composition. This would be logical because, as we have already seen, a very important role of dietary protein is to supply essential amino acids and, if any of these are deficient in the diet, then the biological value of the protein would be reduced. To quantify this concept, it was necessary to select, as a standard, a protein which would supply all essential amino acids in the correct proportion, and then to relate other proteins to this one. Egg protein was selected as the standard and the crucial parameter was considered to be the percentage composition of the essential amino acid in the lowest concentration in relation to egg protein.

This percentage is described as the 'chemical score' which is defined by:

$$\frac{\text{Percentage limited essential amino (x) acid in protein}}{\text{Percentage of the amino acid (x) in egg protein}}$$

After many investigations and analyses, it became clear that egg protein was not ideal and, in fact, contained a higher percentage of some amino acids than was really necessary. The percentage composition of the theoretical ideal protein was calculated from many nutritional experiments and used as the standard. The method of calculating the chemical score is best understood from the practical examples shown in *Table* 12.9.

Casein is, compared with ideal proteins, surplus in its lysine content, deficient to a small extent in tryptophan, but more deficient in sulphur-containing amino acids. The chemical score is, therefore, calculated in relation to sulphur amino acids, especially cysteine, giving a percentage of 80 (calculated from 215/270). It is interesting that early on it was noted that casein is deficient in cysteine. The 'ideal protein' as standard is clearly important in this calculation, because if egg protein had been used the ratio, 215/242, would have given a much lower value. Egg protein, in fact, contains more sulphur amino acids and lysine than are really required.

Wheat gluten is deficient in sulphur amino acids, tryptophan and lysine. It is however more deficient in lysine (107/270) than in the other amino acids and thus the chemical score is calculated on the basis of lysine analyses. It is interesting to note

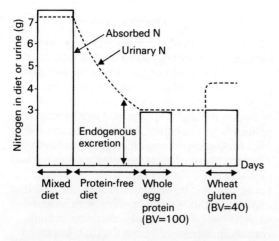

Fig. 12.9. **Schematic representation of the procedure for measuring the biological values (BV) of dietary proteins.**

Table 12.9 **Use of 'chemical scores' to measure the nutritional values of proteins**

Protein	Concn (mg per g N)			Chemical score, %
	Sulphur amino acids	Tryptophan	Lysine	
Ideal	270	90	270	100
Egg protein	342	106	396	>100
Casein	(215)	85	497	$\frac{215}{270} = 80$
Beef muscle	237	(75)	540	$\frac{75}{90} = 80$
Wheat gluten	223	60	(107)	$\frac{107}{270} = 40$

that both casein and beef meat contain large surpluses of lysine and so can easily make up the deficiency of lysine in bread, when bread and cheese or bread and meat are eaten together. If chemical scores are plotted against biological values, a very good straight line is obtained (*Fig.* 12.10), so that we may predict the biological value of a protein simply on the basis of amino acid analysis and without carrying out any animal experiments. Very low values for chemical scores may, however, be misleading, because such proteins do possess a

significant biological value when measured by nutritional experiments. The relationship appears to depend on the essential amino acid which is limiting (*Fig.* 12.10).

The analyses of essential amino acids and measurements of chemical scores provided us with an important scientific basis for good protein nutrition. They do, for example, provide good support for the cult of eating 'sandwiches'. Wheat protein is deficient in lysine and tryptophan, but these deficiencies can be supplied by cheese or meat fillings. As a consequence, the effective biological value of the wheat protein of bread is considerably enhanced. The effect is lost if the bread and meat are eaten separately, say 2–5 hours apart.

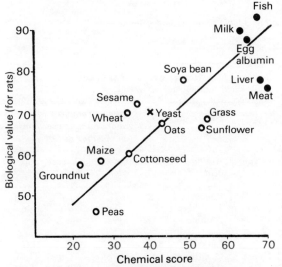

Fig. 12.10. **Correlation of biological and chemical values.** Correlation diagram of the biological value and the chemical score for 16 animal protein sources (full circles) and vegetable protein sources (open circles).

12.8 Simulated meat foods

A complete understanding of the nutritional value of different proteins has led to the search for plant proteins which have very similar amino acid compositions to that of meat. One of the best of these is prepared from the soya bean. As extracted, this protein is a rather unappetizing powder and great efforts have been made to convert it into fibres, to give a 'chewiness' and to make it marketable. Techniques based on those used in the textile industry have been found useful for making food fibres. The amino acid compositions of proteins prepared from soya beans are shown in *Table* 12.10. A study of this table should convince the reader that it is now no longer

Table 12.10 **Chemical analysis on moisture-free basis (per 100 g)**

Food	Protein, g	Carbohydrate, g	Fat, g	Ash, g	Fibre, g	Energy, kcal
Whole milk	26·1	37·0	28·4	5·3	0	506
Whole egg	48·5	3·3	43·2	3·7	0	595
Raw beef steak	60·3	0	32·8	3·1	0	536
Texgran*	56·6	32·5	1·0	6·5	3·3	365
TVP†	54·3	34·7	1·0	6·5	3·3	364
Bontrae†	60	17	20	3	0	488

* Texgram is the tradename of Swifts, marketed in the UK by Malga Products Ltd.
† TVP is the tradename of Archer Daniel Midlan, marketed in the UK by British Arkady Co. Ltd.
† Bontrae is the tradename of General Mills Ltd.

essential to eat meat, because proteins prepared from these vegetable proteins can readily supply all the essential amino acids and all the nitrogen required in the diet. They also possess an important advantage in that their fat content is very low, much less than in meat proteins.

12.9 Consequences of large intakes of protein

If the dietary intake of protein is low or only just adequate, most of the amino acids taken into the plasma are utilized for protein synthesis. This is because the K_m values of the tRNA synthetases which produce aminoacyl-tRNA derivatives for protein synthesis have very low K_m values and thus have a high affinity for all amino acids. If, however, protein intake is high, then a large proportion of the amino acids is catabolized. The amino acid catabolizing enzymes possess high K_m values, so that they will not metabolize any amino acids in short supply (*Fig.* 12.11).

Catabolism of amino acids is regulated by the induction of amino acid catabolizing enzymes, occurring mainly in the liver. Other tissues respond much less effectively (*Table* 12.11), examples in animal experiments being shown in *Fig.* 12.12. Phenylalanine hydroxylase and tryptophan pyrolase activities increase rapidly only 2 hours after a protein meal and the enzyme activities increase as the percentage of protein (casein) in the diet is increased. The enzymes involved

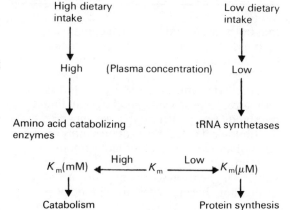

Fig. 12.11. **Fates of dietary amino acids.**

Table 12.11 **Effect of variation of diet composition on amino acid catabolizing enzymes**

Organ	Amino acid	Activity of amino acid α-ketoglutarate amino transferase after feeding diet of			
		Stock	Egg white	Glucose, 20%	Fat
Liver	Leucine	6·5	11	2·4	4·1
	Ornithine	68	134	9·1	26
	Alanine	755	1560	109	610
Heart	Leucine	87	94	79	107
	Ornithine	4·5	1·9	6·8	3·0
	Alanine	172	15	368	0·5

From Krebs H. A., 1972, *Adv. Enzyme Regul.*, Vol. 10, p.397.

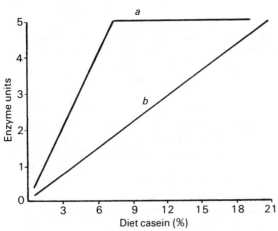

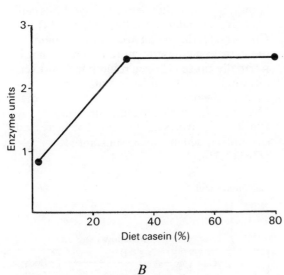

Fig. 12.12. **Responses of amino acid metabolizing enzymes to high protein diets.**
(*A*) Enzyme activities in chick livers 2 h after feeding meal; (*a*) phenylalanine hydroxylase, (*b*) tryptophan pyrolase. (*B*) α-Ketoacid dehydrogenase activity for branched-chain amino acids in rat liver 50 days after feeding high protein diets.

in the oxidation of the ketoacids formed from the branched-chain amino acids leucine, isoleucine and valine also markedly increase. The increased activity of all enzymes is believed to be a result of new enzyme synthesis and is analogous to the induction of enzymes observed in bacteria, although it is unlikely that the mechanisms of induction are identical with those which are known to occur in bacteria.

These induction systems provide an important method of conserving amino acids in short supply and also demonstrate the futility of excess protein consumption, because the induced catabolizing enzymes form α-keto-acids which are also easily produced from carbohydrates.

12.10 Nitrogen balance

One of the most useful methods for assessing the status of protein nutrition of any individual subject is to evaluate the nitrogen balance. The measurement also provides valuable information concerning overall protein metabolism. The nitrogen balance evaluates the relationship between the nitrogen intake (mainly protein) and the nitrogen excretion. In normal stable adults:

Nitrogen intake = nitrogen excreted when the subject is in balance

There are, however, two other possible situations:

a. *Nitrogen intake > nitrogen excretion.* This is called positive balance. It indicates that the nitrogen is being retained in the body which usually means that protein is being laid down. This occurs during growth or during recovery after a serious illness.

Or the opposite situation:

b. *Nitrogen intake < nitrogen excretion.* This, if prolonged, is a serious situation and can ultimately lead to death, because clearly we cannot go on indefinitely excreting nitrogen if insufficient is being taken into the body. It occurs during serious malnutrition, for example in children suffering from kwashiorkor or marasmus, but is also common during serious illness or injury. It can result if a patient is unwilling or unable to take sufficient protein on account of digestion or absorption problems. Tissue degradation caused by disease can also lead to greatly increased nitrogen excretion. As explained above, subjects normally go into positive nitrogen balance during the recovery period.

The mechanism of adjustment to a particular level of nitrogen intake is not fully understood. It has been found experimentally that adjustment to a different level of intake may take several weeks to stabilize. Thus, human subjects on a daily intake of 12 g nitrogen per day who were then given 36 g nitrogen per day, took about 20 days to stabilize to a new enhanced level. Conversely, subjects who had been used to 36 g nitrogen per day went rapidly into severe negative nitrogen balance when transferred to a diet of 12 g nitrogen per day which is an adequate intake (*Fig.* 12.13). It is, therefore, clear that the regulatory mechanism operates very slowly. In the condition of nitrogen balance, the rate of synthesis of protein equals the rate of breakdown of body protein (*Fig.* 12.14).

When an individual goes into negative nitrogen balance, it is generally assumed that this is a result of increased catabolism of protein (B in *Fig.* 12.14), but it should be noted that depressed synthesis can also give a similar result (S in *Fig.* 12.14). Conversely, during recovery from protein deficiency, increased synthesis of protein will normally cause nitrogen retention. However, although protein synthesis is markedly increased when protein is restored to the diet after a period of deprivation, increased synthesis is always accompanied by increased catabolism which may be substantial (*Fig.* 12.15).

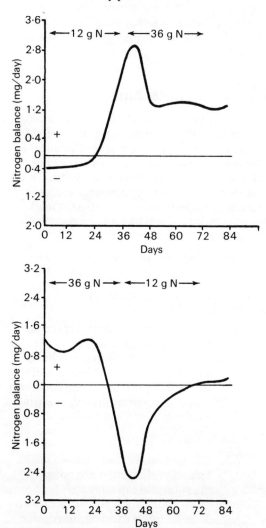

Fig. 12.13. **Regulation of nitrogen balance in human subjects.**

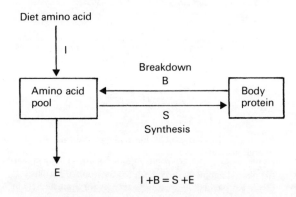

Fig. 12.14. **Nitrogen balance.**
I = amino acid intake; E = amino acid degraded and excreted; B = breakdown of body protein; S = synthesis of body protein.

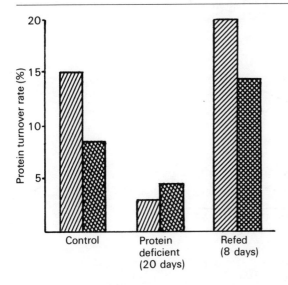

Fig. 12.15. **Muscle protein turnover in rats fed protein deficient and adequate diets.**
(▨) Synthesis; (▨) breakdown.

Efficiency of utilization of protein is also closely related to the nitrogen balance. Many of the original values for biological value of protein were obtained on young growing rats who used protein very efficiently. However, when similar experiments are carried out on young human adults and the nitrogen balance is plotted against the protein intake, it is observed that only under conditions of severe negative balance is egg protein utilized with an efficiency of 100 per cent, the theoretical value. As the nitrogen balance is approached, the efficiency of utilization of the egg protein decreases to approximately 50 per cent (*Fig.* 12.16).

If the intake of protein is increased still further to attempt to move the subject into positive nitrogen balance, then the efficiency of protein utilization becomes much less and extensive loss of nitrogen results even after feeding large quantities of high quality proteins (*Fig.* 12.17). These experiments demonstrate

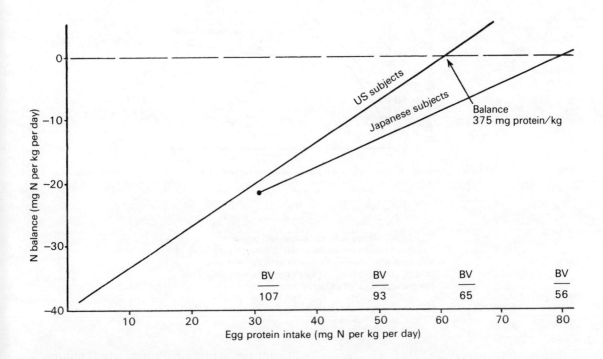

Fig. 12.16. **Nitrogen balance with egg protein.**
Nitrogen balance with egg protein as sole source of protein in human subject from USA and Japan.

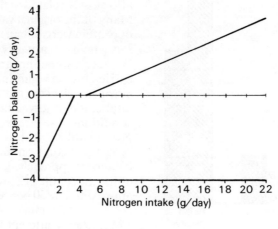

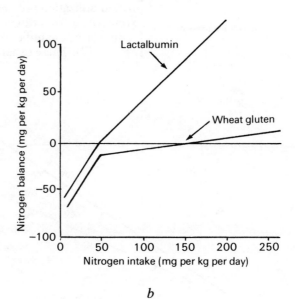

Fig. 12.17. **Nitrogen balance—protein intake.**
In (*a*) the results are mean values of 8 investigations of
different proteins. (*b*) Comparison of the effects of a
protein of high biological value (lactalbumin) and one of
low biological value (wheat gluten).

that, contrary to popular belief, it is very
difficult to lay down much extra protein in the
tissues by eating diets containing large
quantities of protein. Protein is only utilized
efficiently at or below the point of nitrogen
balance.

Chapter 13 Nutrition: dietary fats

13.1 Introduction: why eat fats?

About 45–50 per cent of the energy in the diets of European and US societies is now derived from fat (*see* Chapter 11). Based on a typical intake of 3000 cal/day, we can therefore calculate that an average adult will consume approximately $1500/9 = 166$ g fat per day, which will be mainly in the form of triacylglycerols.

There are several reasons for eating fat:

a. Fat represents the most concentrated form of energy available, supplying about 9 cal/g as compared with approximately 4 cal/g for carbohydrates and protein.

b. It imparts palatability to food and lubricates it. It thus makes eating and swallowing much easier and more acceptable. A fat-free diet tends to be 'dry' and unattractive.

c. Dietary fat is necessary for absorption of the fat-soluble vitamins. These normally occur associated with fat, but even if they are supplied in adequate quantities in a diet devoid of fat, absorption of the vitamins is very poor, since the vitamins must be transported across the intestinal mucosa during fat absorption (cf. Chapter 16).

d. Ingestion of fat is necessary to obtain an adequate supply of essential fatty acids which, like essential amino acids, cannot be synthesized in the human tissues.

Despite the desirability of eating fat, the tendency in many societies has, over the past few decades, been to include far too much fat in the diet. The quantity of essential fatty acids necessary for health, forms, at most, only 5 per cent of the total calories and, apart from fat-soluble vitamins, no other fat is theoretically necessary in the diet.

Modern man, however, appears to eat more fat as he becomes wealthier and can afford more exotic foods, although modern nutritional opinion is strongly against the inclusion of such high quantities of fat in the diet. Many foods currently available, however, contain large quantities of fat and once eating habits are formed it is very difficult to change them.

The main intake of fat in the human diet consists of saturated fatty acids or monounsaturated fatty acids, mainly oleic acid, from animal sources. Polyunsaturated fatty acids are obtained from plant sources such as corn or sunflower oil. Most margarines contain saturated fat which is often produced by hydrogenation of unsaturated fat. This treatment was believed to be essential to stabilize the unsaturated fats and prevent them undergoing autoxidation. It is, however, now realized that an intake of polyunsaturated fats is very desirable and polyunsaturated margarines are now widely marketed.

13.2 Essential fatty acids

Just over 50 years ago, two nutritionalists working in the USA observed that rats fed on a fat-free diet failed to grow at the rate of the controls, and lost large amounts of water through their skin. These abnormalities were not cured by all forms of fat, but only by certain polyunsaturated fatty acids or

Non- essential $\{$

Oleic acid $(18:1)(\omega-9)$: $CH_3(CH_2)_7-\overset{9}{CH}=\overset{10}{CH}-(CH_2)_7-COOH$

Essential (EFA) $\{$

Linoleic acid $(18:2)(\omega-6)$: $CH_3(CH_2)_7-\underset{\underline{\quad}}{\overset{6}{CH}=\overset{7}{CH}}-CH_2\overset{9}{CH}=\overset{10}{CH}-(CH_2)_7-COOH$

Linolenic acid $(18:3)(\omega-3)$

$\omega-3, 6, 9\ CH_3CH_2\underset{\underline{\quad}}{\overset{3}{CH}=\overset{4}{CH}}\ CH_2\ \underset{\underline{\quad}}{\overset{6}{CH}=\overset{7}{CH}}\ CH_2-\overset{9}{CH}=\overset{10}{CH}-(CH_2)_7-COOH$

Arachidonic acid $(20:4)(\omega-6)$

$CH_3(CH_2)_4-\underset{\underline{\quad}}{\overset{6}{CH}=\overset{7}{CH}}\ CH_2\overset{9}{CH}=\overset{10}{CH}-CH_2\overset{12}{CH}=\overset{13}{CH}-CH_2\overset{15}{CH}=\overset{16}{CH}-(CH_2)_3-COOH$

Fig. 13.1. **Essential fatty acids.**
Double bonds which are underlined cannot be synthesized in animal tissues; other double bonds can be synthesized.

triacylglycerols containing them. These fatty acids are linoleic acid, linolenic acid and arachidonic acid (see Fig. 13.1), and they became known as the 'essential fatty acids'.

Subsequently many other symptoms of essential fatty acid deficiency were discovered and these are listed in Table 13.1.

Table 13.1 **Symptoms of essential fatty acid deficiency**

---

1. Decreased growth rate
2. Increased permeability of the skin to water
3. Spontaneous swelling of liver mitochondria
4. Decreased prostaglandin synthesis
5. Changed fatty acid patterns in many tissues, for example in liver mitochondria and endoplasmic reticulum
6. Decrease in the immune response

These three 'essential fatty acids' contain either $\omega-6$ double bonds, in the case of linoleic acid and arachidonic acid, and $\omega-3$ double bonds in the case of linolenic acid. Double bonds in the $\omega-3$ or $\omega-6$ position cannot be synthesized by mammals and thus have to be obtained, initially, from plant sources. For some reason, presumably associated with extensive availability of vegetation for food, the ability to desaturate in the $\omega-3$ and $\omega-6$ positions was lost very early in metazoan evolution. Desaturation is, however, possible in mammalian tissues amongst pairs of carbon

atoms in the part of the molecule between the double bond in the $\omega-9$ position and the terminal —COOH group. Thus, oleic acid can be converted to a triene and this is formed in relatively large quantities under conditions of essential fatty acid deficiency (Fig. 13.2). Similarly, linoleic acid can be metabolized to arachidonic acid by chain elongation and desaturation (Fig. 13.3).

In view of this metabolism it has been suggested that, if sufficient linoleic acid is taken in the diet, arachidonic acid is not required, although in some animal species (especially the cat species), the rate of conversion of linoleic acid to arachidonic acid is very slow because the desaturase enzymes are very weakly active. At present the efficiency of conversion of linoleic acid to arachidonic acid in man is unknown and, because arachidonic acid is a very important cellular component, it is considered desirable to include some of this fatty acid in the diet.

The status of linolenic acid as an essential fatty acid is still unresolved. Early experiments in which linolenic acid was compared with linoleic for their effects on growth rate and skin water loss always showed that linoleic acid was much more effective than linolenic acid (Fig. 13.4). As a result linolenic acid tended to be discounted as an effective essential fatty acid. However, analysis of several lipids of the body, especially those of brain and nervous tissue, show that $\omega-3$ fatty

$$CH_3-(CH_2)_7-CH=CH-(CH_2)_7-COOH$$

oleic acid (C$_{18:2}$)

↓

$$CH_3(CH_2)_7-CH=CH-CH_2-CH=CH-(CH_2)_4-COOH$$

↓

$$CH_3-(CH_2)_7-CH=CH-CH_2-CH=CH-(CH_2)_6-COOH$$

↓

$$CH_3-(CH_2)_7-CH=CH-CH_2-CH=CH-CH_2-CH=CH-(CH_2)_3-COOH$$
Eicosatrienoic acid (C$_{20:3}$)

Fig. 13.2.   **Metabolism of oleic acid to the triene eicosatrienoic acid.**

$$CH_3-(CH_2)_4-CH=CH-CH_2-CH=CH-(CH_2)_7-COOH$$
Linoleic acid (C$_{18:2}$)

↓

$$CH_3-(CH_2)_4-CH=CH-CH_2-CH=CH-CH_2-CH=CH-(CH_2)_4-COOH$$
γ-Linolenic acid

↓

$$CH_3-(CH_2)_4-CH=CH-CH_2-CH=CH-CH_2-CH=CH-(CH_2)_6-COOH$$

↓

$$CH_3-(CH_2)_4-CH=CH-CH_2-CH=CH-CH_2-CH=CH-CH_2-CH=CH-(CH_2)_3-COOH$$
Arachidonic acid (C$_{20:4}$)

Fig. 13.3.   **Metabolism of linoleic acid to arachidonic acid.**

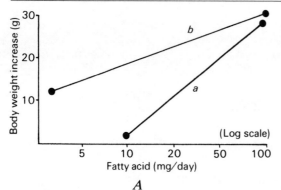

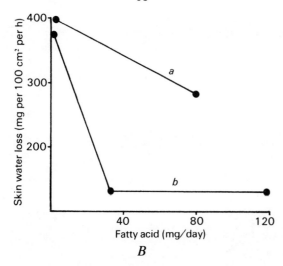

Fig. 13.4.   Comparison of effects of linoleic acid (*b*) and linolenic acid (*a*) on body weight and skin water loss.
(*A*) Effects on body weight of young rats fed fat-free diets.
(*B*) Effects on skin water losses of rats fed fat-free diets.

Table 13.2 **Polyunsaturated fatty acids of phosphatidylethanolamine in brain and liver**

Acid		Concentration (mg/g phospholipid) in	
		Liver	Brain
ω−6 acids	$C_{18:2}$	120	12
	$C_{20:3}$	11	7
	$C_{20:4}$	130	120
	$C_{22:4}$	10	63
	$C_{22:5}$	5	12
ω−3 acids	$C_{18:3}$	21	5
	$C_{20:5}$	23	6
	$C_{22:5}$	54	7
	$C_{22:6}$	98	220

Values are the mean for 25 species.

acids occur commonly (*Table* 13.2). The mammalian tissues cannot synthesize double bonds in the ω−3 position and it is, therefore, virtually certain that linolenic acid or some ω−3 fatty acids are essential in the diet. However, it is of interest that the mammalian body retains ω−3 unsaturated fatty acids very efficiently and thus the daily requirement may be quite low.

ω−3 fatty acids are abundant in fish oils and it has been proposed that the efficiency in the retention of ω−3 acids arose during the migration of animals from the sea when they moved from a situation of great abundance of fatty acids of this type to relative poverty.

13.3 Human response to essential fatty acid deficiency

Many seriously ill patients, often in a coma, are kept alive by receiving nutrients intravenously. This is described as 'parenteral nutrition'. The nutritional solution used contains the correct proportion of amino acids, glucose electrolytes and vitamins, but often no fat. Fats or fatty acids are not easily dispensed in aqueous solutions. Patients kept for only a few days on these diets show marked changes in their plasma fatty acids. The proportions of linoleic acid and arachidonic acid decline sharply and the triene $C_{20:3}$ eicosatrienoic acid formed from oleic acid rises sharply (*Fig.* 13.5).

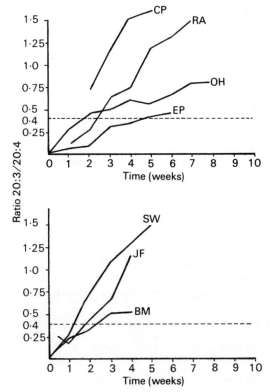

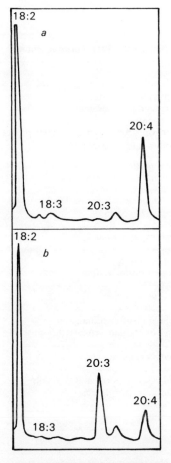

Fig. 13.5. **Analysis of fatty acids in plasma lipids of normal patients** (*a*) **and those suffering from essential fatty acid deficiency** (*b*) **by gas–liquid chromatograms.**

Fig. 13.6. **Triene : tetraene ratios of patients receiving 'parenteral nutrition' with no added fat.**
Seven individuals are shown, to whom the initials relate.

The ratios of the concentration of the triene ($C_{20:3}$) to the tetraene arachidonic acid in the plasma can be used as a measure of essential fatty acid deficiency. It should normally be very small but it rises sharply when the diet is deficient in linoleic acid (*Fig.* 13.6). The importance of this ratio can be seen in infant feeding. Cow milk contains much less linoleic acid than does human milk and infants fed on cow milk show a much greater triene : tetraene ratio than those fed on human milk.

The synthesis of the triene is clearly an attempt by the body to replace the essential fatty acids. It appears to be partially successful and the triene is incorporated into the cellular constituents, mainly membranes, but the replacement causes impaired function of organelles, for example of mitochondria.

13.4 Chain elongation and functions of essential fatty acids

Chain elongation and further desaturation of all unsaturated fatty acids can occur, the processes being summarized in *Fig*. 13.7. As a consequence of this metabolism, long-chain fatty acids containing 20 or 22 carbon atoms are formed and these acids may possess 4, 5 or 6 double bonds (*Fig*. 13.7). Recent research has shown that the B vitamins, folate and B_{12}, may be involved in the formation of the long-chain unsaturated fatty acids, probably in the synthesis of double bonds, but the mechanism of their action is not yet established.

The majority of these long-chain polyunsaturated fatty acids are incorporated into the phospholipids of membranes of nearly all tissues of the body (cf. Chapters 2, 9). If these fatty acids are deficient in the diet, the membrane pattern of fatty acids is changed. It is generally believed that the polyunsaturated fatty acids are required in the membranes, to enable the membranes to attain the correct fluidity or conformation or both. Defects in either of these can lead to defective functions.

A very important group of regulators, the prostaglandins, thromboxanes and prostacyclins, are also formed from polyunsaturated fatty acids. Linoleic acid is an important precursor and one group of prostaglandins is formed from the triene, α-linolenic acid, which is itself produced by desaturation of linoleic acid. However, the tetraene arachidonic acid is the most important precursor of the main series of compounds which play very important roles in the

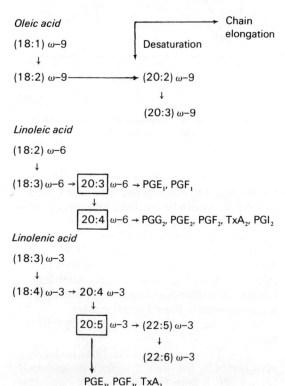

Fig. 13.7. **Chain elongation and desaturation of unsaturated fatty acids.**
PGE, PGF = prostaglandins; PGG = endoperoxides; TxA = thromboxanes.

Table 13.3 **Biological effects of prostaglandins**

1. *Reproductive system*

   i.   Contraction of uterus
         Abortion (2nd trimester)
  ii. 'Luteolytic'
         Failure of corpus luteum to secrete progesterone
 iii. Labour—raised plasma PGs

  Miscellaneous effects
  a. Hypothalamic-releasing factor
  b. Follicles—ovulation
  c. Spermatogenesis—maturation
  d. Transport of sperm

2. *Functional hyperthermia*

  Organ blood supply
  Mediation of vasodilatation

3. *Central nervous system*

  Release in brain after stimulation
  Modulates or regulate synaptic transmission

4. *Inflammatory response*

  Increased vascular permeability in skin
  Causes prolonged erythema
  Induces wheals
  Contained in exudates

5. *Water-electrolytes*

  Bladder Na^+ transport increased
  Ca^{2+} required for effects on transport

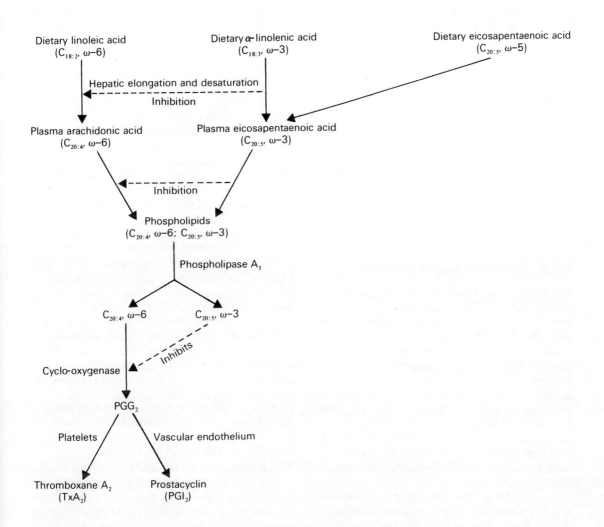

Fig. 13.8. **Inter-relationships of dietary ω–3 and ω–6 polyunsaturated fatty acids.**
Effects on thromboxane synthesis.

regulation of blood clotting (cf. Chapter 26). Prostaglandins are also involved in many other processes which are summarized in *Table* 13.3. These compounds are formed from the ω–6 series of fatty acids, but it has been shown recently that the ω–3 series of acids, such as fatty acids derived from fish oil, can antagonize or completely inhibit some of these regulators, especially the thromboxanes which play an important role in platelet aggregation and, therefore, of blood clotting.

These investigations were initiated following the observation that Eskimos who consume large quantities of fish oils suffer very little arterial disease or abnormal blood clotting. Possible mechanisms of action of ω–3 fatty acids are shown in *Fig*. 13.8.

13.5 Diet and heart disease

Cardiovascular disease has recently been described as the 'Western way of death'. At least 1·5 million are killed by the disease each year, and 150 000 deaths are caused by it annually in England and Wales. The disease appeared to be relatively uncommon at the beginning of the century with few cases being described, but it has grown to very serious proportions during the past 60–70 years. Many factors are no doubt involved, especially smoking, lack of exercise, genetic susceptibility and high blood pressure, but in this chapter we shall concentrate on the possible importance of diet.

The fact that the diet may be involved was first suggested by a Russian investigator, Anitischkow who, in 1913, fed large amounts of cholesterol to rabbits. He found that they developed lesions in the arteries which were very similar to those seen in humans who had arterial disease. The experiments did not attract great attention because the rabbit was not regarded as a good model and cholesterol was not a normal dietary constituent for rabbits.

The possible importance of the relation of diet to heart disease was resurrected as a result of studies of populations of European countries, such as Norway and Holland, which had been occupied during the 1939–1945 World War. Although most of the inhabitants had existed on a very poor diet, cardiovascular disease sharply declined during the 1940s–1950s. Studies of the diets of these populations showed that the proportion of fat in their diets, especially saturated fat, had declined to much lower values than was previously consumed. Subsequent epidemiological studies, mainly of Third World countries, showed the populations consumed much less fat than Western societies and much of the fat ingested was of vegetable origin and, therefore, unsaturated fat. The incidence of cardiovascular disease was much less in these countries than in Europe and the USA. These studies indicated that both reducing the total intake of fat and also increasing the proportion of unsaturated vegetable fat were beneficial.

Cholesterol which was originally indicated to be of great importance by the early work of Anitischkow, was shown to play an important role in arterial disease. First, the majority of patients suffering severe cardiovascular disease had significantly raised cholesterol levels (*Fig*. 13.9) and, secondly, diets high in saturated fats tended to raise the concentration of blood cholesterol (*Fig*. 13.10). Based on these cases a simplified concept of the role of dietary fat in heart disease was developed. It was concluded that raised concentrations of cholesterol in the plasma were directly associated with deterioration of the arteries and the development of arterial disease (*Fig*. 13.11).

The evidence for this relationship is very strong, but exactly how cholesterol causes the deterioration has been the matter of debate for many years and is not yet resolved.

Most of the cholesterol in the plasma is transported on the low-density lipoprotein fraction of the plasma proteins and it is possible that it is this component of the plasma, rather than cholesterol alone, which is primarily involved. Controlled laboratory experiments have clearly demonstrated, in support of epidemiological studies, that polyunsaturated fats, such as corn oil, containing a high percentage of linoleic acid cause a sharp decline in the concentration of

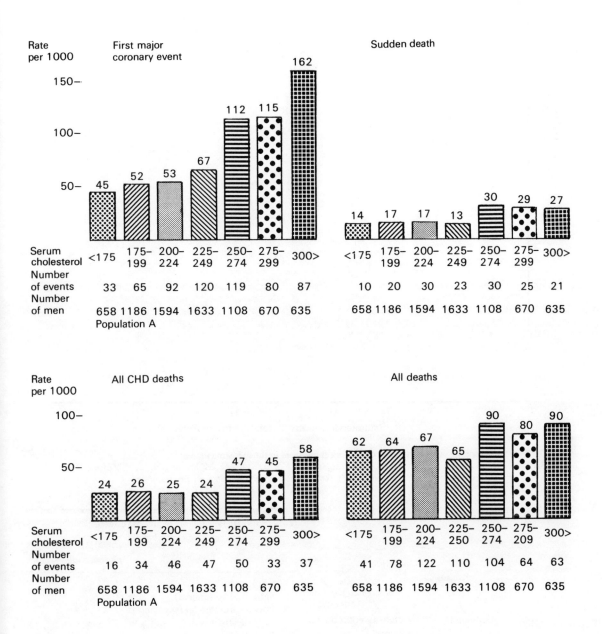

Fig. 13.9.   **The relation between serum cholesterol and the incidence of coronary heart disease (CHD).**
A 10-year study of males aged 30–59 years old.

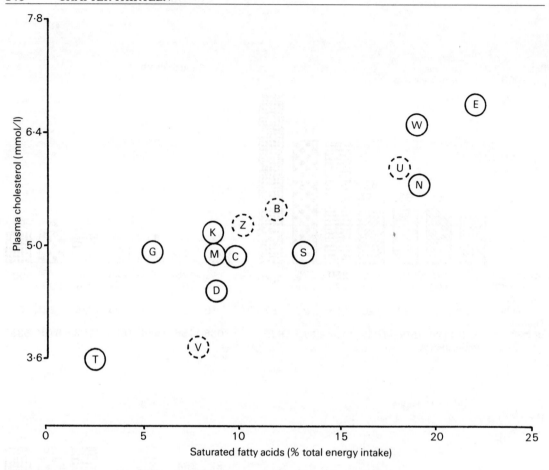

Fig. 13.10. Effect of intake of saturated fats on plasma cholesterol in the following places:
B, Belgrade; C, Crevalcore; D, Dalmatia; E, East Finland; G, Corfu; K, Crete; M, Montegiorgio; N, Zutphen; S, Slavonia; T, Tanushimaru; U, US railroad; V, Velika Krsna; W, West Finland; Z, Zrenjanin.

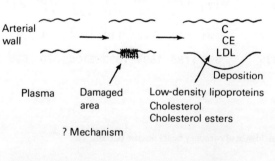

Arterial wall

Plasma   Damaged area   Low-density lipoproteins
                          Cholesterol
                          Cholesterol esters

? Mechanism

C
CE
LDL

Deposition

Fig. 13.11. Hypothesis of arterial disease.
Deposition of cholesterol (C), cholesterol esters (CE) and low-density lipoproteins (LDL) at sites of arterial damage.

Table 13.4 **Dietary factors in arterial disease**

	Effect on plasma cholesterol	
1. Cholesterol	↑	
2. Saturated fat (animal)	↑	
3. Polyunsaturated fat (vegetable)		↓
4. Sucrose (fructose)	↑	
5. Fibre		↓
6. Animal protein	↑	
7. 'Yoghurt factor'		↓
8. Milk		↓
9. Vitamin C		↓ ?
10. Ca (milk)		↓ ?
11. Silicates		↓ ?

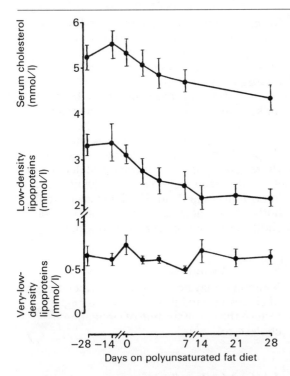

Fig. 13.12. **Effect of feeding a polyunsaturated fat diet to human volunteers on serum cholesterol (30 ml per day of Flora margarine was fed to 7 subjects).**

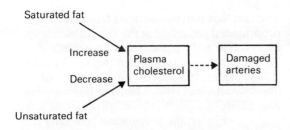

Fig. 13.13. **Role of dietary lipids in cardiovascular disease.**

Cholesterol

Increases in dietary cholesterol do cause some increase in plasma cholesterol, but the effect is complex for two reasons. Firstly, only a fraction of the dietary cholesterol is absorbed (*Fig*. 13.14) and, secondly, dietary cholesterol is effective in inhibiting the *de novo* synthesis

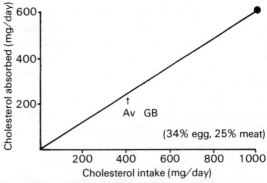

Fig. 13.14. **Cholesterol absorption.**
Mean absorption is: vegetarians, 52%; non-vegetarians, 52%. Av GB = average intake of cholesterol in the United Kingdom = 405 mg/day.

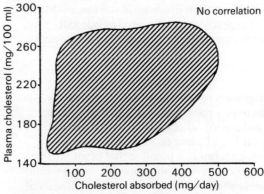

Fig. 13.15. **Effects of dietary cholesterol on plasma cholesterol.**
Increased dietary cholesterol does not necessarily cause increases in plasma cholesterol and wide differences are observed between different individuals.

plasma cholesterol (*Fig*. 13.12). The general concept of the role of dietary fats in cardiovascular disease can therefore be summarized in *Fig*. 13.13. In addition to the effects of saturated and polyunsaturated fats on plasma cholesterol, many other dietary components may also play a role. They are listed in *Table* 13.4.

of cholesterol in the tissues. The mechanism of the feedback inhibition is discussed in Chapter 19. The net effect of increase of dietary cholesterol on plasma may, therefore, be relatively small (*Fig*. 13.15), although many experts recommend that foods containing large quantities of cholesterol, such as eggs, should be kept to a minimum.

Sucrose

Experiments with animals have shown that high concentrations of sucrose in the diet tend to cause increase in the plasma triacylglycerols and small increases in the concentration of plasma cholesterol. The effects are much larger than observed after feeding similar quantities of glucose or starch and the effect is believed to be caused by the fructose component of sucrose. This sugar is very efficiently metabolized by the liver to pyruvate and thence to fat, after incorporation into plasma lipoproteins, mainly the very-low-density lipoproteins.

The importance of this effect in normal human diet is still controversial, but it could be significant if the sucrose intake is very large.

Fibre

As discussed in Chapter 16, dietary fibre tends to reduce the concentration of circulating cholesterol. This is because it binds bile salts which are then excreted in the faeces. More cholesterol is thus diverted to bile salt formation and then excreted in this form.

Animal and plant protein

Ingestion of animal protein significantly increases the concentration of circulating cholesterol, as compared with a similar quantity of plant protein. The reason is unclear, but it is possible that plant proteins supply high concentrations of certain amino acids which are required for the synthesis of certain plasma lipoproteins, for example those of the high-density lipoproteins involved in cholesterol metabolism and transport (cf. Chapter 19).

Yoghurt factor or milk factor

The interest in the 'yoghurt factor' started from observations on the Masai tribesmen in South Kenya and Northern Tanzania. It was found that their serum cholesterol levels were very low, about 135 mg/ml which is close to half the value found for male adults in Western society. They consumed large quantities of milk in the form of yoghurt, about 4 litres per day. This led to careful studies of the value of yoghurt and significant reductions of cholesterol were observed when it was tested on American adults.

It has also been observed that fresh milk, despite its high fat content, does not increase the plasma cholesterol. The reason for the yoghurt and milk effects is unclear. A specific milk factor which interferes with cholesterol synthesis has been postulated, but it has also been suggested that calcium is the important component. Calcium may inhibit both fat and cholesterol absorption because cholesterol is normally absorbed with fat.

Vitamin C

Vitamin C may be involved in the metabolism of cholesterol to bile salts and thus help to reduce the concentration of circulating cholesterol.

13.6 Mode of action of polyunsaturated fats in reducing the concentration of the plasma cholesterol

The fact that polyunsaturated fats cause a pronounced reduction in the concentration of cholesterol in the plasma is well established, but the mechanisms by which this is effected are still not clear. It is possible that cholesterol plays two distinct roles which are illustrated in *Fig.* 13.16.

Firstly, the polyunsaturated fatty acids become incorporated into plasma lipoproteins changing their structures so that they incorporate more cholesterol. This will lead to an enhanced transport from the tissues. Secondly, the increased quantity of poly-unsaturated fatty acids entering the liver leads to the synthesis of phospholipids containing a much higher proportion of these fatty acids. These phopholipids are capable of transporting much more cholesterol and bile salts in the bile than phospholipids containing saturated fatty acids. As a consequence, the excretion of cholesterol is significantly increased.

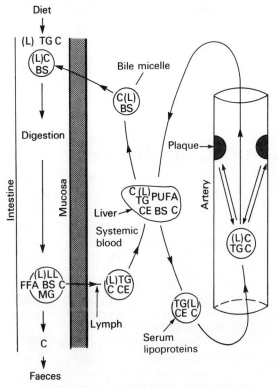

Fig. 13.16. **Effects of polyunsaturated fatty acids on the excretion of cholesterol in the bile and faeces.**
Large amounts of cholesterol may be excreted through the bile and faeces in animals fed polyunsaturated fatty acids. Intake of polyunsaturated fatty acids makes the phospholipids of the plasma lipoproteins, especially high-density lipoproteins, highly unsaturated which enables them to accommodate more cholesterol and thereby prevents deposition of cholesterol in the arteries. The bile phospholipids are also highly unsaturated so that they can hold more cholesterol in the micelles, thus facilitating elimination of cholesterol. Key: (L) lecithin; LL lysolecithin (phosphatidylcholine); TG triacylglycerol; MG 2-monoacylglycerol; FFA free fatty acid; C cholesterol; CE cholesteryl ester; BS bile salt.

These hypotheses are in accord with the observations that sterol excretion in the faeces is increased when the intake of polyunsaturated fats is increased.

13.7 Value of dietary changes in the prevention of heart disease

For many years a controversy has raged on the role of diet in heart disease. All agree that it plays a role and, whilst some believe that this role is of major importance, others believe that it is relatively minor. Furthermore, there is no final agreement as to whether the whole population would really benefit from the dietary changes indicated in *Table* 13.4 or whether only selected individuals seriously at risk should take new diets. It has been suggested that the population may be represented by two extreme groups:

Group I
100 males aged 35 with no hypertension and who are non-smokers. Of this group, 6 would avoid coronary heart disease by adhering to the best diet and the dietary change would have no effect on the remaining 94.

Group II
100 males aged 35 with moderate hypertension and who are moderate smokers. Of this group 29 would avoid coronary heart disease by dieting.

Chapter 14 Nutrition: vitamins

14.1 Introduction

The historical story of the vitamins stretches through 4000 years of man's development from 2000 BC to the present day. About 2000 BC, a serious disease called 'beri-beri' was first described in China, which was characterized by:

Numbness of legs
Pain in muscles
Exhaustion—paralysis
Difficulty in breathing
Bradycardia
Heart failure
Death

It was a serious condition causing many thousands of deaths and, not until the early 1920s, was it clearly understood to be caused by a nutritional deficiency.

Another early observation concerning a disease associated with vitamins was made by Hippocrates who, in about 500 BC, described the condition of 'night blindness', i.e. the lack of ability of adapting the eyes to poor illumination after leaving a well-lighted environment. Many years elapsed before other observations in the field were made, but the voyages of exploration in the fifteenth and sixteenth centuries led to serious outbreaks of scurvy on many ships embarking on long voyages. This is described in Chapter 32. The condition was also ultimately traced to a dietary deficiency, which was eventually shown to be of ascorbic acid.

In the seventeenth century, two other diseases were described which were caused by vitamin deficiencies: rickets, caused by vitamin

D deficiency (described in Chapter 22) and pellagra, which is characterized by a severe dermatitis, diarrhoea, mental disturbances and eventual death. Cases were originally found in Europe especially in Spain, Italy and Rumania, but also in Egypt. However, the most serious outbreaks of the disease occurred in the Southern States of the USA in the early parts of this century, 170 000 cases being described in 1917 and 120 000 cases in 1927.

Towards the end of the nineteenth century five diseases, beri-beri, night blindness, scurvy, rickets and pellagra had been described. At that time it was very difficult to distinguish between these 'diseases' and those caused by micro-organisms or viruses, such as septicaemia or influenza. Gradually, however, the concept of inadequate nutrition began to be considered as a possible cause.

A classic advance was made by a surgeon-admiral in the Japanese Navy, Admiral Takaki, who had been trained at Guy's hospital in London. He was extremely concerned about the very high incidence of beri-beri in the Japanese Navy which, in 1880, was still very serious. Whilst in England, he observed that the sailors of the British Navy did not suffer from beri-beri and he considered that this might be due to the fact that they ate a diet rich in protein rather than rice, the main staple diet of the Japanese. He therefore surmised that beri-beri was caused by a protein deficiency and, when he changed the diet in the Japanese Navy to include much more meat and fish, a dramatic reduction in the incidence of beri-beri occurred. He was able to reduce the incidence from 61 per cent to zero in the space of two years.

A little later some animal experiments were initiated in the Dutch East Indies which showed that birds fed on milled rice developed symptoms very similar to those of human beri-beri, but they were cured if the millings were added to their diet. As described in Chapter 32, in the mid-eighteenth century it had been discovered, by a British Naval Surgeon, that scurvy could be cured by a dietary change, by adding fresh fruit to the diet. Thus, at the beginning of the twentieth century, there were indications that several diseases resulted from inadequate or unsuitable diets, but the concept of the 'vitamin' had not yet developed.

The concept of 'vitamin' followed from the classical experiment of Hopkins in 1912 at Cambridge. He fed groups of rats on synthetic diets containing adequate quantities of purified carbohydrates, proteins, fats and minerals. He discovered that when fed the synthetic diet, the rats ceased to grow, although their energy intake was more than sufficient. If, however, he added a very small quantity of milk, about 2–3 ml, to the daily diet, the rats grew at a rapid rate. The process could be reversed by withdrawing the milk from the group or by re-adding to the group fed on the synthetic milk (*Fig*. 14.1). He calculated that the milk could not have significantly

affected the energy content of the diet and that it must have been effective because it contained very small quantities of certain food which were essential for the health and growth. He called these components the 'accessory food factors'. The idea that the diet must contain very small quantities of certain essential chemicals was completely novel at that time. It was difficult to believe an idea that very small quantities of organic chemicals could play such a significant role in health.

The name 'vitamin' was invented several years later by Funk who assumed that these essential dietary components were all amines, i.e. vital-amines which become contracted to 'vitamine'. Gradually, the 'e' was omitted to give the name 'vitamin'. The concept that all vitamins are amines is not true, but the name passed into general usage and is today well established.

14.2 Classification of the vitamins

A modern classification of vitamins appears to be very confusing and it is useful to consider how it arose. The basis for our description of the vitamins with letters, A, B, C, was first instituted by two workers in the USA in 1915, McCollum and Davis. They pointed out that there appeared to be at least two vitamins, a fat-soluble component, described as A, and a water-soluble component, B. It was very soon shown that 'water-soluble B' must be divisible into two components, the anti-beri-beri factor, B, and another component, described as C, the anti-scurvy factor. Shortly afterwards, the fat-soluble component A was shown to be divisible into two components, A, which was required for rat growth and was subsequently discovered to be the anti-night blindness factor and the anti-rickets factor D (*Table* 14.1).

Of this original list, C was subsequently found to be a single pure compound or ascorbic acid, vitamins A and D to be a small group closely related in structure, but 'B' turned out to be a very complex mixture. These factors accounted for all the known vitamin-deficiency diseases except pellagra.

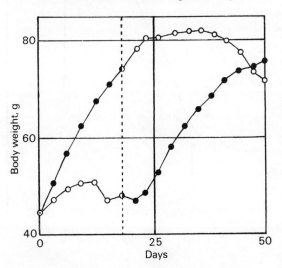

Fig. 14.1. **Growth of rats on synthetic diets with and without vitamins.**
(○) Artificial diet alone; (●) artificial diet plus milk.

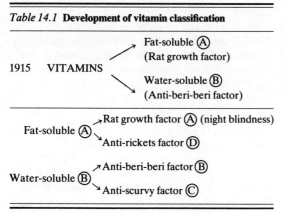

Table 14.1 **Development of vitamin classification**

Nutritionalists then began to consider carefully the possibility of animal feeding experiments and these, in fact, showed that other vitamins existed. The first was discovered by two workers in the USA, Evans in California and Matill in Iowa. They fed rats a correct diet containing all the known vitamins and, although the animals appeared healthy, they failed to reproduce properly. The embryos grew in the uterus until a relatively late stage in pregnancy, but then died and were rapidly resorbed. The condition was described as 'resorption sterility' (*Fig.* 14.2). Evans and Matill found that the condition was caused by a fat-soluble component which could be

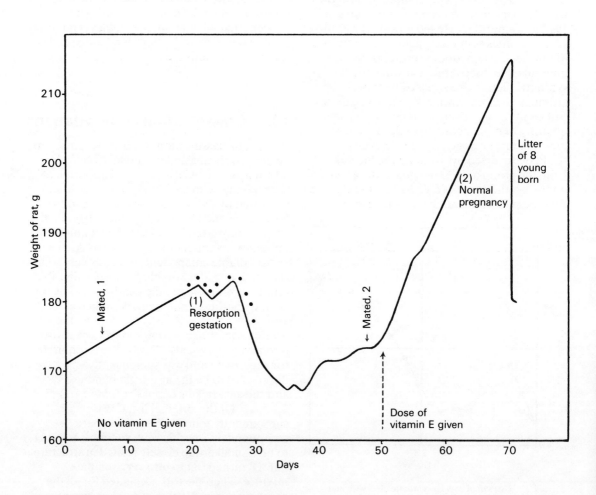

Fig. 14.2.   **Effect of vitamin E on sterility in the female rat.**

Table 14.2 **Fat-soluble vitamins**

Symbol	Name	Human deficiency disease	Minimum daily intake	Important food source
A	Retinol (Carotene)	Night blindness and disorders of mucous membranes, bones, nerves, reproduction	750 μg	Animal Fish Vegetable } Oils
D	Calciferol	Rickets	2·5 μg	Fish oils
E	Tocopherol	Reproduction failure Muscular dystrophy	10–20 mg	Plant oils
K	Pylloquinone	Blood-clotting failure	? (Bacterial synthesis)	Plant oils

extracted from fresh green leaves and that it was quite different from vitamins A and D. It was, therefore, given the letter E and called 'tocopherol' which derives from the Greek word for birth. Subsequently, many other serious consequences followed vitamin E deficiency which are discussed below.

In the mid-1930s a second fat-soluble vitamin was discovered by Dam in Denmark as a result of feeding chicks on special diets. Its deficiency led to defective blood clotting; it was consequently called the 'Koagulation Faktor' and given the letter K, not F as might have appeared in logical sequence. The action of this vitamin is described in Chapter 26. A summary of the fat-soluble vitamins is shown in *Table* 14.2.

Work on separation of vitamin 'B' components proved to be very difficult

although, early on, it was shown that it could be readily divided into two components by heat treatment. These components were called 'B_1' and 'B_2'.

B_1: Heat labile—anti-beri-beri factor
B_2: Heat stable—anti-pellagra factor

The heat-labile factor was eventually isolated in a pure form and given the name thiamin or aneurin. Thiamin is generally used today and is proved as the anti-beri-beri factor. It is mainly concentrated in the germ of cereals of wheat and rice which is removed during milling and polishing.

Vitamin B_2 was, however, found to be a very complex mixture, containing four main components (*Fig.* 14.3). The anti-pellagra factor was eventually shown to be nicotinamide, but the complex contained three

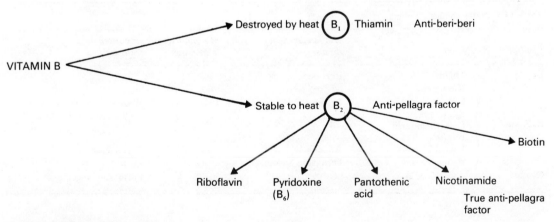

Fig. 14.3. **The vitamin B complex.**

other vitamins, riboflavin, pyridoxine and pantothenic acid. All vitamins of this group are required by humans, but no clear-cut deficiency diseases have been described for deficiencies of riboflavin, pyridoxine or pantothenic acid, although some minor symptoms are recorded. The reason for this is that diets selected by man in most parts of the world will normally contain each of these vitamins in adequate quantities.

Biotin was discovered in a strange way. When rats were fed on diets containing relatively large quantities of raw egg white, they became ill and had very serious deterioration of the skin. The condition was traced to the presence of a protein in the egg white, avidin, which possesses a very powerful binding affinity for biotin, rendering it unavailable for the animal. Humans can also suffer from a similar condition if fed large quantities of raw egg under similar conditions as used for the experimental rats. Much later, in 1945–1955, two more vitamins of the B group were discovered, folic acid and vitamin B_{12} which are involved in the prevention of anaemia and are described in Chapter 24.

The water-soluble vitamins are listed in *Table* 14.3, and it will be observed that the nomenclature of vitamins in the B group has become very complex. Some, such as thiamin, have a name and number (B_1) whereas others, such as nicotinamide, have a name only. Riboflavin, the first member of the vitamin B_2

complex, is sometimes called vitamin B_2, but it is more usual to use the term B_2 complex to describe the group of vitamins. The numbering of pyridoxine as B_6 and cobalamin as B_{12} leads to the question of the whereabouts of vitamins B_3, B_4, B_5, B_7, B_8, B_9, B_{10} and B_{11}. The answer is not simple but arose mainly because many investigations in the early period of intensive study, in the 1920s–1930s, described what they believed to be new vitamins called B_3 or B_4 which were ultimately shown to be impure mixtures of already known vitamins or of new vitamins and known vitamins.

When reviewing *Tables* 14.2 and 14.3 which summarize human vitamin requirements, the following should be noted: these vitamins are required in small doses because they cannot be synthesized by the human body; the daily requirement is generally small but can vary from a few microgrammes (B_{12}) to nearly 100 mg/day (vitamin C); the relatively large requirement for vitamin C means that the quantity is close to the minimum daily requirements for one of the essential amino acids, tryptophan (approximately 250 mg/day).

Vitamins are, however, distinct in several ways. They are utilized for specialized purposes, often to form a coenzyme, and are excreted if ingested in large quantities, whereas tryptophan is used for several purposes and excess can even be used to provide energy.

Table 14.3 **Water-soluble vitamins**

Symbol	Name	Human deficiency disease	Minimum daily intake, mg	Important food sources
B_1	Thiamin	Beri-beri	1·2	
	Riboflavin	—	1·7	
	Nicotinamide	Pellagra	18	Widely distributed in many animal and plant foods
	B_6 pyridoxine	—	1·5	
B_2 complex	Pantothenic acid	—	?	
	Biotin	—	?	
	Folic acid	Anaemia	0·4	Green leaves
	B_{12} cobalamin	Pernicious anaemia	5 μg	Liver
C	Ascorbic acid	Scurvy	50–75	Fresh fruit, vegetables

14.3 Vitamin synthesis by intestinal bacteria

Although, by definition, vitamins required by man cannot be synthesized within the tissues, it is important to appreciate that bacteria within the gut can synthesize many of the vitamins required by man and animals. This was observed during early investigations of the sulphonamide drugs (cf. Chapter 43). When these drugs were fed to young rats it was noted that they failed to grow at a normal rate. This was traced, not to a toxic effect of the drug, but to the fact that the animals were vitamin deficient. The sulphonamide drug had killed most of the bacteria in the gut which normally synthesize a large number of the vitamins required and, when the animals were given the drug and a vitamin supplement, they grew at a normal rate.

These observations were a little surprising because most of the bacterial population in rats and man live in the colon region where absorption of all molecules in animals and man is very poor. The solution of the problem was provided when it was observed that rats and many other animals, such as deer, in the wild, eat their own faeces— the practice of coprophagy. They appear to have learnt through the course of evolution that the faeces can provide a good source of vitamins and that ingestion will present the vitamins to the small intestine where they are efficiently absorbed.

In the rat, several members of the vitamin B complex are synthesized including thiamin, nicotinamide, riboflavin, pyridoxine, biotin and vitamin B_{12}. In addition vitamin K is synthesized. It is uncertain whether the synthesis of any of these vitamins is adequate for their needs, but the amounts of biotin, vitamin B_{12} and vitamin K may be sufficient.

Humans do not normally indulge in coprophagy, so that absorption of the vitamin must take place in the gut if it is to be available to the host. It is generally believed that bacterial synthesis and absorption of vitamin K and biotin are adequate and that small but inadequate quantities of other B vitamins synthesized by bacteria may be absorbed.

14.4 Storage of vitamins

It is important to appreciate that the human tissues possess the ability to store vitamins, but that the capacity varies considerably from one vitamin to another. Generally, the fat-soluble vitamins are effectively stored, but little or limited storage of water-soluble vitamins occurs.

Vitamin A is stored very extensively in the liver so that rats fed on diets high in vitamin A can deposit a supply of this vitamin in their livers sufficient for several years, in fact, much more than is required for their normal life span! Vitamins D, E and K are also stored in tissue lipids but to a much smaller extent than vitamin A. Of the water-soluble vitamins, only vitamin B_{12} is extensively stored bound to special proteins in the liver (cf. Chapters 24, 29).

The problem of vitamin storage is of particular importance when patients are put onto a restricted diet or suffer a prolonged period of malabsorption.

14.5 The water-soluble vitamins: the B groups of vitamins and vitamin C

The functions of the B vitamins are now well established and all are known to be components of coenzymes which play vital roles in the metabolism of all cells; they are summarized in *Table* 14.4. Coenzyme formation may simply involve phosphorylation, for example the formation of thiamin pyrophosphate from thiamin or pyridoxal phosphate from pyridoxine, or the vitamin may form part of a larger molecule and take part in the formation of nucleotides such as those formed from riboflavin and nicotinamide.

Most of the coenzymes formed from the vitamins are involved in very specific functions, for example oxidation–reduction reactions, but vitamin B_6 is involved in many different types of amino acid reactions. The roles of folic acid and vitamin B_{12} are described

Table 14.4 **Vitamin B complex**

Name	Formula	Coenzyme form	Metabolic role of coenzyme
B₁ Thiamin		Cocarboxylase Thiamin pyrophosphate T—CH₂—CH₂—O—℗—℗	1. Decarboxylations of ketoacids, e.g. pyruvate → acetyl-SCoA 2. Transketolase
B₂ Riboflavin		1. Flavin mononucleotide (FMN) = R—℗ 2. Flavin adenine dinucleotide (FAD) = R—℗—℗—Ribose—Adenine	Prosthetic group of flavoproteins, e.g. succinic dehydrogenase, cytochrome reductase, xanthine oxidase.
Nicotinamide (Niacin)		NAD⁺ NADP⁺	Hydrogen acceptors for dehydrogenations, e.g. malic, lactic, isocitric dehydrogenases.
B₆ Pyridoxine Pyridoxal Pyridoxamine		Pyridoxal phosphate 	1. Transamination 2. Decarboxylation of amino acids 3. Cysteine desulphydrase 4. Aminolaevulinic acid synthesis 5. Tryptophan degradation
Pantothenic acid		Coenzyme A Adenosine 3′-phosphate 	1. Carrier of acetyl group 2. Oxidation of pyruvate as acetyl-SCoA 3. Oxidation and synthesis of fatty acids and cholesterol
Biotin		Prosthetic group of carboxylases	Carrier 'active CO₂', e.g. 1. Pyruvate → oxaloacetate 2. Acetyl-SCoA → malonyl-SCoA

Table 14.4 Continued

Folic acid 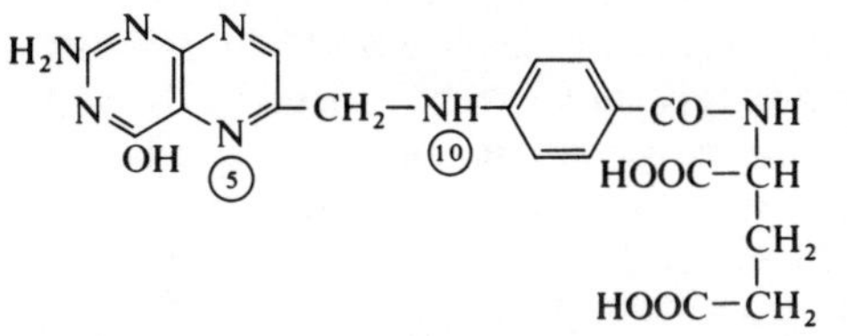	Coenzyme F (tetrahydrofolic acid) 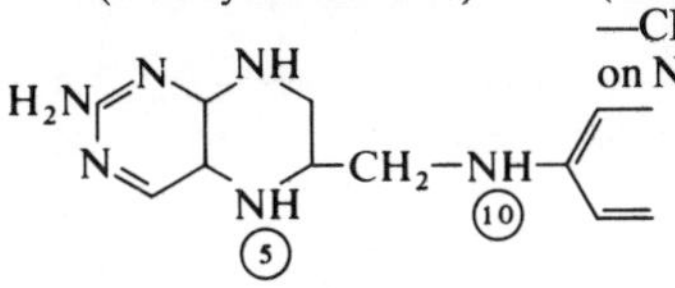	Carrier of 1 carbon units ($-CHO-CH_2OH$, $-CH_3$ $-CH=NH$) on N–5 or N–10

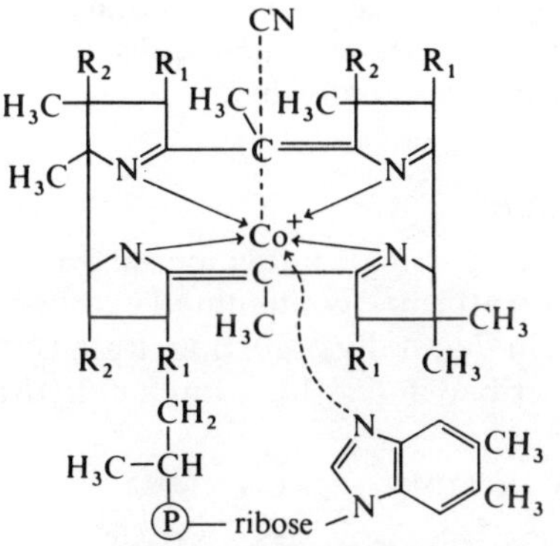 B_{12} cobalamin	(Vitamin B_{12} coenzyme) (schematic representation) 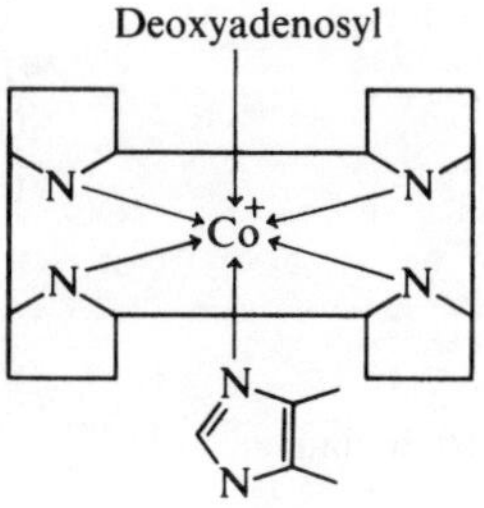	Several functions known in bacteria In mammals, only established roles as cofactor for: 1. Methylmalonyl-SCoA isomerase 2. Transfer of CH_3 from folate especially in methionine synthesis

$R_1 = -CH_2-CH_2-CO$
$\qquad\qquad\qquad NH_2$

$R_2 = -CH_2-CO-NH_2$

Deoxyadenosyl

in Chapter 24. Vitamin C, which is fully discussed in Chapter 32, does not, however, form a coenzyme and is involved in conjunction with iron in oxidation and hydroxylation reactions.

14.6 The fat-soluble vitamins

Four fat-soluble vitamins are known: A, D, E and K. Unlike the water-soluble vitamins the functions of which are well understood, the functions of the fat-soluble vitamins have been much more difficult to elucidate. It is clear that they do not form coenzymes in a manner similar to the vitamins of the B group, but they are involved in much more complex regulatory processes.

Another important difference is that two of the fat-soluble vitamins, A and D, are toxic in high doses, even to the extent of being lethal. Water-soluble vitamins are excreted in high doses. Vitamin D is intimately concerned with calcium and phosphate metabolism and is discussed in Chapter 22. Vitamin K is involved in blood clotting and is discussed in Chapter 26. This chapter will therefore be devoted to a discussion of vitamins A and E.

Vitamin A

Vitamin A is the dietary factor now known to be the antidote to the condition of night blindness first described in 500 BC by Hippocrates. More careful studies of the vitamin in animal experiments demonstrated,

160 CHAPTER FOURTEEN

however, that many complex symptoms develop as a result of vitamin A deficiency and that, in serious cases, death ultimately results (*Table* 14.5).

Table 14.5 **Vitamin A deficiency and excess**

Symptoms of vitamin A deficiency

1. Defective dark adaption
2. Failure of growth
3. Xerosis and keratinization of mucous membranes; xerophthalmia; secondary infections
4. Faulty bone modelling, with the production of thick, cancellous bones instead of thinner more compact bones
5. Nerve lesions, often associated with bone lesions
6. Increased pressure of the cerebrospinal fluid, developed independently or in association with deformed skull bones; hydrocephalus
7. Abnormalities of reproduction, including degeneration of the testes and abortion, or the production of malformed offspring
8. Certain forms of skin disease
9. Death

Symptoms of hypervitaminosis A

1. Cessation of growth (rat)
2. Skin abnormalities (rat, man)
3. Increased pressure of cerebrospinal fluid (human infant); nausea and vomiting (adult man)
4. Bone abnormalities (man); bone fractures (rat)
5. Profuse, fatal internal haemorrhage (rat); secondary vitamin K deficiency (rat)
6. Congenital malformations (rat)

It also became clear that vitamin A is toxic in excess and this is true in man as well as in rat. Toxicity was first observed by Arctic explorers who ate polar bear liver and became ill, suffering severe sickness and nausea. This was ultimately shown to be caused by the very high content of vitamin A in polar bear liver, which was a result of the consumption by the bears of cod and seal, both of which concentrate vitamin A in their tissues and especially in the liver. In one recorded case in the UK, every day a man consumed several pints of carrot juice which contains high concentrations of the precursor of vitamin A, β-carotene. He also took vitamin A pills. He became ill and eventually died.

Structure

Vitamin A contains a single 6-membered ring with a side-chain attached containing 11 carbon atoms. The main, naturally occurring, form is vitamin A$_1$ described in *Fig.* 14.4, but a dehydro

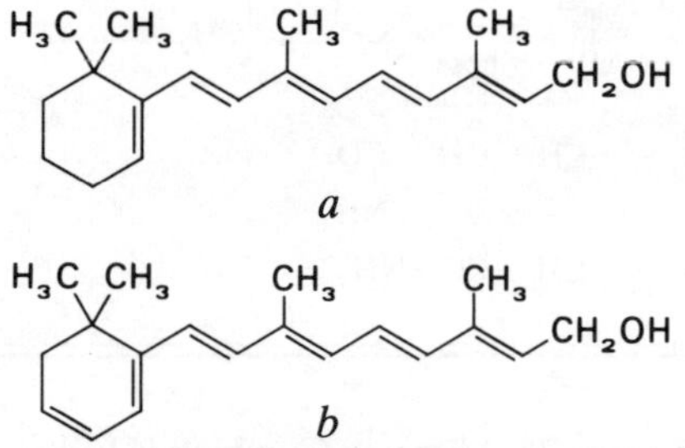

Fig. 14.4. **Structures of vitamin A.**
(*a*) Vitamin A$_1$;
(*b*) vitamin A$_2$.

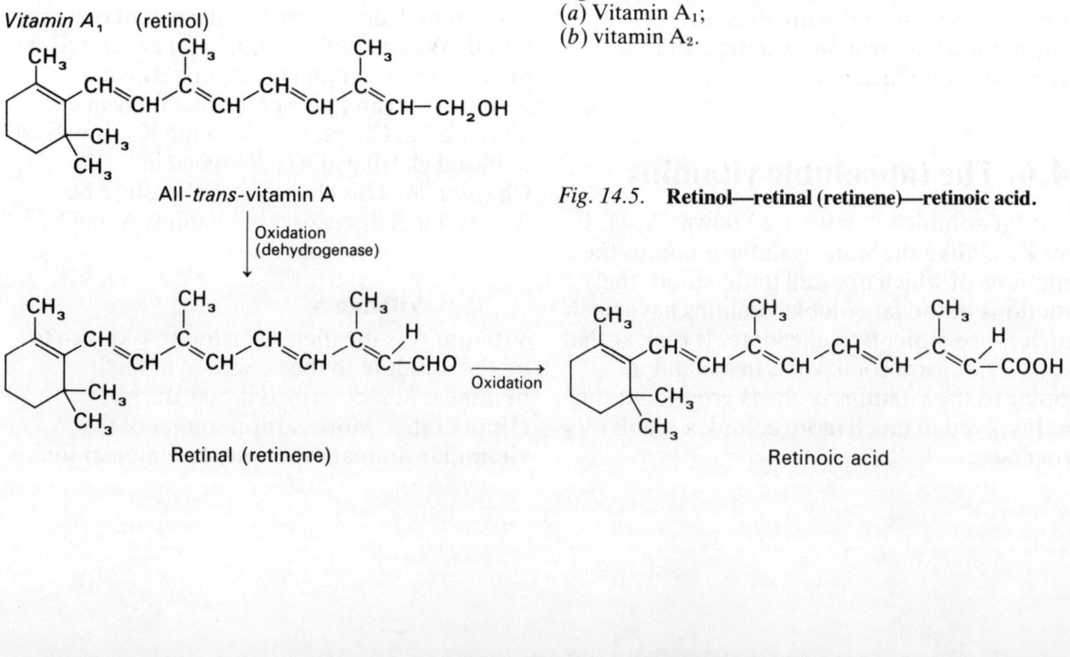

Vitamin A$_1$ (retinol)

All-*trans*-vitamin A

Fig. 14.5. **Retinol—retinal (retinene)—retinoic acid.**

form with 2 double bonds in the ring is also known and called vitamin A_2. Vitamin A_2 is found in several sea fish liver oils, but generally it forms only a small percentage of the total vitamin A. Freshwater fish liver generally contains much more vitamin A_2.

In humans and most mammals, vitamin A_1 is the effective form of vitamin A but, in some freshwater fish, vitamin A_2 performs a similar role. Vitamin A is an alcohol (retinol), but it can be converted into an aldehyde (retinal or retinene) or an acid retinoic acid (*Fig.* 14.5).

Carotene

An important precursor of vitamin A is β-carotene which occurs widely in plants and is responsible for the red colour of carrots. β-Carotene is effectively 2 molecules of vitamin A, joined end to end, and can be converted to vitamin A in the intestinal mucosal cells. It is a two-stage process. The vitamin A is first cleaved by an oxygenase enzyme to form the aldehyde and then it is reduced to the vitamin by retinal reductase in the presence of NADH or NADPH (*Fig.* 14.6).

H_3C CH_3 CH_3 CH_3 H_3C CH_3

CH_3 H_3C H_3C

CH_3

β-Carotene

(1) CHO

[Intestinal mucosa]

(2)

H_3C CH_3 CH_3 CH_3

CH_2OH

CH_3

Vitamin A (2 mol)

Fig. 14.6. Conversion of carotene to vitamin A.
The process in the intestinal mucosa occurs in two stages:
(1) dioxygenase (O_2) which forms the aldehyde and
(2) retinal reductase (NADH or NADPH) which reduces the aldehyde to the alcohol (retinol).

Absorption–transport–storage

Vitamin A esters are hydrolysed in the intestine and β-carotene is converted to vitamin A so that the latter is transported into the plasma. Fat absorption is essential for

Table 14.6 Vitamin A binding protein

Properties of serum protein

Pre-albumin fraction
M_r 20 000

Binds { Retinol / Retinal / Retinoic acid

Effect of protein deficiency as

Protein/day for 12 days, g	Retinol-binding protein, mg/dl ($t_{1/2} = 12 h$)
80	5·7
40	4·4
20	3·3

adequate vitamin A absorption and it is believed that vitamin A, like other fat-soluble vitamins, is absorbed together with monoacylglycerol micelles (cf. Chapter 16).

In the plasma, vitamin A is transported on a specific protein, the retinol-binding protein, and then transferred to the liver for storage. Recently, it has been discovered that the synthesis of the retinol-binding protein is very sensitive to protein deficiency in the diet. Failure to synthesize the protein can also lead to vitamin A deficiency in the tissues and the widespread incidence of this vitamin deficiency in India is primarily due to low concentrations of the retinol-binding protein and not to a lack of carotene or vitamin A in the diet, which is usually adequate (*Table* 14.6). In the liver, fatty acid esters are formed from the retinol and vitamin A is stored in this form (*Fig.* 14.7).

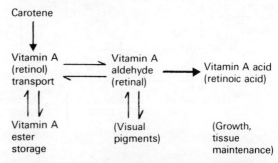

Fig. 14.7. Storage, transport and utilization of vitamin A.
Note that retinoic acid cannot be reduced to retinal.

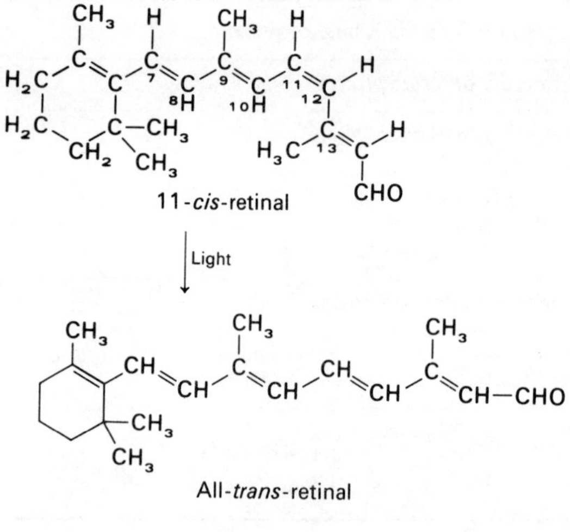

11-*cis*-retinal

Light

All-*trans*-retinal

Fig. 14.8.   Conversion of 11-*cis*-retinal to all-*trans*-retinal by light

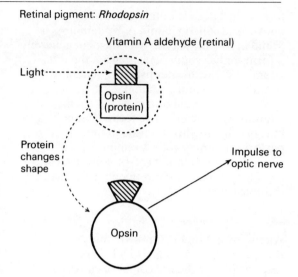

Retinal pigment: *Rhodopsin*

Vitamin A aldehyde (retinal)

Light

Opsin (protein)

Protein changes shape

Impulse to optic nerve

Opsin

Fig. 14.9.   Vitamin A and vision.
Schematic representation of change in *cis*-retinal to *trans*-retinal and its change in the conformation of opsin.

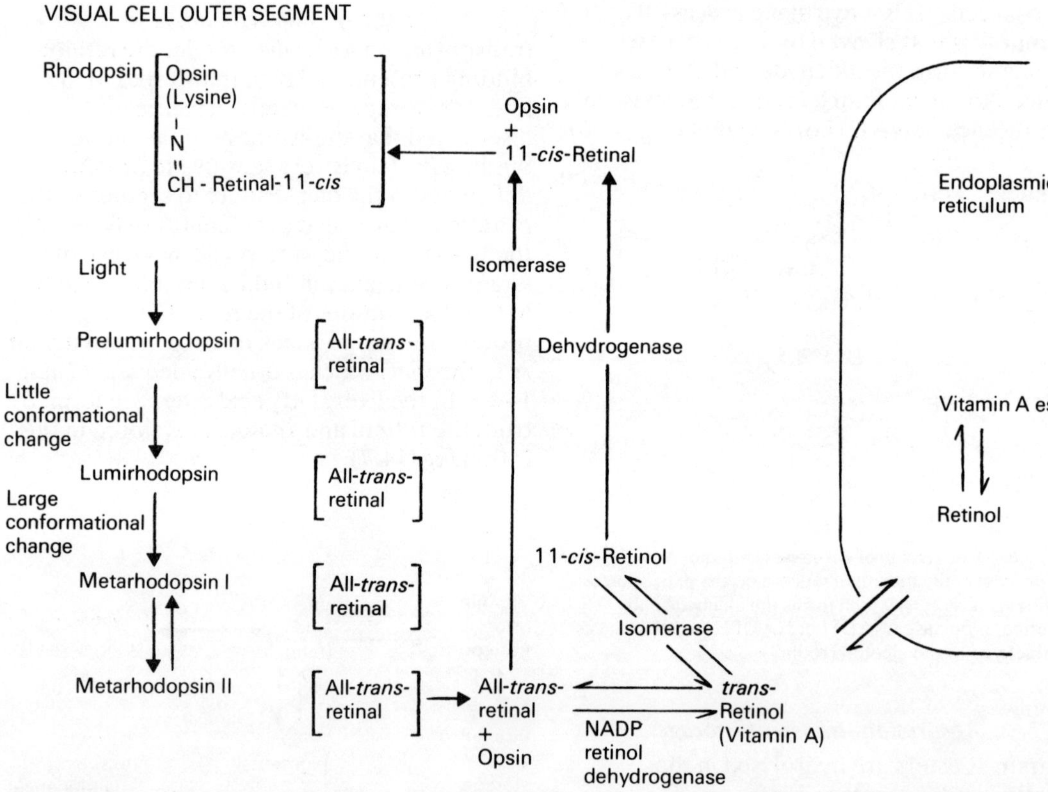

VISUAL CELL OUTER SEGMENT

Rhodopsin [Opsin (Lysine) | N ‖ CH - Retinal-11-*cis*]

Opsin + 11-*cis*-Retinal

Endoplasmic reticulum

Light

Prelumirhodopsin — [All-*trans*-retinal]

Little conformational change

Lumirhodopsin — [All-*trans*-retinal]

Large conformational change

Metarhodopsin I — [All-*trans*-retinal]

Metarhodopsin II — [All-*trans*-retinal] → All-*trans*-retinal + Opsin

Isomerase

Dehydrogenase

11-*cis*-Retinol

Isomerase

trans-Retinol (Vitamin A)

NADP retinol dehydrogenase

Vitamin A ester

Retinol

Fig. 14.10.   Vitamin A in vision.
Details of conformational changes in opsin are reconversion of *trans*-retinal to 11-*cis*-retinal.

Vitamin A in vision

Early observations on the relation of vitamin A deficiency to night blindness led to investigations of the role of vitamin A in vision. This led to the discovery that the visual pigment of the retina of all species investigated contained light-sensitive pigments composed of an apoprotein and the aldehyde formed by oxidation of vitamin A (retinene or retinal). The main pigment is called 'rhodopsin'.

Very important progress was made in understanding the role of retinal in vision by Wald in the USA. He pointed out that retinal could exist in several *cis–trans* isomers, but showed that only two forms were of major importance: 11-*cis*-retinal and all-*trans*-retinal (*Fig.* 14.8). Furthermore, the effect of light was to convert the 11-*cis* form into the all-*trans* form. This caused a conformational change in the opsin constituent, so that an impulse was transmitted to the optic nerve and the visual signal relayed to the brain (*Fig.* 14.9). The rhodopsin went through a series of changes to finally split off opsin from all-*trans*-retinal (*Fig.* 14.10). It is known that retinal is bound by a Schiffs base to a lysine residue on opsin (*Fig.* 14.11), but the mechanism of relay of the conformational change is still uncertain.

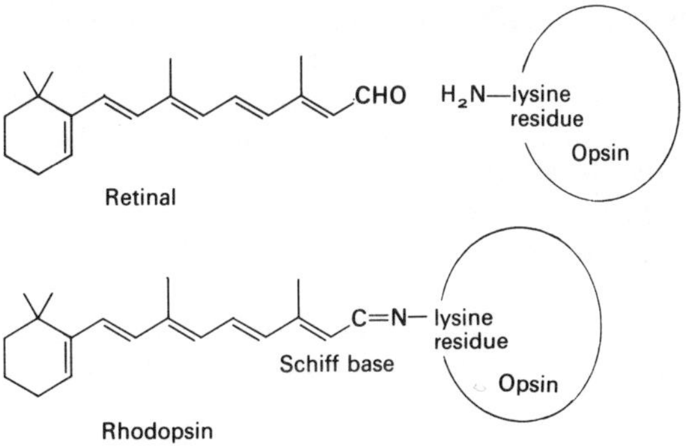

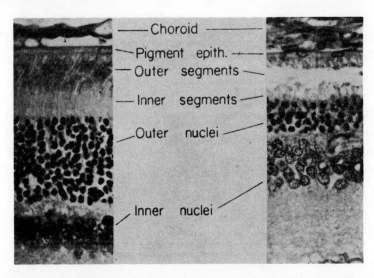

Fig. 14.11. **Binding of retinal to opsin by the formation of a Schiff base.**

Imidazole and —SH groups may be exposed in the process.

To reform rhodopsin after all-*trans*-retinal has been split off, the retinal must first be converted to retinol by a dehydrogenase enzyme, isomerized to 11-*cis*-retinol and then reoxidized to 11-*cis*-retinal which combines with the opsin. *Trans*-retinal is in equilibrium with a cellular store of vitamin A (*Fig.* 14.10).

Investigations of the visual action of retinal were greatly aided by the discovery of vitamin A acid (*Fig.* 14.5). Retinal is

Fig. 14.12. **Effect of vitamin A deficiency on visual pigments.**
Sections through the retina are shown. Animals were maintained on vitamin A acid (retinoic acid) which fulfills the metabolic functions, but cannot form visual pigments.

converted to vitamin A acid or retinoic acid by either a dehydrogenase enzyme or chemically in the body. When fed to animals, retinoic acid will protect against all deficiency symptoms of vitamin A lack, but it cannot be converted to retinal (*Fig*. 14.7). If animals are fed retinoic acid they, therefore, become blind with severe alteration of the rods of the retina (*Fig*. 14.12), but are otherwise healthy. The process can be reversed by feeding with vitamin A (retinol).

Metabolic functions of vitamin A

It has been known for many years that vitamin A must play an important role in cellular metabolism in addition to that in vision. Animals deprived of vitamin A die from serious metabolic disturbances that are unrelated to the visual process. Despite nearly 50 years of intensive research, the real metabolic functions of vitamin A and whether the vitamin exerts one or several functions still remain unsolved. A serious problem in vitamin A research is that animals deprived of the vitamin suffer many serious bacterial infections and thus most of the effects observed can often be secondary, caused by infections, and not be directly dependent on vitamin A deficiency.

Metabolic processes which have been described as dependent on vitamin A are listed in *Table* 14.7. The depression of gluconeogenesis observed in vitamin-A-deficient animals is likely to be a result of the depression of steroid hormone synthesis, especially corticosterone and cortisol. The defective stage is believed to be the hydroxylation of pregnenolone to progesterone which may be vitamin A dependent.

The evidence for the dependence of mucopolysaccharide synthesis on vitamin A is based on observations that, in vitamin A deficiency, many of the mucosal membranes of the body tend to keratinize and fail to synthesize the mucopolysaccharide. However, it is possible that vitamin A plays a much more fundamental role in regulating cell differentiation and cell division, causing the stem epithelial cells to form mucus-secreting cells rather than keratinizing cells (*see Table* 14.7). A role associated with cellular differentiation is also observed in spermatogenesis which is impaired in vitamin A deficiency.

Vitamin A also appears to play a role in cholesterol synthesis, in the final stage of cyclization of squalene to cholesterol (*Fig*. 14.13). In vitamin A deficiency, mevalonate tends to be diverted to the

---

Table 14.7 **Possible metabolic roles of vitamin A**

1. Gluconeogenesis (↓) ? Secondary to (2)

2. Glucocorticoid synthesis (Corticosterone syn.)
   (Cortisol syn.) (↓)

   Pregnenolone $\xrightarrow{\text{3}\beta\text{-Hydroxylation}}$ Progesterone

3. Mucopolysaccharide synthesis (↓)

4. Cholesterol synthesis (↓)

   Mevalonate $\nearrow$ Squalene (↑)
   $\searrow$ Coenzyme Q (↑)

5. Cell differentiation
   Differentiation in: Spermatogenesis
   Epithelial cells
   (↓) Mucous   (↑) Keratinizing
   cells   cells

6. Membrane glycoprotein synthesis (↓)

↓ Decreased activity and ↑ increased activity observed in vitamin A deficiency.

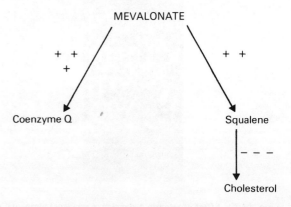

Fig. 14.13. **Vitamin A—lipid metabolism.**
+ increase in vitamin A deficiency; − decrease in vitamin A deficiency.

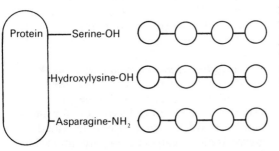

Fig. 14.14. **Membrane glycoproteins.**
The —OH groups of serine or hydroxylysine or the —NH₂ of asparagine are the sites of linkage to the carbohydrate chain (○—○).

synthesis of coenzyme Q which substantially increases in concentration. In fact, coenzyme Q was first discovered in vitamin-A-deficient animals.

Recently, however, it has been discovered that vitamin A may play an important role in the synthesis of membrane glycoproteins. All glycoproteins in membranes possess short-chain oligosaccharides which are attached to —OH groups of serine or hydroxylysine or to the —NH₂ group of asparagine (*Fig.* 14.14). It has been shown that vitamin A is first phosphorylated and then it acts as a carrier for a carbohydrate, such as mannose, for synthesis into the oligosaccharide of a membrane glycoprotein (*Fig.* 14.15).

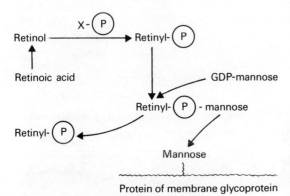

Fig. 14.15. **Role of vitamin A in glycoprotein synthesis in membranes showing interaction of mannose residues.**

It is, as yet, uncertain whether this function of vitamin A could explain all the known metabolic roles, but membrane glycoproteins play very important roles in cellular recognition and impairment of their synthesis could clearly have wide-ranging and serious consequences in the animal body.

Vitamin A excess

Excess of vitamin A in the diet is extremely harmful and can lead to death as explained earlier. Vitamin A *in vitro* has been shown to labilize the lysosomal membrane and cause enzyme release. Lysosomes prepared from animals given high doses of vitamin A release their enzymes into the supernatant very readily.

As discussed in Chapter 3, release of a large battery of hydrolases, such as nucleases or proteases, can cause serious damage to all cells and it is likely, but not certain, that the toxic effects of vitamin A observed *in vivo* are a result of attack on the lysosomal membrane.

By synthesizing an analogue of vitamin A which contains an α-ionone ring in place of a β-ionone ring, it has been demonstrated that the toxic effects of vitamin A are likely to be non-specific pharmacological actions. This is because the compound with an α-ionone ring has a similar effect to natural vitamin A on membranes in excess, but it possesses less than one-hundredth the activity of vitamin A when tested for normal vitamin A functions.

Vitamin E

As described previously, vitamin E was discovered as a result of feeding rats diets deficient in plant oils. The rats suffered from reproduction failure and no live young were born. Further investigations showed that degeneration of germ cells occurred in males and that other symptoms could be demonstrated, sometimes in one species and not in another. These included haemorrhages of the cerebellum, increased capillary

Table 14.8 **Symptoms of vitamin E deficiency**

1. Resorption of fetus
2. Degeneration of germ cells (male)
3. Encephalomalacia (chicks) (haemorrhage of cerebellum)
4. Exudative diatheis (chick) (increased capillary permeability—oedema)
5. Muscular dystrophy (guinea-pig, rat, rabbit)
6. Liver necrosis (rat)

permeability, muscular dystrophy and liver necrosis (*Table* 14.8).

Structure

The structure of vitamin E or tocopherol was not elucidated until 15 years after its discovery. Several tocopherols have been identified known as α, β, γ, δ (*Fig.* 14.16). They differ only in the number and position of the methyl groups on the ring, but the biological activity varies, α being the most active form.

Mode of action

Exactly how vitamin E protects against all the symptoms listed in *Table* 14.8 is not absolutely clear, but it is quite certain that vitamin E is an antioxidant, i.e. it protects readily oxidizable molecules *in vitro* and in the cells. The most vulnerable lipophilic molecules in the cells are the polyunsaturated fatty acid components of

the phospholipids of cell membranes and, to understand how vitamin E functions, we must first consider the process of peroxidation of unsaturated fatty acids.

Peroxidation of unsaturated fatty acids All unsaturated fatty acids, and especially those containing more than two double bonds, are very sensitive to attack by oxygen, forming 'lipid peroxides'. This process, described as 'peroxidation', can occur in pure fatty acids, fatty acids as components of triacylglycerols or as components of phospholipids. It does not require an enzyme but usually takes place by a free radical mechanism and is catalysed by traces of metals, such as Fe or Cu, or by haemoglobin or u.v. or ionizing radiations. Unsaturated fats used in foods are thus liable to peroxidation, but fatty acid components of tissue can also form peroxides. As all membranes in all cells contain unsaturated fatty acids as phospholipid components, they are all potential sites of attack.

The process (*Fig.* 14.17) is usually initiated by radicals, such as OH·, many of which can be formed during the course of metabolism or in the presence of metals or by radiation. It is important to note that, once started, the process can proceed as a 'chain reaction' because the peroxy radical formed at stage C can also abstract a H atom from another fatty acid molecule which enables the process to continue and to form the

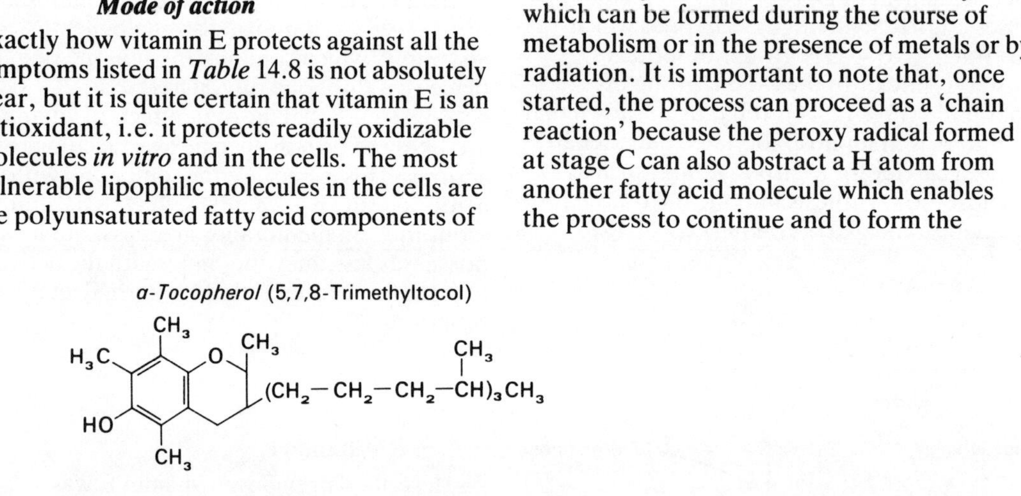

Fig. 14.16. **Vitamin E (tocopherols).**

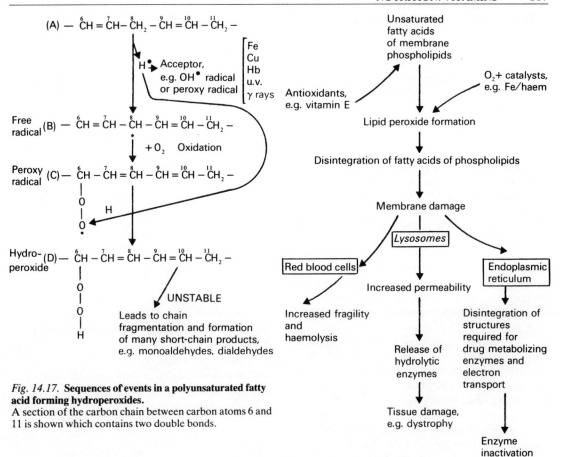

$$\text{(A)} - \overset{6}{C}H = \overset{7}{C}H - \overset{8}{C}H_2 - \overset{9}{C}H = \overset{10}{C}H - \overset{11}{C}H_2 -$$

H• Acceptor, e.g. OH• radical or peroxy radical

Fe
Cu
Hb
u.v.
γ rays

Free radical (B) $- \overset{6}{C}H = \overset{7}{C}H - \overset{8}{C}H - \overset{9}{C}H = \overset{10}{C}H - \overset{11}{C}H_2 -$

$+ O_2$ Oxidation

Peroxy radical (C) $- \overset{6}{C}H - \overset{7}{C}H = \overset{8}{C}H - \overset{9}{C}H = \overset{10}{C}H - \overset{11}{C}H_2 -$
          |
          O
          |
          O•

H

Hydro-peroxide (D) $- \overset{6}{C}H - \overset{7}{C}H = \overset{8}{C}H - \overset{9}{C}H = \overset{10}{C}H - \overset{11}{C}H_2 -$
          |
          O
          |
          O
          |
          H

UNSTABLE

Leads to chain fragmentation and formation of many short-chain products, e.g. monoaldehydes, dialdehydes

Fig. 14.17. **Sequences of events in a polyunsaturated fatty acid forming hydroperoxides.**
A section of the carbon chain between carbon atoms 6 and 11 is shown which contains two double bonds.

Unsaturated fatty acids of membrane phospholipids

Antioxidants, e.g. vitamin E

O_2+ catalysts, e.g. Fe/haem

Lipid peroxide formation

Disintegration of fatty acids of phospholipids

Membrane damage

Lysosomes

Red blood cells

Increased permeability

Endoplasmic reticulum

Increased fragility and haemolysis

Release of hydrolytic enzymes

Disintegration of structures required for drug metabolizing enzymes and electron transport

Tissue damage, e.g. dystrophy

Enzyme inactivation

Fig. 14.19. **Lipid peroxide formation as a cause of membrane damage and the role of antioxidants.**

Catalysts of peroxidation

P — C
Phosphatidyl-choline

P — C
Lyso-phospholipid

Saturated and unsaturated mono- and dialdehydes

TOXICITY

LOSS OF MEMBRANE INTEGRITY

Fig. 14.18. **Basic hypothesis of consequences of lipid peroxidation in membranes.**

hydroperoxide ('lipid peroxide'). A very minor event can thus initiate peroxidation of a large quantity of fatty acids in presence of oxygen.

Consequences of peroxidation The overall process is a potential source of damage to fatty acids and, if it occurs in cells, of serious damage to membranes and other cell components.
    This is because:

i.   The hydroperoxides formed are powerful oxidizing agents and can oxidize vital components, e.g. the —SH groups of proteins

ii. The free radicals formed in the process, e.g. R—OO· are very reactive and can attack cell components such as proteins or DNA

iii. The hydroperoxides formed are very unstable and rapidly disintegrate into fragments which include short-chain aldehydes, dialdehydes and ketones, several of which are potentially toxic

Peroxidation in food lipids This will, as a consequence of (iii) above, lead to the loss of essential fatty acids from the diet and the formation of undesirable short-chain fragments. Cooking in unsaturated oils causes extensive peroxidation.

Peroxidation in lipids of cell membranes As a consequence of (iii) above, the fragmentation of fatty acid hydroperoxides leads to loss of fatty acids from the membrane phospholipids (*see Fig.* 14.18).

Many membranes have been shown to be damaged by this process, including the red-blood-cell membranes, membranes of the lysosomes and those of the endoplasmic reticulum (*Fig.* 14.19).

Antioxidant action Antioxidants such as vitamin E and synthetic antioxidants such as BHT and propyl gallate inhibit peroxidation of lipids by blocking the chain autoxidation process.

We can summarize the normal process of lipid peroxidation as follows, where R = hydrocarbon chain of a polyunsaturated fatty acid:

Initiation: $\quad$ R-H → R· + H· → Acceptor

$$R· + O_2 → R\text{-}OO·$$

Chain $\quad$ R-H + R-OO· → R-OOH + R·
propagation $\qquad\qquad\qquad$ Hydroperoxide

If we then add an antioxidant such as vitamin E, to the system represented by AH, this then supplies a H to enable a new radical, A·, formed from the antioxidant, rapidly

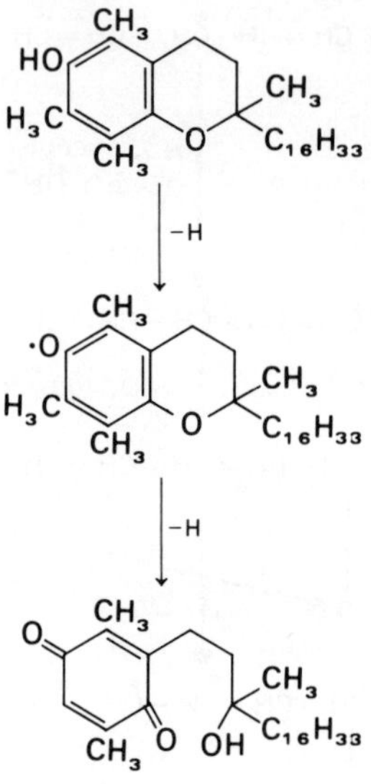

Tocopheryl quinone

Fig. 14.20. **Vitamin E as an antioxidant: formation of quinones.**

giving rise to a stable compound which does not propagate the chain reaction (*Fig.* 14.20).

$$AH + R\text{—}OO· \longrightarrow R\text{—}OOH + A·$$

Chain termination $\qquad$ Does not propagate reaction and gives rise to stable compound, e.g. tocopheryl quinone from vitamin E

In some cell membranes, vitamin E is believed to be embedded in the membrane in close proximity to polyunsaturated fatty acids, such as arachidonic acid, which are very likely to undergo peroxidation (*Fig.* 14.21).

Although this concept of an antioxidant does not explain all the biologically observed consequences of vitamin E deficiency, many may be explained. For example, red blood cells in animals and man become more fragile in vitamin E deficiency,

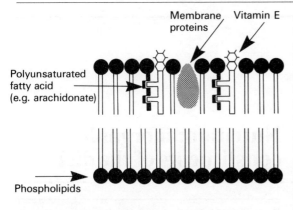

Fig. 14.21. **Vitamin E protecting arachidonic acid in a typical membrane.**
When supplies of the antioxidant vitamin E are adequate, the fatty acids are well protected, but if the supply of antioxidant is inadequate, peroxidation with its consequences is likely to occur.

the permeability of the lysosomal membranes increases, releasing lysosomal enzymes, and the activity of enzymes bound to the endoplasmic reticulum decreases as lipid peroxidation increases (*Fig.* 14.6).

Nevertheless the original observations on vitamin E deficiency, i.e. absorption sterility, have yet to be explained satisfactorily.

Selenium

In the 1960s it was observed by Schwarz in the USA that all animals required minute traces of selenium in their diet. If selenium was omitted by feeding very special diets, the animals showed signs of vitamin E deficiency, for

example lipid peroxidation increased in the tissue. Furthermore, some of the symptoms of selenium deficiency could be cured by feeding high concentrations of vitamin E and some of the symptoms of vitamin E deficiency were cured by feeding high concentrations of selenium. Although the deficiency symptoms were not identical, there was clearly a relationship between the two factors. This was ultimately traced to the enzyme glutathione peroxidase. This enzyme also protects cells against peroxide formation by destroying peroxides after they are formed, utilizing glutathione which becomes oxidized (*Fig.* 14.22). Glutathione peroxidase is thus another factor which protects cells against peroxides and it was eventually discovered that selenium forms part of the active centre of this enzyme, presumably as a seleno-amino acid, probably selenocysteine.

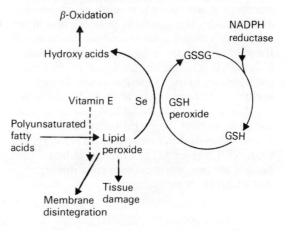

Fig. 14.22. **Glutathione peroxidase–Se–vitamin E.**

Chapter 15 Nutrition: inorganic constituents of the diet

15.1 Introduction

The inorganic constituents of the animal tissues, including human, are determined by the process of ashing, i.e. incinerating the tissue until all the organic matter is burned off. The material left behind contains large numbers of different metals and smaller numbers of anions such as sulphate, carbonate and chloride. It is generally believed that many metals play major roles in cellular function and these have been investigated very thoroughly, although several metals are found in the tissues for which no known function exists. These metals are believed to be absorbed, mainly in the diet, and are unavoidable contaminants of the environment. Some probably do little harm, whereas others such as lead are very toxic (cf. Chapter 35) and metals such as copper are important cellular constituents, but which are toxic in excess.

15.2 Metals found in the human body

Metals in the human body are listed in *Table* 15.1 and it will be seen that they are divisible into two groups, the major metals forming the main proportion of the total metal content of the body and trace metals which are found in low, or even minute, quantities. Iron, although usually placed in the trace metal group, really falls into an intermediate group between the major and trace metals. Typical metal contents in the human adult are shown in *Table* 15.2.

Table 15.1 **Metals in the human body**

| Major metals | Trace metals | |
	Established function	No established function
Na	Fe	Ni
K	Cu	Al
Ca	Mn	Sn
Mg	Co	Ti
	Zn	Pb
	Mo	Li
	Cr	Ba
		Sn
		V
		Ag
		Au
		Ce

Table 15.2 **Metal content of the human body**

Metal		Weight of metals, fat-free tissue, g/kg	Total body content, g
Major metals	Na	1·84	105
	K	3·12	245
	Ca	22·4	1050 (90% in skeleton)
	Mg	0·47	35
		(mg/kg fat-free tissue)	
Minor metals	Fe	74	3
	Cu	1·7	0·3
	Zn	28	0·1
	Mn	1	0·2
	Co	3 μg	120 μg

170

Calcium is the most abundant metal in the whole body, mainly on account of the high concentration in the skeleton, but potassium is more concentrated than other metals in the soft tissues (*Table* 15.1). Of the trace metals, only seven have been shown to possess clearly defined biological functions and these metals play vital roles, forming components of active enzymes. No function has yet been assigned to any of the long list of metals shown in the third column. For several of the metals in this category, it is extremely difficult to be certain whether or not a true function exists; the quantities in the tissues are very small and it is very difficult to carry out feeding experiments with all traces of the metal under study removed from the diet and removed from all materials which are likely to come into contact with the experimental animals. Unless a clearly defined deficiency disease can be shown to be associated with lack of a particular metal, it cannot be stated to be essential for life.

It should also be noted that several metals which are essential for life, such as copper and zinc in small quantities, are toxic when they gain access to the body in large quantities.

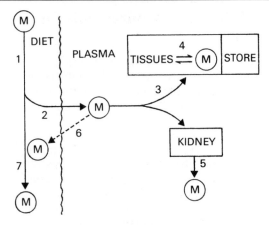

Fig. 15.1. **Factors affecting metal requirements.**
M = metal; (1) intake; (2) percentage absorption; (3) transfer to tissues, release from tissues; (4) store $\rightleftarrows$; (5) urine excretion; (6) re-excretion by bile; (7) excretion in faeces (7 = 1 − 2 + 6).

15.3 Factors affecting metal requirements

The daily dietary requirement of any metal is directly related to losses and the factors involved are illustrated in *Fig.* 15.1. Metals taken in the diet are not absorbed with equal efficiency. For some, such as sodium and potassium, the efficiency is very high but for others it is very low (*Table* 15.3).

After absorption, the metals are transported in the blood, often carried on a special protein, and then transferred to the tissues for use or kept in a stored form, sometimes bound to protein (*see Fig.* 15.1). Circulating metals absorbed from the digestive tract or released from the tissues can be lost in two ways, either by biliary excretion through the liver, or through the kidney.

Table 15.3 **Faecal loss of ingested metals**

Metal	Faecal loss, % intake
Na } K }	1–5
Ca	35–80
Mg	65
Fe	30–95
Zn	50
Co	80
Mn	97
Cu	99

Extensive biliary excretion occurs

The type of excretion varies with the metal. Most of the sodium and potassium is lost through the kidneys and a large proportion, normally about 50 per cent of calcium and magnesium, is also lost by this route. Very small amounts of trace metals, such as iron, copper or zinc, are excreted in the urine but the major proportions of these metals are lost in the faeces.

15.4 Dietary requirements for metals

A steady intake of all essential metals is required to balance the losses from the body. Under some circumstances, for example those of hot humid conditions, losses of sodium can increase substantially and must be compensated by increased dietary intake. Potassium is very widely distributed in many foods and no special measures are normally necessary to ensure adequate intake. Sodium chloride is incorporated into many foods during preparation, cooking or during eating. It is now believed that many individuals consume excess salt which must be excreted and, in older individuals, retention of salt can lead to oedema and high blood pressure. Excessive consumption of sodium chloride is not, therefore, desirable. Magnesium is also widely distributed in the diet and is a constituent of chlorophyll. Deficiency of this metal in the human diet occurs very rarely, if at all.

Some cases of zinc deficiency in humans have been occasionally reported. Zinc plays several vital roles in most cells because it is an essential component of several important enzymes, such as carbonic anhydrase, alcohol dehydrogenase, lactic dehydrogenase, carboxypeptidase and alkaline phosphatase. Rats deprived of zinc die within 2–3 weeks on a deficient diet. Taking all these aspects into consideration, only two metals are of major concern in the human diet both in children and adults: calcium and iron. These metals are discussed fully in Chapters 22 and 28.

15.5 Anions in the diet

The diet must contain certain essential anions. Of these, the most important quantitatively are chloride and phosphate, but smaller quantities of sulphate, carbonate, fluoride and iodide also occur. *Chloride* normally accompanies sodium as sodium chloride and *phosphate* is widely distributed in food. *Sulphate* and *carbonate* are metabolic end-products and not really essential in the diet although consumed incidentally.

Iodide is essential for the biosynthesis of the hormones triiodothyronine and thyroxine and is oxidized to iodine before being incorporated into the thyroid hormones (cf. Chapter 17).

In early periods, before the need for iodide was understood, many people living well away from the sea such as in Derbyshire, UK and in Switzerland, suffered severe iodine deficiency. This was because the main source of iodine was from the sea, either concentrated in sea weed or from fish which had stored it as a result of a natural food chain. Iodides are normally added to table salt and very little deficiency occurs currently in modern societies.

Fluoride is not an essential constituent of the diet but was accidentally found to be effective in the prevention of tooth decay. Studies, mainly in the USA and Canada, showed that children in areas where the fluoride content of the water was relatively

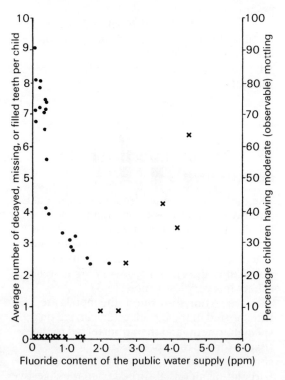

Fig. 15.2. **Relation of fluoride content of public water supply to dental caries and fluorosis.**
(× × × ×) Percentage fluorosis; (·····) mean DMF.

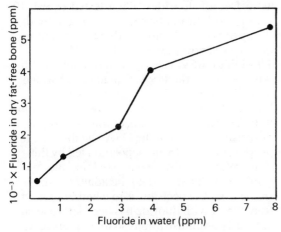

Fig. 15.3. **Relation of fluoride in drinking water to fluoride content of human bones.**

Table 15.4 **Effects of fluoride in different concentrations on man and experimental animals**

Concentration of fluoride, ppm	Vehicle	Effect
1 and over	Water	Dental caries reduction (in man)
2 and over	Water	Fluorosed enamel (in man)
8 and over	Water	Osteofluorosis (in man)
50 and over	Food or water	Thyroid changes (experimental animals)
100 and over	Food or water	Growth retardation (experimental animals)
125 and over	Food or water	Kidney changes (experimental animals)

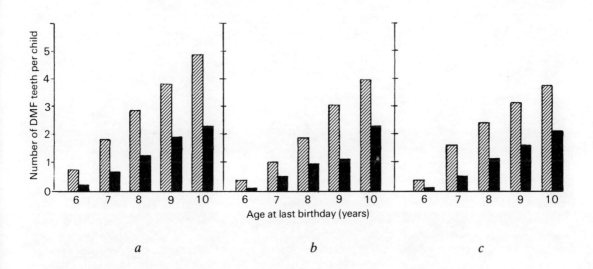

Fig. 15.4. **Effects of controlled fluoridation of water supplies on decay of teeth in children (6–10 years old) in three towns in the USA and Canada.**

1944–1945 is before fluoridation (▨) and 1953–1954 is 10 years after fluoridation (■).
(*a*) Grand Rapids; (*b*) Newsburgh; (*c*) Brantford.

high had far fewer decayed teeth than those in areas where fluoride was absent. If, however, the fluoride content of the water was high, a type of discoloration in the teeth occurred, called 'fluorosis', but this did not appear to have adverse effects on teeth or health (*Fig*. 15.2). The concentration of fluoride in bones also increased as the fluoride content of the water increased (*Fig*. 15.3). In both teeth and bones the F^- ion was believed to replace the OH^- in the crystal lattice. Experiments on aminals showed that fluoride could be toxic if added to the drinking water in concentrations of 50–100 ppm or more, but these were very much greater than the very small additions of 1 ppm required to prevent tooth decay (*Table* 15.4). These observations led to controlled experiments in which fluoride was added in a concentration of 1 ppm to the water supply of towns in the USA and Canada which

were free of fluoride. When tested over a 10-year period, dramatic reductions in the tooth decay of young children within the age group of 6–10 years were observed (*Fig*. 15.4).

This led to the advice by Government Health Authorities in the USA, Canada and Europe that all water supplies should be treated with fluoride to bring the concentration up to 1 ppm if the existing natural concentration was less. Surprisingly, both in the USA and the UK, this proposal led to strong opposition led by the 'Pure Water' lobby who insisted that the drinking water should not be adulterated.

In the UK, the decision as to whether or not to treat water with fluoride has been left to local authority administrations so that, at present, approximately half the water supplies are treated and half are not. Addition of fluoride is strongly supported by the dental profession and most scientific authorities.

Chapter 16 Digestion and absorption of foodstuffs

16.1 Foods digested and absorbed in man

Natural or prepared foods consumed by the average human adult are very complex mixtures, although the majority of the components of a typical human diet can be subdivided as shown in *Table* 16.1. Ingested foods can also be classified depending on whether they are absorbed unchanged, e.g. glucose, or whether they require extensive enzymic degradation as for muscle proteins.

Table 16.1 **Dietary components and absorption**

Dietary intake	Absorbed as
Protein	Amino acids Dipeptides
Carbohydrate	Monosaccharides (? Disaccharides)
Fat (triacylglycerols)	Monoacylglycerols Fatty acids
Vitamins	Vitamins (esters of vitamins)
Electrolytes	Electrolytes
Trace metals	Trace metals (complexes)

Details of most of the processes involved are described later in this chapter, but the absorption of metals, e.g. iron, vitamins and electrolytes, is discussed in later chapters specifically devoted to these topics.

Enzymic digestive processes are by now well understood, but the detailed mechanisms of the absorption of the digested fragments from the intestinal lumen into the blood have only been partially elucidated.

16.2 Biochemical changes in ingested foods and the role of the digestive organs

The digestive processes can be considered from two different viewpoints, biochemical or physiological. In the biochemical approach the processes of digestion and absorption of an ingested food, e.g. starch, are discussed; the physiological approach focuses attention on the role of each organ, e.g. the stomach or the pancreas, in the process.

It is desirable that the student should acquire a coordinated view of the whole digestive process and both approaches will be used in this chapter.

16.3 Enzymic processes involved in digestion

The majority of enzymes involved in the digestive process are hydrolases, i.e. they split bonds of esters, glycosides or peptides by addition of the elements of water:

Ester : $R\text{—}CO\text{—}OR' \xrightarrow{H_2O} R\text{—}COOH + R'\text{—}OH$

Glycoside : $Glucose\text{—}O\text{—}X \xrightarrow{H_2O} Glucose + X\text{—}OH$

Peptide : $\underset{R}{—CH—}NH—CO—\underset{R}{CH—} \xrightarrow{H_2O} \underset{R}{—CH—}NH_3^+ + {}^-OOC—\underset{R}{CH—}$

The digestive enzymes are included in Class 3 of the enzyme classification and catalyse reactions very similar to those occurring in the lysosomes (cf. Chapter 3). The two groups of enzymes are not, however, identical since most lysosomal enzymes have an optimum pH on the acid side of neutrality (pH 4·0–5·0), whereas most digestive enzymes are active within the pH range 6·5–7·5. It was noted in Chapter 3 that the lysosomes constitute a primitive digestive system and have a function in unicellular organisms identical to that of the much more complex digestive tract in mammals.

Many of the digestive enzymes have trivial names, such as pepsin, trypsin or amylase, because they were the first enzymes to be discovered, and their discovery occurred long before systematic nomenclature had come into existence. Most of these names have been retained in common use, since they are of such long standing.

The powerful hydrolytic enzymes of the digestive tract catalyse the degradation of large macromolecules of food, e.g. starch or protein, into small molecules, e.g. glucose or amino acids, that can then be readily absorbed.

16.4 The digestive secretions

The enzymes involved in digestion are present in a number of secretions released at specific locations into the digestive tract.

a. Saliva

Saliva is secreted into the mouth from three main pairs of glands: the submandibular pair, the parotid pair and sublingual pair. Approximately 1500 ml saliva are secreted daily in man. It is a dilute secretion containing only 0·3–1·4 per cent solid material that is composed mainly of protein, proteoglycans, glycoproteins and electrolytes.

The composition of the secretions from different glands is not identical, a greater proportion of glycoproteins and proteoglycans being secreted by the submandibular glands. One important enzyme contained in the secretion is salivary amylase and this hydrolyses food starch into maltose. The major function of the saliva is, however, to lubricate the food by means of its proteoglycans and glycoproteins and thus render swallowing easier. It also protects the epidermal cells.

Saliva, by digesting starch particles that may be lodged in the teeth, may play an important role in cleansing the teeth. In support of this concept, the secretion of saliva by patients who have received X-ray treatment for tumours in the mouth region is often severely impaired and this can rapidly lead to severe tooth decay. The buffering action of saliva is also important in neutralizing acids produced by bacteria which also helps to prevent caries.

b. Gastric juice

Gastric juice is a composite secretion from the parietal cells; these cells both secrete hydrochloric acid and are the chief cells for pepsinogen secretion. The juice normally contains about 1·1 per cent non-aqueous material; of this 1·1 per cent hydrochloric acid of concentration 170 mM makes up about one-half by weight. Consequently, gastric juice is the most acid secretion in the body. The remaining solids are pepsinogen, lipase and glycoproteins.

Some of the glycoproteins present are large molecules of molecular weight $1·5 \times 10^6$–$2·0 \times 10^6$ and contain 80–85 per cent carbohydrate. Amongst the smaller glycoproteins (mol. wt 45 000–60 000) is the intrinsic factor which plays a very important role in the absorption of vitamin B_{12} (cf. Chapter 24).

The main function of the gastric juice is to provide both the powerful proteolytic enzyme, pepsin, and an acid environment for its action. Pepsin is extremely unusual in having such an acid optimum pH (≈ 1.5–2.0) and it will attack nearly all proteins in their native state.

The mechanism of hydrochloric acid secretion has been controversial for many years and is still not fully understood. An early mechanism, based on the observation that carbonic anhydrase occurs in high concentrations in parietal cells, proposed that catalysis of the $H_2O + CO_2 \rightarrow H_2CO_3$ reaction was of fundamental importance, the H^+ then being released by the ionization $H_2CO_3 \rightarrow H^+ + HCO_3^-$.

More recently, attention has focused on the possible role of oxidative energy metabolism in the generation of acidity, H^+ secretion being considered as directly coupled to a redox system. In essence the theory is based on a concept very similar to that proposed for the coupling of oxidation and phosphorylation in the chemiosmotic theory (Chapter 6), but the mechanism must also involve the transfer of H^+ from the mitochondria to the stomach lumen.

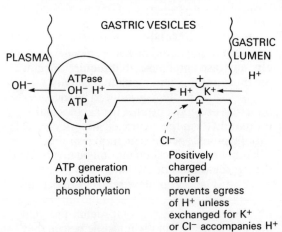

Fig. 16.1. **Possible mechanism of the generation of acidity in the stomach utilizing a membrane ATPase.**

Alternatively, a membrane ATPase may be involved as illustrated in *Fig.* 16.1. The ATPase could be activated by the exchange of H^+ for K^+. The activation of ATPase couples ADP with phosphate, generating ATP with the release of H^+ and OH^-; this reaction will require energy from the electron transport system (cf. Chapter 6). The positively charged barrier will prevent egress of H^+ unless exchanged for a K^+ or accompanied by a Cl^-. Chloride ions may be provided by exchange with HCO_3^- at the oxyntic cells and generation of HCO_3^- by carbonic anhydrase could also be a role of this enzyme.

c. Bile

Bile secreted by the human liver is a clear brownish-yellow or green fluid and about 500 ml are produced daily by the human adult. It is stored in the gallbladder where water is removed and the total solid content proportionally increased (*Table* 16.2).

Table 16.2 **Constitution of bile**

Constituent	Percentage composition	
	Liver	Gallbladder
Water	96.5–97.5	83–90
Total solids	2.5–3.5	10–17.5
Bile salts	1.0–1.8	6–11
Mucopolysaccharides	0.4–0.5	1.5–3.0
Bile pigments Cholesterol Phospholipids	0.2–0.4	0.5–5.0
Electrolytes	0.7–0.8	0.6–1.0

The bile serves two functions: an excretory function and a digestive function.

i. Excretory function

The major excretory products are the bile pigments formed by the degradation of haemoglobin and are described in Chapter 27. In addition, insoluble solid material, such as sand, soot or metal particles that accidentally enter the blood stream, are disposed of through the bile. The particles are transported to the liver where they are taken up by the Kupffer cells; they then enter the lysosomes, any digestible material is hydrolysed by the hydrolytic enzymes and undigested material is then excreted into the bile by reversed pinocytosis (Chapter 3).

Cholesterol, in excess, may also be regarded as an excretory product and several drugs and their metabolites are excreted in the bile (cf. Chapters 29 and 36).

ii. Digestive function

Bile plays two important roles in digestion. Firstly, due to its bicarbonate content it is an alkaline secretion with a pH as high as 8·0, and it is therefore important in neutralizing the acidity of the gastric juice and in increasing the pH of the intestinal contents to a value close to pH 6·5–7·0. The enzymes of the pancreatic juice and the intestine can then function under optimal conditions. Secondly, bile is essential for the digestion and absorption of fats. Bile salts, together with the phospholipids, emulsify the fats into small droplets so that they are readily attacked by pancreatic lipase. They then aid absorption by taking part in micelle formation together with the monoacylglycerols formed by digestion.

d. Pancreatic juice

The pancreatic juice is secreted from the pancreas by the duct joining the common bile duct at the ampulla of Vater, so that both the pancreatic juice and the bile mix before entering the intestinal tract.

The pancreatic juice contains about 1·8 per cent solid material; it is, in fact, really composed of two separate secretions: (a) the electrolyte secretion ($\approx 1\cdot2$ per cent solid) and (b) the organic and mainly enzyme protein secretion ($\approx 0\cdot1$ per cent solid). The separate identity of these secretions is indicated by the fact that they are under entirely different hormonal controls, as described below.

The positive ions of the juice are mainly Na^+, K^+ and Ca^{2+} at concentrations close to that of the plasma. The concentration of HCO_3^- is normally about 80 mM, but it can increase to 135 mM; it has an important regulatory effect on the pH of the intestinal contents, pancreatic juice being capable of neutralizing an equal volume of gastric juice.

Table 16.3 **Digestive secretions**

Pancreatic juice	Substrate
Chymotrypsin(ogen) Trypsin(ogen) Elastase Carboxypeptidase	Proteins
Amylase	Starch
Lipase Esterase Phospholipase Cholesterol esterase	Glycerols/Esters
Ribonuclease Deoxyribonuclease	Nucleic acids

The pancreatic juice supplies a large range of powerful hydrolytic enzymes capable of digesting the majority of ingested foodstuffs and these are listed in Table 16.3. The enzymes are synthesized in the rough endoplasmic reticulum and are then transported to the Golgi apparatus where carbohydrate residues are added to form glycoproteins and proteoglycans. Metals, such as Ca^{2+}, Mg^{2+} and Zn^{2+} are sequestered by some enzymes, e.g. amylase and lipase, that require calcium for their stability and maximum activity.

The proteins produced by the pancreas are concentrated in vacuoles and ultimately form the 'zymogen granules'. Lipoproteins, special structural proteins, calcium and magnesium may all be involved in the aggregation process.

These granules are so called because of the inactive state of most of the enzymes, i.e. they are 'zymogens' or 'proenzymes' that must be activated in the intestine before they can carry out their catalytic functions. The granules move along a microtubule system and then pass out of the cells by exocytosis.

16.5 Control of digestive secretions: the gastrointestinal hormones

The gastrointestinal hormones are polypeptides that play an important part in the regulation of the secretions of gastric and pancreatic juices, and of bile. During the passage of a meal they are released from the gastrointestinal tract, absorbed and passed through the blood stream to act on their respective target organs. The distribution of the hormones through the length of the gastrointestinal tract is shown in *Fig.* 16.2.

a. Secreted in the stomach

Gastrin

Gastrin release is stimulated by the presence of food in the stomach. It is secreted by the antral region of the gastric mucosa and by the duodenal and jejunal mucosae. Several different forms of gastrin exist: big gastrin (34 amino acid residues), little gastrin (17 amino acid residues) and minigastrin (14 amino acid residues). The secretagogue activity resides in the terminal five amino acid

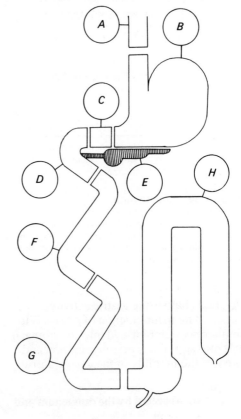

Fig. 16.2.   **The distribution of hormones in the gastrointestinal tract.**

Hormone	Amount in tissue, pmol/g, at						
	A	B	C	D	E	F	G
1. Gastrin	0·1	23·5	2342	1397	190	62	0·1
2. Secretin	0·1	0·1	0·1	73	32	5	0·1
3. Choleocystokinin–pancreozymin	0·6	0·6	2·5	62·5	26	3	0·6
4. Enteroglucagon	0·6	0·6	0·6	10	45·7	220	35
5. Gastric inhibitory polypeptide	0·2	0·2	404	71	62	2·4	0·2
6. Somatostatin	0·1	89	310	210	110	40	20
7. Vasoactive intestinal polypeptide	94	44	46	106	61	78	136
8. Pancreatic polypeptide			Mainly in the pancreas				

A = oesophagus.   E = pancreas.
B = stomach.   F = jejunum.
C = pyloric sphincter.   G = ileum.
D = duodenum.   H = colon.

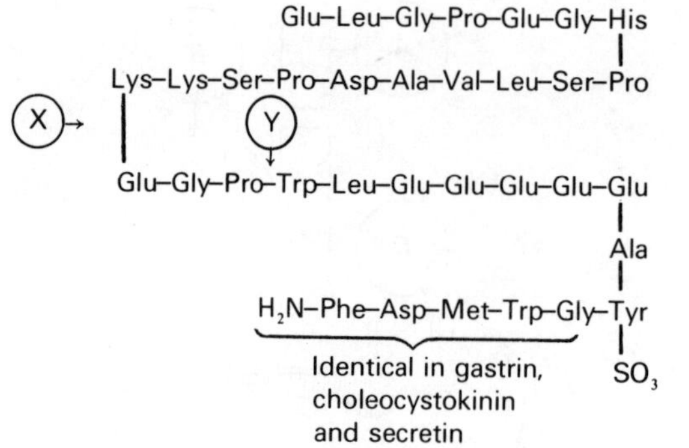

```
                    Glu–Leu–Gly–Pro–Glu–Gly–His
                                              |
           Lys–Lys–Ser–Pro–Asp–Ala–Val–Leu–Ser–Pro
    (X)→        |            (Y)
                    Glu–Gly–Pro–Trp–Leu–Glu–Glu–Glu–Glu–Glu
                                                            |
                                                           Ala
                                                            |
                H₂N–Phe–Asp–Met–Trp–Gly–Tyr
                 _____/    |
                    Identical in gastrin,      SO₃
                    choleocystokinin
                    and secretin
```

Fig. 16.3. **Structure of gastrin.**
Big gastrin = whole sequence (34 residues); little
gastrin = sequence from X to terminal —NH₂;
minigastrin = sequence from Y to terminal —NH₂. Note
the folding shown in the molecule is solely to permit fitting
sequence into space on the page.

sequence and synthetic pentagastrin
(*Fig.* 16.3) has strong gastric activity.

The main function of gastrin is to
stimulate secretion of acid into the stomach,
but it also stimulates pepsin secretion and
increases the mobility of the gastric antrum.

b. Secreted by the duodenum and jejunum

i. Secretin

Secretion of secretin is stimulated by the
presence of acid in the duodenum and
jejunum and is released from the duodenum
and the jejunal mucosa. It is a polypeptide
composed of 27 amino acid residues and its
major functions are to stimulate pancreatic
secretion of water and electrolytes, to
potentiate the effect of cholecystokinin–
pancreozymin on the pancreas, to increase the
secretion of bicarbonate by the liver, and to
delay gastric emptying.

Its structure is very similar to that of
glucagon. It possesses no active fragments
(*Fig.* 16.4) and must, therefore, depend on the
tertiary structure for activity.

ii. Choleocystokinin–pancreozymin

This hormone is produced by the duodenal and
jejunal mucosae; it is released by distension of
these organs and by the digested products of
fat and protein in the duodenum and jejunum.
It is composed of 33 amino acids, and the
entire activity is found in the C-terminal eight
amino acids. Five of these amino acids are
identical with the five terminal amino acids of
gastrin.

The main function of this hormone is
to stimulate enzyme secretions from the
pancreas and to cause contraction of the
gallbladder.

iii. Enteroglucagon

Enteroglucagon is secreted to a limited extent
by the stomach mucosa, but it is also secreted
from the mucosa of the small intestine and
colon. As its name implies, it is very similar in
structure (or even identical) to glucagon itself
(Chapter 17). Its main functions are to delay
the passage of food from the stomach, through
the intestine and to release pancreatic insulin.
It also serves to stimulate the growth of
mucosal cells.

Position:	1	2	3	4	5	6	7	8	9	10	11	12	13	14
VIP	H-His	Ser	Asp⁻	Ala	Val	Phe	Thr	Asp⁻	Asn	Tyr	Thr	Arg⁺	Leu	Arg⁺
Secretin	H-His	Ser	Asp⁻	*Gly*	Thr	Phe	Thr	Ser	Glu⁻	*Leu*	*Ser*	Arg⁺	Leu	Arg⁺
Glucagon	H-His	Ser	Gln	*Gly*	Thr	Phe	Thr	Ser	Asp⁻	Tyr	*Ser*	Lys⁺	Tyr	Leu

Position:	15	16	17	18	19	20	21	22	23	24	25	26	27	28	29
VIP	Lys⁺	Gln	Met	Ala	Val	Lys⁺	Lys⁺	Tyr	Leu	Asn	Ser	Ile	Leu	Asn	(NH₂)
Secretin	Asp⁻	Ser	Ala	Arg⁺	*Leu*	Gln	*Arg⁺*	Leu	Leu	*Gln*	Gly	*Leu*	Val	(NH₂)	
Glucagon	Asp⁻	Ser	Arg⁺	Arg⁺	Ala	Gln	Asp⁻	*Phe*	*Val*	Gln	Trp	*Leu*	Met	*Asn*	Thr

Fig. 16.4. **Structures of secretin, glucagon and vasoactive intestinal octacosapeptide (VIP).**

iv. Gastric inhibitory peptide

Gastric inhibitory peptide, a polypeptide composed of 43 amino acids, is secreted by the duodenal and jejunal mucosae as a result of the action of digested products in the duodenum and jejunum. It inhibits gastric acid secretion, gastric emptying and causes the release of insulin.

v. Pancreatic polypeptide

Pancreatic polypeptide is produced by the pancreas and inhibits pancreatic enzyme secretion and gallbladder contraction. These hormones are all distributed by the bloodstream but some hormones are produced locally in the gastrointestinal tract, possibly at nerve terminals and they act locally. In this group are the following:

Somatostatin (growth hormone release inhibiting hormone)

This hormone is found in the hypothalamus, the pancreas and in the gastrointestinal tract and is discussed in Chapter 17. It has many complex actions on the gastrointestinal tract, including inhibition of release of gastric acid, pepsin and gastrin from the stomach, and it thus acts as a powerful anti-acid. It also inhibits the release of glucagon, insulin and exocrine secretions from the pancreas and the release of motilin, enteroglucagon and gastric inhibitory peptide from the duodenum and jejunum.

Vasoactive intestinal polypeptide

Vasoactive intestinal polypeptide is a polypeptide composed of 28 amino acids and is secreted along the whole length of the gastrointestinal tract. Like gastric inhibitory peptide and somatostatin, it inhibits gastric secretion, but stimulates intestinal blood flow and secretion from the small intestine (*Fig.* 16.4).

In summary, it will be seen that this group of hormones is very important in the regulation of digestion and absorption over the whole gastrointestinal tract. This regulation is accomplished by two main methods:

a. By regulating the quantities of the various secretions of the gastrointestinal tract and, therefore, the concentrations of enzymes to which the various foodstuffs are exposed.

b. By regulating the rate of movement of foodstuffs along the gastrointestinal tract, this in turn controlling (i) the time for which any particular foodstuff is exposed to the enzymes and (ii) the time available during which digestive products are exposed to the intestinal mucosa for absorption.

16.6 Methods of studying absorption from the intestine

a. Animal experiments

Short lengths of intestine of small animals, e.g. rats or guinea-pigs, can be tied at each end to form small sacs which can be used to study absorption of many digested fragments, e.g. glucose or amino acids. The technique is greatly improved by 'everting' the sacs (*Fig.* 16.5), so that the substance under study can be absorbed into a relatively small volume.

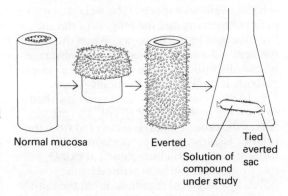

Normal mucosa   Everted   Tied everted sac

Solution of compound under study

Fig. 16.5. **Method of studying intestinal absorption using 'everted sacs'.**

b. Human experiments

Although experiments using animal tissues *in vitro* are valuable, it is much more important to study absorption in man under normal physiological conditions. This can be carried

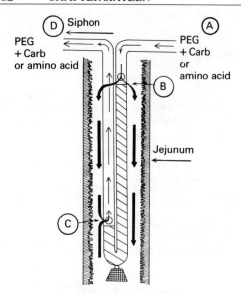

Fig. 16.6. **Method of studying intestinal absorption in man.**
PEG = polyethylene glycol.

$$\text{Amount absorbed} = \text{concn in} - \text{concn out} \times \frac{\text{PEG}_{in}}{\text{PEG}_{out}} \times \text{flow rate}$$

out by the method developed by Dawson and Holdsworth and is shown in *Fig.* 16.6. The subject swallows a special tube weighted with a pellet of mercury and the progess of the end of the tube is followed by X-ray. When the tube has reached a suitable location, the substance whose absorption is to be studied, for example glucose, is pumped in at the point A (*see Fig.* 16.6), mixed with an inert non-absorbed carrier, e.g. polyethylene glycol. The mixture leaves the tube at B and is recovered from the inlet C. Clearly, only a proportion of the infused mixture will be regained at exit D, but this proportion will be measured by the polyethylene glycol regained. From the initial and final concentration of the substance under study, the absorption can be calculated.

16.7 Absorption mechanisms

The absorption of digested food components follow the general pattern described in Chapter 2. These components thus reach the plasma by one of the following three processes.

a. Simple diffusion

A large concentration of a foodstuff may reach the plasma by simple diffusion down a sharp concentration gradient; no special absorption mechanism is necessary. This occurs, for example, in the stomach when a relatively concentrated solution of glucose is imbibed.

b. Facilitated diffusion

Studies of intestinal absorption have shown that this is an important phenomenon. This process, the absorption process, is selective, but no energy is needed for the system since the substance moves down a concentration gradient from a high concentration in the gut lumen to a lower concentration in the mucosal cells.

Selective carrier molecules may be involved and are clearly important in regulating the uptake of molecules into the plasma (*Fig.* 16.7).

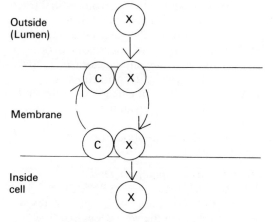

Fig. 16.7. **Concept of carrier molecules.**
X molecule to be absorbed; C carrier. Properties of the carrier molecules: (i) specificity; (ii) a system that can be saturated; (iii) competitive inhibition; (iv) active transport and facilitated diffusion; (v) show Michaelis–Menten kinetics.

c. Active transport

This process is highly specific but also requires energy, usually in the form of ATP, so that the low concentration of the molecules can be absorbed against a concentration gradient. A typical example is the absorption of glucose which is described later.

16.8 Carbohydrate digestion and absorption

A typical daily intake of carbohydrate for a human adult is shown in *Table* 16.4.

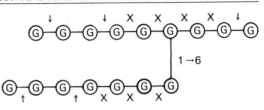

Table 16.4 Typical daily carbohydrate intake for a human adult consuming 2600 cal/day

	Intake	
Carbohydrate	g/day	%
Polysaccharides		
Starch	200	64
Glycogen	1	0·5
Disaccharides		
Sucrose	80	26
Lactose	20	6·5
Maltose	≈0	≈0
Monosaccharides		
Fructose	10	3
Glucose	≈0	0

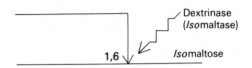

Fig. 16.8. **Hydrolysis of branched-chain polysaccharides by amylase and isomaltase.**
Note that the bonds in the vicinity of the 1,6-branching points are resistant to amylase, but hydrolysis of the 1,6-link by isomaltase releases unbranched chains. ↓ Split by amylase; X resistant to amylase; Ⓖ glucose unit.

For practical purposes, it will be noted that the human digestive system is concerned mainly with the digestion and absorption of starch, sucrose, lactose and, usually, some fructose.

The ingested starch is first attacked by an amylase in the saliva and later by a very similar amylase secreted in the pancreatic juice. The amylase attacks and almost completely digests the unbranched component of starch (amylose) which contains only 1:4 linkages to maltose; the branched component (amylopectin) is resistant to digestion around the 1:6 branching links (*Fig.* 16.8).

The value of salivary amylase in digesting starch has often been considered to be negligible, its role being primarily involved with teeth cleansing as described earlier. This view is based on the fact that, after swallowing, food rapidly comes into contact with the acid environment of the stomach, so that the salivary amylase, which is most active at pH 7·0, is inactivated. In practice, however, the role of the enzyme may be more important in the digestion of starch than is generally assumed. Its activity will depend on the time and extent to which the food is chewed and on

the type of bolus swallowed. If this is relatively compact, the acid of the stomach may take some time to penetrate it so that digestion of the starch can continue after the food has entered the stomach. Experiments on volunteers have demonstrated that the percentage of starch digested may reach 40 per cent, but is usually within the range 15–20 per cent.

After the food leaves the stomach, the remaining starch is attacked by pancreatic amylase. In addition, the highly branched starch fragments formed by amylase, sometimes known as 'limit dextrans', are hydrolysed by isomaltase which splits the 1:6 glucose linkages (*Fig.* 16.8). Digestion of short fragments containing 1:4 linkages may then be continued by amylase so that nearly all the original starch is converted to maltose. A small quantity of glucose is also formed.

After a mixed meal, the intestine will, therefore, contain a mixture of disaccharides, maltose, lactose and sucrose, although some non-enzymic hydrolysis of sucrose can occur in the acid conditions of the stomach. The disaccharase enzymes located in the brush border of microvilli (sucrase, maltase, and β-galactosidase or lactase) hydrolyse the disaccharides into glucose + fructose, glucose and glucose + galactose, respectively. It is not absolutely clear whether hydrolysis of the disaccharides occurs prior to absorption or whether the disaccharides are absorbed and then hydrolysed immediately after absorption. It is, nevertheless, clear that the intestine possesses specific absorption mechanisms for the monosaccharides glucose, fructose and galactose and this can be demonstrated by feeding or infusing the free monosaccharides.

The rate of absorption of glucose, fructose or galactose depends on their infused concentrations in the intestine and the kinetics of their absorption rates are very similar to those described for enzyme actions described by Michaelis–Menten kinetics (Appendix 13a). A translocation rate (equivalent to ν_1) is related to the concentration infused as shown

in *Fig.* 16.9 and a constant equivalent to the Michaelis constant (K_m) can be calculated. This will effectively give a measure of the affinity of the sugar for the transport system. The glucose transport system has been intensively studied and two important concepts have been described which are probably of more general application to other absorbed foodstuffs.

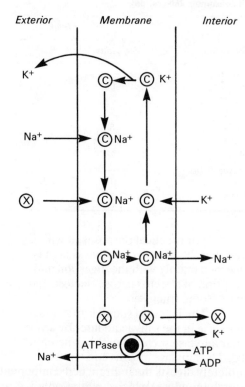

Fig. 16.11. **Coupling of absorption of glucose with electrolyte transport.**
In the model the carrier (ⓒ) is assumed to have a low affinity for the solute transported, for Ⓧ (glucose), until bound to sodium when the affinity increases to a marked extent so that Ⓧ is bound to ⓒ. The Ⓧ—ⓒ—Na^+ complex is then transferred to the membrane adjacent to the interior of the cell. It is not at all certain that the carrier actually moves across the membrane and a conformational change may occur, so that Na^+ and X can pass into the interior of the cell where both Na^+ and the substrate are released. The carrier in the new conformation readily accepts K^+ from the interior of the cell, transferring it to the exterior. The membrane ATPase will transport K^+ back to the interior of the cell and Na^+ to the exterior and the energy derived from hydrolysis of ATP is the driving energy for the absorption of the substrate Ⓧ.
Note: ATP may also be involved in activation of the carrier ⓒ or the transport process. Model adapted from proposal by Crane for glucose transport.

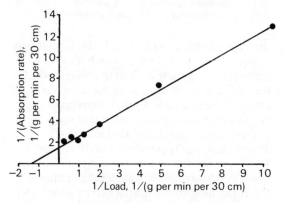

Fig. 16.9. **Application of Michaelis–Menten kinetics to glucose absorption in man using the Lineweaver–Burke plot.**

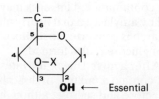

Fig. 16.10. **Specificity of glucose transport systems.**
Note effect of size of X (transport depends on size of group). 3-Methylglucose—active; 3-ethylglucose and 3-propylglycose are inactive.

a. The transport or carrier system has a high degree of specificity as shown in *Fig*. 16.10.

b. The transport of glucose is linked to Na^+ and K^+ transport; transport of a molecule of glucose from the lumen is accompanied by a parallel transport of Na^+ and an antiparallel transport of K^+. The importance of ion transport for glucose transport has been demonstrated by the action of the drug ouabain which specifically inhibits the ATPase responsible for Na^+–K^+ exchange across membranes.

These observations have led to the proposal by Crane of a coupled transport system for glucose, sodium and potassium as shown in *Fig*. 16.11.

16.9 Protein digestion and absorption

Proteins undergoing digestion

As discussed in Chapter 12, adult man requires only 40–45 g protein per day, but typically ingests about 90 g. This forms the exogenous intake, but it should be remembered that the endogenous protein in the gastrointestinal tract, the digestive enzymes themselves, forms a substantial proportion of the total protein undergoing digestion. It is especially large in some animals, such as the dog or rat, but in man is normally about 10–12 g protein for an intake of 30 g protein in a typical meal.

Characteristics of proteolytic enzymes

All proteolytic enzymes show the following characteristics:

i. They are secreted as inactive zymogens

ii. They hydrolyse the peptide linkage liberating free NH_3^+ and COO^- groups

iii. They are classified as 'endo' or 'exo' peptidases depending on whether they attack peptide linkages within the protein molecule (endo) or split off terminal amino acids (exo) (*Fig*. 16.12)

iv. They are specific for peptide linkages which form part of certain amino acids.

Gastric mucosa→Pepsin

Intestinal mucosa→ { Dipeptidases / Aminopeptidases

Pancreas { Trypsin / Chymotrypsin / Carboxypeptidases (A, B) / Aminopeptidases

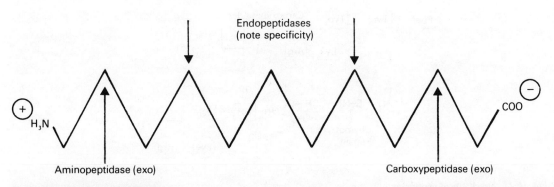

Fig. 16.12. **Endo- and exopeptidases.**
Pepsin and chymotrypsin split peptide bonds adjacent to aromatic amino acids. Trypsin splits peptide bonds adjacent to 'basic amino acids', such as arginine or lysine and carboxypeptidase splits off the terminal amino acid possessing a free carboxyl group. Type A carboxypeptidase has a similar specificity to pepsin and type B carboxypeptidase has a specificity similar to trypsin.

Zymogens

Pepsin is secreted as inactive pepsinogen in the stomach, which is converted to active pepsin first by the acidity of the stomach and then by an autodigestive effect of pepsin itself. Several large peptides are liberated. Not all have been identified, but one peptide contains 29 amino acids and another contains 12 amino acids.

$$\text{Pepsinogen} \rightarrow \text{Pepsin} + \text{Peptides}$$
$$M_r = 40\ 000 \qquad M_r = 35\ 000$$

Chymotrypsinogen and trypsinogen, secreted in the pancreatic juice, are activated by a special enzyme of the intestinal mucosa, enterokinase, and also by trypsin itself. The activation of chymotrypsinogen has been studied in detail and involves the splitting off of two dipeptides enabling an active site to be formed. Trypsinogen is converted to trypsin with the splitting of a hexapeptide, the process being autocatalytic. The precursors of the carboxypeptidases are described as pro-carboxypetidases. Their activation is much more complex than that of chymotrypsinogen or trypsinogen and involves a molecular weight reduction of about 10 000 from 44 500 to 34 000.

Endo- and exopeptidases

There are two types of exopeptidase. Aminopeptidases split off amino acids at the N-terminal part of the peptide chain, whereas carboxypeptidases split the terminal amino acid with a free COO^- group. Endopeptidases attack peptide links within the main peptide chain (*see Fig.* 16.12).

Specificity

All endopeptidases, and also some exopeptidases, exhibit a high degree of specificity. This is, directed by the side chain of the amino acids adjacent to the peptide linkage being hydrolysed and is illustrated in *Fig.* 16.13.

Mechanisms of protein digestion

The action of pepsin in the stomach followed by that of trypsin, chymotrypsin and exopeptidase in the intestine which rapidly digest the ingested proteins into a mixture of amino acids and many short-chain molecules, e.g. dipeptides. The intestinal mucosa possesses many dipeptidases, although it is not clear how many enzymes exist and how specific each dipeptidase is. Dipeptidases so far studied usually appear to be specific for one amino acid and will, for example, hydrolyse leucine aminopeptides. Theoretically, considering just two amino acids, 380 different dipeptides exist and it is not impossible that a separate peptidase exists for each dipeptide. These dipeptidases can thence finally convert all the ingested protein into free amino acids.

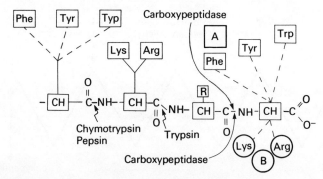

Fig. 16.13. **Specificity of peptidases.**
Pepsin and chymotrypsin will attack peptide links adjacent to aromatic residues, whereas trypsin attacks peptide bonds close to positively charged lysine and arginine residues. Analogous specificity is shown by the carboxypeptidases that are described as belonging to the A or B group.

Absorption of amino acids

The problem of amino acid absorption involving 20 different amino acids is clearly much more complex than that of monosaccharides, where only three are of major importance and, despite intensive studies on animals and man, the precise mechanisms of amino acid absorption remain obscure.

The following facts have, however, been established:

i. Absorption of amino acids obeys Michaelis–Menten kinetics and V_{max} and K_m can be established (*Fig.* 16.14).

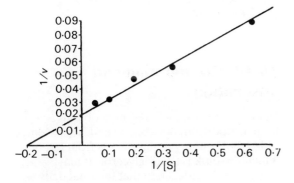

Fig. 16.14. **Lineweaver–Burke plot for the absorption of lysine in man.**
$K_t = 4\cdot9$ mM (equivalent to K_m for an enzyme);
$V_{max} = 51\cdot3\ \mu mol \cdot min^{-1}$ per 30 cm.

ii. Amino acids are absorbed at different rates and some examples are given in *Table* 16.5.

iii. The natural or L series of amino acids are absorbed much more quickly than the unnatural D series, but the stereospecificity of absorption is not absolute.

iv. For those amino acids which have been studied, Na^+ and K^+ translocations occur as described for glucose (*Fig.* 16.11).

v. Some amino acids compete with each other for absorption. This has led to the concept of a limited number of different absorbing sites each of which is specific for a number of amino acids. Although this concept is so far not precisely defined there are likely to be several different sites for amino acid absorption.

a. 'Neutral site'—monoamino, monocarboxylic acid site

b. Dibasic site—specific for lysine, arginine, ornithine and cystine

c. Proline and hydroxyproline site

d. Glycine site
The requirements of amino acid structure for absorption by the neutral site are illustrated in *Fig.* 16.15.

Table 16.5 **Protein absorption in man**

	Amino acid	Time, min*	Amino acid	Time, h*
Free state	Arginine	66	Alanine	11·6
	Lysine	72	Histidine	11·9
	Methionine	104	Isoleucine	13·0
	Tyrosine	108	Serine	13·6
	Valine	175		
	Leucine	358		
			Threonine	23·6
	Unlikely to be released in free state		Glutamic acid	45·5
			Aspartic acid	71·0
			Proline	74·5
			Glycine	98·5

* Time measured is the time for release of all amino acids from a milk protein meal.

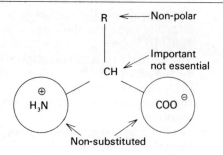

Fig. 16.15. **Structural requirements for absorption of neutral amino acids.**

Although no detailed mechanisms for absorption of amino acids from the intestinal lumen have been proposed, a mechanism has been suggested for amino acid absorption from the kidney tubules. This is known as the γ-glutamyl cycle and is described in Chapter 30.

There are many similarities between absorption from the intestinal lumen and from the kidney tubules and it is possible that a mechanism of this type may occur in the intestinal wall.

Absorption of peptides

Although it is well established that the small intestine is capable of absorbing all amino acids, it has recently been established that, in addition, dipeptides are also absorbed.

The evidence for this came from two directions. Firstly, studies of the rate of digestion of food proteins in the gut and analysis of digestion products indicated that it was very unlikely that complete digestion of amino acids occurs in normal digestion. Release of free amino acids varied among the amino acids (*Table* 16.5). Secondly, conclusive support for the view that absorption of dipeptides was an important pathway was obtained from the study of patients with a genetic disease known as 'Hartnup disease'. These patients were unable to absorb free histidine, tryptophan, leucine, serine, phenylalanine or methionine, but they were able to absorb dipeptides containing these amino acids, e.g. alanyl-histidine or glycyl-tryptophan, at a normal rate. This proves

that there must be separate and independent absorbing sites for dipeptides distinct from the amino acid sites. It is uncertain how many dipeptide sites exist, but different rates of transfer have been established for dipeptides such as glutamyl-leucine and leucyl-glutamate. Hartnup patients show similar absorption defects in both the intestine and the kidney.

Absorption of intact proteins

Under certain conditions and in certain individuals, intact proteins can be absorbed. These proteins often cause undesirable immunological responses and are responsible for the symptoms of food allergies (cf. Chapter 42).

16.10 Fat digestion and absorption

In most Western societies, man obtains a very large proportion, between 40 and 50 per cent, of his daily energy requirements from fat, of which the bulk is triacylglycerol. It may, therefore, be calculated that the total daily intake of triacylglycerol is about 150 g. Other lipid-soluble components form a much smaller proportion of the total lipid intake (*see Table* 16.6). As discussed in Chapter 13, cholesterol is not desirable in the food, but cannot be completely avoided. The fat-soluble vitamins A, D, E and K are, however, very important dietary constituents and if fat absorption is impaired, symptoms of vitamin

Table 16.6 **Typical dietary fat consumed by an adult taking in a total of 3200 cal**

Fat	Daily intake, g
Triacylglycerols	150
Cholesterol	0·5–1·0
Phospholipids	Small quantities
Fat-soluble vitamins (A, D, E, K)	Small quantities

Fat normally provides about 45–50% of the total calorie intake and each gramme oxidized will supply approximately 9·0 calories.

deficiences may follow as a consequence (cf. Chapters 13 and 14).

The digestion of triacylglycerol presents a problem not posed by other constituents of the diet, since these molecules are completely immiscible with water. Churning of the food in the stomach helps to disperse the fatty globules and, although it is generally believed that a gastric lipase exists, it is very doubtful if it plays any significant role in fat digestion.

Before fat digestion can occur, it must be dispersed in fine droplets as an emulsion and this is accomplished by the secretions of the bile. After concentration in the gallbladder, human bile contains between 5 to 10 per cent bile acids. These are composed of a mixture of trihydroxy acids, the cholic acid group and the dihydroxy acids, the chenocholic or deoxycholic acid group. Conjugates of glycine or taurine are formed in the liver from these acids (*Fig.* 16.16). The biosynthesis of bile is described in detail in Chapter 29.

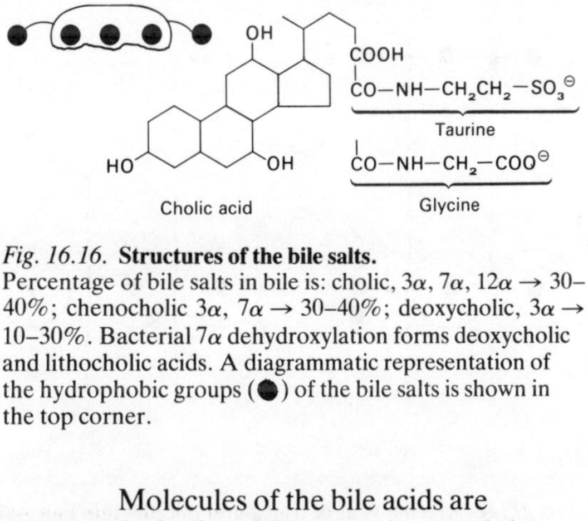

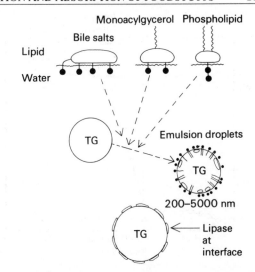

Fig. 16.17. **Role of bile salts, monoacylglycerols and phospholipids in the emulsification of triacylglycerols.**

Molecules of the bile acids are composed of a lipophilic part, the sterol ring, and a hydrophilic part, the three hydroxyl groups and the side chain. As a consequence, bile acids are amphipathic molecules and powerful emulsifying agents. Phospholipids composed of two lipophilic fatty acid chains, hydrophilic phosphate and a hydrophilic molecule, e.g. choline, are also secreted in the bile and, in conjunction with the bile acids, emulsify the triacylglycerols into small droplets of 200–5000 nm in diameter (*Fig.* 16.17).

Fig. 16.16. **Structures of the bile salts.**
Percentage of bile salts in bile is: cholic, 3α, 7α, $12\alpha \rightarrow 30$–40%; chenocholic 3α, $7\alpha \rightarrow 30$–40%; deoxycholic, $3\alpha \rightarrow$ 10–30%. Bacterial 7α dehydroxylation forms deoxycholic and lithocholic acids. A diagrammatic representation of the hydrophobic groups (●) of the bile salts is shown in the top corner.

Emulsified triacylglycerols are readily attacked by pancreatic lipase secreted in the pancreatic juice. Lipase is a remarkable enzyme in that it is only effective at the lipid aqueous phase interface. This may be demonstrated by the use of a special triacylglycerol, triacetin, composed of glycerol linked to three molecules of acetic acid as substrate. On account of its short fatty acid chains, this triacylglycerol is soluble in water in dilute solutions, but in more concentrated solutions it comes out of aqueous solution in the form of globules. As a typical triacylglycerol, lipase will hydrolyse triacetin; hydrolysis of a solution of triacetin is, however, very slow, although as the substrate concentration is increased and globules are formed with a large surface area, the rate of hydrolysis increases dramatically (*Fig.* 16.18).

Lipase hydrolyses fatty acid in the 1 and 3 positions of the triacylglycerols, rapidly producing, from each molecule of triacylglycerol, 2-monoacylglycerols and two molecules of fatty acid. Subsequent slow isomerization of the 2-monoacylglycerol to 1- or 3-monoacylglycerols occurs and these are then hydrolysed to glycerol and a third molecule of fatty acid (*Fig.* 16.19). However, under most conditions in the digestive tract, minimal formation of glycerol occurs, most of

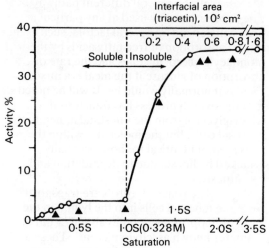

Fig. 16.18. **Effect of concentration on the rate of hydrolysis of triacetin (triacetylglycerol).**
S is the saturation concentration; $1 \cdot 0$ S $\equiv 0 \cdot 328$ M.

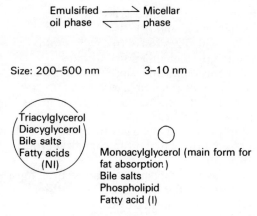

Fig. 16.20. **Relation between emulsified fat droplets and micelles.**
Fatty acids are ionized (I) in the micelles but are not ionized (NI) in the emulsion.

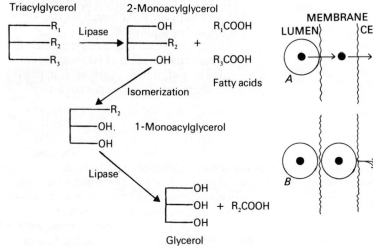

Fig. 16.19. **Formation of monoacylglycerols by hydrolysis of triacylglycerols by lipase.**

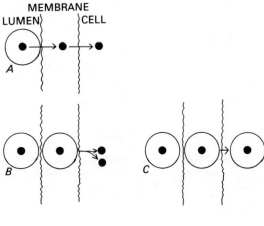

Fig. 16.21. **Mechanisms of transfer of micelles into mucosal cells.**
● Monoacylglycerol.

the hydrolysed triacylglycerols forming monoacylglycerols.

The formation of monoacylglycerols in the presence of bile acids, fatty acids and phospholipids causes the production of micelles that are of very much smaller dimensions than emulsion globules, having a diameter within the range 3–10 nm, i.e. comparable to the dimensions of a large molecule (*Fig.* 16.20). Micelles are stable in a relatively clear solution and are transported into the mucosal cells.

Exactly how the process occurs is not clear, but possible mechanisms are set out in *Fig*. 16.21. According to scheme A, the monoacylglycerol is released at the membrane surface, in scheme B at the interface of the membrane and the cell cytosol, and in scheme C within the mucosal cell.

Whatever process is involved, the net result is the transfer of monoacylglycerol and fatty acid molecules into the cell. On the endoplasmic reticulum, the monoacylglycerols are reconverted to triacylglycerols. The fatty acids required for this synthesis can arise from three sources:

a. Absorbed from the lumen
b. Produced by hydrolysis of absorbed monoacylglycerols
c. Synthesized in the mucosal cells.

Fatty acids are converted to acyl-SCoA derivatives by the action of thiokinase in the presence of coenzyme A and ATP; these are then utilized to form triacylglycerols (cf. Chapter 5). Instead of utilizing the monoacylglycerols for fat synthesis, the mucosal cells can metabolize absorbed glucose to glycerol phosphate and synthesize phosphatidic acid by utilization of two fatty acyl-SCoA (FA-SCoA) derivatives. Phosphatidic acid can then be converted either to phospholipids or to triacylglycerols according to the following scheme:

mucosal cells by several different pathways. The precise pathway used at any particular time will depend on several factors, such as the rate of absorption of free fatty acids by the monoacylglycerol micelles and the rate of absorption of glucose if the meal is a mixed one, as it normally would be. It will be noted that the overall process has transferred triacylglycerol from the intestinal lumen to the mucosal cells, but the molecules within the cells are not identical to those originally attacked by lipase, due to degradation and resynthesis.

The triacylglycerols are transported from the mucosal cells into the lymph in the form of lipoprotein globules about 75 nm in diameter known as 'chylomicrons'. These are composed of a very large (85–90 per cent) proportion of triacylglycerols with small amounts of cholesterol (3 per cent), protein (2 per cent) and phospholipid (5 per cent). The phospholipid and protein surround the globule and are essential for its stability in the lymph and plasma. Although the amounts of protein and phospholipid in the chylomicrons are small they are, nevertheless, essential for the liberation of fat from the mucosal cells and, if their synthesis is impaired, serious reduction in the transport of fat will occur.

The chylomicrons pass from the lymph into the blood through the thoracic duct

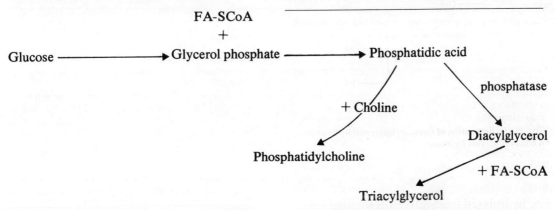

This pathway is believed to be concerned mainly with the biosynthesis of phospholipids and not of triacylglycerols.

It is, therefore, apparent that triacylglycerols may be synthesized in the

and, after a fatty meal, the plasma is distinctly milky in appearance due to the presence of these particles. The fate of the chylomicrons is discussed in Chapter 19.

16.11 Dietary fibre

During recent years, much attention has been devoted to the important role that dietary fibre may play in the digestive tract. Crude fibre has been known for well over a hundred years and described as that portion of any food which remains after extraction with organic solvents, dilute acid and alkali. It has thus been regarded as indigestible and of no food value.

The main constituents of crude fibre are the many complex polysaccharides of plant foods and these, together with lignin, are listed in *Table* 16.7. Although a measure of crude fibre is useful, we really need to know the content of 'dietary fibre' in each food, i.e. the portion of each food not digested and not absorbed. This is difficult to measure but some estimates of crude and dietary fibre are shown for typical foods in *Table* 16.8.

Although for many years fibre has been considered to be of no dietary value, recently, mainly on account of the epidemiological surveys of Burkitt, it is now considered to be of major importance. Burkitt studied the fibre intake, stool volume, and weight and food transit time of many different populations in many parts of the world. From

Table 16.7 **Structural features of the components of dietary fibre**

Fibre	Major groupings	Principal structural types
Structural components of the plant cell wall	Non-cellulose polysaccharides	Galacturonans Arabino- and glucuronoxylans Gluco- and galactomannans Arabinogalactans β-D-Glucan
	Cellulose Lignins	Aromatic polymer
Non-structural polysaccharides	Pectin	Galacturonans
	Gums, mucilages	Great variety including arabinoxylans and gluco- and galactomannans
	Algal polysaccharides Modified celluloses	Sulphated galactans and gulurono-mannuronans Esters, ethers

Table 16.8 **Dietary fibre in some cereals and raw vegetables**

Cereal/vegetable	Total dietary fibre, % dry wt	Composition of the dietary fibre, %		
		Non-cellulose polysaccharides	Cellulose	Lignin
White flour (72%)	3·45	80	19	1
Brown flour (90–95%)	8·70	72	18	10
Wholemeal flour (100%)	13·51	74	20	6
Bran, coarse	48·0	74	18	7
Oatmeal, for porridge	7·66	82	12	6
Rice, long grain	2·74	78	22	Tr
Brussel sprouts	35·5	72	25	3
Cabbage, winter	29·4	62	25	13
Pease, frozen	37·1	69	27	2
Runner beans	26·4	52	42	7
Carrots	28·4	60	40	Tr

Tr = trace.

these studies, he demonstrated that a large fibre intake produced a large stool volume and a fast transit time, whilst a small fibre intake had the opposite effects. In the UK, it has been estimated that an adult typically takes in 20 g fibre per day which compares with nearly 200 g per day for many African countries. Burkitt further showed that low fibre intake could be correlated with many modern diseases (listed in *Table* 16.9) and that it was clearly desirable for Western societies to make large increases in their fibre intake.

Table 16.9 **Consequences of lack of diet fibre**

1. Diverticular disease of colon
2. Hiatus hernia
3. Haemorrhoids
4. Varicose veins
5. Appendicitis
6. Cancer of colon and rectum

From Burkitt, 1976, *Dig. Dis.,* Vol. 21, p. 104.

These epidemiological investigations have stimulated many scientific studies on the effects of dietary fibre and, in general, the main theories of Burkitt's observations and deductions have been confirmed. How and why does fibre in diet exert its beneficial effects? This is not entirely clear, but it is established that molecules of the fibre polysaccharide are strong adsorbing agents. They adsorb large quantities of water and also possible toxic molecules which may be produced by the bacteria in the digestive tract. For example, bile acids, before or after

---

Table 16.10 **Beneficial effects of fibre**

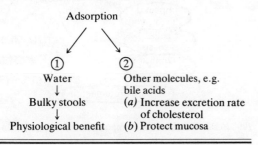

bacterial degradation, are slowly adsorbed and potential carcinogens may also be bound.

The adsorption of large amounts of water is also very important and this leads to a much larger faecal mass and thus to easier transit through the intestines and easier expulsion (*Table* 16.10).

16.12 Bacterial flora in the gastrointestinal tract

The human gastrointestinal tract contains at least 150 different species of bacteria. Most are anaerobic or facultative anaerobes, the species being listed in *Table* 16.11.

Table 16.11 **Species of bacteria* in human gastrointestinal tract**

Group	Bacteria
1	Facultative or aerobic (Gram-negative)
	Enterobacter
	Pseudomonas
2	Facultative streptococci
3	Lactobacilli
4	Anaerobic cocci
5	*Bacteroides*
6	*Eubacterium*
7	*Bifidobacterium*
8	Clostridia

* 150 species, mainly anaerobic.

There are very few bacteria in the stomach or duodenum, more in the lower regions of the small intestine, whilst the large intestine contains a very large population. These bacteria have a profound effect on the foodstuff in the intestinal lumen, on the mucosal cells and particularly on the brush border. Their effects can be demonstrated by the study of gnotobiotic animals, which are born and kept under completely sterile conditions so there are no bacteria in their intestines. These animals show many differences from control (normally bred) animals, and these are listed in *Table* 16.12.

Bacteria can have desirable or undesirable effects on human metabolism.

Table 16.12 Consequences of removal of microflora (gnotobiotic animals)

1. Wall of intestine thinner—reduced in weight
2. Villi longer—slender
3. Surface area of intestine reduced
4. Rate of epithelial cell renewal—reduced
5. Animals able to withstand ionizing radiation more effectively
6. Enzyme complement of mucosa altered
   Alkaline phosphatase increased
   α-Glucosidase increased
7. Absorption of glucose increased
   Vitamins increased

   Greatest response is in small intestine when bacterial count is low

They synthesize some vitamins, e.g. biotin and vitamin K, that can be absorbed and utilized by the hosts. They can be harmful by utilizing valuable nutrients so preventing absorption and by producing toxic compounds that damage the mucosa and inhibit absorption; bacteria also possibly metabolize bile acids to carcinogenic compounds.

The study of the effects of bacteria in human intestine is extremely complex, since there are many variations between population groups, types of diet consumed, times of day and year and even between individuals.

16.13 Malabsorption syndromes

Defects of absorption can occur as a result of:
  i.   Deficiencies of enzyme secretions
  ii. Bile deficiency on account of blockage of the bile duct
  iii. Damage to the brush border and mucosal cells

Deficiency of gastric or pancreatic juices can result in a reduced efficiency of the digestion of proteins, lipids and carbohydrates, but the effects of lack of bile are most clearly shown in the reduced efficiency of fat digestion and absorption. Lactase deficiency is an example of enzyme deficiency, deficiencies of this enzyme in the adult being widespread in many parts of the world. It appears to be

genetically linked and is very common in American Negroes, Africans and in Asia. The lactase which is not absorbed in these populations is fermented by the gut bacteria and frequently causes extensive diarrhoea. The removal of milk from the diet of these patients is clearly desirable. Damage to the intestinal mucosa occurs as a result of coeliac disease or gluten enteropathy. This condition is relatively common and between 1:1000 and 1:2000 of the population suffer from this condition in the UK. Serious damage is caused to the mucosal brush border by the gliadin of wheat protein, but the exact mechanism of the damage is still obscure. Immunological reactions with the mucosal cell may be involved. As a consequence of this condition, the absorption of most digested foods, particularly fat and vitamins, e.g. folate or vitamin B_{12}, is severely impaired and deficiency symptoms will develop (cf. Chapter 24).

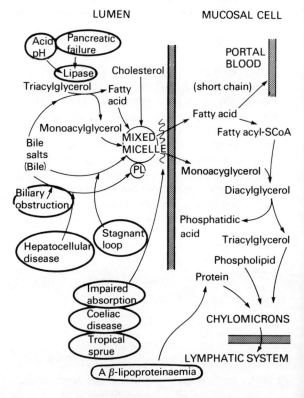

Fig. 16.22. Possible sites of impairment of fat absorption. ◯ indicates the sites.

The absorption of fat can be impaired by many different factors, shown diagrammatically in *Fig*. 16.22 and *Table* 16.13 summarizing many of the characteristic means by which malabsorption can occur. Any of these conditions will lead to steatorrhea, the excretion of fat in the faeces.

Table 16.13 **Some consequences of steatorrhea and associated absorption defects**

i.   Poor absorption of carbohydrate—protein—fat
ii. K^+ depletion (diarrhoea)
iii. Hypocalcaemia (vitamin D)
iv. Haemorrhages (vitamin K)
v.   Anaemia (B_{12}–folate) occurs in Tropical Sprue

Chapter 17

Hormones: a summary of their structures and functions

17.1 Introduction

The discovery of the existence of 'hormones', or 'chemical messengers' as they were originally called, stems from the classic experiments of Bayliss and Starling at the beginning of the century. They showed that extracts of the duodenal mucosa, upon injection into the blood stream, caused a copious flow of pancreatic juice. The factor extracted from the duodenum must clearly have been a chemical carried in the blood, i.e. a chemical messenger.

This discovery was of great and far-reaching importance because, up until that time, control between organs had always appeared to be mediated by the nervous system which is a system of electrical control. Bayliss and Starling, therefore, laid the foundations, probably without realizing it, of a vast area of important biochemistry. Subsequent to the definition of 'hormones', the glands secreting these substances were described as 'endocrines' and the subject as 'endocrinology'.

In more recent years another group of hormone-like substances has been discovered that affects only those cells adjoining the tissues producing them or localized parts of these tissues. Substances of this type are known as 'paracrines'.

Several general aspects of hormones should be noted. Firstly, all hormones have 'target cells' on which they act and these may be located in a specific organ, such as the kidney, or be more widely distributed in the body, forming special cell populations in various organs. Secondly, some hormones function solely to bring about the release

196

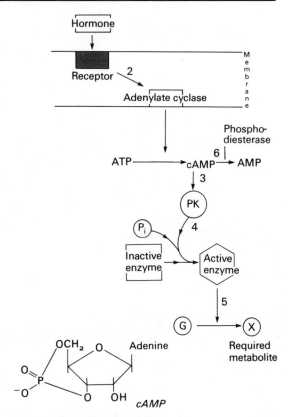

Fig. 17.1. **The role of cAMP.**
PK = protein kinase.

of other hormones from different endocrine glands. The most important amongst this group are the secretions of the pituitary and these are called 'trophic' hormones or 'trophins'. Thirdly, it is now known that many hormones act by means of a second messenger and quite often this is cyclic AMP (cAMP) which is formed from ATP (*Fig.* 17.1). On reaching its receptor in the cell membrane, the hormone causes the release of cAMP, which is the actual regulator of the metabolic process.

Some hormones, such as many of the steroid hormones and those produced by the thyroid, act on the nucleus, usually regulating the control of the synthesis of a specific mRNA or a group of mRNAs. This requires the intervention of a protein receptor on the membrane and of a carrier protein which will transfer the steroid to the DNA receptor protein.

The action of some hormones is regulated by calcium or calcium-binding proteins and the form of control can be in addition to the regulation exerted by cAMP.

For the majority of hormones the chemical structure falls into one of two groups; they are usually either small peptides or steroids, the most important exceptions being adrenaline and the thyroid hormones.

In this chapter it is not intended to describe all the hormone actions in detail, since they are discussed in the chapters concerning the metabolic processes involved. This chapter is designed to present a summary of the structures and major actions of the hormones for reference and to clarify aspects of other chapters.

2. The binding of the hormone activates the enzyme adenylate cyclase which catalyses the conversion of ATP to cAMP.

3. cAMP activates a protein kinase within the cell.

4. The protein kinase catalyses the phosphorylation of an enzyme in the cell which, as a consequence, is often converted from an inactive to an active form. In some cases, phosphorylation leads to inactivation of the enzyme which usually occurs if the enzyme is involved in a synthetic process, e.g. glycogen synthetase.

5. The active enzyme must then catalyse a required metabolic reaction, for example the phosphorylation and degradation of glycogen to form, firstly, glucose 1-phosphate and, ultimately, glucose for release into the blood (cf. Chapter 5). Alternatively, phosphorylation of a second enzyme may take place so that a cascade mechanism becomes involved.

6. A further enzyme, phospho-diesterase, catalyses the conversion of cAMP to AMP thereby rendering the cAMP inactive. The activity of this enzyme can also be regulated by hormones and is clearly important in the overall control process.

17.2 The role of the second messenger: cyclic AMP

One of the great advances in the understanding of the action of hormones has been the development of the 'Second Messenger' concept and the role of cyclic nucleotides, particularly cAMP.

The importance of cAMP was first demonstrated by Sutherland in the USA who showed that many hormones caused the liberation of cAMP in target cells. The typical sequence of events is illustrated in *Fig*. 17.1.

1. The hormone binds to a receptor in the cell membrane, this receptor being specific for the hormone.

17.3 Structural relationships of the hypothalamus and the pituitary gland

Hypothalamus

The hypothalamus is the basal part of the diencephalon or interbrain and is closely associated with the floor and lower parts of the third ventricle. Of special importance are the supraoptic and paraventricular nuclei of the hypothalamus which connect with the pituitary.

Pituitary gland

The pituitary is a very small gland weighing only 600 mg in man which may increase in weight to 1 g in women during pregnancy. It is located very close to the hypothalamus and forms a stalk-like structure linked to the brain.

It is divided both structurally and functionally into two distinct regions, the anterior pituitary or adenohypophysis and the posterior pituitary or neurohypophysis. Structural relationships of the hypothalamus and the pituitary gland are shown in *Fig.* 17.2.

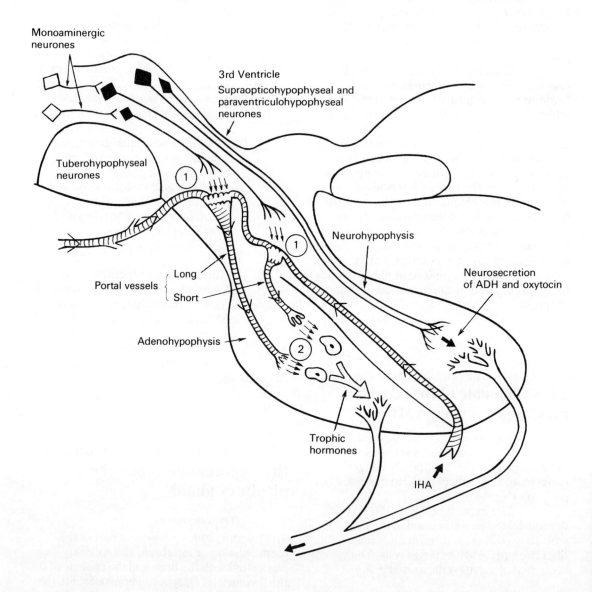

Fig. 17.2. **Structural relationships of the hypothalamus and pituitary.**

17.4 The hypothalamus—anterior pituitary—target organ relationships

This complex system is the most important in the regulation of metabolic activity in the body, control being mediated by a series of steps. Typically, a secretion is released from the hypothalamus which, in its turn, stimulates release of a trophic hormone from the anterior pituitary. This, in turn, travels to the target endocrine organ, for example the adrenal, where release of the active hormone, e.g. aldosterone, is triggered. The term 'trophic hormone' is used to describe those hormones which act as intermediaries: they exert no direct metabolic control, but act only by stimulating the release of the active hormone. In addition to the stimulation of hormone release, some factors can act as inhibitors so depressing the release.

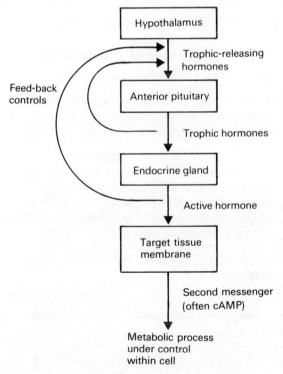

Fig. 17.3. **Relationships of the hypothalamus, anterior pituitary and target organs.**

A summary of the inter-relationships of the system is shown diagrammatically in *Fig.* 17.3 and the structural relationships of the glands are shown in *Fig.* 17.2.

17.5 Hormones of the hypothalamus

The hormones of the hypothalamus are described as 'releasing factors' or 'inhibiting factors'. The hormones known at present are listed below.

a. Thyrotrophin-releasing hormone

Thyrotrophin-releasing hormone (TRH) is one of the simplest peptides, being a tripeptide:

$$\boxed{Glu}-\boxed{His}-\boxed{Pro}-NH_2$$

It is responsible for stimulating the release of thyrotrophin (or thyroid-stimulating hormone, TSH) from the anterior pituitary.

b. Corticotrophin-releasing hormone (β-endorphin releasing factor)

The structure of corticotrophin-releasing hormone (CRF) has recently been established. It is a peptide composed of 41 amino acid residues with a C-terminal serine and an N-terminal alanine. Some of the amino acid sequences are identical to those of vasopressin which possesses CRF activity and was originally believed to be the substance active *in vivo*, but the existence of a separate CRF is now established. Its function is to stimulate the release of corticotrophin (adrenocorticotrophic hormone, ACTH) from the anterior pituitary and to cause the release of β-endorphin.

c. Luteinizing hormone-releasing hormone (LHRH)

This hormone is a peptide containing ten amino acids:

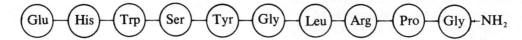

The hormone is responsible for the release of luteinizing hormone (LH) from the anterior pituitary, but it is possible that LHRH also releases folliculotrophin (or follicle-stimulating hormone, FSH). Whether there are two separate releasing hormones or whether LHRH carries out both functions is, as yet, unresolved.

d. Prolactin-releasing factor and prolactin-inhibiting factor

There is, at present, uncertainty surrounding both of these factors. It has been suggested that prolactin-inhibiting factor is not a peptide but is identical to dopamine. Furthermore, it is possible that control of lactotrophin (prolactin) release from the pituitary may be entirely under the control of prolactin-inhibiting factor (PIF) and that prolactin-releasing factor (PRF) has no real existence.

e. Somatotrophin (SRF—growth hormone-releasing factor) and somatostatin (SS—growth hormone-inhibiting factor)

Although it is generally agreed that a growth hormone-releasing factor exists, it has not yet been characterized. Much more extensive progress has, however, been made in the study of somatostatin (SS) which is the inhibiting factor, and its structure has been determined as a peptide containing 14 amino acids:

This hormone is synthesized in many parts of the brain, in addition to parts of the hypothalamus and the gastointestinal tract, and the whole pancreas. In addition to its effect on the release of growth hormone, it has many other effects on metabolism. These include:

i.   Inhibition of synthesis and release of TSH.
ii. Inhibition of secretion of insulin and glucagon.
iii. Inhibition of secretion of gastric acid, pepsin, gastric inhibitory peptide (GIP) and vasoactive intestinal peptide (VIP) (cf. Chapter 16).
iv. Control of cell activity in various parts of the brain.

17.6 Hormones of the anterior pituitary (adenohypophysis)

These peptide hormones fall into the three main groups shown below.

a. ACTH-related peptides

These hormones possess a single peptide chain and their amino acid sequences are often closely related to one another. This group of peptides may all be derived from 'big ACTH', a large polypeptide composed of 120–130 amino acids which could be the precursor of several small biologically active peptides (*Fig.* 17.4).

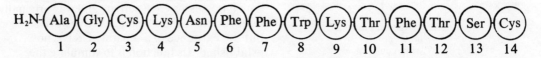

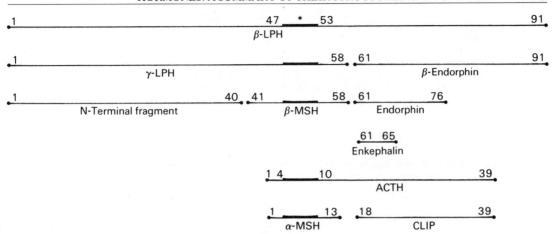

Fig. 17.4. **Structural relationships of hormones of the anterior pituitary.**
LPH lipotrophin; MSH melanocyte-stimulating hormone; CLIP corticotrophic-like intermediate lobe peptide. Regions of structural similarity in several peptide hormones. The hexapeptide sequence ▬▬ is identical in all hormones.

Corticotrophin

Corticotrophin (ACTH) is a peptide possessing 39 amino acids. The first 13 of these 39 form the peptide which is essential in producing the biological effects of the whole molecule, but for production of total activity an additional 11 amino acids (forming a larger peptide of 24 amino acids) are essential (*see Fig.* 17.5). This hormone acts on the cortex of the adrenal gland to stimulate the release and synthesis of a group of steroid hormones, the glucocorticoids. The effect is very rapid, 0·5 ng ACTH per kg body weight causing a release of the steroids within 2–3 minutes and, since the action of ACTH can be prolonged, steroid synthesis must also be affected. The action of ACTH is believed to be mediated by cAMP. Provision of cholesterol and increased concentrations of NADPH, both of which are required for steroid hormone synthesis, are regulated after ACTH has attached to the receptors on the adrenal cortex.

Melanophore (melanocyte) – stimulating hormone

Melanophore-stimulating hormone (MSH) is more important in lower vertebrates than in man, where it regulates the dispersion of dark-coloured melanin granules, and two forms of the hormones, α and β, have been described (*see Fig.* 17.4). β-MSH is a peptide synthesized by the human pituitary (*see Fig.* 17.6 for structure) and, as mentioned, its only known function in man is to increase melanin synthesis by the skin. Increased secretion of MSH, which occurs in certain pathological conditions, e.g. Addison's disease, causes increased pigmentation of the skin.

```
1                                        13                17
Ser-Tyr-Ser-Met-Glu-His-Phe-Arg-Trp-Gly-Lys-Pro-Val-Gly-Lys-Lys-Arg-
```

18	25		31	33		39	ACTH
Arg-Pro-Val-Lys-Val-Tyr-Pro-Asn-Gly-Ala-Glu-Asp-Glu-Ser-Ala-Gln-Ala-Phe-Pro-Leu-Glu-Phe							Sheep
			Ser	Gln			Ox
			Ser	Glu			Human
	Val		Ser	Glu			Rat
			Leu	Glu			Pig

Fig. 17.5. **The structure of ACTH in man and some other species.**

```
                        41              45                  50                  55
Human:    H-Ala-Glu-Lys-Lys-Asp-Glu-Gly-Pro-Tyr-Arg┤Met-Glu-His-Phe-Arg-Trp-Gly├Ser-Pro-Pro-Lys-Asp-OH
Macacus:        H-Asp-Glu-Gly-Pro-Tyr-Arg┤Met-Glu-His-Phe-Arg-Trp-Gly├Ser-Pro-Pro-Lys-Asp-OH
Ovine; Porcine; Bovine:   H-Asp-Glu-Gly-Pro-Tyr-Lys┤Met-Glu-His-Phe-Arg-Trp-Gly├Ser-Pro-Pro-Lys-Asp-OH
Equine:         H-Asp-Glu-Gly-Pro-Tyr-Lys┤Met-Glu-His-Phe-Arg-Trp-Gly┤Ser-Pro-Arg-Lys-Asp-OH
Ovine; Porcine; Bovine:   H-Asp- Ser-Gly-Pro-Tyr-Lys┤Met-Glu-His-Phe-Arg-Trp-Gly┤Ser-Pro-Pro-Lys-Asp-OH
Camel:          H-Asp-Gly-Gly-Pro-Tyr-Lys┤Met-Glu-His-Phe-Arg-Trp-Gly┤Ser-Pro-Pro-Lys-Asp-OH
Camel:          H-Asp-Glu-Gly-Pro-Tyr-Lys┤Met-Gln-His-Phe-Arg-Trp-Gly┤Ser-Pro-Pro-Lys-Asp-OH

Scyliorhinus caniculus:   H-Asx-Glx-Ile-Asx-Tyr-Lys┤Met-Gly-His-Phe-Arg-Trp-Gly┤Ala-Pro-Met-Asp-Lys-OH
Squalus acanthias:        H-Asp-Gly-Asp-Asp-Tyr-Lys┤Phe-Gly-His-Phe-Arg-Trp-Ser┤Val-Pro-Leu-OH
```

Fig. 17.6. **The structures of melanocyte-stimulating hormones (MSH) of several species.**
The amino acid numbering relates to the sequence shown in *Fig.* 17.4.

```
              1              5                  10                 15                 20
Human:    H-Glu-Leu-Thr-Gly-Gln-Arg-Leu-Arg-Gln-Gly-Asp-Gly-Pro-Asn-Ala-Gly-Ala-Asn-Asp-Gly-
Ovine:    H-Glu-Leu-Thr-Gly-Glu-Arg-Leu-Glu-Gln-Ala-Arg-Gly-Pro-Glu-Ala-Gln-Ala-Glu-Ser-Ala-
Porcine:  H-Glu-Leu-Ala-Gly-Ala-Pro-Pro-Glu-Pro-Ala-Arg-Asp-Pro-Glu-Ala-Pro-Ala-Glu-Gly-Ala-

              21             25                 30                 35                 40
Human:    Glu-Gly-Pro-Asn-Ala-Leu-Glu-His-Ser-Leu-Leu-Ala-Asp-Leu-Val-Ala-Ala-Glu-Lys-Lys-
Ovine:    Ala-Ala-Arg-Ala-Glu-Leu-Glu-Tyr-Gly-Leu-Val-Ala-Glu-Ala-Glu-Ala-Ala-Glu-Lys-Lys-
Porcine:  Ala-Ala-Arg-Ala-Glu-Leu-Glu-His-Gly-Leu-Val-Ala-Glu-Ala-Gln-Ala-Ala-Glu-Lys-Lys-

              41             45                 50                 55                 60
Human:    Asp-Glu-Gly-Pro-Tyr-Arg┤Met-Glu-His-Phe-Arg-Trp-Gly┤Ser-Pro-Pro-Lys-Asp-Lys-Arg-
Ovine:    Asp-Ser-Gly-Pro-Tyr-Lys┤Met-Glu-His-Phe-Arg-Trp-Gly┤Ser-Pro-Pro-Lys-Asp-Lys-Arg-
Porcine:  Asp-Glu-Gly-Pro-Tyr-Lys┤Met-Glu-His-Phe-Arg-Trp-Gly┤Ser-Pro-Pro-Lys-Asp-Lys-Arg-

              61             65                 70                 75                 80
Human:    (Tyr-Gly-Gly-Phe-Met-Thr)-Ser-Glu-Lys-Ser-Gln-Thr-Pro-Leu-Val-Thr-Leu-Phe-Lys-Asn-
Ovine:    (Tyr-Gly-Gly-Phe-Met-Thr)-Ser-Glu-Lys-Ser-Gln-Thr-Pro-Leu-Val-Thr-Leu-Phe-Lys-Asn-
Porcine:  (Tyr-Gly-Gly-Phe-Met-Thr)-Ser-Glu-Lys-Ser-Gln-Thr-Pro-Leu-Val-Thr-Leu-Phe-Lys-Asn-

              81             85                 91
Human:    Ala-Ile-Ile-Lys-Asn-Ala-Tyr-Lys-Lys-Gly-Glu-OH    (1)
Ovine:    Ala-Ile-Ile-Lys-Asn-Ala-His-Lys-Lys-Gly-Gln-OH    (2)
Porcine:  Ala-Ile-Val-Lys-Asn-Ala-His-Lys-Lys-Gly-Gln-OH    (3)
```

Fig. 17.7. **The structure of β-lipotrophin (LPH).** Parentheses indicate opiate receptor sequence.

```
              61             65                 70                 75                 80
Human:   H-(Tyr-Gly-Gly-Phe-Met)-Thr-Ser-Glu-Lys-Ser-Gln-Thr-Pro-Leu-Val-Thr-Leu-Phe-Lys-Asn-
Camel:   H-(Tyr-Gly-Gly-Phe-Met)-Thr-Ser-Glu-Lys-Ser-Gln-Thr-Pro-Leu-Val-Thr-Leu-Phe-Lys-Asn-
Bovine:   —   —  —   —   —   —   —   —   —   —   —   —   —  Val-Thr-Leu-Phe-Lys-Asn-

              81             85                 91
Human:   Ala-Ile-Ile-Lys-Asn-Ala-Tyr-Lys-Lys-Gly-Glu-OH    (14)
Camel:   Ala-Ile-Ile-Lys-Asn-Ala-His-Lys-Lys-Gly-Gln-OH    (15)
Bovine:  Ala-Ile-Ile-Lys-Asn-Ala-His-Lys-Lys-Gly-Gln-OH    (16)
```

Fig. 17.8. **The structure of β-endorphins.**
The numbering of the amino acids is based on the chart shown in *Fig.* 17.4. Parentheses indicate common sequence.

Lipotrophins

Two peptides, β- and γ-lipotrophin (β- and γ-LPH), have been isolated from sheep pituitaries. Their function is in the mobilization of fatty acids from the fat depots, but their role in normal human physiology, if any, is not yet known. The structure of β-LPH is shown in *Fig*. 17.7.

Endorphins and enkephalins

These peptides have powerful analgesic and 'morphine-like' actions. They bind to the opiate receptors in the brain and are, on a molar basis, 50–200 times more potent than morphine itself. A pentapeptide sequence (Tyr-Gly-Gly-Phe-Met) is common to this group and appears to be essential for the analgesic activity which is also found in ACTH and in β-LPH. These hormones may, therefore, be considered to be the natural analgesics of the body. The structure of β-endorphin is shown in *Fig*. 17.8.

b. Glycoprotein hormones

This group includes TSH (thyrotrophin), LH (luteinizing hormone or luteotrophin), and FSH (follicle-stimulating hormone or folliculotrophin). HCG (human chorionic gonadotrophin) which is produced by the placenta, has a related structure.

These hormones are composed of two peptide chains, the α chains and the β chains. All contain carbohydrate components and are, therefore, glycoproteins. The α chains of these hormones have no biological activity and show close similarities in each hormone, whereas the β chains differ markedly and confer the biological activity.

Thyrotrophin (thyroid-stimulating hormone, TSH)

This hormone has a molecular weight of 20 000, with an α chain composed of 96 amino acids and a β chain of 113 amino acids. The molecule also contains 18 per cent carbohydrate.

The main function of TSH is to control the release of the thyroid hormones, thyroxine (T_4) and tri-iodothyronine (T_3) from the thyroid gland, but, in addition, it influences many aspects of thyroid function including iodide uptake, synthesis and release of T_3 and T_4 and, subsequently, increase of the thyroid mass.

TSH binds to receptor sites, activates adenylate cyclase which produces cAMP and so controls the synthesis and release of T_3 and T_4. A feedback control of TSH secretion, either directly or by inhibiting TRH, is exerted by T_3 and T_4.

Folliculotrophin (follicle-stimulating hormone, FSH)

The molecular weight of FSH is 30 000, the β chain contains the hormonal specificity and the hormone stimulates the development of the follicle in the ovary.

Luteinizing hormone (LH) and/or intestinal cell-stimulating hormone (ICSH)

Luteinizing hormone has a molecular weight of between 26 000 and 34 000 depending on the species. Unlike FSH the carbohydrate component contains very little sialic acid. The function of LH is to stimulate the formation and secretion of the corpus luteum and acts with FSH to cause development of the follicle and subsequent ovulation.

Human chorionic gonadotrophin (HCG)

This hormone is produced by the trophoblast soon after conception. The β chain is similar to that of LH and the action also resembles that of LH. It is concerned with the enlargement of the corpus luteum in pregnancy and HCG is produced in the placenta.

c. Somatomammotrophins

Growth hormone, human chorionic somatomammotrophin (HCS) and prolactin, produced by the placenta, have closely related structures. They are all composed of a single peptide chain with two or three disulphide bridges.

Prolactin

Prolactin of different species, e.g. bovine prolactin, has been purified and shown to contain 199 amino acids with two disulphide bridges. Human prolactin is believed to be similar in structure and is responsible for the control of milk secretion following oestrogen and progesterone priming.

Growth hormone (somatotrophin)

Human growth hormone contains 191 amino acids with a molecular weight of 21 500; the peptide is linear with two disulphide bridges. The hormone is not species specific in most mammals, but the human derivative is essential for activity in man. The hormone has several complex biochemical actions on many tissues including those listed below.

Protein—amino acid metabolism
Growth hormone causes:

i.   Stimulation of amino acid uptake into cells.
ii.   Stimulation of protein synthesis.
iii. Decrease in urea excretion.
iv.   A positive nitrogen balance.

Fat metabolism
Growth hormone causes breakdown of triacylglycerols in the fat depot and release of fatty acids. Activation of the hormone-sensitive lipase is responsible for this effect (cf. Chapter 19).

Carbohydrate metabolism
Growth hormone causes:

i.   A diabetogenic effect and exacerbation of clinical diabetes.
ii.   Increase of the blood glucose concentration by decreasing glucose uptake by the liver and increasing liver glucose output.

Growth
Excess of the hormone during the postnatal period results in giantism and acromegaly, whereas deficiency of the hormone results in dwarfism. It should be noted that the effects on growth cannot be accounted for by the known metabolic effects and other factors are, therefore, involved in growth control.

17.7 Hormones of the posterior pituitary (neurohypophyis)

Structure and relation to the hypothalamus
The posterior pituitary gland is a downward growth of the floor of the brain and is connected to the nuclei of the hypothalamus by nerve fibres (see Fig. 17.2).

Hormones and their biosynthesis
The posterior pituitary secretes two peptide hormones, oxytocin and vasopressin. Although generally described as postpituitary hormones, these hormones have been shown to be synthesized in the hypothalamus and transported to the posterior pituitary through the axons of the nerve connecting the hypothalamus to the posterior pituitary. When released into the posterior pituitary, the hormones are bound to proteins (of molecular weight about 10 000), composed of 93–95 amino acid residues, which are called neurophysins (see Fig. 17.9). They are stored in the bound form, sufficient for one-week's supply of hormones usually being present.

Vasopressin (AVP—the antidiuretic hormone)
Structure This hormone, usually described as 'arginine vasopressin' (hence AVP) to stress the importance of the amino acid that differentiates it from oxytocin, is a peptide containing nine amino acids and including one disulphide bridge (see Fig. 17.10).

Mode of action Vasopressin acts primarily on the distal convoluted tubule and collecting ducts of the kidney making them more

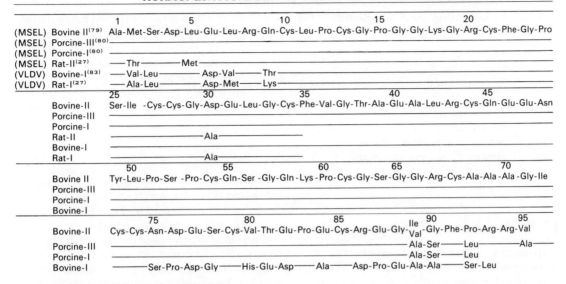

		1		5		10		15		20	
(MSEL)	Bovine II[79]	Ala-Met-Ser-Asp-Leu-Glu-Leu-Arg-Gln-Cys-Leu-Pro-Cys-Gly-Pro-Gly-Gly-Lys-Gly-Arg-Cys-Phe-Gly-Pro									
(MSEL)	Porcine-III[80]										
(MSEL)	Porcine-I[80]										
(MSEL)	Rat-II[27]	——Thr———————Met———									
(MSEL)	Bovine-I[83]	——Val-Leu————————Asp-Val———Thr———									
(VLDV)	Rat-I[27]	——Ala-Leu————————Asp-Met———Lys———									

		25		30		35		40		45	
	Bovine-II	Ser-Ile -Cys-Cys-Gly-Asp-Glu-Leu-Gly-Cys-Phe-Val-Gly-Thr-Ala-Glu-Ala-Leu-Arg-Cys-Gln-Glu-Glu-Asn									
	Porcine-III										
	Porcine-I										
	Rat-II	——————————————Ala———									
	Bovine-I										
	Rat-I	——————————Ala———									

		50		55		60		65		70	
	Bovine II	Tyr-Leu-Pro-Ser -Pro-Cys-Gln-Ser -Gly-Gln -Lys -Pro-Cys-Gly-Ser-Gly-Gly-Arg-Cys-Ala-Ala -Ala -Gly-Ile									
	Porcine-III										
	Porcine-I										
	Bovine-I										

		75		80		85		90		95	
	Bovine-II	Cys-Cys-Asn-Asp-Glu-Ser-Cys-Val-Thr-Glu-Pro-Glu-Cys-Arg-Glu-Gly-$^{Ile}_{Val}$-Gly-Phe-Pro-Arg-Arg-Val									
	Porcine-III	———————————————————————————Ala-Ser———Leu—————————Ala———									
	Porcine-I	———————————————————————————Ala-Ser———Leu———									
	Bovine-I	———————Ser-Pro-Asp-Gly———His-Glu-Asp———Ala———Asp-Pro-Glu-Ala-Ala———Ser-Leu———									

Fig. 17.9. **The structures of some neurophysins.**
Note: the sequence of the rat proteins beyond residue 35 is not known at present.

permeable to water. Consequently, it increases water retention by influencing the osmotic movement of water in the nephron. Conversely, whenever synthesis or release of vasopressin is depressed, a large volume of urine is excreted, this condition being described as 'diabetes insipidus'.

The mode of action of vasopressin is still unclear. It appears to act through the intervention of cAMP by increasing the number of 'water pores'. It must be emphasized, however, that these pores are a hypothetical concept and that their structural existence has not yet been established.

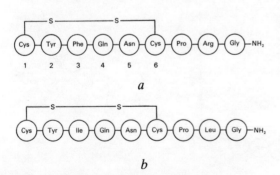

a

b

Fig. 17.10. **Structure of arginine vasopressin (*a*) and oxytocin (*b*).**

Control of secretion The rate of secretion of vasopressin is dependent on the osmolarity of the extracellular fluid, particularly that of the plasma. Increase of solute concentration or decrease in the volume of extracellular fluid promotes vasopressin secretion and, therefore, water retention.

Oxytocin

Oxytocin is a peptide also composed of nine amino acids and is very similar to the vasopressin peptide. It differs only in the amino acids which are present in position 3 (isoleucine) and position 8 (leucine; *see* Fig. 17.10).

Mode of action

a. *Uterus.* Oxytocin causes powerful contractions of the uterus once labour has begun, but it cannot initiate labour. It is the principal hormone involved in the transit of the fetus during birth and is used for facilitating labour. Oxytocin acts by causing a decrease in the resting membrane potential of the cell and, in conjunction with the oestrogens, by increasing the contractibility of the cell.

b.   *Mammary gland.* Oxytocin does not regulate the synthesis of milk, but it is important in the mechanism involved in the release of milk from the gland. This release is mediated by contraction of the myoepithelial cells arranged around the ducts, the contraction being specifically dependent on oxytocin.

17.8 Hormones of the pancreas

Two hormones that are very important in the regulation of metabolism are synthesized in the pancreas: these are insulin and glucagon. Synthesis occurs in small groups of cells, or islets, scattered in the acini of the pancreas, and named 'islets of Langerhans' after their discoverer. Two types of cell may be recognized in the islets, α cells which synthesize glucagon and β cells which synthesize insulin.

Insulin

Insulin is of great historical interest for several reasons. It was the first peptide hormone to be sufficiently well purified by Banting and Best in the 1920s to allow it to be injected into patients in the treatment of the very serious disease 'diabetes'. Diabetes is caused by a deficiency of insulin which can ultimately lead to death. The complete amino acid sequence of insulin was established in the early 1950s by Sanger and it was the first protein to be completely analysed.

Structure

Insulin is made up of two polypeptide chains, the A and B chains, which are linked by disulphide bridges (*see Fig.* 17.11). The hormone is first synthesized on the endoplasmic reticulum of the islet cells as a large single-chain peptide, pro-insulin. At or near the Golgi apparatus, pro-insulin is attacked by a proteolytic enzyme, possibly an enzyme from a lysosome or a lysosome precursor, which splits off a large peptide fragment, peptide C, and releases insulin (*see Fig.* 17.12). This is then stored in granules until its release into the plasma where its half-life is about 10 min. Insulin is destroyed in muscle and adipose tissue by proteolytic action and, in liver, by a glutathione transhydrogenase. Insulin has many effects on metabolism, its main role being involvement in the regulation of the blood glucose concentration; several other metabolic processes are, however, also controlled.

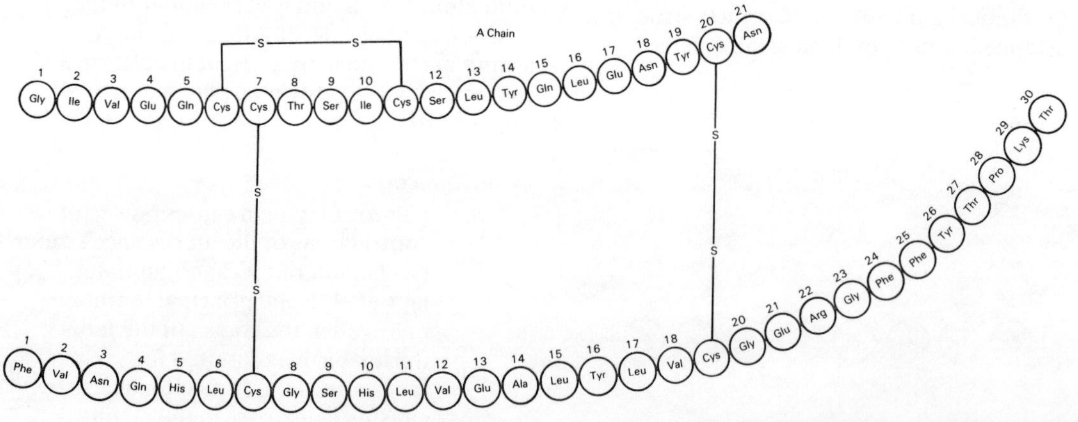

Fig. 17.11. **The structure of bovine insulin.**

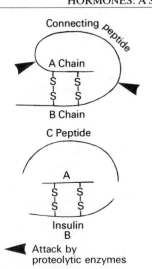

Fig. 17.12. **The formation of insulin from pro-insulin.**

phosphodiesterase which breaks
down cAMP. cAMP is required for
the release of fatty acids (cf. Chapter
19).

b. Insulin also promotes the synthesis of
fatty acids from acetyl-SCoA in the
liver.

Glucagon

This hormone is a polypeptide consisting of a
single chain of 29 amino acids with a molecular
weight of 3485 (*see Fig.* 17.13). Like insulin, it
has a half-life of about 10 min and is degraded
by proteolytic attack. Glucagon also affects
many different metabolic processes.

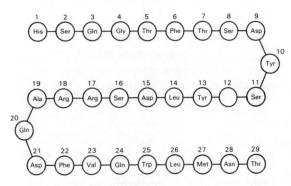

Fig. 17.13. **The structure of glucagon.**

Carbohydrate metabolism

a. Insulin lowers the blood glucose
concentration by increasing glucose
entry into the tissue cells, particularly
those of muscle and adipose tissue,
and by depressing the glucose release
from the liver glycogen.

b. Insulin promotes the production of
glycogen from glucose by induction of
glycogen synthase and depresses
gluconeogenesis from glucogenic
amino acids.

c. Insulin may also inhibit the release of
glucagon.

Protein metabolism

a. Insulin promotes the transport of
amino acids into liver and muscle
cells.

b. Insulin powerfully inhibits protein
catabolism in muscle.

Fat metabolism

a. Insulin inhibits the release of fatty
acids and glycerol from
triacylglycerols stored in adipose
tissue, possibly by stimulating the

Carbohydrate metabolism

a. Glucagon is a very powerful
hyperglycaemic agent and this effect
is accomplished by stimulating
glycogenolysis. It activates adenylate
cyclase to produce cAMP which, by a
stepwise process (cf. Chapter 5,
Appendix 15), activates liver
phosphorylase so liberating glucose
1-phosphate and thence glucose from
the glycogen. The cascade system
(Chapter 5) magnifies the effect of the
hormone so that one molecule of
glucagon, attached to a specific
receptor, can cause the release of
about 3000 000 molecules of glucose.

b. Glycogen synthesis in the liver
is inhibited by the hormone, and this

is also a consequence of cAMP action.

c.　Gluconeogenesis from amino acids is stimulated. Although this occurs in the liver, glucagon may initiate the process by causing an increase in the release of amino acids from muscle proteins.

Fat metabolism

Glucagon stimulates the release of free fatty acids from the triacylglycerols of adipose tissue. This results from stimulation of a hormone-sensitive lipase in the tissue which is also triggered by the release of cAMP (*see* Chapter 19).

Effects on hormone release

Glucagon will also stimulate the release of insulin from the β cells of the pancreas and the release of catecholamines from the adrenals.

17.9　Hormones of the thyroid

In the normal adult the thyroid weighs about 25 g and consists of two connected lobes, closely associated with the trachea. The blood flow to the gland is rapid, and exceeded only by that to the lungs and carotid body. The gland is divided by connective tissue into clusters of follicles or acini. Each follicle consists of a single layer of cells surrounding the fluid containing the protein thyroglobulin; there are about 10^6 follicles in the human gland.

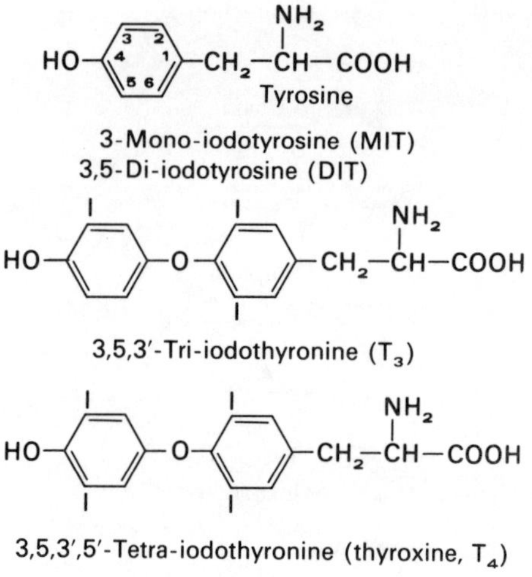

Fig. 17.14.　**Hormones of the thyroid.**

Hormones

Two hormones are produced by the thyroid, thyroxine (T_4) and tri-iodothyronine (T_3), both of which are iodinated derivatives of tyrosine (*see* Fig. 17.14).

Hormone synthesis

Iodide absorbed from the diet or drinking water is transported in the blood to the thyroid, where it is very efficiently transferred into the gland by an 'iodide' pump system dependent on ATP. Uptake is inhibited by certain other ions, such as perchlorate and thiocyanate, which compete for the transport sites (*see* Fig. 17.15).

Fig. 17.15.　**Biosynthesis of thyroid hormones.**
(*a*) Shows thyroxine and tri-iodothyronine synthesis. (*b*) Shows overall synthesis in the thyroid gland from uptake of iodide to secretion of thyroglobulin: (1) trapping of iodide ion from plasma by follicular cells; (2) oxidation of iodide to iodine by membrane-bound peroxidase; (3) synthesis of thyroglobulin. Note reuse of amino acids from proteolytic degradation of thyroglobulin carrying thyroxine in the lysosomes before release into the plasma. (4) Release of thyroglobulin (TG) into the colloid surrounding the follicular cells; (5) iodination of tyrosine residues on thyroglobulin to form MIT and DIT; (6) coupling of iodinated tyrosines to form iodothyronine (T_3 and T_4); (7) pinocytosis—uptake of iodinated thyroglobulin containing T_3 and T_4; (8) uptake of iodinated thyroglobulin by lysosomes followed by proteolysis which releases amino acids, T_3 and T_4; (9) note that T_3 and T_4 may be deiodinated and recycled if not released; (10) secretion of T_3 and T_4 into plasma.

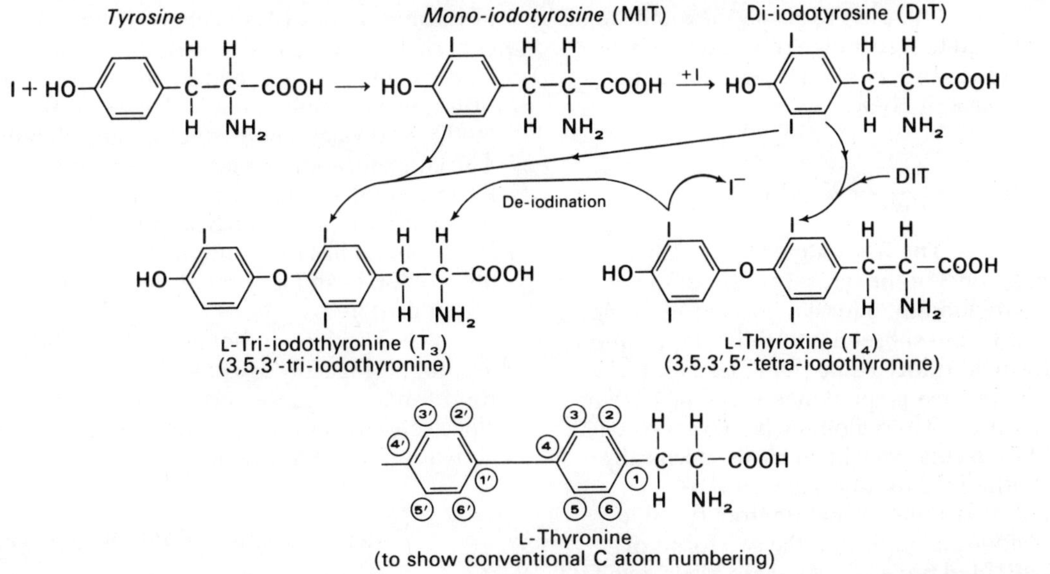

Tyrosine   Mono-iodotyrosine (MIT)   Di-iodotyrosine (DIT)

De-iodination

L-Tri-iodothyronine (T$_3$)
(3,5,3'-tri-iodothyronine)

L-Thyroxine (T$_4$)
(3,5,3',5'-tetra-iodothyronine)

L-Thyronine
(to show conventional C atom numbering)

a

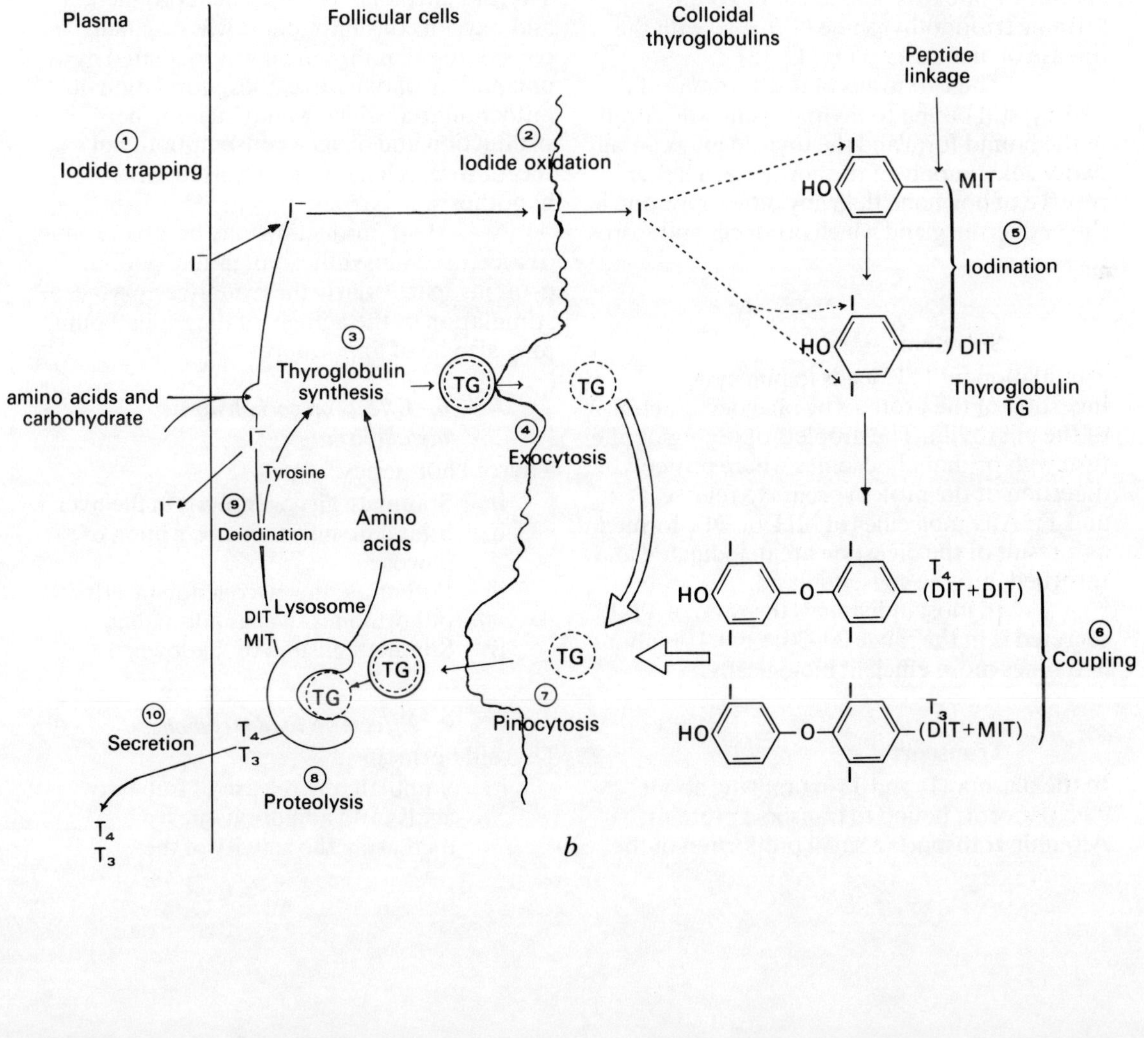

Plasma   Follicular cells   Colloidal thyroglobulins

Peptide linkage

① Iodide trapping

② Iodide oxidation

MIT

⑤ Iodination

DIT

③ Thyroglobulin synthesis

amino acids and carbohydrate

④ Exocytosis

Thyroglobulin TG

Tyrosine

⑨ Deiodination

Amino acids

T$_4$ (DIT+DIT)

⑥ Coupling

Lysosome

DIT
MIT

T$_3$ (DIT+MIT)

⑩ Secretion   T$_4$ T$_3$

⑦ Pinocytosis

⑧ Proteolysis

T$_4$
T$_3$

b

Once in the gland, the iodide ion is oxidized to free iodine, the reaction being catalysed by a peroxidase enzyme in the presence of H_2O_2:

$$2I^- \rightarrow I_2 + 2e^-$$

$$2H_2O_2 \xrightarrow[\text{peroxidase}]{} 2H_2O + O_2$$

The free iodine first rapidly iodinates position 3 of the tyrosyl groups of the protein thyroglobulin to form mono-iodotyrosine (MIT) and subsequently iodinates position 5 to form di-iodotyrosine (DIT; *see Fig.* 17.15). The process is spontaneous and no enzyme is required. Thyroglobulin is a glycoprotein of molecular weight 660 000, specifically synthesized for thyroxine and tri-iodothyronine formation from tyrosine residues. Coupling of the two forms of iodinated tyrosine then takes place, two DIT couples forming thyroxine (T_4) with release of alanine or one MIT and one DIT couple forming tri-iodothyronine (T_3), also with the release of alanine (*see Fig.* 17.15).

The two forms of the hormone, T_3 and T_4, still bound to thyroglobulin are stored in the bound form and the thyroid may contain two-weeks' supply of the hormone, a larger reserve of hormone than any other hormone in the endocrine gland which produces and stores it.

Secretion

Stimulation by TSH leads to pinocytic ingestion of the proteins by phagocytic action of the microvilli. The droplets of thyroglobulin fuse with primary lysosomes where proteolytic digestion of the protein occurs to release T_3 and T_4. Any molecules of MIT or DIT formed as a result of the digestion are deiodinated and resorbed into the cell cytoplasm.

In most individuals the ratio of $T_4:T_3$ secreted is in the range 20–30:1 but T_3 is about 3–5 times more efficient biologically.

Transport

In the plasma, T_3 and T_4 are mostly, about 99·5 per cent, bound to transport proteins. Albumin transports a small proportion of the hormones, about 10 per cent of T_4 and 25 per cent of T_3, but the majority of the hormones, 75 per cent of both T_4 and T_3, are transported by a special α-globulin of molecular weight 60 000 known as a thyroxine-binding globulin. The concentration in plasma is very low, 2 mg/100 ml, but it has one binding site per molecule with a very high affinity for T_4 and T_3. Pre-albumin has a low affinity for T_4, no affinity for T_3 and it normally transports about 15 per cent of T_4.

The protein-bound forms of T_4 and T_3 are not biologically active and should, therefore, be regarded primarily as a reserve, since release of the free forms is essential for activation of the hormones.

Metabolic effects of thyroid hormones

a. *Caloric effects*

Thyroid hormone increases heat production and oxygen consumption. It was originally considered that this effect was mediated by uncoupling of oxidative phosphorylation in mitochondria, which would increase heat production and oxygen consumption, but recent research has not supported this hypothesis.

Heat production may be a by-product of increased biosynthesis of mitochondrial proteins, particularly the oxidative enzymes; stimulation of the activity of the sodium pump may also be of importance.

b. *Effects on carbohydrate metabolism*

Thyroid hormones:
  i.   Stimulate glycogenolysis in the liver
  ii.   Stimulate intestinal absorption of glucose
  iii. Potentiate the glycogenolytic effects of hormones such as adrenaline
  iv.   Stimulate insulin breakdown.

c. *Effects on lipid metabolism*

Thyroid hormones:
  i.   Stimulate the release of free fatty acids from adipose tissue by increasing the activity of the

hormone-sensitive lipase through the intermediate action of cAMP.

ii. Stimulate oxidation of fatty acids

iii. Reduce plasma cholesterol by stimulation of the oxidation of cholesterol to bile acids.

d. Effects on protein metabolism

Thyroid hormones:

i. Stimulate synthesis of proteins, mainly enzymes, involved in oxidative reactions and are necessary for a normal rate of protein synthesis

ii. Stimulate protein catabolism, especially in muscle when present in excess.

Calcitonin

Calcitonin is a peptide hormone which reduces the concentration of blood calcium; it is a single polypeptide chain with a molecular weight of 3421 containing 32 amino acids (*see Fig*. 17.16).

Mode of action

Calcitonin:

i. Reduces the blood calcium concentration by inhibiting bone resorption; osteocytic and osteoblastic activities are reduced

ii. Increases the urinary excretion of calcium and phosphate. These effects

may be mediated by the inhibition of the action of parathyroid hormone.

Parathyroid hormone (parathormone)

The parathyroids, usually four in man, are small glands weighing 30–40 mg each and situated on the dorsal surface of the thyroid, located in the neck. The glands secrete a single hormone, parathyroid hormone (PTH) or parathormone.

Structure of parathormone

Parathormone is a polypeptide composed of 84 amino acids, although the biological activity is associated with the first 20–30 amino acids (*see Fig*. 17.17). The hormone is synthesized as pre-pro-parathormone containing 115 amino acids. This is cleared to produce pro-PTH containing 90 amino acids (molecular weight 12 000), which is then converted into parathormone (molecular weight 9500) and stored in granules. Subsequent to release into the circulation, the large precursor is converted into a large inactive fragment (molecular weight 7000) and a small active fragment (molecular weight 2500).

Mode of action

Parathormone:

i. Promotes bone resorption to mobilize calcium and phosphate; although increased lysosomal enzyme activity

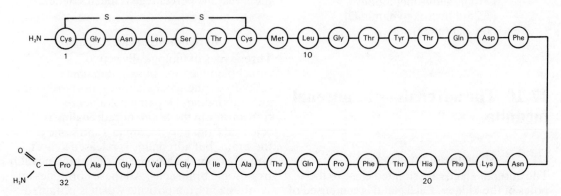

Fig. 17.16. **The structure of human calcitonin.**

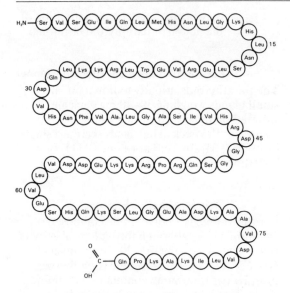

Fig. 17.17. **The structure of human parathyroid hormone.**

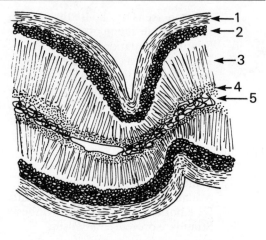

Fig. 17.18. **The structure of the adrenal glands.**
(1) Capsule; (2) zona glomerulosa; (3) zona fasciculata;
(4) zona reticularis; (5) medulla.

may be subsequently involved, the initial process of calcium release is very rapid and is probably dependent on changes in permeability of the plasma membranes or mitochondrial membranes (cf. Chapter 22).

ii.   Facilitates the reabsorption and retention of Ca^{2+} by the kidney

iii. Decreases phosphate reabsorption—the phosphaturic effect. Consequently, the concentration of plasma phosphate is reduced.

iv.   Regulates the conversion of vitamin D to its active metabolite (1,25-dihydroxy-vitamin D_3; cf. Chapter 22).

17.10 The adrenals—the adrenal medulla

Structure of the gland

The adrenal glands are situated close to the poles of the kidney. Each gland is composed of two distinct parts, the medulla which secretes catecholamines and the cortex which secretes steroids (*see Fig.* 17.18).

Adrenal medulla

The medullary cells are arranged in a series of irregular columns close to the venous sinusoids. Each cell contains dense-cored vesicles called 'chromaffin granules' which contain the catecholamines. There are two types of storage cells within the medulla—'adrenaline-storing cells' and 'noradrenaline-storing cells'. In the adult, adrenaline forms 80 per cent of the stored catecholamines but, in the fetus, the percentage is much smaller.

Biosynthesis of catecholamines

Three types of biologically active catecholamines are known: dopamine (dihydroxyphenylethylamine), noradrenaline and adrenaline. Dopamine is a neuro-transmitter in the brain, noradrenaline is released from nerve terminals and some cells in the brain, but adrenaline is released almost entirely from the adrenal medulla, and possibly from some brain cells. All these hormones are synthesized from tyrosine which is obtained from the blood, the sequence of reactions

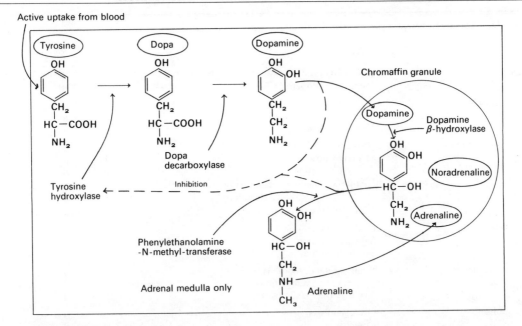

Active uptake from blood

Fig. 17.19. **Biosynthesis of catecholamines.**

being shown in *Fig.* 17.19. It will be noted that dopamine enters the chromaffin granules where it is hydroxylated to noradrenaline, which then leaves the granules to undergo methylation, and is finally returned to the granules.

The catecholamines are stored in the granules which also contain the dopamine hydroxylase enzyme, ATP and a group of poorly defined proteins, the chromagranins.

Control of release

The release of catecholamines is triggered by nerve impulses, which cause release of acetylcholine from the nerve terminals, depolarization of the membrane, followed by entry of Ca^{2+} into the cell. The calcium then causes, by a mechanism not yet understood, a fusion of the granules with the cell membranes and the expulsion of their contents by exocytosis.

Adrenergic receptors

In order to exert their effects, catecholamines must first bind to receptors, known as 'adrenergic receptors', present on the cell surface. There are two types of receptor—α and β. α receptors respond preferentially to noradrenaline and generally involve the contraction of smooth muscle, whereas β receptors respond preferentially to adrenaline, causing relaxation of smooth muscle and metabolic effects, and are involved mainly in the production of cAMP. A summary of the receptors of different tissues is shown in *Table* 17.1.

The catecholamines have powerful effects on the cardiovascular system and on several metabolic processes.

Cardiovascular system

Adrenaline and noradrenaline:

a. Increase the heart rate by their action on the sinoatrial node

b. Increase the blood sugar concentration

c. Stimulate breakdown of triacylglycerols stored in the adipose tissues to form free fatty acids; this effect is mediated by

Table 17.1 **Adrenergic receptor classification**

Tissue	Effect	Receptor
Heart	Increased rate	β
	Increased contractility	β
Blood vessels	Arterial constriction	α
	Arterial dilation (muscle)	β
	Venoconstriction	α
Other smooth muscle	Bronchioles (dilation)	β
	Spleen (contraction)	α
	(relaxation)	β
	Nictitating membrane (contraction)	α
	(relaxation)	β
	Iris radial muscle (contraction)	α
Exocrine glands	Pancreatic secretion	α
	Salivary secretion	β
	Sweat (apocrine)	α
Endocrine glands	β cell: insulin (inhibition)	α
	(stimulation)	β
	α cell: glucagon (stimulation)	β
Metabolism	Glycogenolysis	β
	Lipolysis	β
	Calorigenesis	β

activation of the formation of cAMP

d.   Stimulate gluconeogenesis in the liver.

17.11 The adrenal cortex

The adrenal cortex completely surrounds the medulla but, despite the close relationship, the cortex and medulla are quite separate entities carrying out entirely different functions.

The cortex is composed of three morphologically distinct zones or regions which synthesize different hormones.

a.   Zona glomerulosa—the outermost region which secretes aldosterone
b.   Zona reticularis—close to the medulla and synthesizes oestrogens and androgens
c.   Zona fasciculata—in an intermediate region between the other two zones, producing glucocorticoids, cortisol and corticosterone.

The structural relationships between these zones are shown diagrammatically in *Fig.* 17.18.

Biosynthesis

The major steroids of the adrenals are shown in *Fig.* 17.20, and the synthetic pathways in *Fig.* 17.21. Cholesterol, shown as the starting compound, may be obtained from the blood or synthesized within the gland. The cholesterol is stored, probably in the form of cholesterol esters, and release of cholesterol from these esters by a lipase is vital in controlling the synthetic process. Both the endoplasmic reticulum and mitochondria are involved in the synthetic processes and, since there is no storage of steroid hormones in the gland, the whole process of synthesis, which is triggered by stimulation, must be very efficient.

In the zona glomerulosa, corticosterone is converted to aldosterone and in the zona reticularis, pregnenolone is converted to androstenedione.

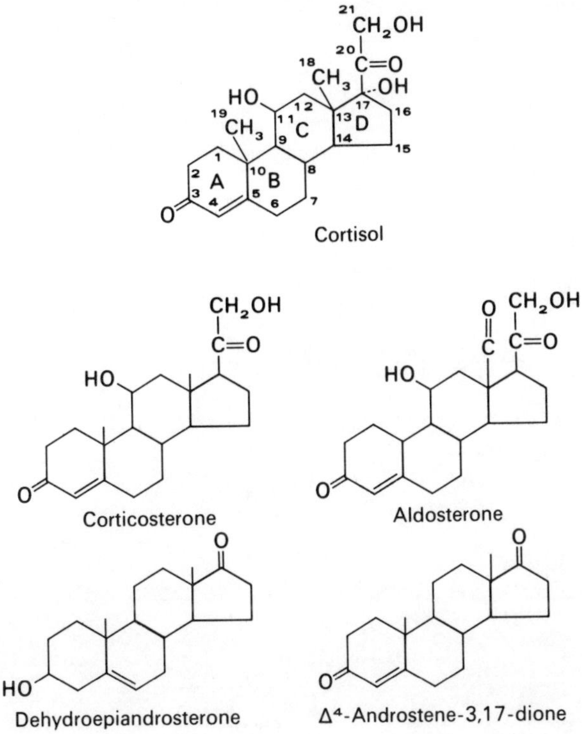

Fig. 17.20. **Steroid hormones of the adrenal cortex.**

The rate of production and release of glucocorticoids is controlled by ACTH. ACTH attaches to receptors on the zona fasciculata and causes a rapid increase in the production of glucocorticoids. After binding of ACTH there is an increase in ion permeability and activation of adenylate cyclase to produce cAMP. A protein kinase activated by cAMP then phosphorylates and activates two proteins, one which splits cholesterol from its esters and another which promotes the conversion of cholesterol to pregnenolone.

The release of aldosterone is controlled by several factors of which the most important are ACTH and the renin–angiotensin system. The liver secretes the peptide angiotensinogen (hypertensinogen) which contains 14 amino acids and is transported with the α_2 globulin fractions. This is hydrolysed by a proteolytic enzyme secreted by the juxtaglomerulosa cells of the kidney to form a decapeptide, angiotensin I (hypertensin). A conversion enzyme, occurring mainly in the lungs, then acts on the decapeptide, removing two further amino acids, to form an octapeptide—angiotensin II (*see Fig.* 17.22). Antiotensin II regulates the release of aldosterone from the zona glomerulosa.

Plasma transport

All steroid hormones, 90 per cent of glucocorticoids and 50–70 per cent aldosterone bind to plasma proteins. Cortisol and corticosterone bind to a special α-globulin—corticosteroid-binding globulin or transcortin. It is not yet established whether aldosterone has a specific binding protein.

The concentrations of hormones in plasma are shown in *Table* 17.2. Cortical hormones show wide variations in

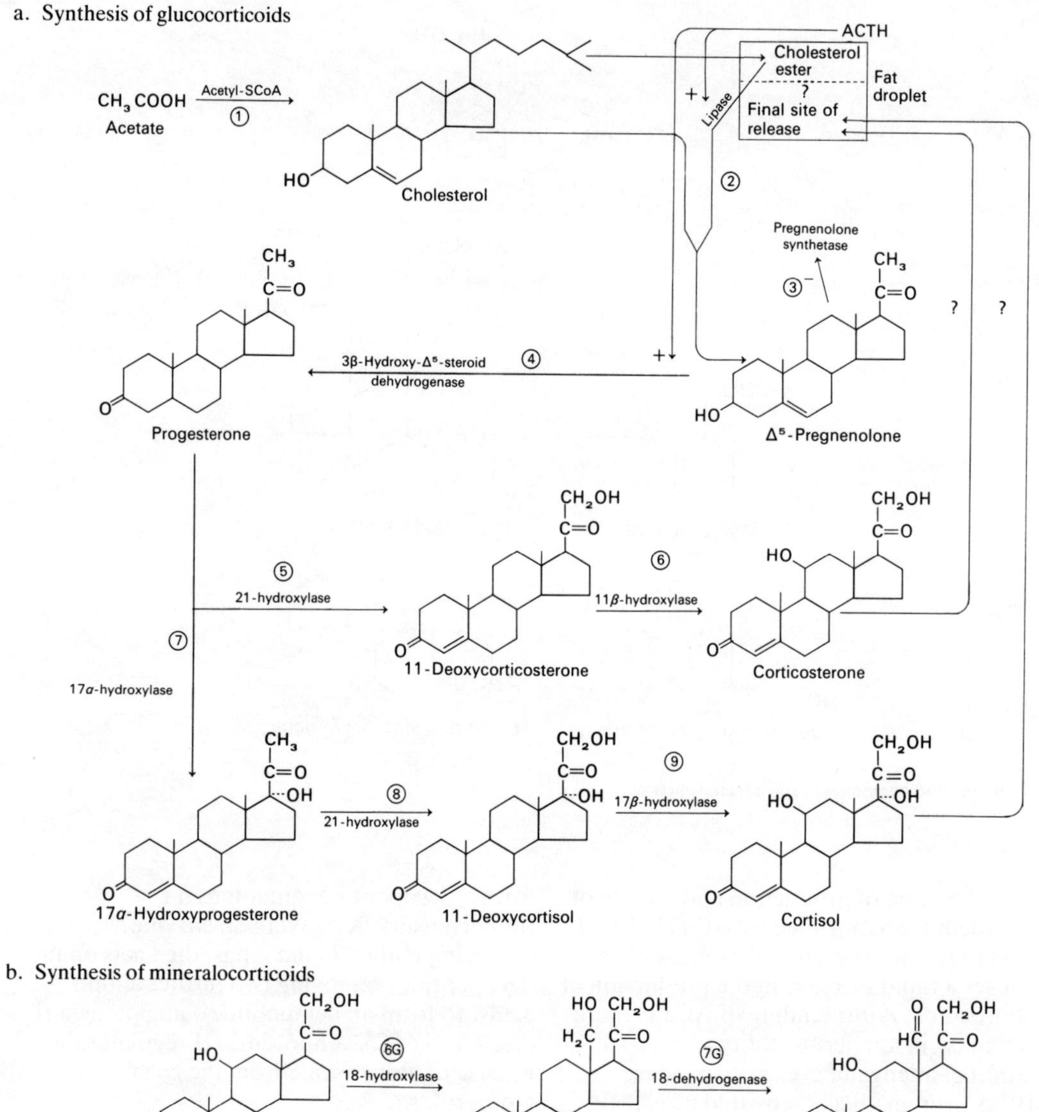

a. Synthesis of glucocorticoids

b. Synthesis of mineralocorticoids

c. Synthesis of androgens

Fig. 17.21. **Biosynthesis of adrenal steroids.**

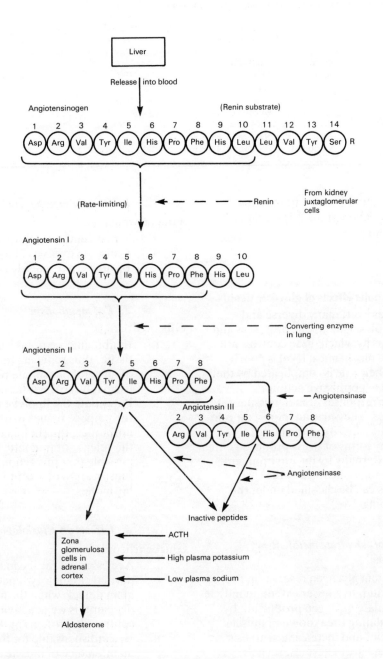

Fig. 17.22. **The renin–angiotensin system showing mechanism of release of aldosterone.**

Table 17.2 **Characteristics of circulating adrenal corticoids in man**

Adrenal corticoid		Total plasma concentration nmol ($\mu g/dl$)	Percentage bound	Half-life, min
Cortisol	08:00	140–170 (5–25)		
	24:00	<140 (<5)	~90	50–90
Corticosterone		~35 (1)		
		pmol/l (ng/dl)		
Aldosterone (mean ± s.e.)	Recumbent	92 ± 75 (3·4 ± 2·8)		
	Ambulent	378 ± 169 (14·3 ± 6·4)		

concentration and this is due both to the degree of stimulation of the release and to circadian rhythms.

Metabolic effects of glucocorticoids

Glucocorticoids exert many diverse and complex controls on metabolic processes and the mechanisms by which these controls are achieved at the biochemical level is poorly understood. Their role is complicated by their ability to exert a 'permissive action', influencing the responses of many tissues to other hormones. They play an important role in increasing the capacity of the body to withstand many forms of stress produced either internally or externally by the environment.

The metabolic roles of the glucocorticoids can be classified under the following headings.

a. Carbohydrate metabolism

Glucocorticoids:

  i.   Maintain glycogen reserves in the liver and, to a lesser extent, in muscle
 ii.   Stimulate glycogen production by stimulating breakdown of muscle proteins and increasing synthesis of some liver enzymes, such as transaminases, required for gluconeogenesis
iii.   Exert an anti-insulin effect on peripheral tissues (including adipose tissues) by inhibiting glucose uptake.

b. Protein metabolism

Glucocorticoids:

  i.   Are essential for normal growth
 ii.   Cause, in excess, catabolism of proteins, particularly in muscle.

c. Fat metabolism

Glucocorticoids:

  i.   Inhibit glucose uptake in adipose tissue, thus reducing the availability of glycerol phosphate required for triacylglycerol synthesis
 ii.   Potentiate the lipolytic response of the adipose tissues to other hormones, thus helping to increase the release of free fatty acids. It is possible that this action results from a sensitization of receptors for other hormones.

d. Effects on physiological functions

Glucocorticoids:

  i.   Are essential for the maintenance of normal circulatory functions; the main defect, when the hormones are deficient, is weak activity of the contractile proteins of the heart
 ii.   Are indispensable for the maintenance of muscle activity; the mechanisms involved are complex and not fully understood, and neither glucose nor electrolytes can compensate for a deficiency of the hormone

iii. Are essential for normal brain function and reduced concentrations of the hormone produce detectable changes in sensations and higher functions, such as memory; the effects may be mediated by the role which the glucocorticoids play in conduction of the nerve impulses

iv. Regulate the development of lymphoid tissue. In deficiency conditions, rapid hyperplasia of lymphoid tissue occurs. It is possible that the glucocorticoids may regulate the activity of lysosomal enzymes in this tissue.

v. Exert anti-inflammatory effects probably by reducing the concentration of mast cells around the site of injury. Mast cells are the main source of histamine which is responsible for the inflammation. The glucocorticoids are effective in stabilizing lysosomal membranes (cf. Chapter 3) and many of the inflammatory responses may be associated with the release of lysosomal hydrolyases. Stabilization of the lysosomal membrane would clearly prevent this occurring

vi. Control the water balance by exerting a control on the glomerular filtration rate

vii. Play a role in parturition by stimulating the conversion of progesterone to oestrogen.

Mineralocorticoids

Synthesis and release of aldosterone

Both the rate of synthesis and the release of aldosterone from the zona glomerulosa are influenced by angtiotensin II, ACTH, high plasma K^+ and low plasma Na^+. Long-term salt deprivation stimulates aldosterone secretion by promoting conversion of corticosterone to aldosterone.

ACTH appears, in man, to be less important than other factors in the regulation of aldosterone secretion. It is likely, however, to exert a permissive effect on the zona glomerulosa by ensuring adequate steroid precursors for aldosterone.

Mode of action

Aldosterone is the principal and most potent mineralocorticoid. Aldosterone:

a. Promotes reabsorption of Na^+ from the urine, sweat, saliva and gastrointestinal tract. The effect on Na^+ reabsorption from the distal tubule of the kidney is most important; reabsorption from the proximal tubule also occurs

b. Promotes Na^+/K^+ exchange across membranes

c. Leads to K^+ depletion if used in excess

d. Causes increased excretion of both Ca^{2+} and Mg^{2+} in the urine. This effect is probably secondary to the increased expansion of extracellular fluid.

The present biochemical action is not yet resolved but it may act on DNA allowing production of an mRNA which codes for the synthesis of a specific protein. The role of the protein is uncertain: it may regulate a form of oxidative phosphorylation which specifically provides energy for the entry of Na^+ or it may be more specifically concerned with the activity of a Na^+ carrier.

Deficiency of aldosterone and glucocorticoids

Adrenal cortical deficiency occurs in Addison's disease, as a result of tubercular or neoplastic damage to the adrenal gland. Complete loss of adrenal function rapidly causes death, but the disease usually develops over a long period and is insidious. The symptoms are all predictable from the known functions of the hormone.

17.12 Control of metabolism by hormones

In summary it is useful to consider the inter-relationships of the hormone actions which regulate metabolism. These are summarized in *Fig*. 17.23.

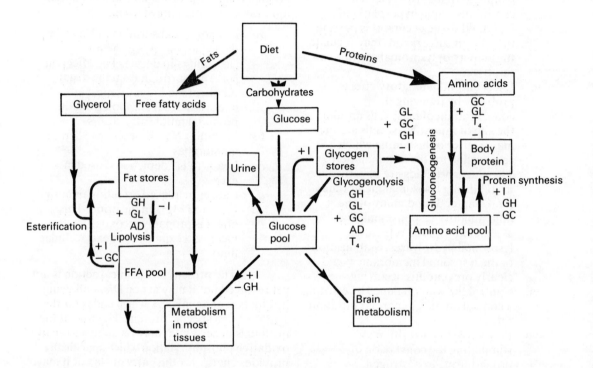

Fig. 17.23. **Summary of metabolic control by hormones.**
I insulin; GH growth hormone, GL glucagon; GC glucocorticoids; AD adrenaline; T$_4$ thyroxine. + stimulates the process; − inhibits the process.

Chapter 18 Plasma glucose and its regulation

18.1 Introduction

The concentration of glucose in blood in all mammals, including man, is maintained at a very constant level, about 90 mg per 100 ml or 5 mM. In normal healthy man, the concentration is kept constant during fasting periods between meals by a steady release of glucose from liver glycogen and the major fluctuations are increases which occur following meals (*see Fig.* 18.1).

  The accurate maintenance of the glucose concentration must possess strong survival value because, during the course of evolution, the control mechanisms have become much more effective and accurate. In reptiles, for example, the fluctuations of blood glucose concentration are much greater. The reason for this very effective homeostasis of glucose is probably associated with the needs of the brain which uses between

60 to 80 per cent of the glucose released in the fasting state. A very serious clinical condition develops if the blood glucose concentration falls drastically, for example to half its normal value. Coma, convulsions and even death rapidly ensue and the nervous system is clearly deprived of adequate glucose.

  Conversely, a raised glucose concentration in blood, for example to double its normal value, is not in itself dangerous over a short period, but over a longer period as in poorly controlled diabetes, glucose, in high concentrations in the blood, reacts with the amino groups of haemoglobin and gives rise to a new form of haemoglobin HbA_{IC}. The measurement of this form can indicate the severity of the diabetes. Glycosylation of other proteins, for example in the blood vessel walls, may also occur and cause significant pathological disturbances. The raised glucose concentration indicates an existence of clinical disease, usually diabetes. The high circulating concentration of glucose causes the renal threshold for glucose to be exceeded, and glucose appears in the urine. Thus it is possible to carry out a simple test for diabetes by testing urine for glucose. In normal subjects a negligible quantity of glucose should appear in the urine. Some other pathological conditions, for example kidney disease, can cause an increase in glucose concentration in the urine; accurate diagnosis of diabetes can, however, be made by the 'glucose tolerance test'. The fasted subject is given a test dose of 50 g glucose, the blood glucose being measured just before the test and at intervals afterwards. In normal individuals, the raised glucose concentration in blood, resulting from the

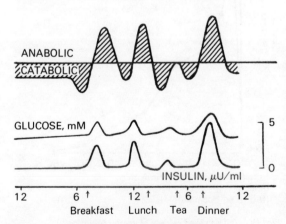

Fig. 18.1. **Changes in glucose and insulin levels during a typical day showing the responses to meals.**

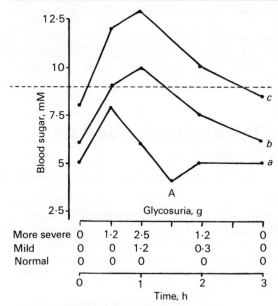

Glycosuria, g				
More severe	0	1·2	2·5	1·2	0
Mild	0	0	1·2	0·3	0
Normal	0	0	0	0	0

Fig. 18.2. **Glucose tolerance test.**
Typical curves are shown for (*a*) a normal individual, (*b*) mild diabetes and (*c*) severe diabetes after ingestion of 50 g glucose. – – – Indicates the renal 'threshold'.

intestinal intake, is rapidly reduced to a normal value but, in diabetic individuals who secrete little or no insulin, the glucose concentration remains at a much higher level for a prolonged period (*see Fig.* 18.2).

18.2 Maintenance of glucose concentration during fasting conditions

Between meals the blood glucose is maintained by the catabolism of liver glycogen by the process shown in Chapter 5. The human liver can produce about 125–150 mg glucose per min or 180–220 g per 24 h. About 100 g glycogen are stored in the liver and this store can, therefore, last about 12 hours. It is unlikely, however, that supplies of glycogen are normally run down to zero during the overnight fasting period and gluconeogenesis (cf. Appendix 24) is switched on to provide blood glucose from amino acids. The major proportion of the glucose, about 70 per cent, is utilized by the central

nervous system and the remainder by the red blood cells, bone marrow and renal medulla. However, unlike the brain where glucose is completely oxidized to CO_2, the other tissues convert it to lactate which is circulated to the liver where it is resynthesized into glycogen.

During fasting, the plasma concentration of insulin is very low and this permits the release of amino acids from muscle proteins for gluconeogenesis and fatty acids from adipose tissue. If, however, the starvation condition is prolonged for several days, the release of amino acids from muscle is depressed. This condition is discussed in Chapter 23.

18.3 Insulin release in the fed condition

One of the most characteristic responses to the ingestion of a carbohydrate meal is the rise in blood glucose concentration which is

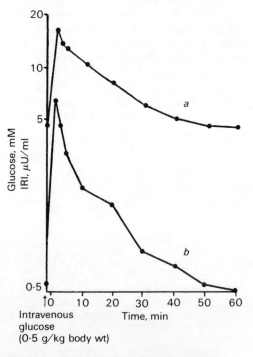

Fig. 18.3. **The rapid release of insulin following intravenous infusion of a dose of glucose.**
(*a*) Glucose; (*b*) IRI (immunoreacting insulin).

accompanied by a rise in the concentration of plasma insulin (*Fig.* 18.1). The response of the pancreatic β cells in secreting insulin is extremely rapid, a fact that can be demonstrated by infusing glucose intravenously. When this is carried out, an increase of insulin concentration can be demonstrated to occur within a minute or even less (*Fig.* 18.3).

It is clear that immediately glucose becomes available from food, it is essential to switch off the production of glucose from liver glycogen. This occurs very rapidly after infusion of glucose and is caused by the rapid response of the liver to very small rises in plasma insulin concentration. The mechanism by which this is achieved is still uncertain.

The rapid release of insulin is much more important in the control of blood glucose than the total quantity of insulin release over a period of an hour. This fact can be demonstrated in patients suffering from severe diabetes who obviously produce very little insulin. The effects observed after injection of insulin into these individuals clearly show that the efficient utilization of glucose correlates most closely to the amount of insulin available during the first 10 minutes of dosing with glucose rather than to the total insulin release. These observations have led to the low-dose infusion or injection procedure for the treatment of diabetic ketoacidosis.

18.4 Tissue response to increased insulin and glucose plasma concentration

Studies of the effects of glucose infusions into man show that only a small quantity of glucose, about 10 per cent, is removed by the liver, 30–40 per cent is removed by muscle and 25 per cent by the adipose tissue; the remaining 25 per cent is believed to be oxidized (*see Fig.* 18.4). During absorption of glucose from the gut these proportions are likely to be similar although a greater proportion of glucose may become incorporated into the liver glycogen.

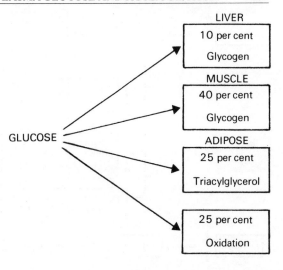

Fig. 18.4.   **Tissue uptake of perfused glucose.**
Note: tissue uptake from ingested glucose is believed to follow a similar pattern, but uptake by the liver may be slightly greater than for perfused glucose.

The uptake of glucose into the tissues is controlled, to a large extent, by insulin which is bound to special receptors in the tissues and stimulates the uptake of glucose into, for example, muscle and adipose tissue. In the muscle, glucose is incorporated into glycogen and, in the adipose tissue, into the glycerol moiety of triacylglycerols. It is first converted to glycerol phosphate through part of the glycolysis pathway prior to incorporation of fatty acids and formation of triacylglycerols for storage (cf. Chapter 19).

In addition to its affect on the uptake of glucose by muscle, insulin has an important regulatory action on the supply of amino acids from muscle protein for gluconeogenesis. The release of amino acids is dependent on the low level of circulating insulin and on the release of glucagon. When insulin concentration is raised as the result of glucose intake, the uptake of amino acids into muscle is promoted, as is their conversion into protein. The release of amino acids therefore ceases immediately.

Insulin also switches off the release of fatty acids from adipose tissue in a similar fashion. It reduces the concentration

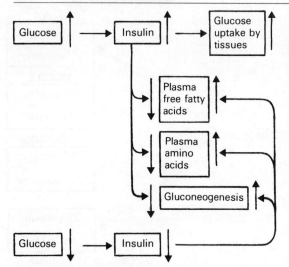

Fig. 18.5. **The role of insulin in the regulation of plasma glucose, amino acids and fatty acids.**
↑ Increase in plasma concentrations; ↓ decrease in plasma concentrations.

of cAMP essential for the activation of lipase which in turn releases fatty acids from stored triacylglycerols (cf. Chapter 19). The essential sequence of events is shown in *Fig.* 18.5.

18.5 The role of glucagon

Glucagon which is released from the α cells of the pancreatic islets, also plays a very important role in the regulation of the plasma glucose concentration. Its effects are virtually opposite to those of insulin, causing liver glycogenolysis, inhibition of glycogen synthesis and stimulation of gluconeogenesis in the liver. Glucagon is, therefore, most active during the fasting condition when the concentration of insulin is low and its release is inhibited by an influx of glucose.

The action of glucagon on glycogenolysis is mediated by the release of cAMP which ultimately causes the activation of phosphorylase to catalyse the phosphorolysis of glycogen producing glucose 1-phosphate (Chapter 5).

It is also likely that glucagon, in addition to and in conjunction with insulin,

plays an important role in the regulation of glucose production from amino acids in muscle. An influx of amino acids from a large protein meal, like glucose causes a stimulation of insulin release which immediately switches off the liberation of glucose from liver glycogen. This could clearly result in a disastrous hypoglycaemia. To overcome this problem, glucagon is believed to stimulate the trapping of amino acids in the liver and these amino acids, by undergoing gluconeogenesis, are able to compensate for the inhibition of glucose release from liver glycogen.

A summary of the inter-relationships between insulin and glucagon is shown in *Table* 18.1.

Table 18.1 **Insulin and glucagon levels and liver glucose production in various metabolic states**

Meal	Insulin level	Glucagon level	Liver glucose production
None	+	+	+
Carbohydrate (small)	+ +	±	0
Carbohydrate (large)	+ + +	±	0
Protein	+ +	+ +	+
Protein and carbohydrate	+ + + +	±	0
Prolonged fast	±	+ +	+
Diabetes	0	+ + + +	+ + +

18.6 Circulating glucose under stress conditions

So far we have discussed the responses of glucose in the blood to varied conditions of nutrition; marked fluctuations can also, however, occur for short periods under conditions of stress which are unrelated to the nutritional state.

One of the best studied hormones to cause this effect is adrenaline. This hormone, together with some other catecholamines, is released in response to a very wide variety of stressful situations or pathological conditions, such as hypoxia, asphyxia, acidaemia, hypoglycaemia, hypothermia, hypotension, haemorrhage, exercise, fear, anger or excitement. The release is triggered by the sympathetic nervous system and was described in 1929 by Cannon as the 'fright, fight or flight' syndrome.

Adrenaline causes a wide variety of important physiological effects particularly on the circulating system, but it is also a powerful stimulator of glycogenolysis and causes a rapid increase in the concentration of blood glucose. The action of adrenaline is mediated by the formation of cAMP in a similar manner to the action of glucagon (Chapter 17). Adrenaline also causes release of fatty acids from adipose tissue (Chapter 19) and the glucose and fatty acids clearly provide the energy for the 'flight' or 'fight'.

In addition to adrenaline, the glucocorticoids released from the adrenal cortex also have important effects on blood glucose (Chapter 17). Their effects are, however, generally responses to a more prolonged stress condition, such as a debilitating illness, rather than the immediate response which occurs following adrenaline release. The glucocorticoids have important effects on gluconeogenesis and these are exerted in two ways. Firstly, they stimulate catabolism of muscle protein releasing amino acids for uptake by the liver, and secondly they stimulate the production of enzymes, such as transaminases, in the liver, these enzymes aiding the conversion of amino acids to glucose in the liver.

In summary, we can state that the concentration of glucose in the plasma is dependent on a wide range of nutritional and hormonal factors, but these are carefully inter-regulated to effect a remarkably accurate control and reliable homeostasis.

Chapter 19 Plasma lipids and their regulation

Whereas glucose and amino acids are transported in aqueous solution in the blood, the transport of the water-insoluble lipids involves much more complex molecules. All lipids are transported as lipoprotein complexes and the majority of these contain triacylglycerols (triglycerides), phospholipids, cholesterol and cholesterol esters.

The roles, biosynthesis and utilization of all the components of these complexes are not yet completely understood, but they are of major importance in health and disease. For example, increased concentrations of some of the plasma lipids are believed to be one of the main causes of arterial disease and thrombosis.

19.1 Classification of plasma lipids

The plasma lipids are usually classified into two main groups:

  a.   Free fatty acids
  b.   Lipoproteins.

a. Free fatty acids

The so-called free fatty acids do not, in fact, exist in the plasma in a free form but are always bound to albumin. Normally about 4–5 molecules of fatty acid are transported on each molecule of albumin and the complex so formed may be described as a special type of lipoprotein.

b. Lipoproteins

Partial separation of the lipoproteins may be achieved by electrophoresis and, as a result, α- and β-lipoproteins can be distinguished depending on whether they fractionate with the α- or β-globulins (*Fig.* 19.1).

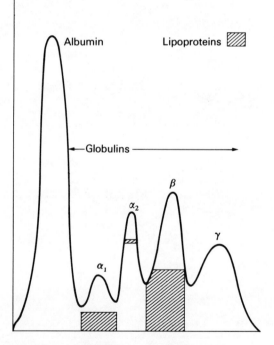

Fig. 19.1. **Electrophoretic separation of plasma lipoproteins.**
Area beneath each component peak and cross-hatched (lipoprotein) areas are proportional to the quantity in the plasma.

However, most of our knowledge of plasma lipoproteins has been achieved by ultracentrifugation or gradient centrifugation of plasma in media of different densities adjusted by varying the concentration of salts, such as sodium chloride or potassium bromide. By this means five distinct types can be separated and described as the very-low-density lipoproteins (VLDLs), low-density lipoproteins (LDLs), high-density lipoproteins (HDLs) and the very-high-density lipoproteins (VHDLs).

To these categories we may also add the chylomicrons which are of comparable density to the VLDLs but present in the plasma only after a fatty meal. It is important to realize that each classification includes a family of closely related proteins and that all lipoproteins of the plasma may not fit neatly into any one of the categories. Thus an intermediate density lipoprotein (IDL) has recently been described.

Constitution of the lipoproteins

Lipids Typical analyses of the lipid composition of the plasma lipoproteins are shown in *Table* 19.1. The following points should be noted:

    i.   All lipoproteins contain all types of lipid, although the quantities and proportions vary widely

    ii.   The density is directly proportional to the protein content and inversely proportional to the lipid content

    iii.   There are close similarities between the compositions of VLDLs and HDLs, but in the latter the protein has replaced most of the triacylglycerols

    iv.   The main bulk of the plasma cholesterol is transported on the β-lipoprotein fraction.

Protein It was originally believed that each lipoprotein type contained one of three apoproteins (A, B or C), but it is now realized that each species of lipoprotein may contain at least seven distinct apoprotein species. This has given rise to the rather confusing nomenclature shown in *Table* 19.2.

Some of these proteins have been purified and characterized (*Table* 19.3) but apoB of the LDL complex is extremely difficult to separate from the associated lipid. The most interesting finding to result from the studies of these proteins is that some have specific activating effects on some of the important enzymes involved in lipoprotein metabolism. These roles will be explained in the following paragraphs.

Structure of the lipoproteins
By X-ray diffraction techniques some progress has been made towards elucidating the very

---

Table 19.1 **Composition of the lipoprotein families**

Lipoprotein type	Composition, % dry wt, of					Total lipid, % dry wt
	Protein	Triacylglycerol	Unesterified cholesterol	Esterified cholesterol	Phospholipids	
Chylomicrons	2	88	2	4	4	98
VLDL	5–12	50–60	3–5	10–13	13–20	88–95
LDL	22	8–10	10	47–55	28–30	78
HDL	50	8	3	14	22	50
VHDL	62	5	0·3	3	28	38

VLDL = very-low-density lipoproteins.
LDL = low-density lipoproteins.
HDL = high-density lipoproteins.
VHDL = very-high-density lipoproteins.

Table 19.2 **Composition and properties of human plasma lipoproteins**

Lipoprotein type	Density, g/cm^3	$10^{-6} \times M_r$	Electrophoretic mobility	Apoproteins*	
				Major	Minor
Chylomicrons	0·95	500–430 000	Origin	apoA-I (1·4%) apoB (22·5%) apoC-I (15%) apoC-II (15%) apoC-III (36%)	apoA-II (4·2%) apoE
VLDL	0·95–1·006	3–13	pre-β	apoB (37%) apoC-I (3%) apoC-II (7%) apoC-III (40%) apoE (13%)	apoA-I apoA-II apoD
LDL	1·006–1·063	2·2	β	apoB (98%)	
HDL	1·063–1·210	330 000	α	apoA-I (67%) apoA-II (22%)	apoC-I (1–3%) apoC-II (1–3%) apoC-III (3–5%) apoD apoE
VHDL	1·210–1·250	175 000	α	apoA-I apoA-II	

* The figures in parentheses show the percentage of total apoproteins in the fraction. For abbreviations *see Table* 19.1.

Table 19.3 **The apoproteins of plasma lipoproteins**

Apoprotein	$10^{-3} \times M_r$	No. of amino acids	Function
A-I	27–28	245 (single chain)	Activator of lecithin : cholesterol acyltransferase
A-II	≈8·5	77 (2 chains)	Inhibitor of lecithin : cholesterol acyltransferase
B	8–275	?	?
C-I	≈6·5	57	Activator of lecithin : cholesterol acyltransferase and lipoprotein lipase
C-II	12·5	100	Powerful activator of lipoprotein lipase
C-III	9·0	79 (single chain)	?

complex structures of the lipoproteins, but the precise structures remain obscure.

It is likely that the lipoproteins containing high concentrations of lipid, such as the chylomicrons and VLDLs are composed of a triacylglycerol core with a surrounding layer of protein and phospholipids to maintain water stability.

The LDLs may be composed of a protein core surrounded by a lipid or trilayer of similar structure to that proposed for membranes (Chapter 2), but protein molecules are also located in the outer lipid layer.

The outer surface of the HDLs is composed of approximately 50 per cent protein and 50 per cent polar heads of phospholipids; a possible representation, obtained by X-ray diffraction analysis, is shown in *Fig.* 19.2.

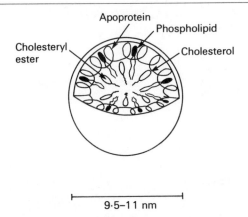

Fig. 19.2. **Schematic representation of a high-density lipoprotein.**

Utilization of plasma lipids

Before discussing the detailed metabolism of the plasma lipids, it is useful to consider the main sites of synthesis and catabolism of the most common ones; these are summarized in *Table* 19.4. Important information about their mode of usage can be obtained from the study of radioactively labelled lipids. The free fatty acids have a very short half-life of a few minutes in the plasma, so that they are rapidly utilized after production, whereas the half-life of the triacylglycerols of the chylomicrons or VLDLs is between 5 and 10 hours.

Cholesterol is much less rapidly turned over in the plasma and the half-life is between 1 and 2 weeks (*Table* 19.4).

19.2 Lipid transport in the fed state

After a fatty meal the main bulk of the ingested triacylglycerols is absorbed into the lymph as chylomicrons, which enter the blood mainly through the thoracic duct.

The adipose tissue is mainly responsible for the removal of chylomicrons from the circulation, although a small proportion may be taken up by other tissues. Whatever tissue is involved, the mechanism of uptake is always the same and involves a special lipase described as 'clearing factor lipase' or lipoprotein lipase. The term 'clearing factor' was originally used because it is the enzyme involved in clearing the milky serum which exists after a fatty meal. The name lipoprotein lipase is used to emphasize the fact that, as distinct from pancreatic lipase, it will act only on triacylglycerols in a lipoprotein complex.

Mode of action of lipoprotein lipase

A precursor of lipoprotein lipase is synthesized in the adipocytes and transferred to an outer compartment, or outer membrane of the cells, where it is activated by heparin or a related proteoglycan. Thence it is transported to the endothelial cells of the adipose tissue (*Figs*. 19.3, 19.4); *Fig.* 19.3 shows that lipoprotein lipase may be controlled

---

Table 19.4 **Sites of synthesis and catabolism of lipoproteins**

Plasma component	Site of synthesis	Half-life	Site of catabolism
Free fatty acids	Adipose	few min	All tissues
Chylomicrons	Intestine	15–30 min	Many tissues, especially adipose
VLDL	Liver + intestine	6–12 h	Many tissues
LDL	Capillary endothelium	2–4 days	Macrophages
HDL and VHDL	Liver	3–5 days	{ Liver Kidney Lysosome

Apoproteins A-I, C-I, C-II and C-III are synthesized in the liver, and A-I, A-II and B in the intestine.
VLDL = very-low-density lipoprotein.
LDL = low-density lipoprotein.
HDL = high-density lipoprotein.
VHDL = very-high-density lipoprotein.

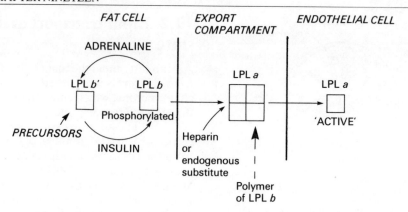

Fig. 19.3. Activation of lipoprotein lipase and transport to the endothelial cells of adipose tissue.
Lipoprotein lipase (LPL) is believed to exist in 3 forms b', b, and a. The active form a is a polymer of b forms, and is produced in the presence of heparin or a related proteoglycan. Before lipoprotein lipase b' is polymerized to a it must first be changed to b. This process, which may involve phosphorylation or a configurational change, is under control of insulin and adrenaline.

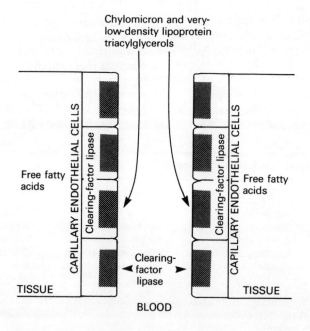

Fig. 19.4. Mode of action of lipoprotein lipase at the endothelial cell.

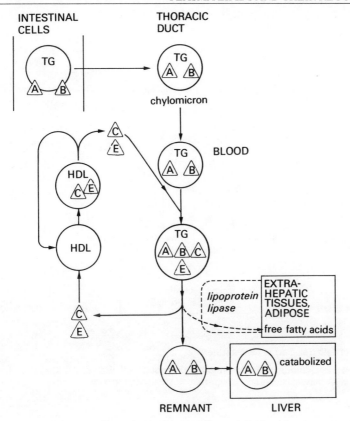

Fig. 19.5. **Disposal of chylomicrons.**

Ⓐ Ⓑ Ⓒ represent the apoprotein components of the lipoproteins.

Chylomicrons are synthesized with apoA and apoB but they pick up apoC from high-density lipoproteins (HDLs) in order to activate the lipoprotein lipase. After hydrolysis of the triacylglycerol (TG), apoC returns to an HDL molecule. Final catabolism of the chylomicron remnant occurs in the liver.

hormonally by insulin and adrenaline. Insulin promotes the activation of the enzyme and adrenaline inhibits it.

As shown in *Table* 19.2, apoprotein C-II is a strong activator of lipoprotein lipase and this apoprotein is transferred from HDL to the chylomicrons to activate the enzyme when they reach the endothelial cells. At the endothelial cell, rapid hydrolysis of the chylomicron triacylglycerols to diacylglycerols and free fatty acids occurs. One molecule of free fatty acid from each molecule of triacylglycerol is released into the circulation

and the diacylglycerols, by a process not yet understood, cross the cell membrane, and one is subsequently completely hydrolysed to glycerol and fatty acids probably by the hormone-sensitive lipase. The glycerol is released into the circulation but the fatty acids are retained and synthesized into triacylglycerols in the adipose tissue cells.

The chylomicron fragment, now stripped of most of its triacylglycerols, releases its C-II apoprotein to HDLs and these are finally taken up by the liver (*Fig.* 19.5).

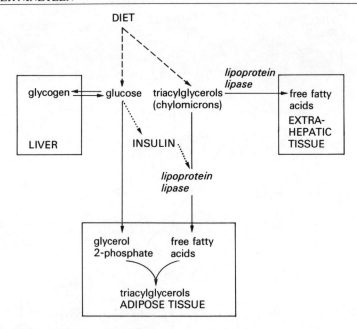

Fig. 19.6. **Summary of lipid transport in the fed state.**

Note that the increased concentration of insulin caused by the raised level of
glucose after a mixed meal causes activation of lipoprotein lipase in adipose tissue
(cf. *Fig.* 19.3).

Role of glucose and insulin

After a normal mixed meal, the plasma glucose
will rise and this will cause a rise in the
concentration of plasma insulin. Insulin, as
shown in *Fig.* 19.3, enhances the activation
of lipoprotein lipase so that the circulating
triacylglycerols are rapidly taken up into the
tissues.

Glucose also plays an important role
in the provision of glycerol phosphates for the
synthesis of triacylglycerol from the fatty acids
taken up from the chylomicrons. The adipose
tissue is unable to convert glycerol to glycerol
phosphates and these must be provided by the
metabolism of glucose through the glycolytic
pathway (Appendices 14, 15).

Summary

The process is summarized in *Fig.* 19.6, which
also shows the roles of glucose and insulin.

19.3 Lipid transport in the fasting state

When the chylomicrons have been cleared
completely from the plasma, a process which
may take from 5 to 10 hours, then the adipose
tissue begins to release its store of
triacylglycerol.

This is accomplished by another
lipase contained within the adipose cell
described as 'hormone-sensitive lipase' and a
second lipase that is not hormonally controlled
(*Fig.* 19.7). The two lipases catalyse hydrolysis
of stored triacylglycerols to release fatty acids
and glycerol.

The hormone-sensitive lipase is
normally in an inactive form but is activated
by phosphorylation requiring a cAMP-
dependent kinase. Triacylglycerol hydrolysis is
thus influenced by the level of cyclic AMP
(adenosine cyclic 3′:5′-monophosphate),

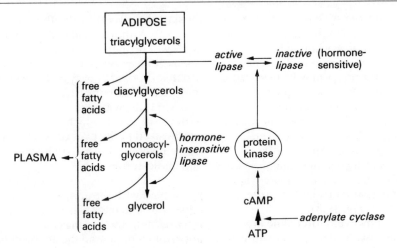

Fig. 19.7.   **Release of free fatty acids from adipose tissue.**

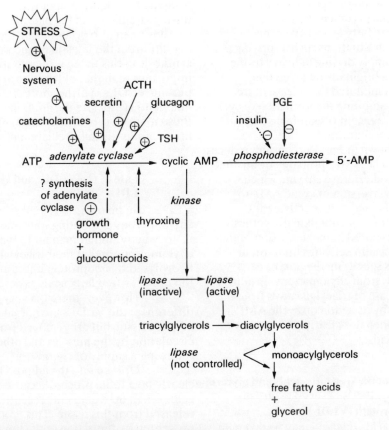

Fig. 19.8.   **Control of release of free fatty acids from adipose tissue.**
⊕ Stimulation of lipase action; ⊖ inhibition of lipase action; –·–→ implies
indirect action through an intermediate, probably a protein. PGE, prostaglandin.

which is generated from ATP by the enzyme adenylate cyclase. It is on this enzyme that the hormones act and studies on animals, especially rats, have demonstrated that no fewer than seven hormones may be involved: catecholamines, adrenocorticotrophic hormone (ACTH), glucagon, thyroid-stimulating hormone (TSH), growth hormone, glucocorticoids and thyroxine. Thyroxine, growth hormone and glucocorticoids may, however, only act by a general stimulation of the synthesis of proteins, including adenylate cyclase, as their action is blocked by protein synthesis inhibitors (*Fig*. 19.8).

In man, however, adrenaline may be the only hormone involved in the activation of adenylate cyclase. The stress situation that can, by increasing the release of adrenaline, increase the amount of free fatty acids circulating in the plasma is shown in *Fig*. 19.8. Increases of free fatty acids have been demonstrated in situations such as driving a racing car or simply driving in heavy traffic.

It was originally believed that prostaglandins inhibited the release of free fatty acids by inhibiting the formation of cyclic AMP, but more recent research has thrown doubt on this mechanism.

As shown in *Fig*. 19.8, insulin reduces the levels of cyclic AMP, probably by an effect on the phosphodiesterase enzyme which catalyses the conversion of cyclic AMP to 5'-AMP. Insulin may either activate the phosphodiesterase or stimulate its synthesis.

It was noted in the discussion of the fed state that insulin activates lipoprotein lipase and thus speeds up the uptake of triacylglycerols from chylomicrons. Insulin also blocks the release of triacylglycerols from the adipose tissue by its action on cyclic AMP in the fed state when it is clearly unnecessary to release fatty acids.

Fate of the free fatty acids and glycerol and synthesis of very-low-density lipoprotein (VLDL)

The free fatty acids released may be taken up into most tissues of the body and used as a source of energy. There is no doubt, however, that a large proportion of the fatty acids and most of the glycerol enter the cells of the liver.

Although some of these free fatty acids will be metabolized by β-oxidation to release energy, depending on the availability of other sources of energy in the liver, the majority wil be converted into coenzyme A (CoASH) derivatives. These then combine with the glycerol (which has been taken up or synthesized in the liver cells) mainly to make triacylglycerols but also to make phospholipids and cholesterol esters.

The whole complex of lipids is assembled with the apoproteins in the Golgi apparatus and endoplasmic reticulum and then released into the plasma (*Fig*. 19.9). It is uncertain whether all the apoproteins normally found associated with the VLDLs (*see Table 19.2*) are synthesized together, because exchanges with other lipoproteins can occur during circulation in the plasma. VLDL synthesis can also occur during fasting conditions in the intestinal mucosa, but synthesis in this tissue is quantitatively less important than that of the liver in provision of plasma VLDLs. Furthermore, the intestine is unable to synthesize the apoB apoproteins and these must be provided by exchange with other lipoproteins during circulation in the plasma.

Fate of the VLDLs and synthesis of LDLs

VLDLs can be regarded as substitutes for chylomicrons in a fasting state and are treated in an exactly similar manner to the chylomicrons in that their triacylglycerols are hydrolysed by lipoprotein lipase and taken into the tissue as free fatty acids as described.

However, there is a very important difference: the VLDLs are not taken up into adipose tissue but are removed from the circulation by the muscles and other tissues requiring a supply of energy.

Uptake into the adipose tissue would clearly be a futile process because the free fatty acids taken up would only recently have been released from this tissue. This process is prevented by the action of the hormones insulin and adrenaline (*Fig*. 19.3). The level of circulating insulin is depressed by the fasting condition, whereas the supply of adrenaline is

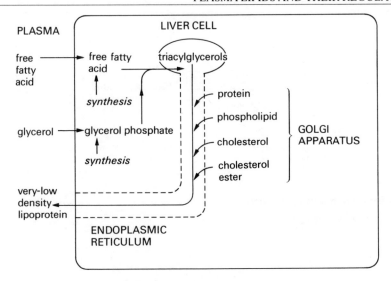

Fig. 19.9. **Synthesis of very-low-density lipoprotein in liver cell.**
Free fatty acids can be provided from adipose tissue, or synthesized glycerol can be
converted to glycerol phosphate; alternatively this may be synthesized from
glucose.

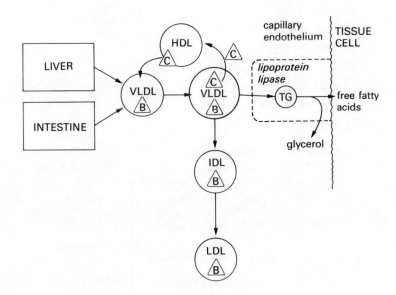

Fig. 19.10. **Formation of low-density lipoproteins (LDLs)
from very-low-density lipoproteins (VLDLs).**
Ⓑ Ⓒ represent the apoproteins apoB and apoC.
IDL, intermediate-density lipoproteins; HDL, high-
density lipoproteins; TG, triacylglycerols.

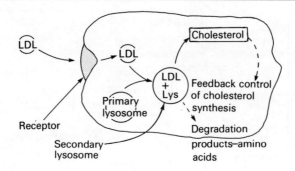

Fig. 19.11. **Catabolism of low-density lipoproteins.**

Familial hypercholesterolaemia: defective receptors.

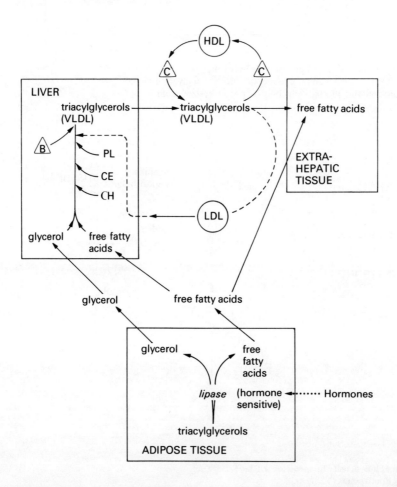

Fig. 19.12. **Lipid transport in the fasting condition.**

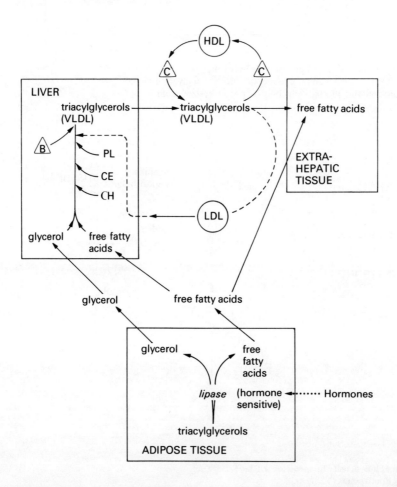, apoproteins B and C; PL = phospholipid; CE = cholesterol ester; CH = cholesterol.

increased. Both these factors tend to reduce the activation of lipoprotein lipase (*Fig.* 19.3) in adipose tissue.

VLDLs take up apoC apoproteins from HDLs to provide an activator for lipoprotein lipase in exactly the same manner as the chylomicrons. After hydrolysis of the triacylglycerols, IDLs and finally LDLs are formed (*Fig.* 19.10). ApoC apoproteins are then recovered by the HDLs.

Relation between free fatty acids and VLDLs

The question arises of why, if the tissues can readily use free fatty acids as a source of energy, it is necessary to utilize this complex procedure of synthesizing triacylglycerols, which are converted back into free fatty acids at the tissues. No complete explanation has been advanced for the phenomenon, but it would appear to be related to the provision of an adequate homeostasis of the plasma lipids.

The half-life of the free fatty acids is very short and it would appear to be unwise to rely on these as an energy source, whereas the triacylglycerols with their much longer half-life (5–10 h) can supply a steady source of energy over a relatively long period.

Fate of LDLs

The half-life of the LDLs is normally within the range of 2·5–3·5 days, but in some individuals suffering from hyperlipoproteinaemia it can be much longer (cf. p. 241). Catabolism of LDL is of great importance because this lipoprotein is the main carrier of cholesterol and raised concentrations of LDL and cholesterol closely correlate with the incidence of coronary heart disease.

Studies of catabolism of LDLs by mammalian cell cultures *in vitro* have demonstrated that specific receptors for LDLs exist on the surface of many cells, and especially macrophages. LDL is taken up by these cells, engulfed in a vacuole and the protein constituents degraded by lysosomal enzymes. This process leads to the release of amino acids and free cholesterol which is an important regulator of cholesterol synthesis (*Fig.* 19.11).

Defective receptors are found in several individuals suffering from hyper-cholesterolaemia and the defect appears to be genetically linked. The lack of receptors for LDLs allows cholesterol synthesis to proceed unregulated and thus to a raised circulating cholesterol. The failure to regulate cholesterol synthesis is believed to be a major factor in genetically linked arterial disease.

Summary of the fasting state

A summary of the tissue inter-relationships in the fasting state is shown in *Fig.* 19.12.

Synthesis and catabolism of high-density lipoprotein

High-density lipoprotein (HDL) is synthesized mainly by the liver, but also by the intestine in the cells' Golgi apparatus and excreted via the endoplasmic reticulum. The main function, as has been described, is to carry important proteins, e.g. apoC-II, that regulate the activity of the enzymes involved in lipoprotein metabolism.

The site of catabolism of HDL in man is unknown. In rats the half-life is 10·5 hours and the liver is the major site of catabolism.

19.4 Fatty livers

Fat, mainly triacylglycerol, can accumulate in the liver to such an extent that the fat is observed by the naked eye. The condition is described as 'fatty liver' or sometimes as 'fatty infiltration of the liver', It is usually a result of defective lipoprotein synthesis and it is logical to discuss the process at this stage.

The disruption of lipoprotein synthesis can be caused by a wide variety of pathological or disturbed physiological conditions. Starvation, diabetes, disturbance of adrenaline, secretion of anterior pituitary hormones, diet deficiency in certain essential amino acids or poisoning by ethionine, orotic acid, ethanol, carbon tetrachloride,

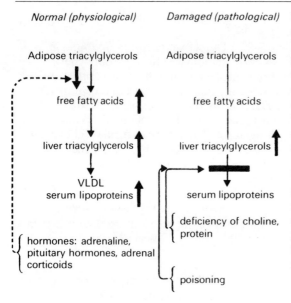

Fig. 19.13. **Elevated fat contents of normal and damaged livers.**

chloroform and several other compounds all cause fatty liver.

Although the exact mechanism of the generation of fatty liver is not established for every situation, the effects can, in the main, be classified into two categories, sometimes described as 'physiological' and 'pathological'. The pattern of effects is illustrated in *Fig.* 19.13. It will be noted that 'physiological' fatty livers can result from an excess supply of fatty acids to the liver, which occurs in starvation, diabetes or excess hormone (e.g. adrenaline) release. These fatty acids are either metabolized, often extensively to ketone bodies (Appendix 20) or used for synthesis of triacyglycerols which are then exported as lipoproteins. Triacylglycerol synthesis is normally very rapid, but should the rate of synthesis of other components of the lipoproteins, such as phosphatidylcholine and the apoproteins, not keep pace with the synthesis of the triacylglycerols, then they tend to accumulate and cause the production of fatty liver.

The importance of phospholipid and apoprotein synthesis in the synthesis of lipoproteins can be seen by studying the effects of protein-deficiency diets or diets deficient in choline or methionine. If choline itself is lacking in the diet or if methionine, which supplies methyl groups for choline synthesis (Appendix 6), is lacking then synthesis of phosphatidylcholine will be strongly depressed. A general deficiency of amino acids will cause a depression of apoprotein synthesis and thus either situation will lead to an accumulation of triacylglycerols in the liver.

Ethionine inhibits protein synthesis because it traps ATP, required for protein synthesis, by forming adenosylethionine. The modes of action of carbon tetrachloride and chloroform are much more complex and may result from their metabolism to free radicals that then attack membranes of the endoplasmic reticulum.

19.5 Interaction and interchange between lipoproteins

The protein and lipid components of the various lipoproteins do not remain constant during their transport through the plasma and it has already been noted (pp. 234–236) that the triacylglycerol component of the VLDL is hydrolysed by lipoprotein lipase.

In addition, however, a number of complicated exchanges of other lipids and protein components can occur and all plasma lipoproteins must be regarded as though they were in a state of dynamic equilibrium with other lipoproteins and with tissue membranes (*Fig.* 19.14).

Protein transfer

The activator of lipoprotein lipase, apoC-II, is transferred from HDLs to VLDLs or to chylomicrons; this is in order to allow maximum activity of the lipoprotein lipase when the VLDLs or chylomicrons reach the tissues where uptake of their fatty acids occurs. When the triacylglycerols are degraded and VLDLs converted to LDLs (p. 236), the apoprotein C-II reassociates with HDLs.

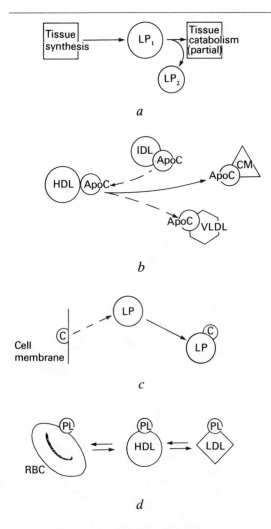

Fig. 19.14. **Lipoprotein interactions in plasma.**
(*a*) Tissue metabolism of lipoprotein; (*b*) apoprotein
transfer; (*c*) cholesterol transfer; (*d*) phospholipid transfer.
Key: LP, LP$_1$ and LP$_2$ are lipoproteins; CM, chylomicrons;
PL, phospholipid; RBC, red blood cells.

Phospholipids

Rapid exchange of labelled phospholipids
between LDLs and HDLs occurs in the plasma
but, in addition, phospholipids of the
lipoproteins will exchange with phospholipids
of the red-blood-cell membranes; *in vitro* the
phospholipids of lipoproteins will also
exchange with lipid components of the
mitochondria and endoplasmic reticulum.
In man, approximately 15 per cent of the
phospholipid of the red-blood-cell membrane
is believed to exchange with lipoprotein
phospholipids every 12 hours.

Cholesterol exchange: role of lecithin and cholesterol acyltransferase

Cholesterol exchange is catalysed by an
enzyme lecithin:cholesterol acyltransferase,
which transfers fatty acids from the β position
of phospholipid to cholesterol (*Fig.* 19.15).
In Appendix 6 it is shown that the β position
of the phospholipid is normally occupied by an
unsaturated (and usually polyunsaturated)
fatty acid, such as linoleic acid. The cholesterol
esters formed will thus be of this type and, in
fact, cholesterol linoleate forms the major
proportion of cholesterol esters in plasma.
    The activity of the enzyme
lecithin:choline acyltransferase in the plasma
is controlled by at least three proteins. ApoA-I
is a strong activator and so also, to a lesser
extent, is apoC-I; apoA-II is a strong
inhibitor.
    Cholesterol and phospholipid
substrates are both components of all the
lipoproteins, but the product, cholesterol
ester, tends to be taken up by the HDLs.

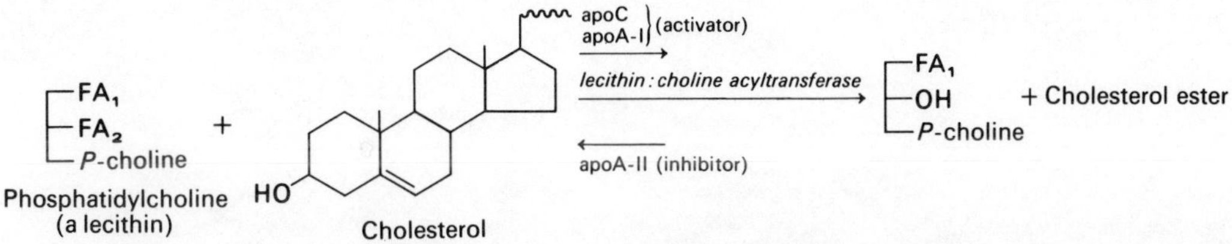

Fig. 19.15. **Mode of action of lecithin:cholesterol
acyltransferase.**
FA$_2$ is an unsaturated fatty acid.

Cholesterol (in the lipoproteins) is in equilibrium with membrane cholesterol, and particuarly that of the red-blood-cell membrane, but cholesterol esters do not equilibrate with membrane components. The esters are therefore effectively trapped in the HDL. Cholesterol esters are believed to be removed during transfer of the HDL through the liver, so that the overall process ensures that excessively high concentrations of free cholesterol do not build up in the plasma. There is thus a reduction of free circulating cholesterol (*Fig.* 19.16).

The function of lecithin:choline acyltransferase is not entirely clear but, by reducing the concentration of free cholesterol in the plasma, it appears to play an important role in the regulation of the composition of the plasma lipoproteins and, indirectly, of the cholesterol content of membranes in equilibrium with the lipoproteins.

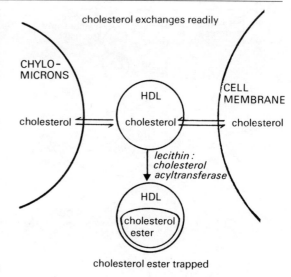

Fig. 19.16. **Role of lecithin:cholesterol acyltransferase.** HDL, high-density lipoprotein.

19.6 Hyperlipoproteinaemias

During recent years much attention has been devoted to the study of the relationship of plasma lipoprotein concentration to disease, particularly arterial disease. Arterial disease, ultimately resulting in coronary thrombosis or brain embolism (stroke), is of major concern in nearly all the well-developed countries of the world and it accounts for a very large proportion of all deaths over 60 years of age (*Table* 19.5; *Fig.* 19.17).

Numerous studies in many parts of the world have provided substantial evidence that there is a strong correlation between raised plasma lipid levels especially cholesterol and death from arterial disease (*Table* 19.6; *Fig.* 19.18). Possible mechanisms involved are discussed below.

Table 19.5 **Arterial disease as a cause of death in the UK**

Cause of death	Death rate per million	
	65–74 years	*75–84 years*
Total neoplasms (cancer)	19 235	29 248
Total vascular disease	36 617	98 272
Heart disease	23 418	55 439
Vascular lesions of central nervous system	10 623	33 487
Other disease of arteries and veins	2576	9356

Raised concentrations of plasma lipids are classified according to the scheme shown in *Table* 19.7. Types I, III, and V are relatively rare inherited defects. Type I is caused by lack of active lipoprotein lipase, that

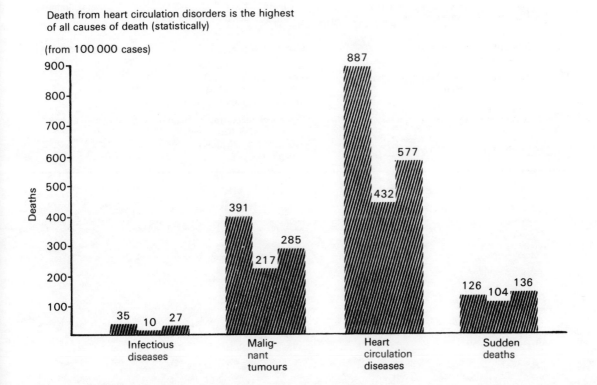

Death from heart circulation disorders is the highest of all causes of death (statistically)

Fig. 19.17. **Death from different causes in three countries.**

Left block = West Germany; central block = Switzerland; right block = Austria.

Table 19.6 **Relation between heart and arterial disease and plasma cholesterol**

| Patient | Disease frequency, % total population | | | Total plasma lipid, mg/100 ml | Plasma cholesterol, mg/100 ml |
	Coronary	Possible coronary	Hypertension		
White	26·0	5·5	46·0	850	235–245
Bantu	1·1	0·5	35·0	393	145–165

Results from patients of the Groote Scheer Hospital, Cape Town.

Table 19.7 **Hyperlipoproteinaemias**

Type	Symptoms	Incidence	Genetic control
I	Inability to clear dietary fat in form of chylomicrons Serum triacylglycerols greatly increased	Familial Rare	
II	Raised β-lipoproteins of normal type Cholesterol elevated	Common Closely associated with coronary artery disease	If genetic, dominant
III	Abnormal β-lipoproteins Cholesterol elevated	Uncommon	If genetic, recessive
IV	Increased levels of triacylglycerols and pre-β-lipoproteins	Very common later in life Often associated with obesity May be secondary to other disorders	
V	Serum triacylglycerols and cholesterol markedly elevated	Often secondary to acute metabolic disorders, such as diabetic acidosis, pancreatitis, alcoholism or nephrosis	Probably genetic

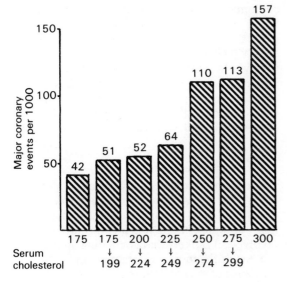

Fig. 19.18. **Plasma cholesterol and coronary events in males in the age group 30–59 years from a 10-year study.**

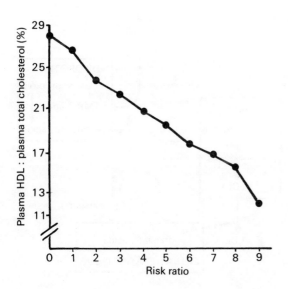

Fig. 19.19. **HDL–cholesterol ratio and coronary high risk factors.**
Risk ratios are based on scores (0, 1 or 2) for:
(1) age; (2) smoking; (3) activity; (4) alcohol consumption; (5) blood pressure.

is by failure to synthesize the enzyme or failure to synthesize the correct apoprotein activators.

Most interest has, however, been paid to type II hyperlipoproteinaemias, in which the level of circulating cholesterol is raised, because this type is closely associated with arterial disease (*Table* 19.7; *Fig.* 19.18). Type IV hyperlipidaemias are very commonly associated with obesity and also with arterial disease. In fact studies in the US and Sweden have shown closer correlation between arterial disease and raised triacylglycerol levels than between the disease and raised cholesterol levels. Recently it has been observed that raised concentrations of HDL are negatively correlated with the incidence of arterial disease (*Fig.* 19.19). This would appear to be related to the supply of cofactor apoproteins carried by HDL.

19.7 Factors leading to raised plasma lipid levels

If raised concentrations of plasma cholesterol or triacylglycerols are important in the development of arterial disease, it is important to attempt to understand what causes lipid levels to increase (*Table* 19.8).

Table 19.8 **Factors controlling plasma lipoprotein concentrations**

1. Solubility
2. Triacylglycerol transport (exogenous and endogenous loads; rate of clearance)
3. Cholesterol (dietary load and rate of clearance)
4. Diet (lipid, carbohydrate, protein)
5. Hormones
6. Age
7. Psyche
8. Stress

It is important to realize that most of the circulating cholesterol is located in the LDLs and the majority of triacylglycerols are transported on the VLDLs (*Table* 19.1). Raised cholesterol levels will, therefore, almost certainly indicate an increase in LDLs and raised triacylglycerol levels indicate an increase in concentration of VLDL. It is very much easier to determine plasma cholesterol or plasma triacylglycerol concentrations than plasma lipoprotein concentration.

a. Age

The concentrations of plasma cholesterol and plasma triacylglycerols increase gradually with age, female values being below those of males until 50 years of age (*Fig*. 19.20). It is uncertain to what extent this is an unavoidable process or how far it is related to the factors listed below.

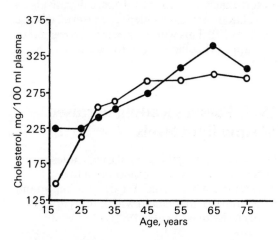

Fig. 19.20. **Changes in plasma cholesterol concentration during the life-span.**
(○−○) Males; (●--●) females.

b. Exercise, activity

It is generally agreed that the more sedentary and inactive the life-style the more likely is the occurrence of arterial disease. One must realize, however, that it may be complex physiological effects which are of paramount importance, especially the dilatation of arteries of the heart and other muscular tissues and the increased rate of blood flow which occurs during exercise. Nevertheless, inactivity, particularly when accompanied by a high calorie intake, tends to cause a rise in the level of circulating triacylglycerols. Raised triacylglycerol levels can occur when the large quantity of ingested fat is converted to

chylomicrons and then deposited in adipose tissue. The obesity which results nearly always causes a raised level of circulating free fatty acids which, as discussed on pp. 233–234, causes the liver to synthesize more triacylglycerol and incorporate it into VLDL. VLDL is catabolized at the tissues to LDL and, because this contains a high concentration of cholesterol, cholesterol levels also tend to rise.

c. Stress

There is good evidence, developed over many years, that stress plays an important role in the development of arterial disease. Although many complex interacting factors may be involved, it is likely that circulating lipids are one of the most important factors. The sequence of events is illustrated in *Fig*. 19.21. The release of adrenaline, triggered by stress situations such as driving in heavy traffic,

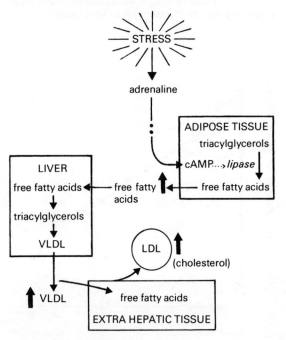

Fig. 19.21. **Effect of stress on plasma lipoprotein concentrations.**
Stress causes raised plasma free fatty acids, very-low-density lipoproteins (VLDL), low-density lipoproteins (LDL) and, if continued, cholesterol, because LDL is the main vehicle for cholesterol transport (cf. *Table* 19.1).

causes a release of free fatty acids from the adipose tissue (*Fig.* 19.21), which can synthesize VLDL in the liver. Repeated stress situations can clearly cause a raised circulating triacylglycerol concentration.

d. Diet

The importance of diet in the causation of arterial disease has been established through many epidemiological surveys (cf. Chapter 13). These studies have indicated that increased plasma cholesterol concentrations, which correlate with arterial disease, may be caused by a diet (i) high in cholesterol content; (ii) high in saturated fat content; (iii) high in sucrose content. Polyunsaturated fats, such as corn oil, containing a high concentration of linoleic acid, depress the concentration of circulating cholesterol.

Although cholesterol in the diet may appear an obvious and important cause of raised plasma cholesterol, the relation between dietary cholesterol and plasma cholesterol is, in fact, complex. This is for two reasons. First, the dietary cholesterol is not completely absorbed and absorption may fall to 50 per cent of the intake with a high-cholesterol diet (cf. Chapter 13). Secondly, there is a mechanism of feedback control (cf. Appendix 22), by which dietary cholesterol switches off synthesis of cholesterol by the liver.

Diets containing large amounts of saturated fat, as would be expected, tend to increase the triacylglycerol concentration and deposit triacylglycerols in the adipose tissue. As mentioned before (p. 243), raised triacylglycerol itself can be correlated with an increased incidence of heart disease, but, as shown for the stress situation (*Fig.* 19.21), raised triacylglycerol, in the form of VLDL, can ultimately lead to a raised concentration of LDL which transports the main bulk of plasma cholesterol.

It is well established that diets high in polyunsaturated fats lead to a reduction in plasma cholesterol concentration (cf. Chapter 13), but their mechanism of action has not been fully elucidated (*see Fig.* 13.16, Chapter 13 for recent hypothesis).

The effect of sucrose on plasma lipids and arterial disease is still somewhat controversial but there is now good evidence that fructose, which is formed from the ingested sucrose, is rapidly metabolized in the liver and converted to fatty acids. These fatty acids are incorporated into VLDL by the liver and thus fructose can be regarded as an important source of plasma triacylglycerols.

Chapter 20

Plasma amino acids and utilization of amino acids by the tissues

The plasma normally contains all the 20 amino acids commonly found in proteins, but other amino acids, such as citrulline, ornithine, taurine, and 3-methylhistidine, are also present.

It is clear that a discussion of the various factors that regulate the concentration of each of these amino acids separately will pose complex problems, and it is not surprising that these factors have only been partially resolved. Fortunately, it is not essential to discuss the metabolism and regulation of the concentration of each individual amino acid, since certain amino acids, e.g. the dicarboxylic amino acids or the branched-chain amino acids, can be considered in groups.

Study of the blood amino acids is very important because it provides valuable information about the state of nutrition of the body and about pathological changes that may be occurring in the tissues. However, this study often presents difficulties.

20.1 How does man utilize amino acids?

A summary of the utilization of amino acids in the adult human body is shown in *Fig.* 20.1. The average adult in Western societies ingests about 90–100 g protein a day. The protein in the digestive tract is, however, substantially increased by the secretion of digestive juices in the form of enzymes, which can add up to 50–70 g protein per day to the intestinal tract. Approximately 160 g amino acids are absorbed each day and 10 g lost in the faeces, either as protein or amino acids.

246

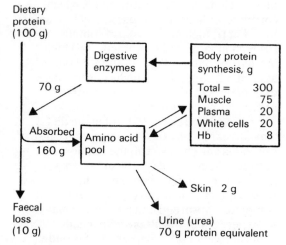

Fig. 20.1. **Daily amino acid flux in man.**

The amino acids are absorbed from the intestine and become incorporated into the 'amino acid pool' which is a useful concept and is represented by the summation of all free amino acids in all cells of the body. If necessary, we can restrict the confines of the 'amino acid pool' to a particular organ, such as the liver, but in this section the 'total body amino acid pool' is discussed. The 'pool' is being constantly replenished by ingested amino acids or by degradation of body proteins, and steadily loses amino acids by incorporation into newly synthesized protein or gains amino acids as a result of the action of proteolytic enzymes. Approximately 300 g body proteins are synthesized and degraded each day, as shown in *Fig.* 20.1, and about 70 g/day lost in the urine after conversion to urea. The overall turnover, 300 g, is 200 g greater than the net intake, 100 g, and this difference emphasizes the large contribution made by the dynamic state of the body proteins to the free amino acid pool.

The concepts of an amino acid pool and a dynamic state of body proteins emphasize the important roles that protein synthesis and degradation play, in different tissues, in the regulation of the nature and concentration of the various amino acids in the blood. The ingested amino acids contribute to this pool and dynamic state, but add only a proportion of the total amino acids involved in the daily turnover in the body.

Four non-essential amino acids, alanine, glutamic acid, glutamine and glycine, form the major proportion of amino acids in the pool.

20.2 The effect of a protein meal on plasma amino acid concentrations

Typical concentrations of amino acids in the blood of normal adults are shown in *Table* 20.1. It will be noted that the concentrations vary considerably, from 32 μmol/l for tyrosine to 145 μmol/l for glycine. The majority of amino acids are found in approximately equal concentrations in the plasma and red blood cells, the main exceptions being aspartic acid which is transported almost entirely in the red cells, and glycine whose cell concentration is much greater than the plasma concentration.

It might be anticipated that the ingestion of a protein meal would cause a large and significant increase in the concentration of all amino acids in systemic blood but, for several reasons, this does not occur.

Mucosal cells of the intestine actively transaminate the dicarboxylic acids, glutamate and aspartate, by utilization of pyruvate to form alanine. Consequently, the portal blood contains a high concentration of alanine but a low concentration of the dicarboxylic acids.

A second tissue which plays an important role in the control of the plasma amino acids is the liver. The fate of the amino acids of a protein meal, as shown in *Fig.* 20.2, has been demonstrated definitively in dogs and is believed to be the same in man. It will be observed that only a relatively small proportion of the absorbed amino acids reach the systematic circulation. The major fraction of amino acids entering the liver from the portal blood after a meal is either catabolized or incorporated into protein.

The liver is relatively inefficient at oxidizing tyrosine, lysine and the branched-chain amino acids, leucine, isoleucine and

Table 20.1 **Typical amino acid concentrations in human blood in the fasting state**

Amino acid	Concentration, μmol/l, in	
	Plasma	Whole blood
Aspartic acid	0	192
Threonine	87	170
Serine	80	171
Proline	122	241
Glycine	145	402
Valine	141	260
Isoleucine	44	81
Leucine	79	144
Tyrosine	32	63
Phenylalanine	30	55
Ornithine	42	100
Lysine	155	192
Histidine	54	93
Arginine	51	64

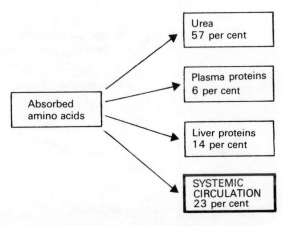

Fig. 20.2. **Fate of amino acids absorbed from a protein meal after passage through the liver.**

valine. These branched-chain amino acids therefore pass through the liver and are readily taken up by muscle and oxidized to the virtual exclusion of any other amino acids. The structural similarity of branched-chain amino acids to fatty acids, the primary fuel of muscle, is the most probable explanation for this.

20.3 The utilization of branched-chain amino acids in muscle and formation of alanine

Rapid oxidation of the branched-chain amino acids in muscle results in the liberation of the amino nitrogen which is incorporated into alanine or glutamine by the reactions shown in *Fig.* 20.3. Transamination of the branched-chain amino acids occurs almost exclusively with α-ketoglutarate which is converted to glutamate. The glutamate can then either donate its amino group to pyruvate, with the formation of alanine, or be converted to glutamine by the incorporation of an additional amino group. The ammonia required for the formation of glutamine can arise from extracellular sources or by deamination of amino acids or purines.

Provision of branched-chain amino acids to muscle, causes the level of intracellular glutamate to rise and production of alanine and glutamine is stimulated. The relative quantities of alanine and glutamine found will depend on the concentration of ammonia within the tissue. Increase of NH_3 concentration will cause increase of glutamine synthesis and decrease of alanine synthesis. The ketoacids formed from the branched-chain amino acids can be completely degraded to CO_2 with the liberation of energy as ATP.

There has been much dispute about the source of the glucose and pyruvate and, therefore, of the alanine released from the muscle. The glucose concentration in muscle is relatively low and, in order to maintain the glucose supply, glucose must be taken up by the muscle from the blood and returned as alanine. The advantages of glucose–alanine conversion seem doubtful, particularly in the fasting animal, where, as discussed in Chapter 23, branched-chain amino acids arise from muscle proteins. It has, therefore, been suggested that pyruvate could be produced by deamination and metabolism of several other amino acids produced by muscle protein degradation. If this occurred, these amino acids, by conversion to alanine and take-up by the liver, could eventually be converted to glycogen and blood glucose. After careful experiments, the extensive formation of pyruvate from other amino acids in muscle

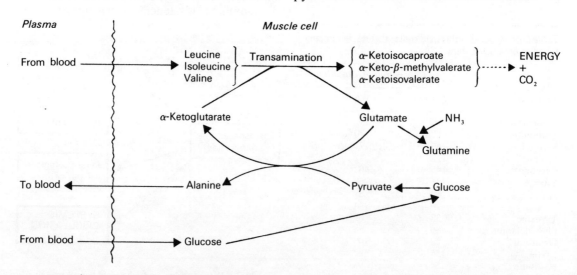

Fig. 20.3.   **Metabolism of branched-chain amino acids by muscle.**

has not been confirmed. The main role of the glucose–alanine conversion therefore appears to be in enhancing deamination of the branched-chain amino acids that can then be completely metabolized to produce energy. There is no resultant net loss of glucose from the blood, as the liver converts the alanine to glucose with the production of urea from the amino group. Alanine production is, therefore, a means of transfer of nitrogen to the liver.

The plasma concentrations of branched-chain amino acids probably play an important regulatory role in nitrogen metabolism, since low concentrations of these amino acids are closely associated with nitrogen conservation. Administration of insulin or glucose causes reduction in the concentrations of branched-chain amino acids in the plasma and significant decreases in the excretion of urinary urea and total nitrogen. Conversely, high concentrations of branched-chain amino acids are associated with increased muscle proteolysis and are observed in diabetes and ketoacidosis.

The regulatory mechanism may depend on the activities of the important catabolic enzymes in muscle and is illustrated in *Fig.* 20.4. Muscle cells cannot accumulate

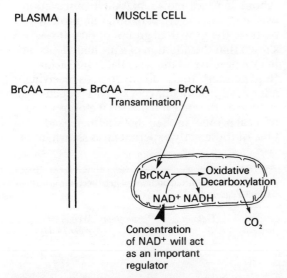

Fig. 20.4. **Role of NAD$^+$ in regulating the rate of catabolism of branched-chain amino acids.**
BrCAA branched-chain amino acids; BrCKA branched-chain ketoacids.

the branched-chain amino acids and the muscle cell concentration reflects the plasma concentration. Transamination of the amino acids is believed to be rapid and uncontrolled, but the second stage of their metabolism in which the enzymes have low K_m values, is regulated by NAD$^+$ availability. In fasting conditions, when NAD$^+$ concentrations in mitochodria are low, the activity of the oxidative decarboxylation stage is inhibited by the unavailability of NAD$^+$. During fasting, fatty acid oxidation in the muscle is increased and the utilization of NAD$^+$ for the β-oxidation process is probably an important factor in reducing the availability of NAD$^+$ for amino acid catabolism.

Reduced rates of decarboxylation of branched-chain amino acids lead to a depression in their catabolism and insulin plays an important role in the mechanism, since reduction in insulin concentration increases the rate of fatty acid oxidation. Consequently, the gradual decrease of the circulating insulin concentration in fasting man is an important factor in the control of protein conservation. This role of insulin is, however, quite different from its role after a meal, when high concentrations of insulin cause increased protein synthesis and inhibition of muscle proteolysis.

This regulatory mechanism only takes into consideration the branched-chain amino acids, but it appears likely that total protein conservation in the body is regulated by the rate of catabolism of these amino acids.

20.4 The induction of amino acid catabolizing enzymes

When the concentrations of amino acids reaching the liver are relatively low, the major proportion of these amino acids is incorporated into protein. However, as the concentrations increase, a proportion of the amino acids is catabolized. The relative K_m values of the two-enzyme system are of major importance in regulating the fate of the amino acids (*Fig.* 20.5), this clearly being a mechanism to prevent the waste of valuable essential amino acids by catabolism.

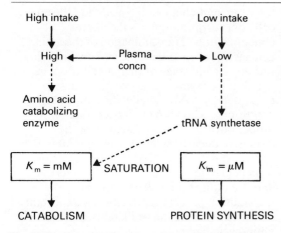

Fig. 20.5. **Fate of high and low concentrations of ingested amino acids in the liver.**

When animals are maintained on a high protein diet, many of the amino acid catabolizing enzymes, e.g. tryptophan pyrolase, phenylalanine hydroxylase, α-keto-acid dehydrogenase and serine dehydratase, are rapidly induced. This effect occurs mainly in the liver and is much less marked in other tissues, e.g. the kidney or heart. Typical responses of two enzymes, tryptophan pyrolase and phenylalanine hydroxylase, to increased casein concentration in the diet are shown in *Fig*. 12.12 (Chapter 12). The branched-chain amino acid leucine has two catabolizing enzymes, an aminotransferase and a dehydrogenase, and their response to a high protein diet is shown in *Table* 20.2, only the latter enzyme being induced.

The induction mechanism for these enzymes has yet to be clarified. It has been demonstrated with some of the enzymes that increased quantities of the enzyme mRNA are synthesized, indicating that the control site is probably at the DNA, and implicating a derepression mechanism similar to the mechanism of bacterial enzyme induction described by Jacob and Monod. It is, of course, unlikely that enzyme synthesis is regulated by such a simple mechanism, since several hormones may also mediate the regulation. Glucose strongly inhibits the induction process, maybe resulting from insulin release, whereas glucagon is a potent stimulator of the induction of several of the amino acid catabolizing enzymes.

Most of the experiments on induction of amino acid catabolism have been carried out on animals and, as yet, there is nothing to contradict the belief that similar regulatory mechanisms operate in the human. On this assumption, the activities of several of the enzymes that catabolize essential amino acids, such as tryptophan and the branched-chain amino acids, will be very low before meals, but will be induced by a meal with a high protein concentration.

20.5 Amino acid imbalance

About 25 years ago a remarkable observation was made on the effect of different protein diets on the growth of groups of rats. It was shown that the addition of a protein, deficient in one or more of the essential amino acids (e.g. gelatin), to the diet of animals receiving adequate protein intake, caused a reduction in their growth rate even though the proportion of total protein in their diet was increased. One of these early experiments is shown in

Table 20.2 **Effect of high protein diets on enzymes involved in the catabolism of leucine in rats**

Diet	Specific activity of leucine α-ketoglutarate amino transferase		Specific activity of α-ketoisocaproate dehydrogenase	
	Liver	Kidney	Liver	Kidney
9% casein	19·21	418	38·9	35·6
80% casein	20·9	423	198	44·5

Table 20.3 **The effect of addition of gelatin on the growth of rats fed a casein diet**

Casein, %	Gelatin, %	Tryptophan, %	Weight gain, g per 14 days
6	—	—	22
6	12	—	13
6	12	0·2	51
8	—	—	53
8	15	—	24
8	15	0·2	77

Table 20.3. The growth rates of the animals fed on 6 or 8 per cent casein diets were strongly depressed by the addition of either 12 or 18 per cent gelatin. The reduction in growth occurred despite the fact that the diet then contained much larger proportions of protein (18 or 23 per cent) than the original diet. Gelatin is deficient in the essential amino acid tryptophan and, if a small amount of tryptophan is also added to the diet with the gelatin, the inhibiting effect of the gelatin is completely overcome and the animals grow at a much faster rate. The explanation of this phenomenon was extensively investigated by Harper at the Massachuschetts Institute of Technology, USA and his work has made important progress towards understanding the mechanism.

The effect can also be demonstrated by feeding incomplete mixtures of amino acids or by addition of groups of amino acids to otherwise complete diets. Very small additions of pairs of essential amino acids drastically reduce the growth rate, as shown in *Table* 20.4.

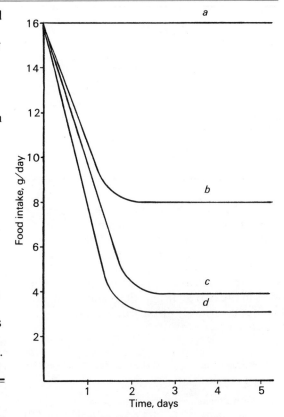

Fig. 20.6. **Effect of feeding incomplete amino acid mixtures on food intake of groups of rats.**
Basal diet was supplemented with all essential amino acids less those indicated. (*a*) Basal diets; (*b*) basal diet + amino acids (− His, − Thr, − Lys); (*c*) basal diet + amino acids (− His); (*d*) basal diet + amino acids (− His, − Thr).

Table 20.4 **The effect of additions of essential amino acids to a 6% fibrin diet on the growth of rats**

	Diet	Weight gain, g per 14 days
Fibrin, %	Added amino acids*	
6	—	33
6	Met (0·4) + Phe (0·6)	17
6	Arg (0·4) + Lys (0·6)	18
6	Arg (0·4) + Thr (0·6)	20
6	Thr (0·4) + Tyr (0·4)	19

*The values in parentheses are the percentage of the amino acid added to 100 g of diet.

The effect in rats of a basal diet that has been supplemented with a mixture of essential amino acids from which certain amino acids are removed is shown in *Fig.* 20.6. This clearly illustrates that the reduction in growth rate is caused by reduction in food intake. The food intake of animals receiving a basal diet was much greater than that of those receiving a supplement of the essential amino acids less histidine. Although there is clear evidence of the effect of an imbalanced diet, the relationships between the various amino acids are complex and, as yet, unresolved. For example, although removal of histidine and threonine from the mixture is more effective than removal of histidine alone, the removal of histidine, threonine and lysine is less effective.

The food intake of the animals has been shown to be indirectly dependent on the concentrations of plasma amino acids, these being directly controlled by the dietary intake of amino acids. For example, when rats feeding on a 6 per cent fibrin diet are additionally fed a mixture of all essential amino acids except histidine, a rapid fall in the concentration of plasma histidine results (*Table* 20.5).

Table 20.5 **The effect of addition of a mixture of essential amino acids lacking histidine on the plasma histidine concentration**

	Plasma histidine concentration, μM, on diet of	
Time after feeding, h	6% Fibrin	6% Fibrin + amino acid mixture
1	4·9	4·9
3·5	4·5	1·8
6·5	3·2	0·9

This demonstration of the fall in concentration of one essential amino acid enabled Harper to develop a hypothesis (*see Fig.* 20.7) to explain the reduction in food intake. It postulates that the appetite centre is very sensitive to a supply of essential amino acids in the correct proportions and, if the balance is seriously disturbed, feeding is switched off.

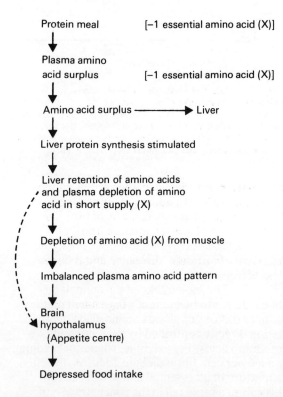

Protein meal   [−1 essential amino acid (X)]

Plasma amino acid surplus   [−1 essential amino acid (X)]

Amino acid surplus ⟶ Liver

Liver protein synthesis stimulated

Liver retention of amino acids and plasma depletion of amino acid in short supply (X)

Depletion of amino acid (X) from muscle

Imbalanced plasma amino acid pattern

Brain hypothalamus (Appetite centre)

Depressed food intake

Fig. 20.7. **Hypothesis of amino acid imbalance.**

This process is clearly of great survival value to the animal, guiding it to the selection of protein foods containing essential amino acids in the correct proportions. Its precise mechanism has not yet been resolved and, although this regulatory process could play an important role in humans, it has not been clearly demonstrated. Most probably the regulation of human appetite is a complex process dependent on many factors.

In Western societies, varied diets are consumed and the problem of reduced food intake resulting from imbalanced amino acids is not likely to prove serious. However, in many Third World countries where the total protein food supply and types of protein are very limited, amino acid imbalance could be of major importance, malnutrition problems arising through poor appetite for the limited supplies available.

20.6 Hormonal regulation of plasma amino acids

Many of the hormones produced in the animal body regulate the concentrations of plasma amino acids. The relationships between them are summarized in *Fig.* 20.8. Discussion of their actions can be simplified by division into two main categories, the anabolic hormones and the catabolic hormones. The anabolic steroids, e.g. testosterone and growth hormone, will be in the former group, and these hormones tend to promote the incorporation of plasma amino acids into muscle protein. Insulin may also be included in this group, although its action appears to be mediated mainly through inhibition of muscle proteolysis rather than by promotion of increased uptake. In fact one of the main effects of dietary glucose is the release of insulin which, in turn, inhibits the muscle protein degradation. Amino acids are clearly not required for energy supply when adequate glucose is available.

The catabolic hormones, particularly catabolic steroid hormones such as corticosterone, act antagonistically to the anabolic hormones. Release of these hormones

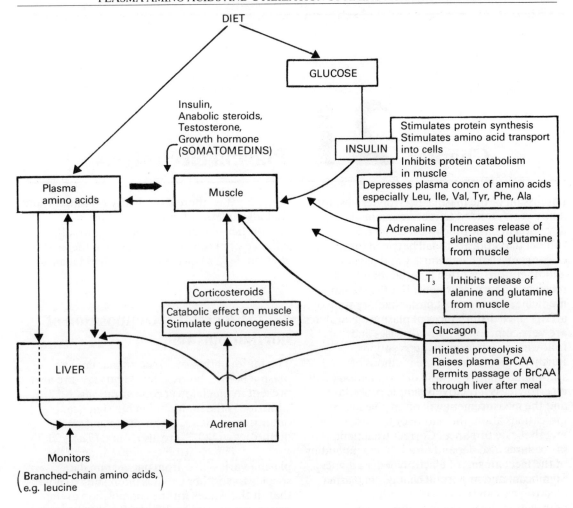

Fig. 20.8. **Hormonal regulation of plasma amino acids.**

BrCAA = branched-chain amino acids.

causes degradation of muscle protein, increase of amino acid oxidation in muscle and amino acid release into the plasma. Glucagon can also act in this manner and, in addition to its action on the muscle, also plays an important role in stimulating the production of amino acid catabolizing enzymes.

The plasma amino acid concentration is, therefore, dependent to an important extent on the delicate balance between these two groups of hormones and the control of their release.

Chapter 21 Plasma electrolytes

Discussion of the plasma electrolytes has, for many years, tended to receive minimal attention in basic preclinical texts. It is a subject that appears to fall between the areas of conventional biochemistry and physiology and, therefore, is often ignored, or at best receives cursory treatment. For the student of medicine, this is a most unsatisfactory state of affairs since disturbances of plasma electrolytes are very common in many diseases and following many surgical interventions. Furthermore, with modern methods of automated analyses, rapid determination of all electrolytes is a relatively simple procedure and the measurement will often give important indications about the nature of the disease or metabolic disturbance. Correct treatment procedures also depend on a full understanding of the mechanisms of electrolyte disturbance. Significant and important changes in plasma electrolytes can be caused by water deprivation, heat effects, acidosis and alkalosis, intestinal and gastric disorders, pulmonary disease, liver disease, circulation failure, diabetes, renal disease, shock, burns and adrenal cortical disease.

It is, therefore, clearly very important that all students of medicine acquire a comprehensive understanding of normal balance of plasma electrolytes, and the factors that can lead to disturbances in their balance.

21.1 Electrolyte composition of normal plasma

The major cation occurring in mammalian plasma is Na^+, but K^+, Mg^{2+} and Ca^{2+} are also present in much lower concentrations. Chloride is the major anion but significant quantities of HCO_3^-, phosphate and small quantities of SO_4^{2-} are also found (*Table* 21.1).

The electrolyte composition of the plasma varies little from one mammalian species to another and it is interesting to note that, if the values for the composition of sea water are divided by 3·8, the figures obtained are very close to those of plasma. The main exception is that of magnesium, whose concentration in sea water is relatively high. This remarkable relation between the

Table 21.1 **Electrolytes in plasma**

Species	Na^+, mmol equiv./l	K^+, mmol equiv./l	Mg^{2+}, mmol equiv./l	Cl^-, mmol equiv./l
Man	135–145	4·0–5·5	1·1–2·1	96–107
Horse	142	4·8	2·1	100
Sheep	148	4·5	2·4	108
Ox	140	4·7	2·4	101
Cat	147	4·3	—	104
Dog	145	4·0	1·5	110
Sea water/3·8	119	4·0	13·1	134

composition of sea water and plasma indicates that the constitution of sea water has changed during the passage of time by becoming more concentrated and especially by increases in its magnesium content and that the electrolyte constitution of the plasma closely resembles that of primitive sea water in which life began.

From that time, it has always been essential for living cells, for reasons that are not clear, to be bathed in the primitive sea water and most cells die rapidly if the electrolyte composition or constitution of the medium differs essentially from that normally found in plasma. It should also be noted that the electrolyte composition of the red blood cells is quite different from that of plasma, esecially in man and some other mammals. Differential concentrations of sodium and potassium ions are maintained across the red-blood-cell membranes (*Table* 21.2).

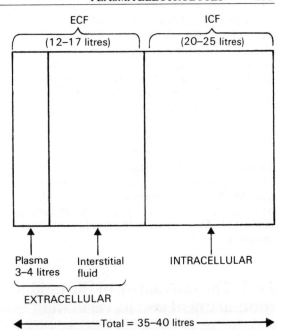

Fig. 21.1.   **Water compartments of the body.**

Table 21.2 **Electrolytes of red blood cells**

Species	Na^+, mmol equiv./l	K^+, mmol equiv./l	Cl^-, mmol equiv./l
Man	19	136	78
Horse	16	140	85
Sheep	98	46	78
Ox	104	35	85
Cat	142	8	84
Dog	135	10	87

Maintenance of these concentration differences requires a special transport mechanism in the red-blood-cell membrane and a source of ATP as an energy supply; it is discussed fully in Chapter 25.

21.2 Water components of the body

A discussion of the roles of the plasma electrolytes is closely related to the water content of the tissues and plasma and it is first essential to describe the 'water compartments' of the body. Although, to a certain extent, these compartments are artificial concepts, they form a very useful basis for the discussion of

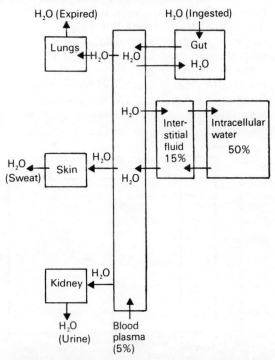

Fig. 21.2.   **Inter-relationships of important organs in fluid balance.**
Percentage figures are based on a total body water concentration of 70 per cent.

water movement in the body and its control by electrolytes. Body water may be conveniently divided into three main compartments: the intracellular water which forms about two-thirds of the total body water of 35–40 litres in the adult, the interstitial fluid water or extracellular water which is normally about 8–13 litres and the plasma water which is normally about 3–4 litres (*Fig.* 21.1).

Water diffuses rapidly into all compartments taking only a few minutes in small animals and 2–3 hours in man. Some ions such as Cl^- also rapidly equilibrate with all compartments. The roles played by the lungs, skin and kidney in the body's fluid balance are shown in *Fig.* 21.2.

21.3 The osmolarity of the fluid compartment and its regulation

Since all water compartments of the body are in contact and water is freely permeable between the compartments, it follows that the osmolarity of all compartments, at equilibrium, must be identical. A useful diagrammatic model for representing the volumes of fluid compartments and the osmolarities was proposed by Darrow and Yannett (*see Fig.* 21.3). The vertical axis represents the osmolarity of the compartment

and the horizontal axis the volume. For simplification it is usual to group the plasma and interstitial fluid together as the 'extracellular fluid'.

It will be noted that the total electrolyte concentration is important in the maintenance of the osmotic pressure and, due to their high concentration in the extracellular fluid, Na^+ and Cl^- play a major role. The use of the Darrow–Yannett model for representing fluids and their osmolarities can best be seen by practical examples, shown in *Fig.* 21.4. In (*a*) the effect of water deprivation is shown: urine continues to be excreted although the volume is greatly reduced. The loss of water from the plasma causes the osmotic pressure of the extracellular fluid to rise and this, in turn, causes water movement from the cells into the extracellular fluid to balance the osmotic pressure. Water loss from the cells can be of serious consequence and death ensues when 20 per cent of cellular water has been lost. The situation is usually made worse by the loss of K^+ from the cells, and serious consequences follow from cellular K^+ deficiency.

The reverse situation is seen as a result of salt deprivation, which can occur when the body sweats excessively during hot weather and the fluid loss is made up of fresh water containing very little salt: the situation is shown in *Fig.* 21.4*b*. The loss of salt solution, as sweat from the plasma, followed by its replacement with pure water causes the osmotic pressure of the plasma to fall. Water will thus move from the extracellular fluid into the cells in order to balance the osmotic pressure and the cells will swell as water enters.

Drinking of sea water which contains about 600 mmol equiv./l electrolyte paradoxically causes dehydration of the cells as shown in *Fig.* 21.4*c*. The high concentration of salt solution entering the plasma from the gut causes the osmotic pressure to rise. As a consequence water moves from the cells to balance the osmotic pressure, thus dehydrating the cells.

These examples show qualitatively the consequence of changes in body water and electrolytes. Quantitative calculations can also be made and these are often important in correct treatment. It will, therefore, be useful

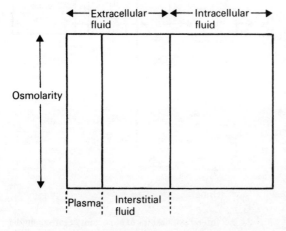

Fig. 21.3. **The water and electrolytes in the body fluid compartments illustrated by the Darrow–Yannett diagram.**

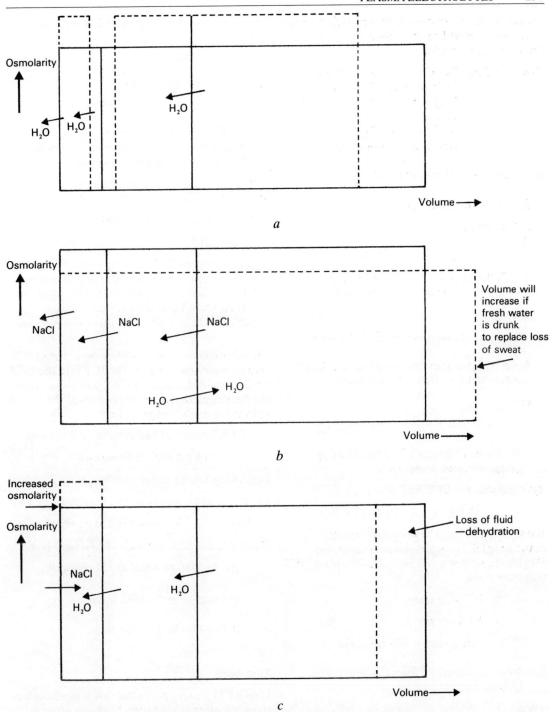

Fig. 21.4. **Use of the Darrow–Yannett diagrams to illustrate changes in body water and electrolytes.**
(*a*) Water deprivation; (*b*) salt deprivation as a result of excess sweating; (*c*) intake of sea water. (——) Initial volume;
(— — —) final volume.

to discuss the following examples of changes in water and electrolytes based on that of a 70-kg man (Mr B) of average build.

Relevant data: Total body water = 60% of body weight = 42 litres
ECF = 16% of body weight = 11·2 litres
ICF = 42–11 = <u>31 litres</u>
Osmolarity of body fluids = 310 mosmol/l

[equivalent to 155 mequiv. salt (NaCl)]

Darrow–Yannett diagram

310 mosmol

	ECF	ICF
	310 × 11 litres = 3410 mosmol	310 × 31 litres = 9610 mosmol

←——11 litres——→ ←—— 31 litres ——→

Total milliosmoles (mosmol) in total body water = 9610 + 3410 = 13 020 mosmol

or

$$11 + 31 = 42 \times 310 = \underline{13\ 020\ \text{mosmol}}$$

i.   *Mr B loses 2 litres ECF (excess loss of gastrointestinal secretions)*

This reduces the ECF by 2 litres:

$$ECF = 11 - 2 = 9\ \text{litres ECF}$$

Removing ECF at 310 mosmol/l leaves remaining ECF unchanged in osmolarity; therefore, no net water transfers occur and ICF stays the same.

ICF = 31 litres

ECF = 9 litres

osmolarity = 310 mosmol/l

ii. *Mr B then drinks 2 litres isotonic saline (310 mosmol/l)*

Then

$$ECF = 9 + 2 = 11\ \text{litres}$$

There is no change in osmolarity of ECF, no net water transfer occurs and ICF remains the same.

ECF = 11 litres

ICF = 31 litres

osmolarity = 310 mosmol/l

iii. *Mr B loses 3 litres of insensible perspiration by hyperventilation which is equivalent to loss of water without loss of salt*

Immediate effect on compartment volumes and osmolarities

Total volume = 42 − 3 = 39 litres

ECF = 11 − 3 = 8 litres

Osmolarity of ECF goes up since 3410 mosmol was in 11 litres and is now in 8 litres; therefore new osmolarity = 3410/8 = 426 mosmol/l and the ECF osmolarity increases from 310 mosmol/l to 426 mosmol/l.

ECF osmolarity is now higher than that of ICF and so water moves from the ICF into the ECF until equal osmolarity is obtained. What will the final osmolarity be, regardless of the volume of each compartment?

13 020 mosmol *had been* in 42 litres and

13 020/42 = 310 mosmol/l.

Following loss of water,

13 020 mosmol is now in 42 − 3 or 39 litres and

13 020/39 = 334 mosmol/l at equilibrium.

What are the new volumes of ECF and ICF?

ECF: 3410 mosmol at 334 mosmol/l;

$$\text{volume} = \frac{3410}{334} = 10\cdot2\ \text{litres}$$

ICF = 39 − 10·2 = 28·8 litres

Summary

Loss of 3 l water causes the following transfers to occur after equilibrium has been attained:

ECF 11 litres → 8 litres → 10·2 litres

ICF 31 litres → 28·8 litres

Total volume: 42 litres → 39 litres

Osmolarity: 310 mosmol/l→426 mosmol/l→ 334 mosmol/l.

21.4 Changes in the plasma concentrations of specific electrolytes

Sodium

Sodium forms the major proportion of the cations in the plasma and thus changes in osmolarity, for the various reasons discussed in Section 21.3, are often closely related to changes in sodium concentration. Examples of such conditions are renal disease, inadequate renal perfusion and excessive aldosterone activity.

Renal disease can cause increased sodium retention, either by reducing the rate of sodium filtration or by increasing the proportion of filtered sodium which is reabsorbed by the tubules.

Inadequate renal perfusion: patients suffering from shock and acutely induced hypotension show a marked reduction in the renal blood flow. As a consequence, the glomerular filtration rate is severely depressed and sodium extraction is considerably reduced. The concentration of sodium in plasma then rises considerably.

In both of these conditions the concentration of chloride in the plasma normally increases with the sodium.

Excess aldosterone activity: as discussed in Chapter 17, aldosterone causes increased sodium retention by its action on the renal tubules. In patients suffering from tumours of the adrenal, thus producing aldosterone in excess, the retention of Na^+ is much greater than normal and this is accompanied by an excessive loss of potassium.

Potassium

The effects of changes of potassium concentration on the cellular metabolism are much more significant and important than those of sodium.

Hypokalaemia

The importance of K^+ for the action of heart muscle was first observed by Ringer in 1883 who showed that, if the medium in which his frog's heart was bathed was made deficient in potassium, it finally stopped beating. Ringer described the heart as stopping in diastole, but it is now known that it is arrested in systole. Both cardiac and skeletal muscle are affected by 'hypokalaemia' as the condition of low plasma K^+ is known.

If the plasma K^+ level falls significantly, then serious symptoms may develop, including arrhythmia, tachycardia and hypotension; changes in the ECG are also observed. Clinical conditions which lead to depletion of potassium include alimentary losses and renal losses.

Alimentary losses of K^+ in the stools can accompany such conditions as steatorrhea. Loss of K^+ by vomiting will not normally lead to serious K^+ deficiency, however, unless it is accompanied by semistarvation.

Renal losses of potassium can occur as a result of loss of cell potassium caused by acidosis, sodium overload, dehydration and excessive protein catabolism. In addition, large doses of aldosterone cause excess excretion of potassium, accompanied by increased sodium retention which leads to depletion of plasma K^+. Clinically, it is commonly associated with tumours of the adrenal which cause a greatly increased secretion of aldosterone. Renal disease does not normally lead to abnormal potassium loss and K^+ is commonly retained in both acute and chronic renal failure.

Hyperkalaemia

High concentrations of potassium are also toxic to muscle causing paralysis of cardiac and skeletal muscle. Early signs of cardiac damage caused by increased concentrations of potassium include arrhythmia followed by a slow rhythm of the heart, ventricular fibrillation and arrest in diastole. Changes in its ECG also occur. Death occurs when the concentration of K^+ reaches 10 mmol equiv./l.

Absolute increases of the total body potassium are not common and hyperkalaemia does not necessarily indicate raised cell concentrations of potassium. Leakage from cells may occur to increase the plasma potassium, leaving the cells depleted and this may be particularly important in acidosis which causes leakage of potassium from cells. It is clear that under conditions such as these, plasma K^+ determination will give misleading information about the intracellular concentration of potassium.

Clinical hyperkalaemia will arise when potassium excretion by the kidney is depressed and in acute renal failure the condition may be serious unless potassium is rigidly excluded from the diet. Insulin and glucose promote the uptake of potassium into cells and this treatment is used for patients suffering from serious hyperkalaemia.

Chapter 22

Plasma calcium and phosphate: regulation by vitamin D and parathyroid hormone

22.1 Introduction: importance of calcium and phosphate in the animal body

Before discussing the regulation of calcium in the plasma, the many varied roles which calcium plays in the animal body are summarized and their importance considered. Some functions of calcium, especially in the bone, are closely related to phosphate, and it is often useful to consider calcium and phosphate metabolism together.

Calcium is known to play an important role in the following processes or systems:

a. Regulation of ionic movements across membranes, particularly of Na^+ and K^+, and *subsequent* development of potential gradients in muscle Regulation of the 'sodium gate'

b. Muscle contraction initiated by binding of calcium to troponin (cf. Chapter 31)

c. Activation of many enzymes which require Ca^{2+} as a specific cofactor

d. Neurotransmission: acetylcholine release at membranes is dependent on calcium concentration

e. Release of hormones, such as insulin, adrenaline and anterior pituitary hormones, from their site of synthesis into the circulation

f. Interactions of hormones with receptor sites

g. Formation of the bone salt, hydroxyapatite.

In some of these processes the mechanism of action of Ca^{2+} ions is well understood (e.g. in muscle contraction), but by no means in all.

22.2 Plasma calcium

In view of its great importance in the regulation of ionic movement, in transmission of nerve impulses and in muscle contraction, the control of the plasma calcium concentration within a very limited range is of great medical importance.

The calcium concentration in normal individuals is approximately 10 mg per 100 ml plasma (5 mmol equiv./l) and it seldom falls outside the range 9·2–10·4 mg per 100 ml. It is important to note that approximately only 45 per cent of the plasma calcium is in the ionic freely diffusible form. The major proportion, 50 per cent, is bound to plasma proteins, and the remaining 5 per cent is complexed with molecules, such as citrate (*Fig.* 22.1).

When the calcium concentration falls to a figure below the normal range, a condition of 'hypocalcaemia' is described. When this happens, the nervous system becomes progressively more and more excitable as the permeability of the membranes increases. The increased excitability causes spontaneous discharges that initiate nerve impulses, and these are transferred to the skeletal muscles so causing tetany. This condition is lethal when the blood calcium concentration falls to 6 mg per 100 ml, but 'latent tetany' can be observed

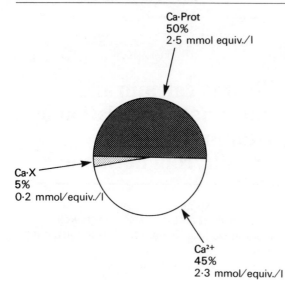

Fig. 22.1. Distribution of calcium in plasma.
Ca·Prot, calcium–protein complexes; Ca²⁺, ionic calcium;
Ca·X, non-protein-bound Ca complexes.

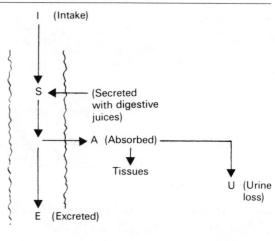

Fig. 22.2.   Relation between dietary intake of calcium and secretion into the gut.
If α = fraction absorbed,

$$A = [I + S]$$
$$E = (1 - \alpha)[I + S]$$

Total daily loss $= E + U = (1 - \alpha)[I + S] + U$
(to be replaced).

when the plasma concentration of calcium is between 6 and 10 mg per 100 ml.

If the blood calcium concentration rises above 15 mg per 100 ml, the condition of 'hypercalcaemia' develops. The nervous system is depressed, reflexes are sluggish, and the muscles become weak as a result of calcium effects on the muscle cell membranes. The serious consequences resulting from disturbances of the calcium concentrations in plasma clearly emphasize the physiological need for elaborate mechanisms to control the calcium level.

22.3 Dietary calcium and phosphate and calcium kinetics

The daily requirement of calcium in the human adult is dependent on three main factors:

a. The percentage of the total intake absorbed from the gastrointestinal tract

b. The quantity of calcium excreted into the tract with the digestive secretions

c. The quantity excreted daily in the urine.

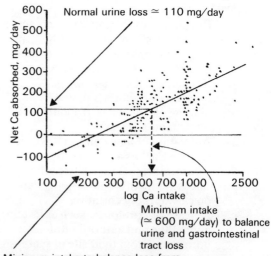

Fig. 22.3. Net absorption of calcium for human adults on varied dietary calcium intakes.
212 balances.

The inter-relationships of the factors are illustrated diagrammatically in *Fig.* 22.2.

The net absorption of calcium is negative if the intake falls below about 200 mg/day (*Fig.* 22.3), because of the large secretion of calcium into the gastrointestinal tract (of which only a fraction is reabsorbed, *Fig.* 22.2). A mean requirement of 200 mg calcium per day is therefore essential to balance gastrointestinal losses. In addition there is an obligatory loss of calcium in the urine of 110 mg/day, making total obligatory losses of calcium about 310 mg/day in the adult. Calcium absorption is never 100 per cent efficient (as discussed below) and it can be seen from *Fig.* 22.3 that a minimum daily intake of about 600 mg calcium is essential to provide sufficient absorbed calcium to make up for these losses.

Phosphate reabsorption by the kidney, unlike that of calcium, is extremely efficient, and an almost linear relationship exists between phosphate intake and urinary excretion (*Fig.* 22.4). Phosphate is widely distributed in foods and problems of lack of phosphate in the diet are almost unknown. A minimum intake of 600 mg/day is usually recommended.

Growing children require larger intakes of calcium and phosphate than adults for deposition as bone.

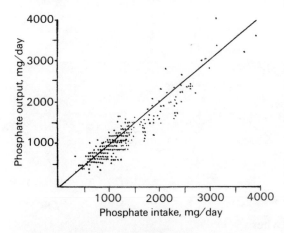

Fig. 22.4. **Urinary excretion of phosphate for human adults on varied dietary phosphate intakes.**
646 balances.

CALCIUM KINETICS

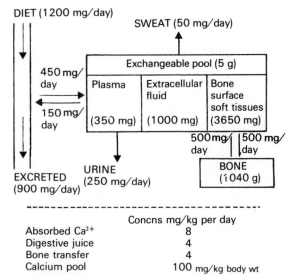

	Concns mg/kg per day
Absorbed Ca²⁺	8
Digestive juice	4
Bone transfer	4
Calcium pool	100 mg/kg body wt

Fig. 22.5. **Calcium kinetics.**
Relationships between calcium absorption from the gut; calcium secretion into the gut; loss of calcium in the urine and sweat, and deposition in the bones in a typical human adult.

The movements of calcium through the main systems of the body during a 24-hour period for a typical adult are summarized in *Fig.* 22.5. It shows that besides the major routes of calcium loss, i.e. in the faeces from non-absorbed calcium and via the digestive juices, and in the urine due to inefficient reabsorption by the kidney tubules, there are smaller quantities excreted in the sweat.

It is important to note that, even in the adult, the large reserves of calcium in the bone are in dynamic equilibrium with plasma calcium. Mechanisms of regulating its equilibrium differ depending on the direction. Calcium leaves the tissue fluids and is incorporated into hydroxyapatite crystals continuously, but calcium leaves the bone at certain localized sites under the control of parathyroid hormone, calcitonin and 1,25-dihydroxy-vitamin-D_3. In proportion to weight, calcium exchange will be much greater in the infant and growing child, than in the adult.

22.4 Factors regulating calcium absorption

The absorption of calcium from the intestinal tract is under the control of many factors, but for several of these the detailed working of their mechanisms is not clearly understood.

The interpretation for tests for calcium absorption depend on the previous dietary intake of calcium, because the body tends to adapt to its normal dietary calcium intake. The percentage of a test dose of calcium absorbed is therefore inversely related to the dietary calcium intake to which the subject has become accustomed. Thus, subjects whose normal intake of calcium is high, will absorb a relatively small percentage of the intake and vice versa. The factors that stimulate calcium absorption are lactose, basic amino acids, vitamin D_3 and parathyroid hormone.

Factors stimulating absorption

a. Lactose

Experiments using young growing animals have clearly established that the inclusion of lactose in the diet will substantially enhance the absorption of calcium. A diet consisting of 12 per cent lactose will increase the absorption of calcium in rats by 30 per cent. This is clearly a very desirable regulating system for the young animal, which will efficiently absorb calcium at the same time as lactose, both being supplied by the milk. Regulation of calcium absorption by lactose has not been demonstrated in young children, but it is believed to occur. The molecular mechanism for this process is not known.

b. Basic amino acids

Several amino acids significantly stimulate calcium absorption: lysine is most effective, but arginine, histidine, tryptophan, methionine, and isoleucine also increase the absorption rate. Calcium uptake from the intestine is, therefore, most efficient when a high-protein diet is consumed.

c. Vitamin D_3

The metabolite of vitamin D_3, *1,25-dihydroxy-vitamin-D_3* is very important in regulating calcium absorption and is discussed in detail in Section 22.7.

d. Parathyroid hormone

It has been suggested that parathyroid hormone may stimulate calcium absorption from the intestine, although the effect may not be a direct one, and is likely to be mediated by vitamin D_3 metabolites. The relationships of parathyroid hormone and vitamin D_3 are discussed in Section 22.10.

Factors inhibiting absorption

On the other hand, calcium absorption is inhibited by corticosteroids, phytates, bile duct obstruction, coeliac disease and chronic renal failure.

a. Corticosteroids

Corticosteroids have been reported to cause depression of calcium absorption in man, but their effects are equivocal.

b. Phytates in the diet

Phytic acid or *inositol hexaphosphate*

where

is a common constituent of many different cereals. It has a strong affinity for several metal ions, particularly calcium, and the binding inhibits their absorption. Phytate is particularly important in many parts of the world where the people exist on a mainly vegetable diet. Calcium is poorly absorbed from these diets on account of the high phytate content and calcium malnutrition, particularly in infancy and pregnancy, frequently develops amongst those populations relying on a vegetable diet. Milk is a very good dietary source of calcium and no phytate is present to inhibit absorption.

c. Obstruction of the bile duct or a defective bile salt supply

It was described in Chapter 16 how fat-soluble vitamins, including vitamin D, require bile salts for adequate absorption. If the supply of bile salts to the intestine is, therefore, depressed, inadequate absorption of vitamin D occurs and, consequently, calcium absorption is impaired.

d. Coeliac disease: malfunction of the intestinal mucosa

Pathological conditions which give rise to damaged intestinal mucosa severely impair absorption of fat and fat-soluble vitamins, and so the absorption of vitamin D is inadequate.

e. Chronic renal failure

Clinical studies established several years ago that one of the characteristic consequences of chronic renal failure was a tendency towards inadequately calcified and weak bones. The reasons for this have only become clear in recent years, during which time the metabolism of vitamin D became more fully understood. An essential stage in the conversion of vitamin D to its active metabolite occurs in the kidneys (see Section 22.6) and if the kidney function is seriously impaired then metabolism of the vitamin is much less efficient. Consequently, the active form of vitamin D that stimulates calcium absorption is not formed in adequate amounts and calcium absorption from the intestine is depressed.

22.5 Vitamin D

The disease 'rickets' has been known in Europe for many hundreds of year, occurring in young growing children and characterized by very poor calcification of bones of the body. The main disease features are illustrated in Fig. 22.6.

Rickets was first accurately described in 1650 by Glisson, Professor of Medicine at Cambridge in his Treatise on Rickets. Strangely the disease was almost unknown in tropical countries, in Scandinavia and amongst Eskimos, the two latter being populations that consume a high proportion of fish in their diets. Both these facts are important in understanding this condition, but a long period elapsed before their significance was appreciated and understood. With the

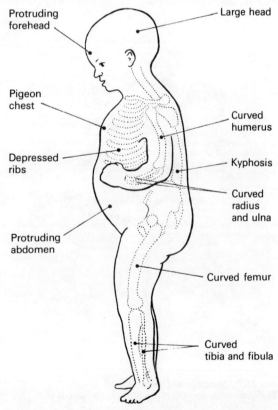

Fig. 22.6. **Diagrammatic representation of symptoms of rickets.**

Table 22.1 **Incidence of rickets in the UK during 1868–1926**

Date	Place	Incidence of rickets, %	Authority
1868	London	33 (with severe rickets)	Gee
1871	Manchester	33 (with severe rickets)	Ritchie
1915	LCC	80 (with some degree of rickets)	Dick (1600 children examined)
1920	Durham	82 (with some degree of rickets)	McIntosh (1300 children examined)
1928	LCC	87 (with some degree of rickets)	Newman (1600 children examined)
1926	England and Wales	50 (with some degree of rickets)	Ministry of Health (1000 children examined)

LCC = London County Council.

development of communities living in the cramped conditions of an industrial society the disease became more prevalent. By the end of the nineteenth century many school children in British cities were suffering from rickets to some degree (*Table* 22.1). In severe cases rickets leads to death but, in the milder cases that were usually encountered, symptoms such as enlarged wrists, ankles and bow legs were frequently seen (*Fig.* 22.7).

During the early part of this century it gradually became apparent that the disease could be treated successfully by incorporating fish oils, such as cod or halibut, into the diet or, most remarkably, by exposing the children to ultraviolet light. The relationship between these two forms of treatment was, at first, completely baffling, but by the mid-1920s it was discovered that in rickets an essential factor for the body was lacking. This factor was present in fish oils or could be formed in the skin by ultraviolet light. It was called vitamin D because it was the fourth vitamin to be discovered (after A, B, and C).

There are two main forms of vitamin D, D_2 and D_3, which differ very slightly in the structure of their side chains. Their biological actions are identical. Vitamin D_3 (cholecalciferol) is the natural vitamin, obtained in fish oils or by the action of ultraviolet light on 7-dehydrocholesterol in the skin, whilst vitamin D_2 can be made in the laboratory by irradiating ergosterol. 7-Dehydrocholesterol is synthesized in the body, and ultraviolet light causes scissoring of the β-ring of 7-dehydrocholesterol to form pre-vitamin-D_3. The rearrangement to form vitamin D_3 is a spontaneous, but slow, process taking 36 h at 37 °C (*Fig.* 22.8). When vitamin D is fed to patients suffering from rickets there is a dramatic improvement in their condition; the plasma calcium level rises and deposition of calcium in the bones is markedly stimulated.

Fig. 22.7.   Symptoms of severe rickets.
This picture shows rickets in Vienna in the post-War famine of 1920. The children, six years of age, show severe rachitic deformities compared with the normally grown child of the same age in the centre. Note enlarged wrists and ankles and bow legs.

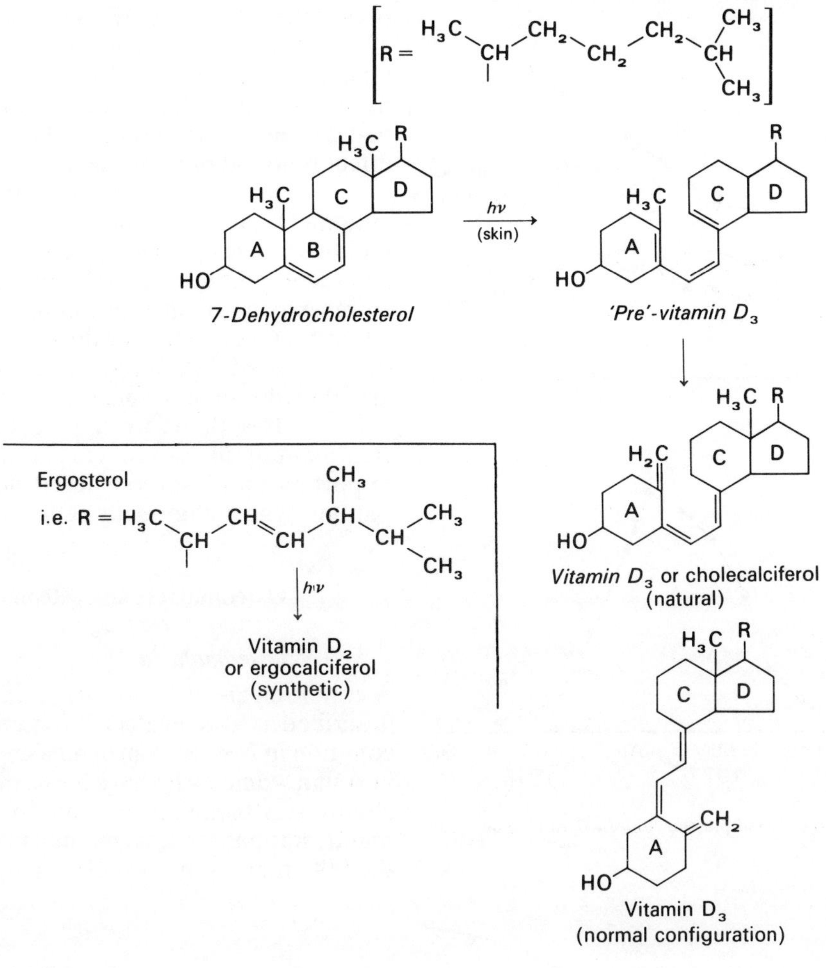

Fig. 22.8. **Synthesis of vitamin D$_3$ by the action of ultraviolet light.**
In the manufacture of vitamin D$_2$, a plant sterol, ergosterol, is irradiated. In the skin, 7-dehydrocholesterol that can be synthesized from cholesterol is converted by ultraviolet light into vitamin D$_3$ (cholecalciferol).

Sources of vitamin D—dietary and synthesized

In view of the fact that vitamin D can be obtained from dietary sources, especially fish oils, and by the action of ultraviolet light on the skin, the question arises as to which will form the major source of the vitamin under normal conditions. In tropical countries with prolonged periods of sunshine, it is clear that ultraviolet light will be of major importance.

In temperate areas of the world, however, ultraviolet light is still important, especially in the summer, but the quantities of vitamin D available depend on the lifestyle of the individual. This is clearly shown by a recent study of local authority workers in Scotland. In both outdoor workers and office workers, the concentration of 25-hydroxy-vitamin-D$_3$ (the metabolite of vitamin D$_3$) in the plasma was much higher at the end of the summer

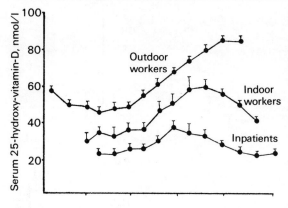

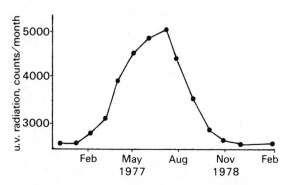

Fig. 22.9. **Variation of plasma vitamin D₃ during one year in Scotland.**

period than during the winter months (*Fig. 22.9*). However, the concentration of the vitamin D_3 metabolite in the serum continued to rise after the intensity of the ultraviolet light had declined. The reason for this delayed effect is not, at present, clear.

In other studies, the concentration of vitamin D_3 metabolite in a group of elderly inpatients in hospital was found to be very low, and the importance of ultraviolet light in increasing the circulating plasma 25-hydroxy-vitamin-D_3 very dramatically was demonstrated by treating the patients with artificial ultraviolet light (*Fig.* 22.10).

It is, therefore, apparent that treatment of this nature is of considerable importance for institutionalized elderly patients who are normally deprived of sunlight.

Osteomalacia and osteoporosis

Osteomalacia

A condition analogous to rickets in adults is described as 'osteomalacia'. It is much more common in women than in men and especially in Asian women who have borne many children. Although severe clinical manifestations are less common in women in the UK, there is, nevertheless, a significant fall

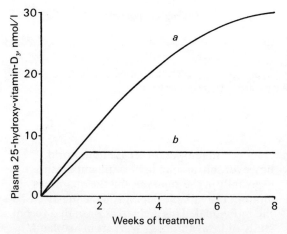

Fig. 22.10. **Ultraviolet treatment in geriatric patients.** Patients (age 70–90 years) suffering from osteomalacia caused by negligible sunlight, poor absorption and defects in 25-hydroxylation.
(*a*) u.v. treated; (*b*) control.

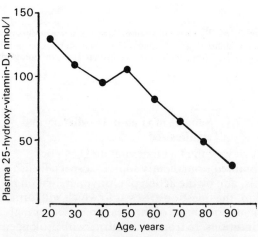

Fig. 22.11. **Plasma vitamin D₃ related to age in British females.**

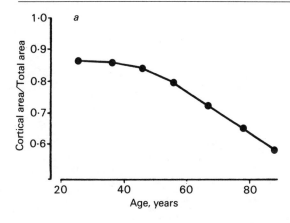

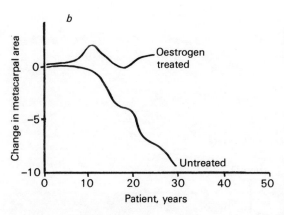

Fig. 22.12. **Osteoporosis.**
Age-related bone loss starting at menopause
(0·5% per year). (*a*) Relationship of metacarpal cortical
and metacarpal total area. Urinary Ca^{2+} is increased, as
are plasma alkaline phosphatase and hydroxyproline.
(*b*) Change in mean metacarpal area in postmenopausal
women, both treated and not treated with oestrogen.

in the circulating 25-hydroxy-vitamin-D_3
during adult life (*Fig.* 22.11). As in children,
vitamin D itself may be deficient and a
contributing cause of the condition, but,
alternatively, the enzymes involved in the
conversion of vitamin D_3 to its active form, the
25-hydroxylase in the liver or the
'1-hydroxylase' in the kidney, may be
diminished in activity.

Osteoporosis

Osteoporosis is an age-related bone loss which
occurs in post-menopausal women, the total
loss of bone salt being approximately 0·5 per
cent per year. Measurements on X-ray
photographs of the hand enable the metacarpal
area of the bone to be calculated and thus
quantitative measurements of loss of bone to
be evaluated (*Fig.* 22.12). Some of the
symptoms, such as increased urinary calcium
and increased urinary hydroxyproline which is
formed from the degradation of bone collagen,
are similar to those observed in osteomalacia
but osteoporosis does not appear to be related
to deficiency of vitamin D_3 or its metabolites.
The fact that it begins at the menopause
indicates that it is hormone related and the
symptoms can be almost completely prevented
by oestrogen treatment. It is uncertain exactly
how oestrogen functions, but it appears that it
protects bone against bone-resorbing agents
and especially parathyroid hormone.

22.6 Conversion of vitamin D_3 to 1,25-dihydroxy-vitamin-D_3

After the discovery of the structure of
vitamin D in the early 1930s, nearly 40 years
elapsed before any real progress was made
towards an understanding of its mode of action
beyond the general concept of elevation of
calcium levels in the serum and improvement
of bone calcification. During recent years,
since *radioactively* labelled vitamin D_3 became
available, a better understanding of its mode of
action has developed; it has become clear that
the vitamin must undergo metabolism before
conversion to an active form, the sequence of
events being illustrated in *Fig.* 22.13.

In the liver, vitamin D_3 undergoes
25-hydroxylation catalysed by an enzyme,
25-hydroxylase, that requires oxygen and
NADPH. The 25-hydroxy-vitamin-D_3 formed
is then transported in the plasma to the kidney,
where it is further hydroxylated in the
1 position to 1,25-dihydroxy-vitamin-D_3.
This hydroxylation, which is NADPH, oxygen
and cytochrome P450 dependent, is an
important site of regulation (*see Fig.* 22.14).

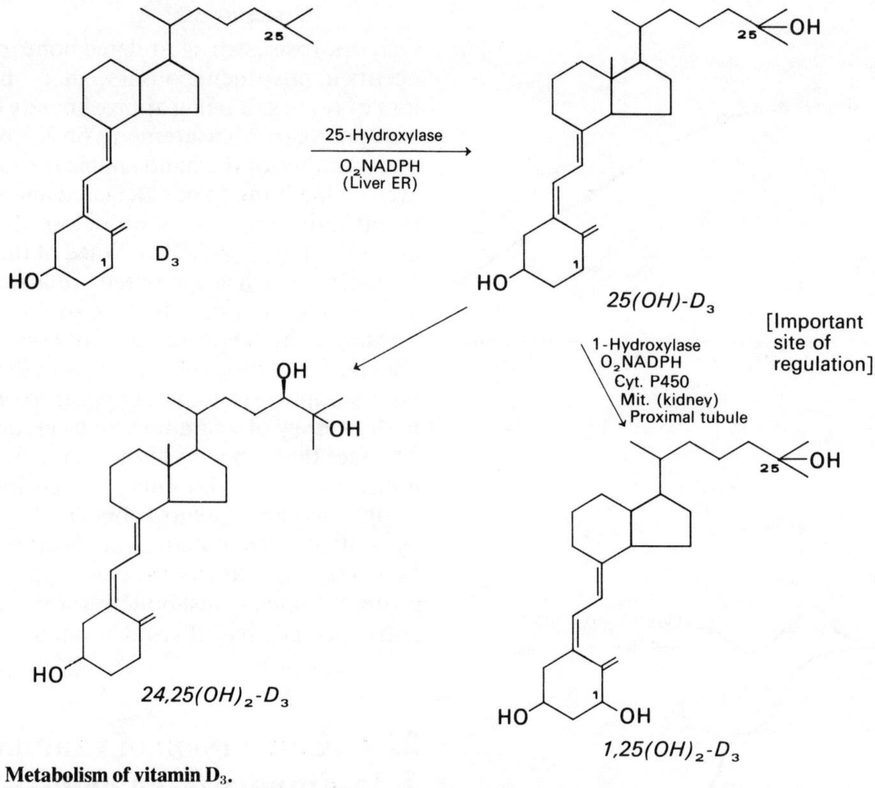

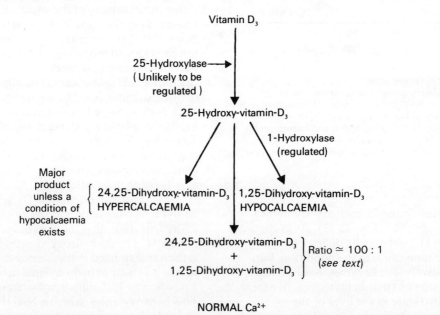

Fig. 22.13. Metabolism of vitamin D₃.
Conversion to 24,25-dihydroxy-vitamin-D₃ and to 1,25-dihydroxy-vitamin-D₃.

Vitamin D₃

25-Hydroxylase⟶
(Unlikely to be
regulated)

25-Hydroxy-vitamin-D₃

1-Hydroxylase
(regulated)

Major
product
unless a   { 24,25-Dihydroxy-vitamin-D₃ | 1,25-Dihydroxy-vitamin-D₃
condition of   HYPERCALCAEMIA | HYPOCALCAEMIA
hypocalcaemia
exists

24,25-Dihydroxy-vitamin-D₃
+   } Ratio ≃ 100 : 1
1,25-Dihydroxy-vitamin-D₃   *(see text)*

NORMAL Ca²⁺

Fig. 22.14. Regulation of metabolism of vitamin D₃ by the level of circulating plasma calcium.

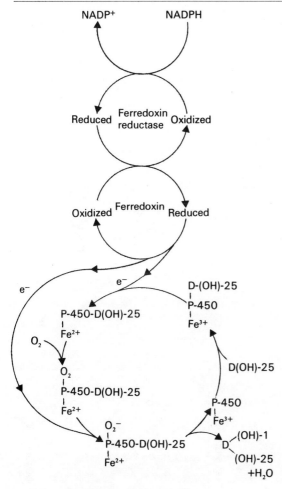

Fig. 22.15. **Mechanism of 1-hydroxylation of 25-hydroxy-vitamin-D₃.**

forming 24,25-dihydroxy-vitamin-D_3. This is not believed to be an active form of vitamin D_3 and, in most tests, it is approximately equal to that of the substrate from which it was formed. The plasma concentration of 24,25-dihydroxy-vitamin-D_3 is, however, approximately 100 times that of 1,25-dihydroxy-vitamin-D_3, so that the formation of 24,25-dihydroxy-vitamin-D_3 must be the normal pathway.

In addition to these major pathways of vitamin D_3 metabolism, a larger number of minor metabolites has been described (*Fig.* 22.16). Whether these are mainly excretory products or whether any exert undiscovered physiological roles is not, as yet, clear. It is, however, certain that none can replace 1,25-dihydroxy-vitamin-D_3 as the active form of the vitamin.

Controls of the activity of the 1-hydroxylase enzyme

It is generally believed that no effective control exists for the activity of the 25-hydroxylase in the liver and that all the vitamin D_3 available in the body is converted to the 25-hydroxy derivative. The 1-hydroxylase in the kidney is, however, an important site of control. 25-Hydroxy-vitamin-D_3 entering the kidney is normally metabolized to 24,25-dihydroxy-vitamin-D_3, unless the 1-hydroxylase is activated. Control of 1-hydroxylase activity is exerted by the parathyroid hormone, the serum calcium and phosphate concentration and by feedback inhibition (*Fig.* 22.16).

As the concentration of serum calcium falls, the 1-hydroxylase is activated but this also requires the presence of parathyroid hormone and does not occur in its absence. Reduced serum phosphate also activates the 1-hydroxylase but parathyroid hormone is not required for this action (*Fig.* 22.17). As the concentration of the product 1,25-dihydroxy-vitamin-D_3 increases, this will cause a feedback inhibition of the 1-hydroxylase allowing more 24,25-dihydroxy-vitamin-D_3 to be formed. The system therefore permits an accurate control to be exerted over the plasma calcium concentration.

The cytochrome P450 systems involved in the 1-hydroxylation of 25-hydroxy-vitamin-D_3 is very similar to that involved in oxidative drug metabolism (cf. Chapter 36), but the main difference is that a non-haem iron-containing protein, ferredoxin, forms part of the electron transport chain transferring electrons to cytochrome P450 (*Fig.* 22.15).

The 1,25-dihydroxy-vitamin-D_3 is the main active form of vitamin D_3. The kidney, however, also has the capacity to hydroxylate 25-hydroxy-vitamin-D_3 in the 24 position

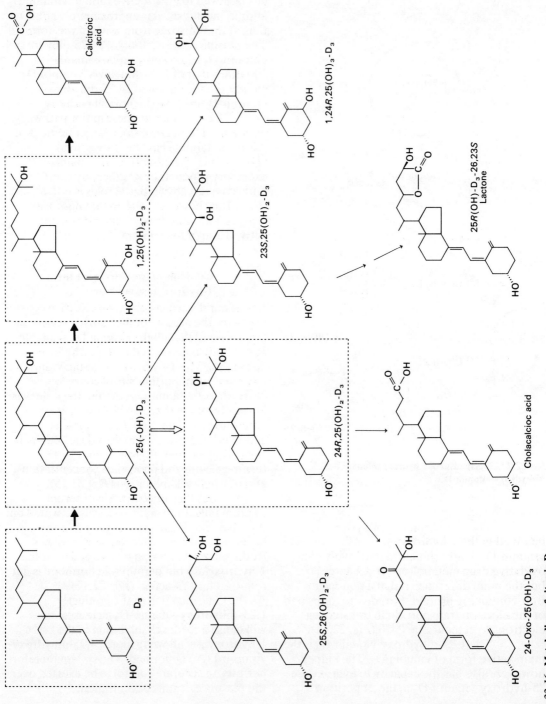

Fig. 22.16. **Metabolism of vitamin D₃.**

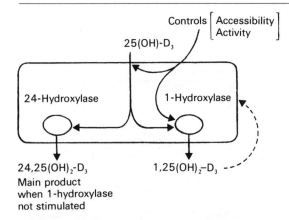

Fig. 22.17. **Regulation of metabolism of 25-hydroxy-vitamin-D₃.**
Main controls of 1-hydroxylation are: (1) PTH (↑);
(2) Ca²⁺ (↓); (3) Pᵢ (↓); (4) vitamin D status (feedback inhibition).

22.7 Mode of action of 1,25-dihydroxy-vitamin-D₃

The metabolite of vitamin D, when injected or fed, soon causes an increased concentration of calcium and phosphate in the plasma. The sequence of events is a complex one. The main effects of 1,25-dihydroxy-vitamin-D₃ are to stimulate the following processes:

a.   Intestinal transport of Ca²⁺
b.   Intestinal transport of phosphate
c.   Renal absorption of Ca²⁺
d.   Renal absorption of phosphate
e.   Mobilization of Ca²⁺ from bone
f.   Mobilization of phosphate from bone.

Inorganic phosphate always accompanies calcium through the membranes of the intestinal or renal tubule cells, but whether this is a passive process—the negatively charged phosphate may simply balance the charges on the positive Ca²⁺ ions—or an active process controlled by vitamin D is uncertain.

When the concentration of calcium in the plasma is raised there is an increased rate of deposition of calcium in bone tissue. In fact, if the intake of vitamin D is maintained at a high level for long periods, the consequent high concentration of calcium in the plasma begins to cause calcium deposits in many tissues, especially in the mucoproteins of synovial membranes, the kidneys, the myocardium, the pancreas, and the uterus. Calcification in these organs can be a serious matter, and may prove lethal. High doses of vitamin D over a long period are, therefore, toxic.

Many investigations have been carried out to establish whether or not vitamin D, or its metabolite, plays a direct role in the deposition of calcium in bone and other tissues, but most of these studies have yielded negative or equivocal results. It appears likely that extensive calcium deposition is simply a consequence of a raised plasma calcium concentration. High concentrations of vitamin D metabolite can also cause mobilization of Ca²⁺ and phosphate from bone, which appears to contradict the deposition effects. However, if mobilization were a direct effect of parathyroid hormone, the role of the vitamin D metabolite would only be to regulate release or activity of the hormone. The inter-relationship of the hormone and the metabolite is discussed below.

The main action of the vitamin D metabolite has been established as increasing calcium absorption from the intestinal lumen (as discussed below).

The effect of 1,25-dihydroxy-vitamin-D₃ on calcium absorption from the intestine

In the mucosal cells 1,25-dihydroxy-vitamin-D₃ stimulates the synthesis of a protein that has a very high affinity for calcium—the calcium-binding protein. The mechanism of action is illustrated in *Fig.* 22.19. A special carrier protein transfers the 1,25-dihydroxy-vitamin-D₃ brought to the mucosal cell to the nucleus. This crosses the nuclear membrane and the vitamin D metabolite, or its protein complex, then causes derepression of part of the DNA. Synthesis of a special mRNA containing the code for the calcium-binding protein can then occur, followed by synthesis of the protein on the ribosomes (*see* Appendices 28, 31, 32).

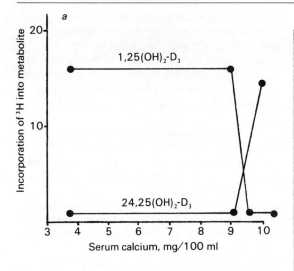

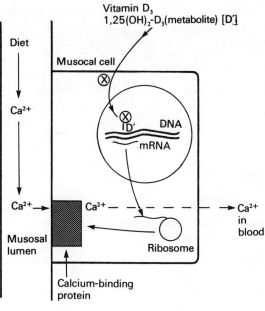

Fig. 22.19. **Control of synthesis of calcium-binding protein in the mucosal cells by 1,25-dihydroxy-vitamin-D₃.**

$[1,25(OH)_2\text{-}D_3] = 1,25$-dihydroxy-vitamin-$D_3$. $\bigotimes$ = carrier protein.

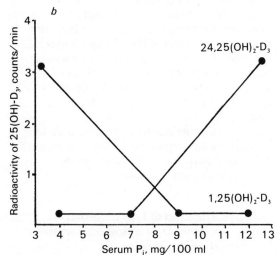

Fig. 22.18. **Regulation of hydroxylation of 25-hydroxy-vitamin-D₃ by serum Ca²⁺ and Pᵢ.**
(a) Requires PTH;
(b) thyroparathyroidectomized.

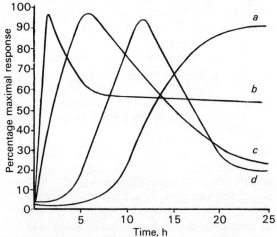

Fig. 22.20. **Sequence of events following dosing with 1,25-dihydroxy-vitamin-D₃ (1,25(OH)₂-D₃) in chick intestine.**
(a) Calcium-binding protein in mucosa; (b) 1,25(OH)₂-D₃; (c) calcium transport; (d) calcium-binding protein synthesis on ribosomes.

The calcium-binding protein has been purified from chick intestinal cells and the properties of the protein are summarized in *Table* 22.2. There are a large number of polar amino acid side chains, mainly from glutamate and aspartate, and it is assumed that calcium binding occurs at the carboxyl groups of these side chains.

Table 22.2 **Calcium-binding protein**

Tissue	Intestine Kidney
Molecular weight	28 000
Number of amino acids	242
Acidic	
Glutamic acid	44
Aspartic acid	34
Polar side chains, % total	53
Calcium binding sites	
High affinity	4
Weak affinity	32
Metal binding affinities	Ca > Mn > Zn > Co > Ni > Mg

The protein is also formed in the kidney tubule cells where its function is probably to retain calcium and prevent excess calcium excretion—this being a well-known consequence of vitamin D treatment.

The hypothesis that vitamin D acts by controlling the synthesis of the calcium-binding protein is attractive and was widely accepted for a few years, but subsequent research has raised many questions about its precise role.

Experiments indicating that the calcium-binding protein was not the main factor involved in the stimulation of calcium uptake from the intestine are illustrated in *Fig.* 22.20. Vitamin D deficient chickens were dosed with 1,25-dihydroxy-vitamin-D_3 and several measurements were made on mucosal cells. 1,25-Dihydroxy-vitamin-D_3 was taken up very rapidly by the cells and, although synthesis of calcium-binding protein did occur, the rate of synthesis was not significant until 8–12 hours had elapsed. However, calcium transport was stimulated much earlier, 2–4 hours after treatment, and the rate of transport of calcium was actually declining when the synthesis of the calcium-binding protein was rising (*Fig.* 22.20). This experiment establishes that the calcium-binding protein cannot be directly involved in transport and therefore poses two important questions: firstly what is the role of the calcium-binding protein and, secondly, how

does 1,25-dihydroxy-vitamin-D_3 regulate calcium absorption?

It is now generally believed that the calcium-binding protein plays a protective role in the cells, effectively removing the calcium ions from the cell organelles where it could easily cause serious disturbances in enzyme action or metabolic function. This still leaves the second question, as to how calcium absorption is regulated by vitamin D and, at present, there is no answer to this problem.

22.8 Parathyroid hormone

The parathyroid hormones are *polypeptides* synthesized by the parathyroid gland. The bovine parathyroid hormones have been most carefully studied, and three have been distinguished: PTH I, PTH II and PTH III. Normal secretions from the gland contain approximately 75 per cent PTH I, 15–20 per cent PTH II and 5–10 per cent PTH III.

Each hormone contains 84 amino acids and differs from the others in only very few amino acids (*see Fig.* 22.21). By careful degradation of the polypeptides, it has been possible to demonstrate that most of the activity resides in the peptide containing residues 1–34 (*Table* 22.3); consequently it is believed that this peptide is the active and effective hormone.

Table 22.3 **Activity of fragments of bovine parathyroid hormone**

Peptide	Metabolic rate, units/mg
1–84 (Native)	3000
1–34	5400
3–34	10
1–28	440

Note: Peptides shorter than 28 residues possess very little hormonal activity.

Human parathyroid hormone also contains 84 amino acid units and only 8 of these differ from those of the main bovine hormone (*Fig.* 22.21). It has, however, been difficult to prove that the 1–34 peptide is the active

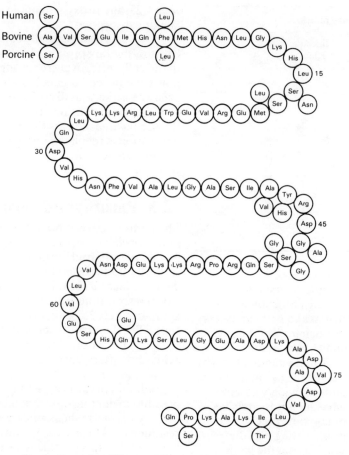

Fig. 22.21. **Structure of bovine I parathyroid hormone in comparison with the human and porcine hormone.**

component of human parathyroid hormones despite the general assumption that it is. This peptide has been synthesized for the human hormone and recent research indicates that it has a much weaker effect on adenylate cyclase than the bovine peptide. The complete (1–84) peptide may, therefore, be required in the human for interaction with the receptor sites, although other tests, for example on chick, show that both peptides are equally effective on hypocalcaemia; the helical structure of the larger peptide appears to be essential in man.

Control of release of parathyroid hormone

In the parathyroid gland, pre-pro-parathyroid hormone (pre-pro-PTH) is first synthesized.

This hormone is composed of 115 amino acids and is converted with the loss of 25 amino acids to the immediate precursor, pro-parathyroid hormone composed of 90 amino acids and it is stored in this form until required. Release of the parathyroid hormone (84 amino acid units) is accomplished by the splitting off of a hexapeptide from the N-terminal group by the action of a protease. The release of the hormone is controlled by the circulating calcium concentration. Response is extremely sensitive and parathyroid hormone is released within 20 seconds following a decrease in the calcium concentration, the hormone released being proportional to the change in calcium concentration. The 1–84 peptide is released into the plasma but is rapidly cleaved, probably at the site of its action, into an N-terminal

1–34 peptide with a very short half-life and a C-terminal peptide with a longer half-life.

Mode of action of parathyroid hormone

Parathyroid hormone acts in a complex manner on three main organs: the intestine, bone and the kidney (*Fig.* 22.22). The net result of its action, like that of 1,25-dihydroxy-vitamin-D_3, is to raise the plasma calcium concentration. There are, however, important differences between the actions of parathyroid hormone and 1,25-dihydroxy-vitamin-D_3.

i. Parathyroid hormone raises the blood calcium level primarily by causing Ca^{2+} release from bone, whereas 1,25-dihydroxy-vitamin-D_3 raises it primarily by causing increase in the rate of intestinal Ca^{2+} absorption.

ii. Parathyroid hormone inhibits absorption of inorganic phosphate by the kidney tubule, whereas 1,25-dihydroxy-vitamin-D_3 enhances it.

Biochemical experiments have established that parathyroid hormones activate adenylate cyclase to produce adenosine cyclic 3′:5′-monophosphate (cAMP) (cf. Chapter 17). The precise sequence of events between cAMP formation

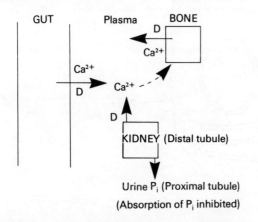

GUT   Plasma   BONE

Fig. 22.22. **Summary of actions of parathyroid hormone on the intestine, bone and kidney.**
D indicates that vitamin D_3 metabolite is also likely to be involved.

and the observed physiological effects have not, as yet, been fully elucidated and it has also not been established whether all the physiological effects exerted by parathyroid hormone are mediated by cAMP or involve other mechanisms.

In the kidney, parathyroid hormone acts at two separate sites. The first involves increased Ca^{2+} reabsorption in the distal tubule. The response of the tubule cells is very rapid but the capacity of the system is limited and activation of adenylate cyclase may be involved. Enhancement of synthesis of the calcium-binding protein may, however, occur and, therefore, 1,25-dihydroxy-vitamin-D_3 could also be involved in the process. In the second, parathyroid hormone acts on the proximal tubule to cause inhibition of inorganic phosphate reabsorption; activation of adenylate cyclase may also be involved in this process.

Parathyroid hormone is believed to have an important catabolic effect on bone with the release of both calcium and phosphate. Recently, however, it has been suggested that the low concentrations of the active N-terminal peptide fraction of parathyroid hormone (10–20 pg/ml, half-life 2–4 min) normally circulating in the human, have an anabolic effect on bone and that catabolic effects only occur when the concentration of parathyroid hormone is abnormally high. If this is correct then the normal physiological role of parathyroid hormone must be moderated by the kidney or through its inter-relationship with the regulation of vitamin D metabolism.

Several questions arise in regard to the action of parathyroid hormone on bone.

i. To what extent is the Ca^{2+} released from bones stored in a readily available ionic form or released from the hydroxyapatite complex?

ii. What roles do the different types of bone cells, the osteoblasts, osteocytes and osteoclasts, play?

iii. Are Ca^{2+} and phosphate always released together in equivalent amounts?

iv. What role does cAMP play in the process?

Although no clearcut answers can be given to all these problems, the catabolic process can be split into two stages.

First stage. This involves a rapid release of Ca^{2+} from the cells, mainly the *osteoblasts* and *osteocytes*. Protein synthesis is not involved and no resorption of the actual bone matrix is involved. The process can be initiated within a few minutes of exposure of bone to parathyroid hormones.

Second stage. This stage is initiated within 15 min–2 h, but it involves extensive RNA and protein synthesis within the *osteoclast* cells. New synthesis of lysosomal enzymes occurs, e.g. collagenase, involved in bone resorption. Calcium and phosphate are gradually released

as a consequence of the increased activity of the lysosomal enzymes (cf. Chapter 3).

The mitochondria probably play a very important role in the process. Several years ago Lehninger demonstrated that mitochondria of several tissues, including liver, kidney and mineralizing tissues, rapidly took up Ca^{2+} and phosphate during active respiration to deposit hydroxyapatite granules. It is postulated that in bone these mitochondrial deposits may be available for bone mineralization. The parathyroid hormone could function by regulation of the activity or permeability of the osteoblast and osteocyte mitochondria, making Ca^{2+} rapidly available for release when required into the plasma (*Fig.* 22.23).

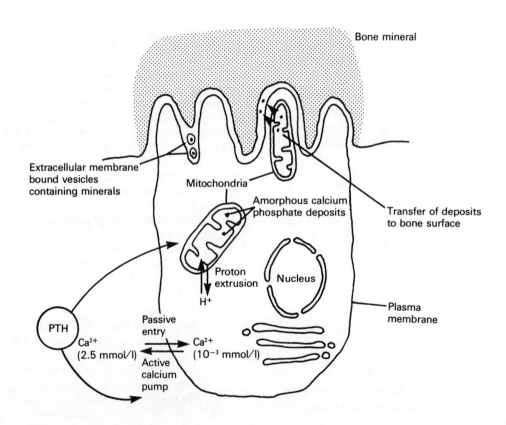

Fig. 22.23. **Possible regulation of permeability of mitochondria and plasma membranes of osteoblasts and osteocytes by parathyroid hormone.**

22.9 Calcitonin

When the plasma concentration of Ca^{2+} rises significantly above normal levels, the secretion of parathyroid hormone virtually ceases and the secretion of another hormone, calcitonin, commences (*Fig.* 22.24). Calcitonin is not formed in the parathyroid gland but in the C cells of the thyroid, and is a polypeptide containing only 32 amino acids (*Fig.* 22.25). The release and synthesis of the hormone is directly controlled by calcium and both respond to elevated concentrations of calcium in the plasma.

      The main effect of calcitonin is to inhibit bone resorption and is, therefore, directly antagonistic to parathyroid hormone. Calcitonin stimulates adenylate cyclase but the mechanism of resorption inhibition is unknown. Calcitonin effects on bone occur independently of parathyroid hormone and vitamin D, are not inhibited by actinomycin, and therefore probably do not require RNA or protein synthesis.

      Calcitonin also has various effects on the kidney tubule causing increased loss of many electrolytes, particularly inorganic phosphates. Consequently the plasma concentration of phosphate, in addition to that of calcium, falls in response to calcitonin.

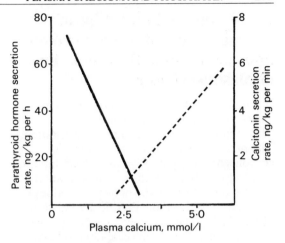

Fig. 22.24. **Relationship between the circulating levels of calcitonin (– – –) and parathyroid hormone (——).**

Calcitonin may possess important regulating effects on mitosis and cell division, since it has been demonstrated that mitotic activity in lymphocytes and RNA synthesis in bone cells, both of which are stimulated by parathyroid hormone, are strongly inhibited by calcitonin. These actions may explain the inhibiting effects of calcitonin on the stimulation of osteoclast activity, but they do not account for the inhibition of calcium release from osteoclasts and osteocytes.

Fig. 22.25. **Structure of calcitonin.**

22.10 Inter-relationships of vitamin D, parathyroid hormone and calcitonin in the regulation of plasma calcium

It is now clear that a complex relationship exists between the different hormones controlling the plasma calcium concentration. The precise mechanism of the control is complex and still partially unresolved, but the relationships between the major regulation organs, the intestine, the bones, the kidneys, the parathyroid and the thyroid are illustrated in *Fig.* 22.26. The sequence of events thought to occur when the plasma calcium level falls is shown and has been summarized below.

a. Release of calcitonin from the thyroid is depressed

b. Release of parathyroid hormone from the parathyroid gland is increased

c. Parathyroid hormone release into the plasma (i) causes a release of calcium and inorganic phosphate from bone, (ii) stimulates the activity of the hydroxylase in the kidney to convert 25-hydroxy-vitamin-D_3 into 1,25-dihydroxy-vitamin-D_3

d. 1,25-Dihydroxy-vitamin-D_3 causes an increase of plasma calcium by three mechanisms:

   i. By stimulating the intestinal uptake of calcium. The synthesis of the calcium-binding protein is also increased.

   ii. By acting in conjunction with parathyroid hormone to release calcium and phosphate from bone

   iii. By increasing calcium and phosphate retention by the kidney.

Plasma phosphate levels are also increased together with calcium:

a. When parathyroid hormone and 1,25-dihydroxy-vitamin-D_3 stimulate release of calcium and phosphate from bone

b. When 1,25-dihydroxy-vitamin-D_3 stimulates intestinal uptake of calcium and phosphate from the intestine

c. When 1,25-dihydroxy-vitamin-D_3 stimulates calcium and phosphate retention by the kidney.

Phosphate can, however, be regulated independently of calcium, based on two mechanisms. Firstly, parathyroid hormone causes an *increase* in excretion of inorganic phosphate and will, in the absence of vitamin D_3, therefore cause a *rise* in plasma calcium, but no change or even a fall in plasma phosphate. Secondly, the 1-hydroxylase is stimulated by a low plasma phosphate concentration independent of parathyroid hormone. The increased activity of this enzyme causes the formation of 1,25-dihydroxy-vitamin-D_3 and, consequently, plasma phosphate concentration increases.

The regulation of plasma calcium is, therefore, not purely dependent on parathyroid hormone or vitamin D_3, but on both and in their interdependence, the parathyroid hormone : vitamin D_3 ratio is thus of major importance. Also recent research has indicated that vitamin D_3 and its metabolite may play an equal or larger part in plasma phosphate regulation, than in plasma calcium regulation.

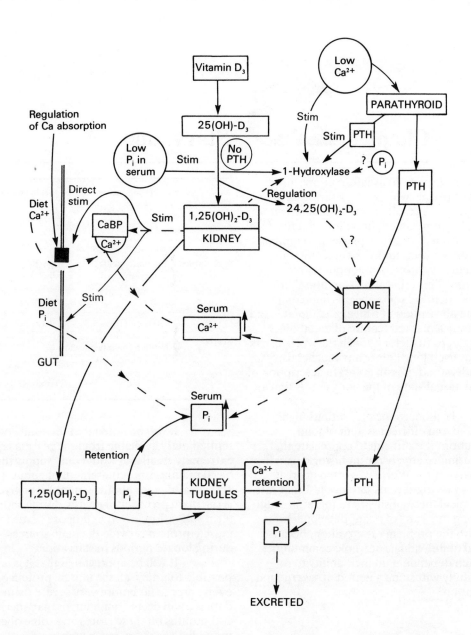

Fig. 22.26. **Calcium homeostasis: response to low serum Ca^{2+}.**
Note: (1) ratio of PTH/vitamin D is vital in regulation of Ca^{2+}; (2) 1,25(OH)$_2$-D$_3$ regulates P$_i$ independently.
Stim—stimulation of enzyme activity, protein synthesis or absorption. CaBP—calcium-binding protein.

Chapter 23 Starvation

The term 'starvation' usually conjures up images of people who are very seriously ill or even dying as a result of serious malnutrition. It must be appreciated, however, that this form of 'starvation' is only one phase of a complex sequence of events and all people normally pass regularly through minor phases of starvation. This is because 'starvation', in a strict biochemical sense, begins immediately after the absorption of a meal is complete, and is certainly a marked feature of metabolism during sleep. In fact, although the word 'breakfast' was used long before any biochemistry was understood, it is an accurate description of the nutritional state of the body at this time of day.

In modern society, with its highly organized and efficient system of food distribution, it is difficult to appreciate that, even during relatively recent history, man in most parts of the World was often in great danger of facing serious starvation. Elaborate biochemical mechanisms were therefore built up during the years of mans' development to deal with the problem of starvation, because survival of individuals or whole communities was often dependent on their ability to successfully withstand a period of severe food deprivation.

23.1 Energy storage

The energy available in the human body may be divided into two categories: that immediately available in the plasma and that stored in the tissues. Typical quantities for a human adult are shown in *Table* 23.1.

282

Table 23.1 **Circulating fuel and stored fuel in adult man**

Fuel		Weight, kg	Energy, kcal
Circulating	Glucose	0·02	80
	Free fatty acids	0·0003	3
	Triacylglycerol	0·003	30
	Total		113
Stored	Glycogen (muscle)	0·15	600
	Glycogen (liver)	0·075	300
	Protein (mainly muscle)	6	24 000
	Fat (adipose tissue)	15	141 000
	Total		165 900

It will be noted that the total energy immediately available from the plasma is extremely small and would only supply the basal metabolism requirement of about 1800 kcal/day for about 80 min. Liver glycogen is the main provider of energy during short-term starvation, whereas adipose tissue fat and muscle protein provide the main sources during longer periods of starvation.

It will be appreciated that it is not possible to utilize all the muscle protein or even adipose fat before very severe tissue damage will occur, but humans may survive 1–3 months on these stores and some obese individuals for very much longer.

23.2 Phases of starvation

It is convenient to divide the periods of starvation into distinct phases:

a. The interprandial phase—the period between meals in a normal day

b. The postabsorptive phase—the overnight fast period of 12 hours; this period may extend to 24 hours

c. The prolonged fast lasting longer than 24 hours and which may be extended into several days or weeks.

It must be appreciated that these are not sharply defined phases. They merge gradually into one another so that, for example, metabolic processes typical of prolonged fast may begin 12–24 hours after taking food.

23.3 Interprandial phase

This, the first phase of starvation, occurs between meals after the absorption of ingested food is complete. This period of absorption normally varies between 2 to 5 hours depending on the quantity and type of food ingested, so that 4–5 hours after a meal the first phase of starvation begins. When the supply of glucose from the intestine ceases, glucose must continue to be supplied from other sources for the energy needs of the brain and other tissues shown in *Fig.* 23.1. The source of blood glucose is liver glycogen which is broken down by the process of glycogenolysis (*Scheme* 23.1). It should be appreciated that not all tissues

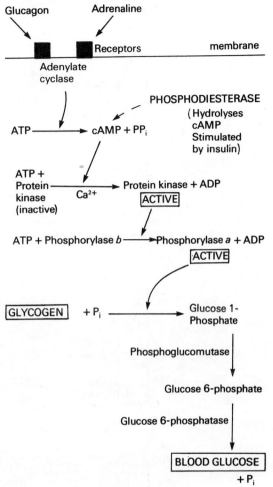

Scheme 23.1. **Glycogenolysis in the liver.**

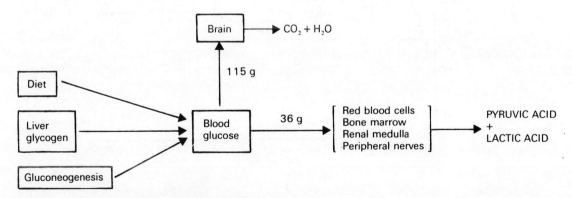

Fig. 23.1. **Tissues for which blood glucose is essential.**

Figures show requirements for 24 h. Note that the brain oxidizes glucose completely to CO_2 and H_2O, whereas the other tissues produce lactic or pyruvic acid as a result of glycolysis.

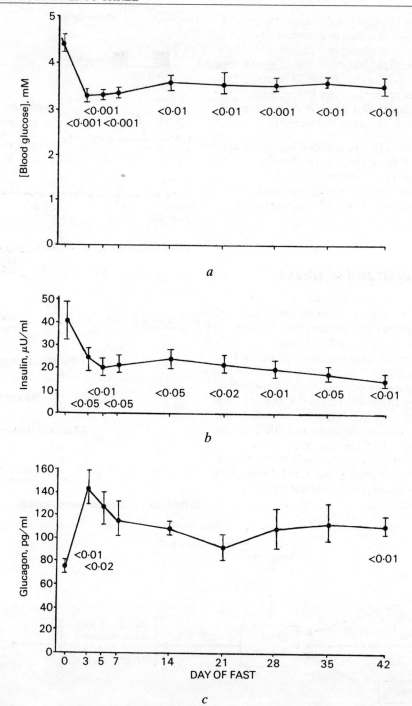

Fig. 23.2. **Changes of the concentrations of (*a*) plasma glucose, (*b*) insulin and (*c*) glucagon during a period of prolonged fasting.**

require glucose as an essential fuel and the muscles of the body can readily use free fatty acids released from adipose tissue. The utilization of fatty acids becomes more important as starvation proceeds, but fatty acids are metabolized by many tissues during the early phase.

Glycogenolysis (cf. Chapter 5) is controlled, mainly during this phase of starvation, by the plasma glucose concentration, by insulin and by glucagon. The system which transports glucose into the liver has a very large capacity, or high V_{max} so that glucose readily crosses the membranes of the liver cell and its concentration in the cell is close to that in the plasma. However, the enzyme mainly responsible for the phosphorylation of glucose in the liver, glucokinase, has a low affinity for glucose with a very high K_m value (10 mM) compared with that for the K_m of hexokinase of 100 μM. This system therefore ensures that, as the concentration of glucose in the portal blood begins to fall, the concentration of glucose 6-phosphate in the liver cell, produced by the action of glucokinase, falls sharply. This in turn helps to stimulate breakdown of glycogen which first forms glucose 1-phosphate and then glucose 6-phosphate (*Scheme* 23.1).

Insulin and glucagon, and particularly the ratio of their concentrations, are also important in the control of glycogenolysis; they probably act by regulating the tissue concentrations of cAMP. Glucagon stimulates cAMP formation which activates the protein kinase essential for phosphorylase activation (*Scheme* 23.1), whereas insulin inhibits cAMP action, probably by increasing the activity of phosphodiesterase which hydrolyses cAMP. The fall of blood glucose concentration and of insulin and the rise of glucagon concentrations during starvation are shown in *Fig.* 23.2.

It should be noted that both these hormones also act on the adipose tissue, the fall of the insulin level and the rise in the glucagon level in the plasma combining to stimulate release of free fatty acids (cf. Chapter 19). The net result of changes in concentrations of these hormones is release of free fatty acids from the adipose tissue and of glucose from glycogen.

As this phase of starvation continues, peripheral muscles and adipose tissue consume progessively less glucose and switch over to fatty acids as their main energy source so that after 8–10 hours more than half the muscles' need for energy is met by free fatty acids.

23.4 Postabsorptive phase (overnight fast)

In normal adult man, the liver glycogen is usually considered to form about 4–5 per cent of the total liver weight, but occasionally it may reach 10 per cent of this weight shortly after a meal. This quantity of glycogen is capable of maintaining the blood glucose concentration at normal values for 12–16 hours. It is interesting to note that this period corresponds closely to the period of overnight fasting.

However, as the glycogen store begins to be depleted, important metabolic changes take place to ensure a continuation of the supply of glucose to the blood which is essential for energy supply for the brain, erythrocytes, bone marrow, renal medulla and peripheral nerves (*Fig.* 23.1). Sources of glucose, other than glycogen, must therefore be used and the body begins to synthesize blood glucose from lactic and pyruvic acids, glycerol released by the degradation of triacylglycerol and from glucogenic amino acids. The process is known as gluconeogenesis (*Scheme* 23.2).

Although the lactic and pyruvic acids are formed from glucose (*Fig.* 23.1), resynthesis into glucose conserves other glucose stores and the energy from glucose synthesis can be derived from the oxidation of fatty acids that are in plentiful supply. About 36 g glucose per day can be synthesized from lactic and pyruvic acids produced by the tissues carrying out active glycolysis and about 16 g from glycerol.

The main source of glucose for gluconeogenesis is, however, amino acids. These arise by proteolytic degradation of tissue proteins, probably as a result of the action of lysosomal cathepsins. Protein degradation can occur in the liver but, as the main store of

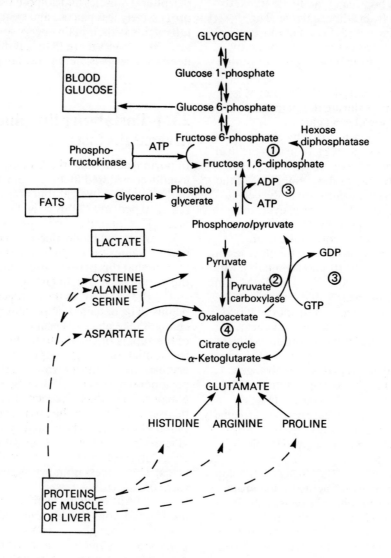

Scheme 23.2. **Gluconeogenesis.**
The following points should be noted: (1) allosteric control of hexose
diphosphatase; (2) allosteric control of pyruvate carboxylase; (3) the need for
energy in the form of ATP and GTP; (4) although it is not shown in the diagram,
pyruvate carboxylase is located in the mitochondria so that pyruvate must cross into
the mitochondria before conversion into oxaloacetate. Oxaloacetate is reduced to
malate that crosses the mitochondrial membrane and is then oxidized back to
oxaloacetate by the cytosolic malic dehydrogenase. Other reactions occur in the
cytosol.

protein is in the muscle, this tissue is the most important source of amino acid supply for gluconeogenesis, particularly if the starvation is at all prolonged.

In the liver, amino acids are deaminated, usually as a result of transamination to form α-ketoacids such as pyruvic acid or one of the intermediates of the citrate cycle. Pyruvate is converted directly to oxaloacetate and α-ketoglutarate which can be formed from several amino acids is also converted to oxalolacetate by the operation of the cycle. This then forms phospho*enol*pyruvate and eventually, by reversal of the glycolytic pathway, blood glucose (*Scheme* 23.2).

The process of gluconeogenesis must be controlled so that it proceeds in the direction of glucose synthesis and not in the direction of glucose degradation. There are two important sites of control: the enzyme converting pyruvate to oxaloacetate, pyruvate carboxylase, is an allosteric enzyme and inactive in the absence of acetyl-SCoA. High concentrations of acetyl-SCoA, produced by the β-oxidation of fatty acids, increase the rate of conversion of pyruvate to oxaloacetate and thus, ultimately, of gluconeogenesis.

A second control point is that involving the inconversion of fructose 6-phosphate and fructose 1,6-diphosphate. Two different enzymes are involved at this stage, phosphofructokinase requiring ATP with the diphosphatase splitting off inorganic phosphate. The diphosphatase is strongly inhibited allosterically by AMP and stimulated by 3-phosphoglycerate and citrate. Precursors of glucose thus increase the activity of the enzyme. Conversely, phosphofructokinase, catalysing the reverse reaction, is inhibited by ATP and citrate but stimulated by AMP. Although some of the liver proteins can be used for gluconeogenesis, muscle protein forms the main mass of protein in the body and is the most important source of supply of amino acids. The muscle, however, cannot synthesize glucose for release into the blood,

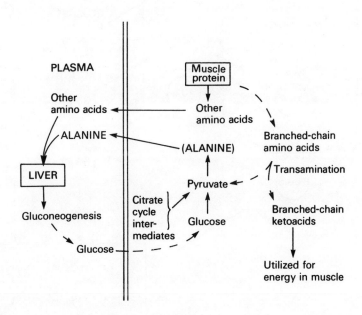

Fig. 23.3. **Alanine formation in muscle.**

primarily because the important enzyme, glucose 6-phosphatase, is absent and the amino acids released from the muscle must be transported to the liver for glucose synthesis to take place.

However, when the composition of amino acids leaving the muscle was carefully measured, it was noted that it did not reflect the amino acid composition of the muscle and that alanine was in much higher concentration in the plasma than in the muscle. Alanine forms about 7–10 per cent of the amino acids of muscle protein, but accounts for 30 per cent or more of the amino acids released from the muscle during starvation. Alanine must therefore be synthesized in the muscle, probably by transamination from pyruvate, and

branched-chain amino acids are believed to be mainly involved in the transaminations. This process forms branched-chain ketoacids that are available for oxidative metabolism, so supplying energy for the muscle (*Fig.* 23.3). Pyruvate can arise from several sources: from muscle glycolysis or from the metabolism of several different amino acids (*Fig.* 23.2). These amino acids are transported to the liver where they are converted into glucose by the process described (*Scheme* 23.2). Alanine, by reason of its ready conversion to pyruvate, is an especially effective substrate for gluconeogenesis in the liver.

The overall process showing approximate quantities of metabolites involved is illustrated in *Fig.* 23.4.

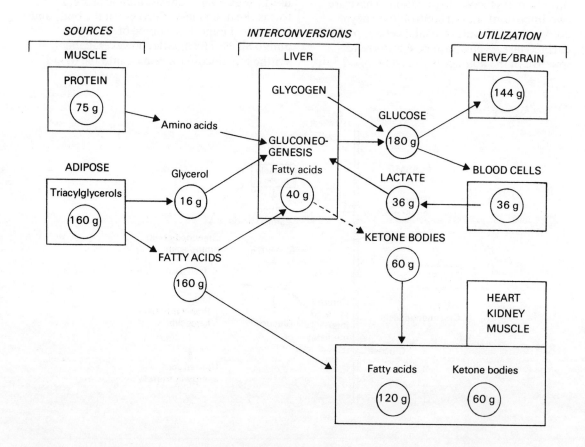

Fig. 23.4. **Early period of starvation in man (24 h–1800 cal utilized).**
Total of sources utilized: 75 g protein; 160 g triacylglycerols.

23.5 Prolonged starvation

Although the processes described for the early stages of starvation efficiently provide for the bodies' energy requirements, it is clear that they cannot be permitted to proceed indefinitely. Extensive depletion of body protein causes serious effects on the tissues and death ensues when 30–50 per cent of the body protein has been lost.

If the starvation is prolonged for more than 24–48 hours, adjustments to metabolism are made to conserve the body protein. This is accomplished by a reduction of glucose production so that the liver glucose output falls from 150–250 g/day after 1–3 days to 40–50 g/day after 4–6 weeks fasting. This drop in glucose production is dependent on the fall in the output of muscle alanine, which may be controlled by the rising concentration of ketones produced as a result of greatly increased fatty acid oxidation.

As the production and utilization of glucose declines, oxidation of fatty acids increases and all tissues capable of oxidizing fatty acids will use these as their main source of energy. The oxidation of fatty acids takes place in the mitochondria and these acids must be transferred across the mitochondrial membrane as carnitine–acyl-SCoA complexes. The concentration of carnitine increases during starvation and carnitine has a stimulatory effect on ketogenesis by increasing the activity of carnitine acyltransferase I, the enzyme which catalyses the synthesis of the carnitine–acyl-SCoA complex. This control may be important *in vivo* although it has not yet been proved under physiological conditions. Malonyl-SCoA may also be an important regulator of transport of fatty acids into the mitochondria. It exerts this control by inhibiting carnitine acyltransferase I. The concentration of malonyl-SCoA in the cells correlates with the rate of lipogenesis and, if

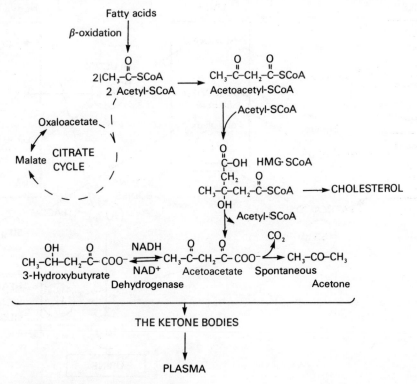

Scheme 23.3. **Formation of 'ketone bodies' in the liver.**
HMG-SCoA = 3-hydroxy-3-methylglutaryl-SCoA.

this is severely depressed, as would occur under starvation conditions, then the inhibitory control of malonyl-SCoA on the transferase is removed and fatty acids readily pass into the mitochondria to undergo oxidation. β-Oxidation of large quantities of fatty acids leads to the production of large quantities of acetyl-SCoA, much greater than can be oxidized in the citrate cycle. A large proportion of the acetyl-SCoA is diverted from complete oxidation in the citrate cycle to produce ketone bodies and this is caused, not only by the high rate of production, but also by a depletion of oxaloacetate which is required for condensation with the acetyl-SCoA. Synthesis of oxaloacetate from pyruvate is greatly reduced on account of conservation of glucose and oxaloacetate tends to be reduced to malate by malic dehydrogenase in the

presence of the large quantities of NADH that are produced to aid glucose synthesis. This process is particularly important in the liver where molecules of acetyl-SCoA are condensed in pairs to form acetoacetyl-SCoA and hence form ketone bodies, i.e. acetoacetic acid, 3-hydroxybutyric acid and acetone are formed (*Scheme* 23.3). The liver is not capable of utilizing the ketoacids and they are liberated in high concentrations into the plasma. As a consequence, the plasma concentration of ketone bodies can increase from 1–2 mmol/l after 3–4 days starvation to 6–10 mmol/l after 14 days without food. The acids produced, acetoacetic and 3-hydroxybutyric, lead to a compensated metabolic acidosis and the serum bicarbonate concentration is reduced. Many tissues, and in particular the heart and skeletal

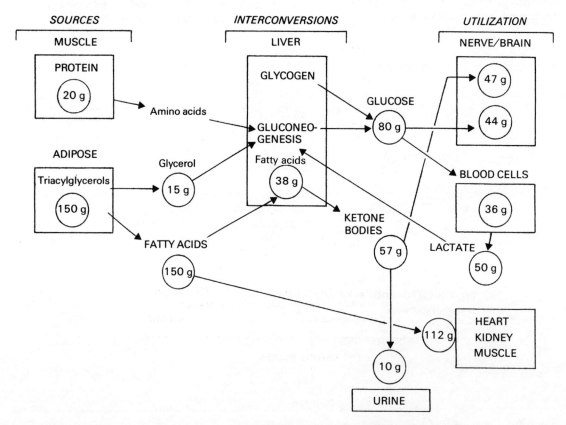

Fig. 23.5. **Late period of starvation in man, i.e. after 5–6 weeks fasting (24 h–1500 cal utilized).**
Total of sources utilized: 20 g protein; 150 g triacylglycerols.

muscle, can oxidize these ketoacids to carbon dioxide and water and thus use them as a source of energy.

During prolonged starvation, however, a remarkable change occurs in the utilization of substrates by the brain. It was noted previously (*Fig.* 23.1) that most of the glucose produced by glycogenolysis was used by the brain but as the supply of glucose diminishes and the supply of ketone bodies increases, the brain reduces its utilization of glucose and increases its oxidation of ketone bodies for its energy supply. As shown in *Fig.* 23.5, under these conditions the brain can consume over 80 per cent of the ketone bodies produced by the liver, although it should be noted that the brain still has a residual glucose requirement, about 40 per cent of that in early stages of starvation. The ketone bodies can, therefore, provide 50–60 per cent of the brain's energy requirements, and the ability of the brain to switch over to ketone bodies as a source of energy must clearly have been of important survival value in serious conditions of food deprivation.

23.6 Clinical aspects of starvation—anorexia nervosa

It is well known that most cases of severe starvation are encountered in countries of the Third World and clinicians working in these countries are constantly surrounded by the problem. It should be realized, however, that many persons who become ill or suffer injury in well-developed Western society may also suffer from mild or even serious forms of starvation.

These will include:

a. Those suffering from infectious illnesses, especially diseases involving the gastrointestinal tract, which leads to a reluctance or inability to eat

b. People, usually elderly, who are unable or unwilling to purchase and prepare adequate meals

c. Those involved in serious accidents which may result in deep coma for several days or weeks; parenteral nutrition is usually essential in these cases

d. People who are reluctant to eat in order to reduce their weight drastically or for psychological reasons. This condition, where psychological problems are involved, is described as anorexia nervosa. It usually affects adolescent females and can cause very serious illness or even death.

Anorexia nervosa is believed to be rooted in certain constitutional and growth factors interacting with psychosocial conflicts of adolescent and postpubertal fatness. It seems to reflect a psychologically necessary stifling of biological maturation through a reversal of the pubertal biological process, back to a child-like biological state. This is generally achieved by the avoidance of intake of dietary carbohydrate and a large proportion of females in the group exhibit this condition to some degree. As an alternative to restriction of dietary carbohydrate intake, some girls will vomit most of their food soon after ingestion.

It is not entirely clear what causes this mental attitude. Over-reaction to mild obesity, fashion, cult of dieting and early menarche have all been suggested as possible causes. It may be significant that extreme dietary restriction can induce amenorrhea.

The hypothalamus, and the extensive subcortical systems, are very important in the control of appetite and, although these systems are likely to play a major role in anorexia nervosa, exactly what this role is remains to be established.

Part 3

Specialized metabolism of tissues

Chapter 24

Blood: erythropoiesis— role of folate and vitamin B_{12}

24.1 Normal erythropoiesis and site of formation of abnormal cells in conditions of folate and vitamin B_{12} deficiencies

In the adult, the active bone marrow forms 3·5–6 per cent of the total body weight, about 1500–3000 g, or approximately the weight of the liver. Only half the marrow is normally functional, but enormous capacity for expansion exists so that, under stress, 6–7 times the normal physiological output of cells can be produced.

A summary of the process of erythropoiesis is shown in *Fig.* 24.1. All cells of the blood originate from primitive stem cells, usually called 'haematocytoblasts'; they are large cells, 18–23 μm in diameter that can differentiate to form polymorphonuclear cells, lymphocytes, monocytes or erythrocytes. The erythrocytes are formed through a sequence of differentiations from proerythroblast and normoblast cells and it is in this sequence that the main abnormalities are seen in folate or vitamin B_{12} (cobalamin) deficiencies. In these conditions, the normoblasts are not formed from the erythroblasts but, instead, distinct large cells described as 'megaloblasts' with diameter of 14–22 μm are produced. These undergo a series of changes in which the cells shrink to 9–18 μm in diameter, then through stages described as 'early megaloblasts', 'intermediate megaloblasts' and 'late megaloblasts'.

These megaloblasts may be released into the blood forming 'megalocytes' or 'macrocytes', instead of the normal erythrocytes. The term 'macrocyte' refers to the large size of these cells, 14–19 μm in diameter, compared to the normal diameter of an erythrocyte, i.e. 7·2 μm. In pernicious anaemia, the disease caused by vitamin B_{12} deficiency, the red cell count in the circulation falls dramatically.

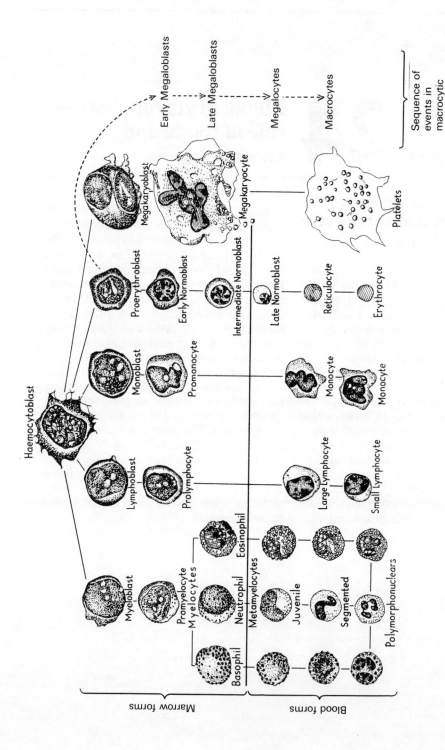

Fig. 24.1. **Normal erythropoiesis and the formation of macrocytes in folate or vitamin B₁₂ deficiencies.**

24.2 Macrocytic anaemias

Pernicious anaemia, although no doubt widespread for many centuries, was first clearly described by Addison in 1855. The disease has a slow insidious onset, usually beginning late in life, after 40 years of age, and being characterized by weakness, loss of weight, vomiting and diarrhoea. Clear nervous symptoms are apparent in most cases with tingling of the extremities and unsteadiness of the gait: the spinal cord and peripheral nerves are clearly involved. In the blood, large numbers of large cells or macrocytes appear and thus the name 'macrocytic anaemia' is given to this type of disease. The macrocytes are formed by the loss of nuclei from the megaloblasts, large immature red blood cells containing nuclei that are discharged from the bone marrow into the blood in this condition. The total blood cell count, however, falls dramatically, often from $4 \cdot 5$–6×10^6 cells/mm^3 to 1–3×10^6 cells/mm^3 and thus a serious condition of anaemia develops. The disease was originally fatal and so the term 'pernicious anaemia' was given to the disease. No treatment was available until 1926, when it was discovered by Minot and Murphy in the USA that feeding patients with raw liver or raw liver extract brought about a considerable improvement in the condition. A dietary factor present in the liver thus appeared to be essential for the prevention of the disease.

A little later, in the 1930s, research on pernicious anaemia was advanced by a remarkable observation: patients in India suffering from tropical sprue, a condition characterized by serious malabsorption from the gastrointestinal tract, were also found to exhibit a form of anaemia very similar to that observed in pernicious anaemia. Studies of the blood cell picture of sprue patients showed that this was virtually identical to that of patients who had contracted pernicious anaemia: the blood contained many macrocytes.

Progress in this field was greatly aided by microbiological investigations. Two strains of bacteria, *Lactobacillus casei* and *Streptococcus lactis*, were found to require liver extracts similar to those used for treatment of pernicious anaemia patients, for adequate growth. It was also observed that these bacteria could utilize, in place of liver extract, extracts of green leaves. This led to intensive investigations of the 'green leaf factor' and, in 1945, it was ultimately purified. It was named 'folic acid', because it was derived from foliage (*Fig.* 24.2). In parenthesis, it should be noted that the *p*-aminobenzoic acid component and the sulphonamide analogues of folic acid were discovered before folic acid itself and this aspect is discussed in Chapter 43.

Very soon after its separation in the pure form, folic acid was injected into patients suffering from tropical sprue and it brought about a complete recovery from the anaemia.

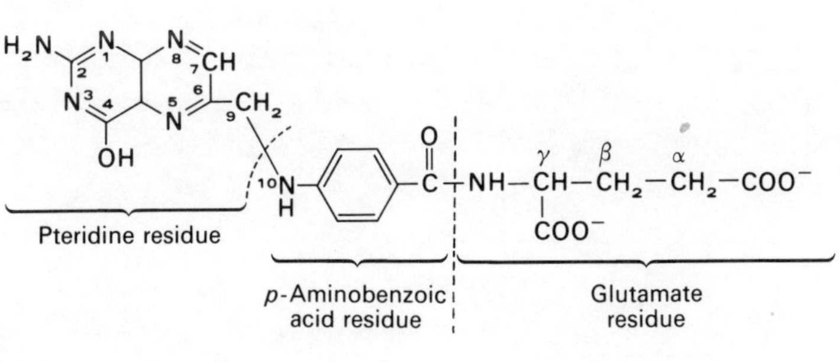

Fig. 24.2. **Folic acid structure.**
Note that this is the monoglutamate form of folic acid. Additional glutamate residues are added by the formation of peptide bonds with the terminal —COO$^-$ group and the amino group of the next glutamate residue. The γ-COOH group takes part in the linkage forming 'γ-glutamyl links'.

Trials on pernicious anaemia patients were, however, less satisfactory: in many patients the anaemia improved, but the neurological symptoms remained. It was clear that folate was not the factor deficient in this disease.

Research once again was concentrated on liver extracts and, in 1956 after 10 years of intensive work, the purification and verification of the structure of a new vitamin, termed 'vitamin B_{12}' or 'cobalamin' was achieved (*Fig.* 24.3). When injected into pernicious anaemia patients it brought about a complete remission of the anaemia and was therefore the factor that was either not absorbed from the diet or completely lacking in it. Vitamin B_{12} is an interesting vitamin in many respects since it was the last vitamin to be discovered, it possesses the most complex structure of all the vitamins, and it is one of the few biological molecules definitely known to require cobalt as part of its structure.

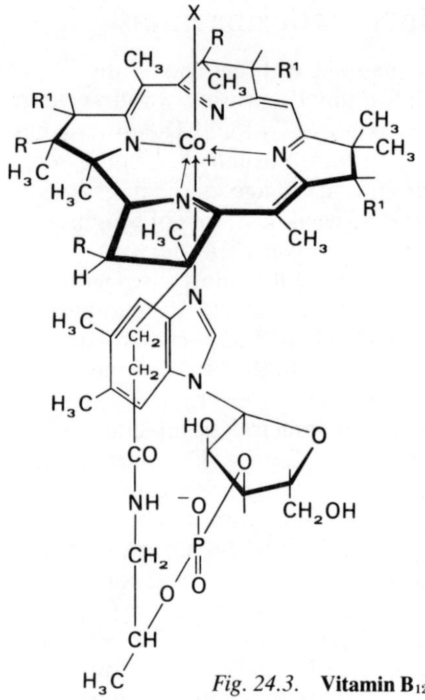

Fig. 24.3.   **Vitamin B_{12}—cobalamin.**

24.3 Causes of folate deficiency

Folate may be deficient in the diet, but this is most likely to occur only in stress conditions, for example in pregnancy, although occasionally the folate in the food may not be readily available. It is, however, more common as in cases of tropical sprue, for the deficiency to be caused by an absorption failure. This can result from the damage of mucosal villi by coeliac disease or tropical sprue, from defective enzymes involved in the transport system or from the effects of certain drugs on this transport. To study absorption failure, it is thus important to consider in detail the transport system.

24.4 Absorption of folate

Folates are present in the food in several forms depending on whether the pteridine ring and p-aminobenzoic acid residue are linked to one, three, five or seven glutamate residues. The glutamate residues are joined through peptide links but, since they are γ-glutamyl links they cannot be attacked by conventional proteolytic enzymes and a special enzyme termed 'folate conjugase' is required (Fig. 24.4), although the name 'conjugase' is somewhat misleading as the main role of the enzyme is to 'deconjugate'. This enzyme is necessary since the monoglutamate form is the only form of folate absorbed into the mucosal cells and thence into the plasma. In humans, the splitting of the glutamate is likely to occur in two stages. A conjugase, found on the brush border of the cell, splits off some of the glutamate residues, whilst a second, found just beneath the villi and possibly in the lysosomes, completes the liberation of folate (monoglutamate). In the mucosal cell, the folate may be converted to methyltetra-hydrofolate and this, together with the monoglutamate form of folate, is transferred to the plasma (Fig. 24.4).

Absorbed folate may be stored in the liver, where it is reconverted to the polyglutamate form, containing either four, five or six residues (Fig. 24.4).

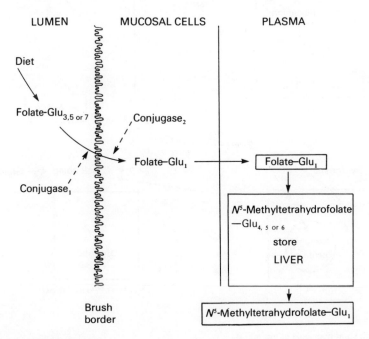

Fig. 24.4. **Absorption of dietary folate.**
Glu = glutamate.

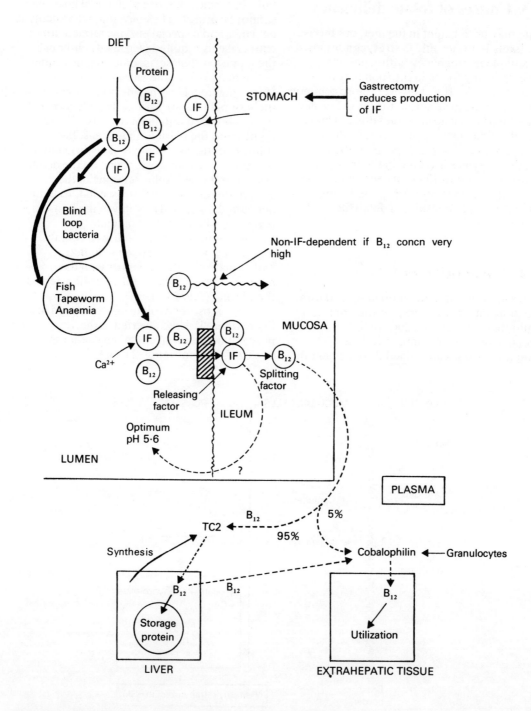

Fig. 24.5. **Absorption and transport of vitamin B$_{12}$.**

IF = intrinsic factor. TC2 = transcobalamin 2.

24.5 Causes of vitamin B$_{12}$ deficiency

Dietary deficiencies of vitamin B$_{12}$ are very rare but have been reported in some parts of the world, such as India or Sri Lanka, where strict vegetarian diets are often eaten. This is because plants do not manufacture or require vitamin B$_{12}$ and animals must rely on micro-organisms to synthesize the vitamin B$_{12}$ they require. Why plants can manage quite well without a vitamin essential for many micro-organisms and animals is not known.

As for folate, absorption failure is the most common cause of vitamin B$_{12}$ deficiency. Coeliac disease or tropical sprue, both of which cause damage to the intestinal villi, will also result in failure to absorb vitamin B$_{12}$ (*Fig.* 24.5), but the most common cause and that which gives rise to pernicious anaemia, is the failure of the stomach to produce sufficient quantities of the 'intrinsic factor', a glycoprotein essential for vitamin B$_{12}$ absorption.

24.6 Absorption and transport of vitamin B$_{12}$

A summary of the processes involved in vitamin B$_{12}$ absorption is shown in *Fig.* 24.5. The vitamin is normally ingested in the food in a form bound to protein, but this is split by the digestive enzymes, initially in the stomach. The stomach secretes a special glycoprotein, or possibly a group of closely related glycoproteins, containing 8–15 per cent carbohydrate with a molecular weight in the region of 50 000–60 000 called the 'intrinsic factor'. This protein, which is resistant to proteolytic digestion, binds very strongly to vitamin B$_{12}$ as it is released by digestion to form a dimer. The binding is relatively specific and small changes in the nucleotide ring of the vitamin drastically reduce the amount bound to the intrinsic factor. One of the roles of the intrinsic factor may be to protect the vitamin against uptake by bacteria, since bacteria and

in particular those in blind or stagnant loops of the intestine, avidly bind vitamin B$_{12}$. Some fish tapeworms that we find in the human gut after the consumption of undercooked fish can also cause anaemia, since these worms utilize vitamin B$_{12}$ leaving little for absorption. The reason for the inadequate production of intrinsic factor in pernicious anaemia, when secretion may fall to only 2 per cent of normal, is not clear. Gastritis is normally associated with this condition, but whether this is a cause or a secondary effect is unknown. The most popular current view is that the failure to produce active intrinsic factor is due to an autoimmune disease, i.e. antibodies are manufactured to one of the bodies' own proteins (cf. Chapter 42).

The vitamin B$_{12}$–intrinsic factor complex is passed to the ileum where it is bound to a receptor site, Ca^{2+} being required for the binding. A releasing factor is necessary for the release of the intrinsic factor from vitamin B$_{12}$ and transfer of the vitamin into the mucosal cell. The released intrinsic factor may be available to facilitate the uptake of additional vitamin B$_{12}$ but it is uncertain whether it can be reused. After passage through the mucosal cell, the majority of the vitamin B$_{12}$ ($\simeq$ 95 per cent) is taken up by a special transport protein, transcobalamin 2 and a small proportion ($\simeq$ 5 per cent) onto another transport protein, cobalophilin. The vitamin B$_{12}$ bound to transcobalamin 2 is transported to the liver where it is stored bound to an intracellular protein. From the liver it can be released into the plasma and here transported on cobalophilin, from which it is taken up by other tissues. The cobalophilin–vitamin B$_{12}$ also acts as a temporary store. As shown in *Fig.* 24.5, transcobalamin 2 (TC 2) is synthesized in the liver, but cobalophilin is synthesized mainly by the granulocytes. In addition to cobalophilin and transcobalamin 2 other transport proteins have also been described: the function of the protein transcobalamin 1 may be to act as an additional vitamin B$_{12}$ transport system to cobalophilin. Cobalophilin is sometimes called 'transcobalamin 3'.

24.7 Inter-relationships of folate and vitamin B_{12}

It was noted previously (*see* Section 24.2) that a relationship appeared to exist between the functions of folate and vitamin B_{12} and before considering their roles it is useful to compare their effects on disease and on metabolism; these effects are summarized in *Table* 24.1 and

Table 24.1 **Relationships of folate and vitamin B_{12} deficiencies**

Symptom	Folate deficiency	Vitamin B_{12} deficiency
Anaemia	← Identical →	
Urine formiminoglutamate	↑	↑
Serum folate	↓	↑
Serum vitamin B_{12}	↑ ↓	↓
Red blood cell folate	↓	↑ ↓
Large dose folate	Responds	May respond
Large dose vitamin B_{12}	May respond	Responds
Neurological damage	——	+ + + +
Methylmalonate excretion	——	+ + + +

explanations presented in the following sections. One of the most significant observations is that serum folate actually rises during vitamin B_{12} deficiency and this has an important bearing on interpretations of the mode of action of the two vitamins.

24.8 Mode of action of folic acid

In consideration of the aspects of the metabolic roles of folate relevant to the development of anaemias, an outline of metabolism is given in *Figs*. 24.6 and 24.7 and details can be found in Appendix 23. Folate is reduced by an enzyme, folate reductase, to dihydrofolate and by a second enzyme, dihydrofolate reductase, to

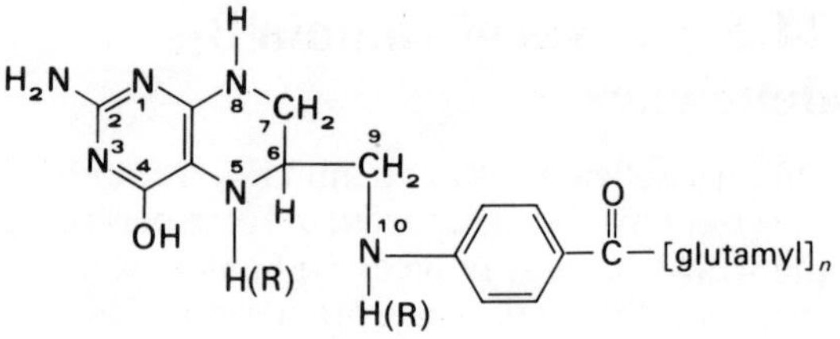

Fig. 24.6. **Tetrahydrofolate and one-carbon adducts.**

One-carbon adducts	R group
N^5-Formyltetrahydrofolate	—CHO
N^{10}-Formyltetrahydrofolate	—CHO
N^5-Formiminotetrahydrofolate	—CH═NH
N^5, N^{10}-Methenyltetrahydrofolate	>CH
N^5, N^{10}-Methylenetetrahydrofolate	>CH$_2$
N^5-Methyltetrahydrofolate	—CH$_3$

tetrahydrofolate which is the active coenzyme form. The tetrahydrofolate coenzyme carries one-carbon units on the N–5 and N–10 nitrogen atoms in the form of formyl (—CHO), methyl (—CH$_3$) or formimino (—CH═NH) groups, or it can also carry a one-carbon unit that bridges the N–5 and N–10 carbon atoms, e.g. methylene (>CH$_2$) and methenyl (≫CH) (*see Fig*. 24.6). The major role of tetrahydrofolate is to receive one-carbon units, for example from the degradation of histidine or in the conversion of serine to glycine.

The one-carbon units can undergo conversion, so for example the formimino group (—CH═NH) accepted from a metabolite of histidine can be converted to the formyl group (—CHO) utilized in purine synthesis (*see Fig*. 24.7). Purine synthesis would then clearly be impaired in a folate deficiency and so would the synthesis of thymidine which is, in turn, required for DNA synthesis. Defective purine synthesis would, naturally, reduce the efficiency of both DNA and RNA synthesis (which are therefore dependent on an adequate supply of tetrahydrofolate).

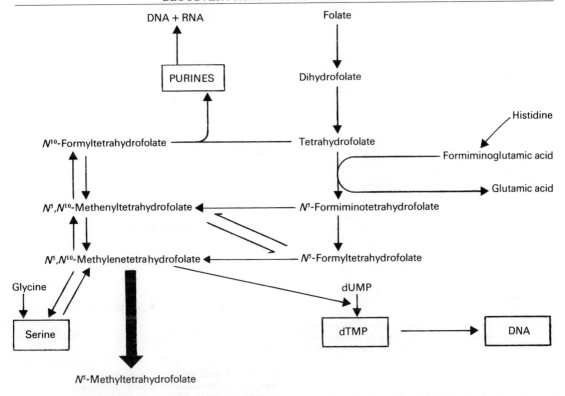

Fig. 24.7. **Tetrahydrofolate as carrier of one-carbon units.**

24.9 Mode of action of vitamin B_{12}

Vitamin B_{12} is known to be involved in many enzyme reactions in various micro-organisms including those of glutamate mutase, ribonucleotide reductase, ethanolamine deaminase. All these reactions either do not occur or do not utilize vitamin B_{12} in mammals. The metabolic roles of vitamin B_{12} in mammals have not been completely elucidated, but it has been established that vitamin B_{12} is involved in two important reactions.

For one of these reactions, the vitamin must first be converted to a coenzyme, a 5′-deoxyadenosyl derivative of vitamin B_{12} (5′-deoxyadenosylcobalamin), the formation of which requires ATP, FAD and NADH (*Fig.* 24.8). This coenzyme (coenzyme B_{12}) is required for the metabolism of methylmalonyl-SCoA; methylmalonyl-SCoA is formed from propionyl-SCoA, in turn produced from the branched-chain amino acid, leucine, or directly from valine (*see Fig.* 24.9). The mutase enzyme that requires vitamin B_{12} coenzyme (5′-deoxyadenosylcobalamin) as a cofactor, converts methylmalonyl-SCoA to succinyl-SCoA which can then undergo normal metabolism to succinate or to porphyrins. If, however, vitamin B_{12} is deficient then further metabolism of methylmalonyl-SCoA is blocked and methylmalonate is excreted, so the formation of large quantities of this metabolite in the urine is characteristic of vitamin B_{12} deficiency and tests for it may be used as an indication of such a deficiency. In practice, however, the diagnosis may be more clearly established: the patient is given a

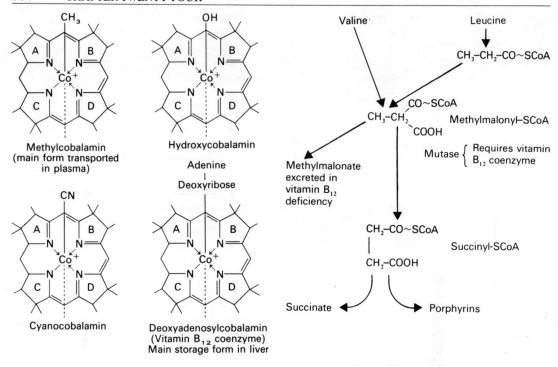

Fig. 24.8. Forms of vitamin B₁₂ (schematic representation).

Fig. 24.9. Role of vitamin B₁₂ coenzyme (deoxyadenosylcobalamin) in the metabolism of methylmalonyl-SCoA.

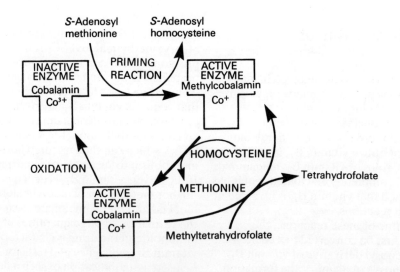

Fig. 24.10. Role of vitamin B₁₂ and folate in methionine synthesis.

Table 24.2 **Methylmalonate excretion in vitamin B$_{12}$ deficiency—loading tests**

	Methylmalonate, mg/day		
	No treatment	+ Valine load	+ Isoleucine load
Normal subjects {	0·2	0·1	0·5
	15·3	15·7	23·1
Vitamin B$_{12}$ deficiency {	0·3	4	2·8
	264	348	730
Folate deficiency {	0·7	0·8	1·0
	6·0	21·6	16·1

loading of valine or leucine and, if deficiency exists, there will be a very large increase in the excretion of methylmalonic acid so providing a clear diagnosis (*see Table* 24.2).

Vitamin B$_{12}$ also acts in methyl group transfer and in this role the vitamin is used without conversion to the coenzyme. It is required in a priming reaction for transfer of a methyl group from tetrahydrofolate to homocysteine forming methionine, the methyl group being transferred to the homocysteine first being taken up by vitamin B$_{12}$ (cobalamin) to form the methyl derivative

(methylcobalamin) and then being attached to the homocysteine forming methionine, and releasing vitamin B$_{12}$ again (*see Fig.* 24.10). This particular reaction is an important linking reaction in tetrahydrofolate and vitamin B$_{12}$ metabolism.

24.10 Metabolic inter-relationships of folate and vitamin B$_{12}$ in the 'methyltetrahydrofolate sink' hypothesis

The fact that transfer of the methyl group from tetrahydrofolate to homocysteine requires vitamin B$_{12}$ has led to the development of a hypothesis that attempts to explain the interdependence of the two vitamins. It was noted in *Table* 24.1 that one of the characteristics of vitamin B$_{12}$ deficiency was the increased concentration of folate in the plasma, present mainly as methyltetra-hydrofolate. It has been proposed that this form of folate is relatively stable, and that it is liable to be released into the plasma

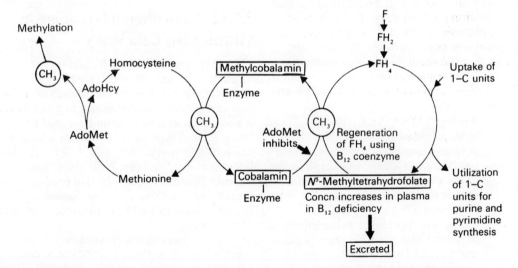

Fig. 24.11. **Inter-relationships of vitamin B$_{12}$ (cobalamin) and folate, dihydrofolate and tetrahydrofolate—the methyltetrahydrofolate sink theory.**

AdoMet = *S*-adenosylmethionine;
AdoHcy = *S*-adenosylhomocysteine.
Note the methyl group release from *S*-adenosylmethionine is used in transmethylations.

and excreted. The folate is, therefore, regarded as 'trapped' or as in a 'sink'. The role of vitamin B_{12} would therefore be to aid in transfer of methyl groups from the methyltetrahydrofolate, so returning the tetrahydrofolate to the metabolic pool from which it takes part in one-carbon metabolism (*Fig*. 24.11). Based on this hypothesis, vitamin B_{12} has the main role of maintaining the concentration of the active tetrahydrofolate coenzyme in the tissues. Most recent research tends to support this concept; it has also been found that *S*-adenosylmethionine may play a role by causing feedback inhibition of tetrahydrofolate formation (*see Fig*. 24.11).

24.11 Possible relationships of folate and vitamin B_{12} deficiencies to anaemias

Although deficiency of folate (*see* Section 19.8) leads to defective purine or pyrimidine synthesis, the anaemic condition that develops cannot be entirely accounted for by this. In this anaemic condition, the red blood cells develop to a certain stage, after which their development is inhibited; it would appear that this condition could be caused by a block in DNA synthesis. Study of the biochemistry of megaloblastic bone marrow cells, however, reveals that the situation is more complex and that many experimental observations do not fit this simplified explanation, some of which are mentioned below.

a.   Although DNA synthesis is impaired in megaloblastic bone marrow, RNA synthesis continues relatively unaffected. It is difficult to explain this effect since folate deficiency should impair production of purines for both DNA and RNA.

b.   There is no imbalance in the selection of deoxyribonucleoside phosphates in the megaloblastic bone marrow. In view of the vital role that tetrahydrofolate plays in dTTP synthesis, it could be expected that the concentration of this pyrimidine nucleotide would be depressed, but

this does not occur. Furthermore, addition of vitamin B_{12} or tetrahydrofolate to a marrow preparation *in vitro* does not cause any increase in the dTTP concentration.

c.   In megaloblastic bone marrow there appears to be an excess of DNA initiation followed by the synthesis of short Okazaki fragments, but gap filling of DNA strands in the synthesis of complete DNA chains is seriously impaired.

From the third observation (*c*), it seems that folate and vitamin B_{12} play a vital role in the DNA elongation process by a mechanism that is as yet not understood. It is conceivable that they are involved in the methylation of other molecules in the nucleus, such as histones, that are known to require methylation during the DNA replication process.

In conclusion, however, it must be stated that the known biochemical functions of folate and vitamin B_{12} cannot so far fully explain the pathological manifestations of the anaemia.

24.12 Neurological damage in vitamin B_{12} deficiency

The symptoms of vitamin B_{12} deficiency are often most clearly demonstrated by neurological manifestations, such as unsteady gait and lack of coordination. For many years, it was not clear how vitamin B_{12} could be involved in this process or whether it was related to anaemia. Recently, however, it has been demonstrated that both folate and vitamin B_{12} could be involved in lipid metabolism and that the neurological symptoms are caused by a disturbance in lipid metabolism.

It has been suggested that if metabolism of methylmalonyl-SCoA were blocked, then either methylmalonyl-SCoA or propionyl-SCoA would be the precursors in fatty acid synthesis. Methylmalonyl-SCoA would give rise to branched-chain fatty acids and propionyl-SCoA to odd-numbered fatty acids with addition of two-carbon units in the usual way. Studies of vitamin B_{12} deficient rats did not demonstrate the existence of branched-chain fatty acids, but small increases in odd-numbered fatty acids were found.

More significant findings have been obtained by analysis of long-chain polyunsaturated fatty acids in the livers and brains of animals deficient in folate or B_{12}. In both these deficiencies, a large fall (approximately 50 per cent) has been demonstrated in the capacity of the animals to elongate and further desaturate fatty acids, such as linoleic acid ($C_{18:2}$) and linolenic acid ($C_{18:3}$). Consequently, the tissue concentrations of fatty acids, such as $C_{20:4}$, $C_{22:4}$, $C_{22:6}$, are much reduced. It is well established that these fatty acids, as components of phospholipids, form vital structures in the brain and nervous tissue and so lack of the ability to synthesize these polyunsaturated fatty acids could be important in the impairment of their function. So far no explanation has been put forward to explain the possible mechanism by which either folate or vitamin B_{12} influence the metabolism of polyunsaturated fatty acids.

Chapter 25 Blood: metabolism in the red blood cell

25.1 Introduction

The erythrocyte or red blood cell is unique in that, unlike all other cells of the body, it is devoid of a nucleus, of DNA, RNA and intracellular organelles, of a cytochrome system and of the capacity for oxidative phosphorylation. The major task of the erythrocytes, the transport of haemoglobin, is performed with neither an expenditure nor a gain of energy.

Energy is required in the red blood cells, however, for maintenance of the correct ion balance, brought about by the pumping out of sodium in exchange for potassium, for protection of haemoglobin against oxidative denaturation, for maintaining the correct conformation of the cell and for protection against the formation of methaemoglobin. A supply of energy is additionally required for the production by the red blood cells of metabolites, such as 2,3-diphosphoglycerate required for haemoglobin function and NADPH which is required for the action of glutathione reductase. Both the energy supply and metabolite production are provided by glycolysis and the pentose-shunt pathway (*see Fig.* 25.1).

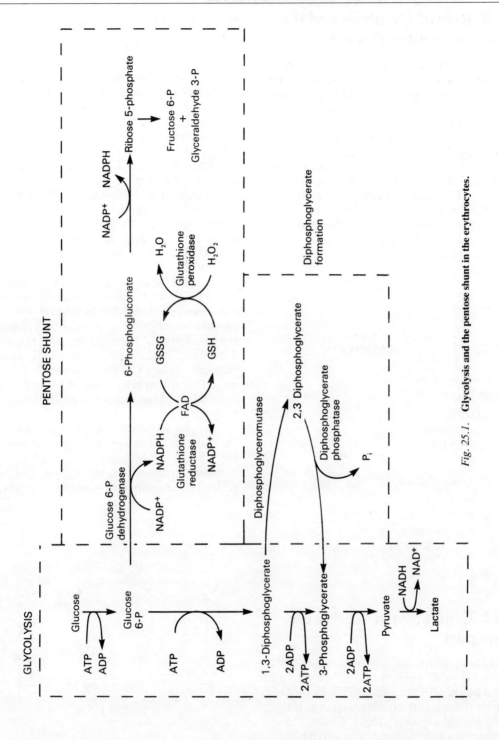

Fig. 25.1. **Glycolysis and the pentose shunt in the erythrocytes.**

25.2 Role of glycolysis and the pentose-shunt pathways

The energy requirements of the cell are met by the conversion of glucose to lactic acid which, as in other tissues, utilizes 2 ATP molecules and produces 4 ATP molecules with a net gain of 2. The production of 2,3-diphospho-glycerate, however, bypasses one of the stages producing ATP so that the net energy produced by the system is zero for each molecule of 2,3-diphosphoglycerate produced.

The main function of the pentose-shunt pathway (cf. Chapter 5) is to produce the NADPH which is essential for the regeneration of reduced glutathione from oxidized glutathione:

i.e.

$$
\begin{array}{c}
\text{Glu—Cys—Gly} \\
| \\
S \\
| \quad\quad\quad + NADPH + H^+ \\
S \\
| \\
\text{Glu—Cys—Gly}
\end{array}
$$

Oxidized glutathione (GSSG)

$\Big\downarrow$ FAD

Glu—Cys—Gly
 |
2   SH $+ NADP^+$

Reduced glutathione (GSH)

25.3 Utilization of ATP in ion transport

In human plasma, the concentration of Na^+ ions is 142 mM, whilst that of K^+ ions is 5 mM; in red blood cells the concentrations are almost exactly reversed, the concentration of Na^+ being 19 mM and that of K^+ 136 mM. The precise reason for the requirement of high concentrations of K^+ in the cell is not clear, but K^+ appears to be essential for the correct configuration and functions of proteins and it cannot be replaced by sodium.

A concentration gradient of both sodium and potassium ions will, therefore, exist across the red blood cell membrane and energy will be required to maintain this gradient; ATP supplies this energy and a membrane ATPase enzyme maintains the differential concentrations. The enzyme has a molecular weight of 270 000 and is a $\alpha_2\beta_2$ tetramer, the larger α unit (mol. wt 95 000) containing an ATP-binding site. The enzyme is located in the membrane so that it is able to bind specifically to Na^+ on the inside of the membrane and to K^+ on the outside. An outline of the mode of action is shown in *Fig*. 25.2.

In the initial stage A, the enzyme E_1, activated by Mg^{2+}, takes up Na^+ and becomes phosphorylated by ATP. In stage B, the enzyme undergoes a conformational change to enzyme form E_2, releasing the Na^+. It is then in the correct conformation to accept K^+ and this happens in stage C with release of P_i. The final stage D involves the release of K^+ and restoration of the original conformation of the enzyme, E_1.

A schematic representation of the sodium–potassium exchange is shown in *Fig*. 25.3: it will be noted that three Na^+ ions are exchanged for two K^+ ions and one action of the pump is to generate an electron

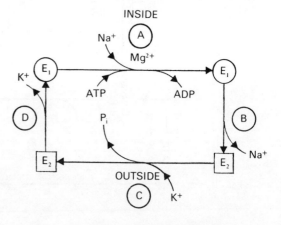

Fig. 25.2. **Utilization of ATP in Na⁺/K⁺ transport.**

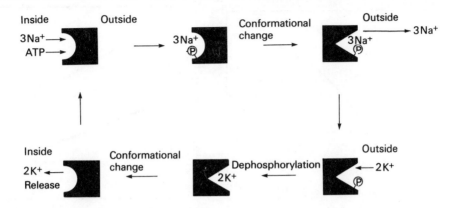

Fig. 25.3.   **Possible mechanism of action of Na$^+$/K$^+$ ATPase by conformational change of the enzyme.**

current across the plasma membrane by disproportionation of the ions. The representation of the action of the enzyme shown in *Fig*. 25.3 assumes that after the Na$^+$ ions have been picked up and phosphorylation of the enzyme has taken place, a conformational conversion of the enzyme occurs transferring the sodium from the inside to the outside of the membrane. The enzyme is then in the correct configuration to accept potassium, and following the dephosphorylation step, the enzyme-bound potassium is transferred to the inside of the cell where K$^+$ is released and the cycle repeated.

If the action of the Na$^+$/K$^+$ membrane ATPase is inhibited the Na$^+$ concentration within the cell will increase due to retention of Na$^+$ within the cells. Plant steroids, such as ouabain or digitoxigenin, are powerful inhibitors of the enzyme and it is interesting that the extract of the foxglove plant, digitalis, that contains digitoxin was one of the earliest drugs known, being described in 1775.

The main physiological effect of these drugs is, however, on the ATPase of the heart muscle cell involved in Ca^{2+} transport, resulting in retention of Ca^{2+} and a consequent increase in the contractibility of the cardiac muscle.

25.4 Metabolic role of 2,3-diphosphoglycerate

2,3-Diphosphoglycerate plays an important role in the binding of oxygen to haemoglobin in the red blood cell. The number of 2,3-diphosphoglycerate molecules in each red blood cell is about 280 millions which is approximately equal to the number of haemoglobin molecules.

Increase in the concentration of 2,3-diphosphoglycerate causes the oxygen equilibrium curve for haemoglobin to shift to the right, meaning that haemoglobin will have a reduced oxygen affinity and so release more oxygen (*Fig*. 25.4). The metabolism of the red blood cell thus switches over to produce more 2,3-diphosphoglycerate in conditions of oxygen deficiency.

The 2,3-diphosphoglycerate molecule forms salt bridges through its negatively charged phosphate groups to the positively charged haemoglobin chains; it is located in a cavity between two β chains and, on oxygenation, the cavity is contracted and the molecule is thus expelled (*Fig*. 25.5). The histidine residues 2 and 143, and the lysine residue 82 of the two β chains are believed to be involved in these salt linkages with the 2,3-diphosphoglycerate molecules.

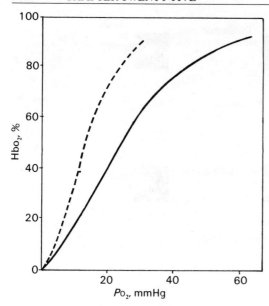

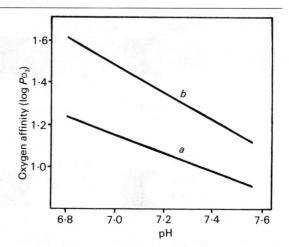

Fig. 25.6. **Effects of pH and 2,3-diphosphoglycerate on the binding of O_2 by haemoglobin.**
(*a*) Without 2,3-diphosphoglycerate; (*b*) with 2,3-diphosphoglycerate.

Fig. 25.4. **The effect of 2,3-diphosphoglycerate on the oxygen equilibrium curve of haemoglobin.**

(– – –) Haemoglobin; (——) haemoglobin + 2,3-diphosphoglycerate.

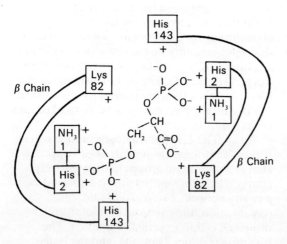

Fig. 25.5. **Binding of 2,3-diphosphoglycerate by human deoxyhaemoglobin β chains.**

It has long been known that the affinity of oxygen for haemoglobin is affected by the pH, affinity decreasing as the pH becomes more acid. This is known as the 'Bohr effect'; it is clearly advantageous, since under conditions of oxygen lack, glycolysis will occur producing lactic acid that in turn lowers the tissue pH and causes release of more oxygen. 2,3-Diphosphoglycerate increases the Bohr effect (*see Fig.* 25.6), since the salt links formed between the molecules increase the p*K* values of the positively charged groups on haemoglobin and correspondingly decrease the p*K* values of the phosphate and carboxyl groups in 2,3-diphosphoglycerate. In fact, the combined effects of 2,3-diphosphoglycerate and increased acidity lead to a marked reduction in oxygen-binding capacity of haemoglobin and, therefore, to oxygen release.

2,3-Diphosphoglycerate plays an important role in blood that has been stored for transfusion. The concentration of 2,3-diphosphoglycerate in the red blood cells stored in a citrate–glucose medium can fall from 4·5 mM to one-tenth of this concentration within a few days. This results in the haemoglobin possessing a very high affinity for oxygen, so that when the blood is transfused it is ineffective in releasing oxygen at the tissues. The problem cannot be solved by incorporating 2,3-diphosphoglycerate into the medium since this polar molecule cannot be transferred across the cell membrane. The most effective method of restoring the 2,3-diphosphoglycerate concentration within the red blood cell is by the addition to the medium of inosine which can be transported into the cell: here the ribose moiety is split off from the purine base and it can then be metabolized via a section of the pentose-shunt pathway and via glycolysis to produce 2,3-diphosphoglycerate.

25.5 The role of glutathione and NADPH

Earlier (*see* Section 25.2 and *Fig.* 25.1), the generation of NADPH and its role in the reduction of oxidized glutathione (GSSG) to reduced glutathione (GSH) was shown.

Haemoglobin in the red blood cell is always at risk since oxygen is an oxidizing agent tending to oxidize vulnerable groups in the protein and the haem group. Oxygen, however, is not a powerful oxidizing agent until it acquires an electron forming the superoxide anion, O_2^-, and this can happen during oxygenation of reduced haemoglobin.

$$Hb + O_2 \rightarrow Methaemoglobin + O_2^-$$

$$(Fe^{2+}) \qquad\qquad (Fe^{3+})$$

About 10^7 superoxide anions are believed to be formed each day in the red cell, and they are extremely toxic to vulnerable groups such as the —SH of proteins and the unsaturated lipids of the cell membranes by initiating lipid peroxidation (*see* Chapter 14). Superoxide must, therefore, be destroyed rapidly and this is accomplished by the enzyme superoxide dismutase, that converts it to hydrogen peroxide:

$$O_2^- + 2H^+ \xrightarrow{\text{Superoxide dismutase}} H_2O_2$$

Hydrogen peroxide is itself a strong and toxic oxidizing agent and must therefore be destroyed rapidly. This is achieved by

means of another enzyme, glutathione peroxidase, that converts reduced glutathione (GSH) to oxidized glutathione (GSSG):

$$H_2O_2 + 2GSH \rightarrow GSSG + 2H_2O$$

The inter-relationships of the metabolic pathways involving glutathione are shown in *Fig.* 25.7.

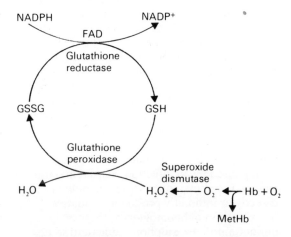

Fig. 25.7. **Metabolism of glutathione and its role in the destruction of O_2^- (the superoxide anion) and H_2O_2 in the red blood cell.**

25.6 Genetic abnormalities: enzyme deficiencies

During the 1939–45 war, many troops were stationed in malaria-infested areas and were treated, regularly, with drugs, such as mepacrine or primaquine, as a prophylactic against malaria. A proportion of those treated suffered from a severe type of haemolytic anaemia which was often fatal and described as 'blackwater fever'. The term described the high concentration of haemoglobin in the urine resulting from haemolysis; the fever was often precipitated by an attack of the malarial parasite.

Not until about 10 years after the war period was the cause discovered. In certain individuals, it was found that the reduced glutathione in their red blood cells was very unstable and this was shown subsequently to be caused by an inherited defect of the first enzyme of the pentose phosphate shunt, glucose 6-phosphate dehydrogenase. It will be clear from *Figs*. 25.1 and 25.7 that a deficiency of this enzyme will lead to a failure of the red blood cell to restore GSSG to GSH, a step essential for the removal of superoxide. Cell damage is likely to result from oxidation of the membranes by the superoxide anion.

Deficiency of glucose 6-phosphate dehydrogenase is now known to be common, more than 100 million people in the world being affected. The incidence appears to be particularly high in black or Caucasian people living around the Mediterranean. Numerous mutant forms of the condition have now been described. In about 2 per cent of affected subjects haemolysis is chronic but, in the remaining 98 per cent, haemolytic episodes are precipitated by an infection, as with malaria referred to above, or by drugs capable of directly or indirectly producing oxidative denaturation of haemoglobin: these include aminoquinolines, sulphonamides and some vitamin K derivatives. In this group normally only the older, more susceptible, cells are destroyed.

Deficiency of glutathione peroxidase can also occur, but is much rarer than that of glucose 6-phosphate dehydrogenase deficiency. The anaemic conditions that result are similar to those observed in glucose 6-phosphate dehydrogenase deficiency. Deficiency of glutathione reductase causes a wide range of haematological and other clinical symptoms. The activity of this enzyme may, however, be impaired because of the lack of availability of the coenzyme, flavin adenine dinucleotide (FAD). This requires riboflavin (vitamin B_2) for its synthesis and nutritional deficiencies of the vitamin are known to cause haemolytic diseases similar to those resulting from the enzyme deficiency.

The most common deficiency of the glycolytic pathway enzymes is that of pyruvate kinase. The anaemia resulting from such a deficiency varies from a very mild form to one necessitating a life-long transfusion. The deficiency is most commonly found in Northern Europeans but is also found widely amongst Negroes, Syrians, Mexicans, Japanese and Italians. The disorder is transmitted as an automosal recessive trait with homozygotes exhibiting a haemolytic syndrome, the heterozygotes being clinically and haematologically normal. In the latter group, enzyme assays on the red blood cells will indicate a partial enzyme deficiency, many mutant forms of the enzyme displaying reduced efficiency. The main metabolic defect, apparent from *Fig*. 25.1, will be the inefficiency of ATP synthesis and an enhanced synthesis of 2,3-diphosphoglycerate resulting from the metabolic blockage and this leads to some of the symptoms described in Section 25.3.

Deficiencies of other enzymes of the glycolytic pathway have also been described but are much less common. These include deficiency of hexokinase, an enzyme that is very active in young reticulocytes but that normally shows decreased activity as the red cells age. In severe enzyme deficiency, cells will, therefore, become old before their normal life span has elapsed.

Glucose phosphatate isomerase, phosphofructokinase, triose phosphate isomerase and lactate dehydrogenase deficiencies (*see* Appendix 15) have also been demonstrated but are relatively rare. It should be noted that in some cases the haemolytic component is relatively severe whereas in others it is mild. Furthermore, because these deficiencies are the result of genetic, chromosomal abnormalities, defective metabolism in other tissues may be of more importance than that in the red blood cell, for example, neurological defects appear early in triose phosphate isomerase deficiency.

These enzyme deficiencies in human red blood cells eloquently support the extensive genetic polymorphism of man. For example, far more than 100 variants of glucose 6-phosphate dehydrogenase have been detected, 2 per cent of this group showing chronic anaemia and 15 mutants have been described.

Polymorphism has also been demonstrated for most other enzymes for which genetic abnormalities have been found and the mutant enzymes are liable to differ from each other in terms of electrophoretic migration, kinetic effects of allosteric modifications, stability to heat and other physicochemical properties. Semirecessively transmitted erythrocyte enzyme abnormalities most often represent double heterozygosity rather than homozygosity.

Chapter **26** Blood: blood clotting

26.1 Introduction

Clotting is the most striking and best known property of blood. Early in history, man must have appreciated the vital and often life-saving necessity of clotting to stem the loss of blood from wounds.

The processes involved in blood clotting have, therefore, been studied for many years, one of the major problems being to attempt to understand how blood remains fluid within the vessels but clots when they are damaged. Although some of the fundamental and correct ideas concerning the mechanisms of the process were first described over a hundred years ago by Schmidt, the details have since been demonstrated to be extremely complex and are only now being unravelled by the most modern research.

During the past few years a strong impetus to a much more intensive study of blood clotting has been provided by the realization that, although the occurrence of blood clotting on the body surface is a vital necessity, if it occurs within the vessels it can cause serious tissue damage and even death. Blood clot formation is extremely serious in the brain, where it may result in a 'stroke', or in the heart where it causes coronary thrombosis.

26.2 Physiological events in blood clotting

a. Role of the platelets

The blood platelets play a vital role in the

clotting of blood. They are small cells, 2–4 μm in diameter and approximately 1 μm thick, and are produced by the explosive disintegration of the very large megakaryocytes that are often up to 40 μm in diameter. The cytoplasm of each megakaryocyte can liberate 2000–3000 platelets. The normal count of platelets in the blood is usually about 250 000 per mm³, although the number can vary greatly even after normal physiological activity such as strenuous exercise.

Despite their small size, platelets have a very complex internal structure containing mitochondria, lysosomes, storage granules and many microtubules and microfilaments (*Fig.* 26.1).

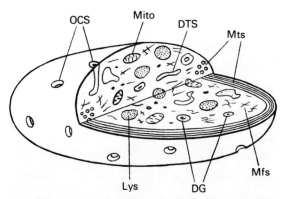

Fig. 26.1. **The structure of the platelet.**
This is a diagrammatic representation of a platelet sectioned through equatorial and transverse planes. Abbreviations: DG dense granules (amine-storage bodies); ÐTS dense tubular system; Lys lysosome-like organelles; Mfs microfilaments; mito mitochondria; Mts microtubules; OCS open canalicular system.

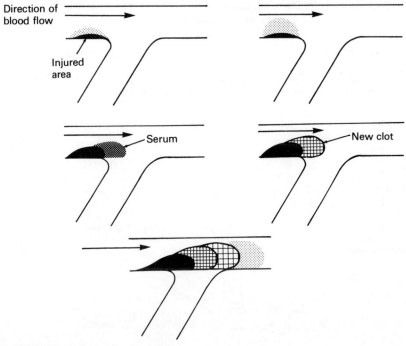

Fig. 26.2. **Diagrammatic representation of the formation of the haemostatic plug from aggregated platelets and fibrin.**

b. Sequence of physiological events

Injury to a blood vessel results in changes to the characteristics of the vessel's surface that cause platelets to activate the coagulation sequences. Contraction of the vessel also helps to reduce blood loss, but the formation of the haemostatic plug is usually vital and is illustrated in *Fig*. 26.2. A more detailed sequence is given below.

i. The platelets adhere to the vessel wall, a process aided by the exposure of the subendothelial structures, such as the basement membrane, collagen and elastin fibres

ii. Activation of the biochemical coagulation processes is initiated

iii. Disruption of the platelets leads to a release of their cytoplasmic constituents, of which ADP and serotonin from the amine-storage granules are believed to be of special importance

iv. Released arachidonic acid from the platelets forms thromboxane (TxA_2) and endoperoxides that accelerate platelet aggregation and thus blood coagulation

v. Fibrin formation provides an additional matrix for more platelet adhesion and thus the blood clot rapidly builds up.

26.3 Platelet adhesion and aggregation

In the previous section the importance of adhesion of platelets to the damaged vessel and aggregation of platelets in the initiation of the blood clot was emphasized and much research has been directed to studying details of the mechanisms involved. Studies *in vitro* have demonstrated that several factors may cause platelet aggregation, including thrombin,

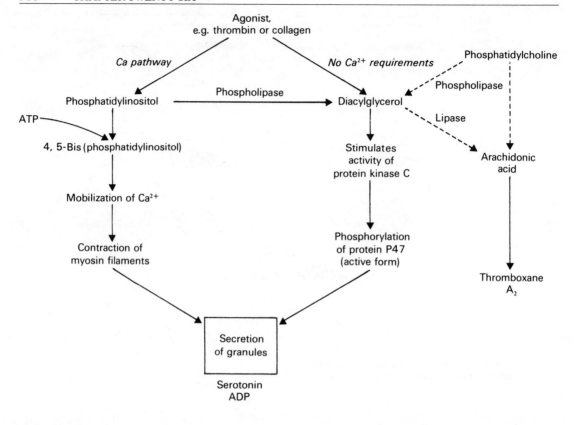

Fig. 26.3   **Initial sequence of events in extrusion of granules from platelets and platelet aggregation.**

collagen and ADP. All of these may also be important *in vivo*, thrombin being produced from plasma components by the action of the cascade mechanism (cf. Section 26.8), collagen from damaged blood vessels and other tissues, and ADP is released from platelets themselves and from erythrocytes. As may be seen in *Fig.* 26.1, platelets contain an elaborate filament system and many granules. Expulsion of these granules containing serotonin and ADP is an important early event in platelet aggregation. The exact sequence of events is not finally established, but current views are illustrated in *Fig.* 26.3.

Granule secretion is caused either by contraction of myosin filaments, a process requiring Ca^{2+}, or by a special protein called 'P47' which requires phosphorylation for activity.

Phosphatidylinositol plays an important role. Exposure of platelets to an aggregating agent such as thrombin initiates the phosphorylation of inositol to form the 4,5-bis(phosphatidylinositol). This mobilizes Ca^{2+} and causes the contraction of myosin which, in turn, causes excretion of the granules. Diacylglycerol, which is formed by the action of a phospholipase C on phosphatidylinositol or phosphatidylcholine, can also play an important role by stimulating the activity of a protein kinase. This, in turn, causes activation of the protein P47 which can also effect the release of secretory granules. Hydrolysis of diacylglycerol by a lipase or of phosphatidylcholine by a phospholipase will also release thromboxane A_2, an important factor in aggregation. Membrane glycoproteins may provide an important site of

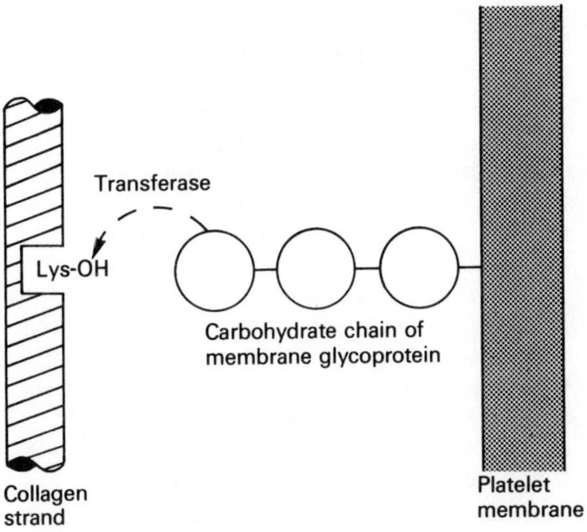
Transferase
Lys-OH
Carbohydrate chain of
membrane glycoprotein
Collagen
strand
Platelet
membrane

Fig. 26.4. **Mechanism of linking of glycoproteins of platelet membrane to collagen strands.**

attachment for the platelets. Three glycoproteins with molecular weights between 90 000 and 150 000 have been isolated from the platelet membranes, and two enzymes, collagen glucosyltransferase and collagen galactosyltransferase, can catalyse the linkage of the glycoprotein carbohydrate chains with the hydroxylysine residues of collagen. This process, illustrated in *Fig.* 26.4 is a mechanism by which platelets could be attached to collagen fibres of the damaged vessels, but some experimental evidence indicates that this process may not be vital for platelet adhesion.

26.4 Roles of prostaglandins, endoperoxides, prostacyclins and thromboxanes in platelet aggregation

During recent years much attention has been devoted to the important roles played by members of the prostaglandin family of compounds in the aggregation of platelets and

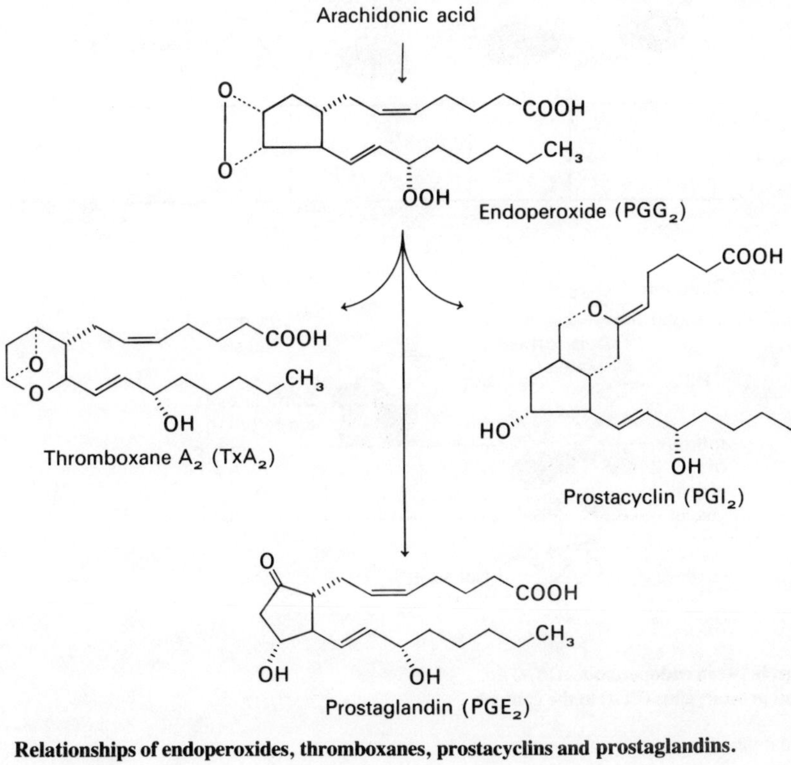
Arachidonic acid
COOH
CH3
OOH
Endoperoxide (PGG2)
COOH
CH3
OH
Thromboxane A2 (TxA2)
COOH
CH3
HO
OH
Prostacyclin (PGI2)
COOH
CH3
OH
OH
Prostaglandin (PGE2)

Fig. 26.5. **Relationships of endoperoxides, thromboxanes, prostacyclins and prostaglandins.**

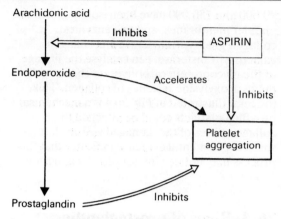

Fig. 26.6. **Effect of aspirin on platelet aggregation and endoperoxide formation.**

blood clotting. The structural relationships between the compounds are illustrated in *Fig.* 26.5 and the biosynthetic pathways are described in Chapter 13.

The importance of this family of compounds in platelet aggregation was demonstrated by the treatment of suspensions of platelets with arachidonic acid which causes rapid and irreversible aggregation. Furthermore, injection of arachidonic acid into animals can cause death by thrombosis. These experiments suggested that arachidonic acid was converted by platelets, and possibly some tissues, into a compound that was a strong activator of platelet aggregation. It was known that arachidonic acid was metabolized to

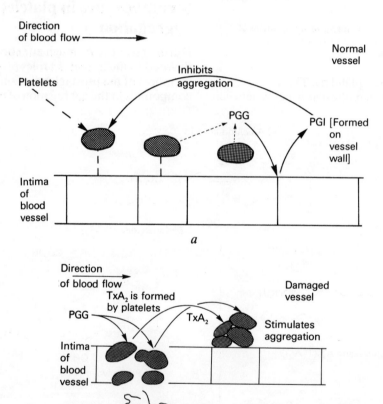

Fig. 26.7. **Relationships between endoperoxides (PGG), thromboxanes (TxA) and prostacyclins (PGI) in the control of platelet aggregation.**
(*a*) Normal vessel; (*b*) damaged vessel.

prostaglandins, but when purified prostaglandins were tested they were found to be strong inhibitors of ADP-induced aggregation. Furthermore, aspirin was shown by Vane to be not only a powerful inhibitor of aggregation, but also of prostaglandin synthesis. It was thus difficult to understand why aspirin should inhibit platelet aggregation when it also inhibited the formation of an aggregation inhibitor, i.e. prostaglandin PGE_1.

The problem ws resolved when it was discovered that aspirin inhibited prostaglandin synthesis because it inhibited the formation of a precursor of the prostaglandins, endoperoxide (*Fig.* 26.6). Endoperoxides are powerful platelet aggregators and thus inhibition of their formation would clearly lead to a decrease in platelet aggregation.

Further research has demonstrated the existence of two other very important groups of compounds that are formed from

arachidonic acid: they are called 'thromboxanes' and 'prostacyclins'. The possible existence of compounds of the thromboxane group was first indicated by the experiments of Vane, who, in 1969, showed in guinea-pigs that a substance was released into the plasma which caused the contraction of rabbit aorta. This compound had a very short half-life of 30 seconds and was ultimately shown to be a member of a new group of compounds, the thromboxanes, that were found to be very powerful platelet aggregators, and at least five times as effective as endoperoxides.

More recently it has been demonstrated that another group of compounds, the prostacyclins, could also be formed from endoperoxides by the vessel walls. These compounds, like the thromboxanes, are very unstable, but, unlike the thromboxanes, they are powerful inhibitors

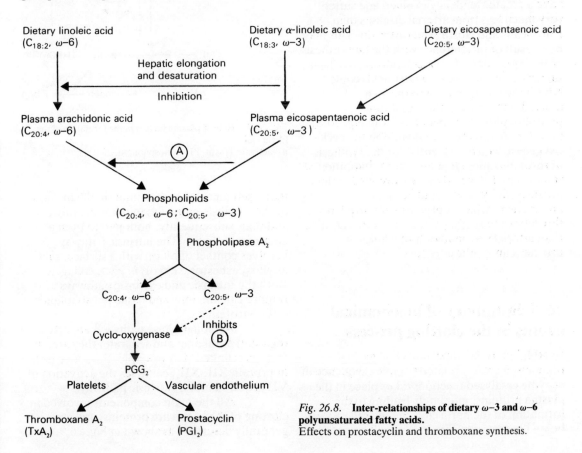

Fig. 26.8. **Inter-relationships of dietary $\omega-3$ and $\omega-6$ polyunsaturated fatty acids.**
Effects on prostacyclin and thromboxane synthesis.

of platelet aggregation and 20–30 times more effective than the prostaglandins.

The endoperoxides, used as substrates by the wall of the arterial vessel, are not formed within the vessel wall but are produced by the platelets. The production of prostacyclins is, therefore, very important in the prevention of clotting within undamaged vessels.

A delicate balance must, therefore, be maintained between the formation of thromboxanes, the platelet aggregators, in the damaged vessels and the formation of platelet aggregation inhibitors that protect the undamaged vessels (*see Fig.* 26.7).

Recently, much interest has been shown in the possible importance of ω–3 polyunsaturated fatty acids in the clotting process. It has been known for some time that Eskimos who consume relatively large amounts of ω–3 polyunsaturated fatty acids have a greater tendency to bleed and suffer very much less from arterial diseases than Europeans and North Americans do. This may be a result of interference with the biosynthesis of thromboxanes. Two mechanisms have been suggested: either the ω–3 fatty acids could inhibit the incorporation of the ω–6 arachidonic acid into the membrane phospholipids of the platelets (*A* in *Fig.* 26.8) or, alternatively, they could inhibit the cyclo-oxygenase which is essential for the synthesis of thromboxanes (*B* in *Fig.* 26.8). Inhibition of stages *A* and *B* would also theoretically inhibit prostacyclin synthesis, but platelet metabolism, which is primarily involved in thromboxane synthesis, may be more sensitive than arterial wall metabolism where prostacyclin synthesis occurs.

26.5 Summary of biochemical events in the clotting process

In addition to the important changes in the platelets, a closely related complex sequence of enzyme catalysed reactions takes place in the plasma proteins, ultimately leading to the formation of the fibrin clot, the overall process being illustrated in *Fig.* 26.9. It will be noted

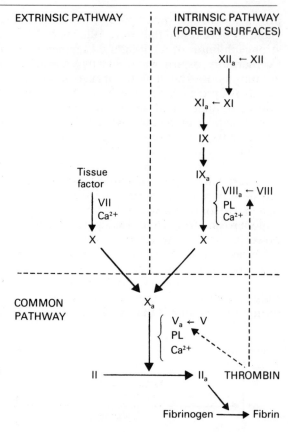

Fig. 26.9.   **Role of plasma factors in the formation of the fibrin clot.**
a – activated form; PL – phospholipid.

that a sequence of events either in the intrinsic or the extrinsic pathways initiates the process and that, subsequently, both join to form a common pathway. The intrinsic pathway involves contact of blood with a surface, such as glass, asbestos, skin or, *in vivo*, collagen. Both the intrinsic and extrinsic pathways require calcium ions and tissue phospholipids or lipoproteins.

Although these pathways are often regarded as distinct and separate, they are, in fact, interlinked. For example, the active form of protein XII, XII_a, catalyses the activation of VII required in the extrinsic pathway.

All the main components involved in clotting of the blood are proteins, and are generally described as shown in *Fig.* 26.9, by

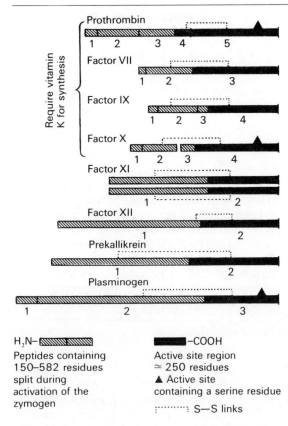

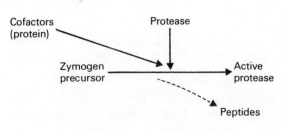

H₂N– ▨▨▨▨

Peptides containing
150–582 residues
split during
activation of the
zymogen

▬▬ –COOH

Active site region
≃ 250 residues
▲ Active site
containing a serine residue

⌐······⌐ S—S links

Fig. 26.10. **Zymogens of the proteinases involved in blood coagulation.**

Cofactors
(protein)

Protease

Zymogen
precursor

Active
protease

Peptides

Fig. 26.11. **Activation of blood clotting factors.**

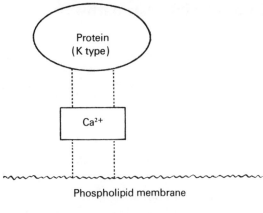

Fig. 26.12. **Roles of calcium.**

the use of Roman numerals, i.e. factors II, XII etc. The process itself is often described as a cascade and involves a series of proteolytic splitting processes of the large protein molecules to produce peptides, each of which converts an inactive zymogen precursor (factors II, XII etc.) into an active enzyme (II_a, XII_a etc.). Structures of the main proteins are shown in *Fig.* 26.10. Splitting of large peptide sequences is usually necessary to release the active protease. At several stages, cofactors such as calcium ions or phospholipids are required. Activation often requires a protein cofactor which is not directly involved in the protease action (*Fig.* 26.11). Calcium is also frequently required and may play several different roles. It stabilizes factor V and fibrinogen, activates factor XIII, but more commonly it is involved in linking proteins to the phospholipid membrane (*Fig.* 26.12). Special binding sites for this purpose are synthesized in the presence of vitamin K (cf. pp. 325–328).

26.6 The intrinsic pathway

a. Activation of factor XII

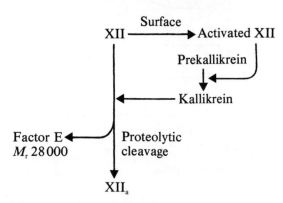

$$\text{XII} \xrightarrow{\text{Surface}} \text{Activated XII}$$

When plasma protein factor XII comes into contact with a surface it becomes activated. Only a conformational change must be involved since the molecular weight (75 000) does not change. The activated form of factor XII converts the protein prekallikrein to kallikrein, which has a molecular weight of 88 000 and is composed of light and heavy chains. This protein has proteolytic enzyme activity and splits off a fragment (E) with a molecular weight of 28 000, so converting XII to active XII_a.

b. Activation of factor XI

$$\text{Factor XI} \xrightarrow[\text{Trypsin}]{\substack{XII_a \\ +\text{ Protein factor (No Ca}^{2+}\text{ required)}}} XI_a$$

Activation of factor XI which is composed of two chains, each of mol. wt $= 80\,000$, depends on a proteolytic action that must involve splitting of internal bonds in the molecule since no change in molecular weight has been detected, and this stays constant at about 160 000. A protein factor in plasma, not yet clearly identified, is required for this process. Factor XI can also be activated by trypsin, a fact which supports the role of a proteolytic split in the molecule.

c. Activation of factor IX

$$\text{IX} \xrightarrow[\text{Ca}^{2+}]{\text{XI}_a} \begin{cases} \text{Light chain, } M_r = 16\,600 \\ \text{Heavy chain, } M_r = 38\,000 \end{cases}$$

$$\xrightarrow{\text{Ca}^{2+}} IX_a$$

$$\text{Peptide, } M_r = 9000$$

Factor IX is a glycoprotein of molecule weight 55 400 composed of a single polypeptide chain and containing 26 per cent carbohydrate. Activation is in two stages: in the first stage, IX is cleaved by XI_a into two proteins, a light chain (mol. wt 16 600) and a heavy chain (mol. wt 38 000). A peptide (mol. wt 9000) is subsequently split from the larger protein molecule to form the active IX_a which is a serine protease. Calcium ions are required for the activation.

d. Activation of factor X

Note that factor X can be activated as a result of the operation of the intrinsic or extrinsic pathways. Although the factors involved in the activation are different in the two pathways, X is converted to X_a by either pathway. In the intrinsic pathway IX_a (a protease), a phospholipid, Ca^{2+} and factor VIII are required.

$$\text{X} \xrightarrow[\substack{\text{Ca}^{2+} \\ \text{VIII}_a}]{\substack{IX_a \\ \text{Phospholipid}}} X_a$$

Factor VIII, which is a large protein (mol. wt $= 0 \cdot 1 - 2 \times 10^6$), requires activation to VIII_a but the process is, as yet, unresolved. The protein has no enzyme function but is essential for the activation of X by factor IX_a. A great deal of attention has been devoted to factor VIII which can vary in concentration in the plasma under different pathological conditions. Increases are observed in liver necrosis, burns, following surgery, in fever and during pregnancy. Furthermore, a genetic deficiency of the factor VIII is well known and is described as 'haemophilia' (*see* Section 26.11).

26.7 The extrinsic pathway

Activation of factor X in the extrinsic pathway is accomplished by a 'tissue factor', factor VII_a and Ca^{2+}.

$$X \xrightarrow[\text{Ca}^{2+}]{\underset{\text{VII}_a}{\text{'Tissue factor'}}} X_a$$

Tissue factor, present in most tissues, has been known for many years to be a lipoprotein complex. It is inactive when dissociated into lipid and protein components, but the lipid component may be replaced by synthetic lipids. Two proteins, of molecular weight 220 000 and 300 000, have been isolated from the complex.

Factor VII is a single chain glycoprotein (mol. wt = 45 000–50 000) and is activated by XII_a or by thrombin (II_a) after its activation. This involves splitting into two chains, α and β, but no change in the overall molecular weight:

$$VII \xrightarrow[]{\overset{XII_a \qquad \text{Thrombin}}{}} VII_a \begin{cases} \alpha VII_a \\ \beta VII_a \end{cases}$$

It is a serine protease with reactive serine residues, but it is unable to bind to its substrate, factor X, unless tissue factor and Ca^{2+} are present. The stages of activation of factor X are identical, whichever pathway is involved in the activation:

$$
\begin{array}{c}
\text{Glycoprotein} \\
M_r = 10\,000 \\
\nearrow \\
X \xrightarrow{} \alpha X_a \xdashrightarrow{Ca^{2+}} \beta X_a \\
M_r = 55\,000 \quad M_r = 45\,000 \quad M_r = 42\,600 \\
\downarrow \\
\text{Peptide} \\
M_r = 2700
\end{array}
$$

Factor X is a glycoprotein, of molecular weight 55 000, composed of a heavy chain (mol. wt 35 000) and a light chain (mol. wt 17 000) that are joined by disulphide bonds. In the first stage of the activation, an Arg–Ile bond is split releasing a glycoprotein (mol. wt 10 000) containing most of the carbohydrate residues. The product, αX_a, is slowly converted, in the presence of Ca^{2+} and phospholipid, into βX_a by the release of a small peptide (mol. wt 2700). The significance of this transformation from the α to the β form is not clear since the anticoagulant activity of the two forms is approximately equal.

26.8 The common pathway of blood clotting

In this part of the clotting process, two well-established reactions occur, the conversion of prothrombin to active thrombin and the formation of the fibrin clot from fibrinogen.

a. Prothrombin and vitamin K

Prothrombin is a single-chain glycoprotein of molecular weight 70 000 and the exact amino acid sequence has been established, the structure being represented in *Fig.* 26.13. A characteristic property of the molecule is the series of calcium-binding sites that are close to the terminal alanine. Progress in the understanding of the structure and action of prothrombin has been closely linked with the study of the role of the fat-soluble vitamin K (phylloquinone).

In 1929, H. Dam, working in Denmark, observed that chicks fed on fat-free diets were, within three to four weeks, severely affected by impaired blood clotting. He further showed that the disease could be cured by the addition of a plant lipid extract to the diet and he correctly surmised that the natural food contained a new vitamin: he termed this 'vitamin K' ('koagulation vitamin').

Chemical analysis of pure vitamins demonstrated that two groups of K vitamins existed, one series being isolated from plants,

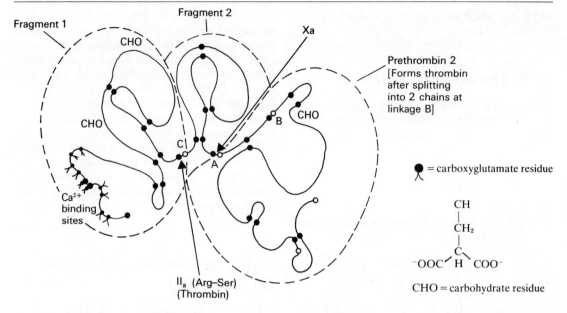

Fig. 26.13. **The structure of prothrombin.**

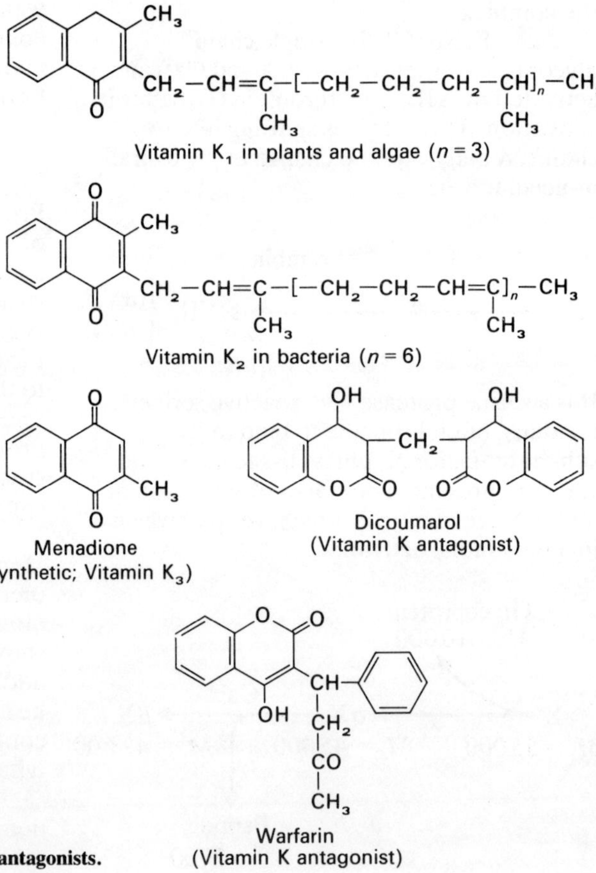

Vitamin K$_1$ in plants and algae ($n = 3$)

Vitamin K$_2$ in bacteria ($n = 6$)

Menadione
(synthetic; Vitamin K$_3$)

Dicoumarol
(Vitamin K antagonist)

Warfarin
(Vitamin K antagonist)

Fig. 26.14. **The structures of the K vitamins and their antagonists.**

or the K_1 series and the other from micro-organisms, the K_2 series. A simple synthetic compound, menadione, was also shown to have vitamin K activity, although it was subsequently demonstrated that this was alkylated *in vivo* before expressing its biological activity (*Fig.* 26.14).

It is noteworthy that there is extensive synthesis of the vitamin by gut micro-organisms and sufficient can be absorbed into the blood to render dietary intake unnecessary for most animals. Adult humans are also believed to be able to synthesize and absorb adequate amounts from their gastrointestinal tract. Very young children, however, are often deficient and supplements are usually given to the mother in later stages of pregnancy or to the infant soon after birth.

The plasma of vitamin-K-deficient animals was found to contain very low concentrations of active prothrombin and some other blood clotting factors, VII, IX and X. It was, therefore, believed for many years that vitamin K was essential for prothrombin synthesis. However, it was discovered recently that the plasma of vitamin-K-deficient animals contained a protein precursor of prothrombin of very similar molecular weight to prothrombin itself. Vitamin K was involved, therefore, not in the actual synthesis of prothrombin, but in some subtle post-translational modification. Careful comparison

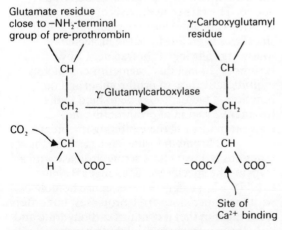

Fig. 26.15. **Formation of γ-carboxyglutamyl residues in prothrombin.**
Note: *see also Fig. 21.9.*

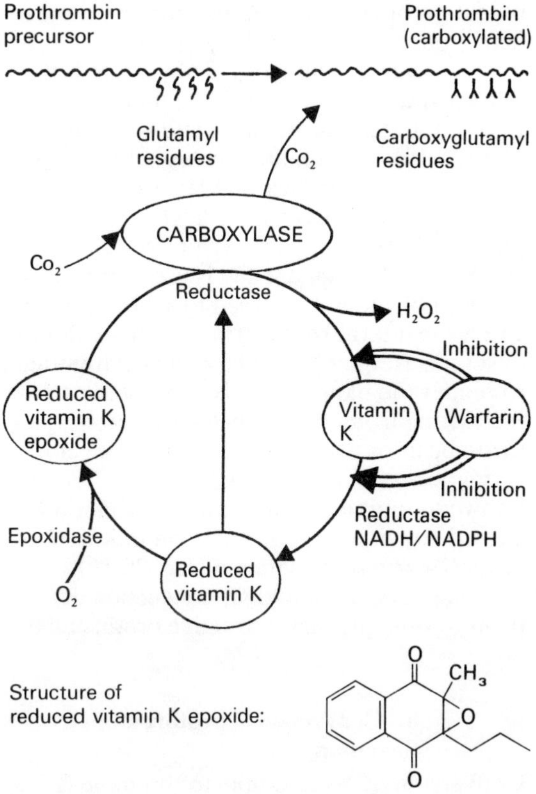

Prothrombin precursor

Prothrombin (carboxylated)

Glutamyl residues

Carboxyglutamyl residues

CO_2

CARBOXYLASE

Reductase

CO_2

H_2O_2

Inhibition

Reduced vitamin K epoxide

Vitamin K

Warfarin

Inhibition

Epoxidase

O_2

Reduced vitamin K

Reductase
NADH/NADPH

Structure of reduced vitamin K epoxide:

Fig. 26.16. **Mode of action of vitamin K in the formation of prothrombin.**

of the structures of the prothrombin precursors found in the plasma of vitamin-K-deficient animals and of normal prothrombin has established that the vital change is the conversion of several glutamyl residues into carboxyglutamyl residues (*Fig.* 26.15). These residues have a powerful affinity for Ca^{2+} and are the site of Ca^{2+} binding in normal prothrombin and, in fact, the lack of activity of prothrombin precursor is entirely due to its inability to bind calcium ions.

Vitamin K is essential for the introduction of carbon dioxide into the glutamyl residues to form the carboxyglutamyl residues, but the exact mechanism by which it functions is not yet established. Several points have, however, been demonstrated: oxygen is also necessary for carboxylation to occur, an epoxide of vitamin K is formed and a reversible oxidation of vitamin K occurs during the cycle. The process is illustrated in *Fig.* 26.16.

Several antagonists of vitamin K are known and two of those most commonly encountered are dicoumarol and warfarin (*Fig.* 26.14). Dicoumarol occurs naturally in sweet clover and has caused serious haemorrhagic disease in cattle grazing on pastures containing such clover. Warfarin is a synthetic anticoagulant and is used extensively as a rat poison; it is believed to act by causing extensive internal haemorrhages. In moderate doses it is frequently used for clinical treatment to render the blood less liable to clot in patients recovering from, or very liable to, coronary thrombosis or other conditions in which clot formation would be very serious. Warfarin is known to inhibit the conversion of vitamin K epoxide to vitamin K and of vitamin K to the reduced form of the vitamin (*Fig.* 26.16) and thus, overall, it inhibits conversion of prothrombin precursor to active prothrombin.

b. Conversion of prothrombin to thrombin

Conversion of prothrombin to thrombin is catalysed by the protease factor X_a in the presence of phospholipid, Ca^{2+} and factor V

$$II \xrightarrow[\text{V, Phospholipid}]{Ca^{2+},\ X_a} II_a$$

Prothrombin   Thrombin

Factor V is a large protein (mol. wt 300 000–350 000) composed of a heavy chain (mol. wt 125 000) and two light chains (mol. wt 73 000). It is activated by thrombin to form three peptides, but it is not clear whether all three peptides or only one or two of them possess activity:

$$V \xrightarrow{\text{Thrombin}} V_a \qquad \text{3 peptides}$$

$$M_r \begin{cases} 70\ 000 \\ 100\ 000 \\ 150\ 000 \end{cases}$$

Factor V_a has no catalytic activity of its own but greatly accelerates the rate of II_a formation in presence of factor X_a.

The activation process of prothrombin has been studied in detail and four portions of the molecule can be distinguished: fragment 1, fragment 2, prethrombin and thrombin (*Fig.* 26.13). Factor X_a hydrolyses prothrombins at an Arg–Thr linkage (*A* on *Fig.* 26.13) releasing prethrombin 2 and fragment 1–2:

$$\text{Prothrombin} \xrightarrow{X_a} \text{Fragment 1–2} + \text{Prethrombin 2}$$

Prethrombin 2 is then split by X_a at an Arg–Ile linkage (*B* on *Fig.* 26.13) to form two-chain thrombin.

The activation of prothrombin is, however, complicated by the fact that it is also hydrolysed by thrombin as the latter is formed. This occurs at an Arg–Ser linkage (*C* on *Fig.* 26.13) liberating fragment 1 and prethrombin 1 which is composed of fragment 2 + prethrombin 2.

$$\text{Prothrombin} \xrightarrow{II_a\ \text{(Thrombin)}} \begin{array}{l} \text{Fragment 1} \\[4pt] + \text{Prethrombin 1} \\[4pt] \begin{bmatrix} \text{Fragment 2} \\ + \\ \text{Prethrombin 2} \end{bmatrix} \end{array}$$

X_a then hydrolyses prethrombin 1 at *A* to form fragment 2 and prethrombin 2 which subsequently forms thrombin as described above. The rate of hydrolysis of prethrombin 1 is, however, slow since the calcium-binding sites have been lost from the molecule following splitting off of fragment 1 by thrombin. In fact this fragment 1 completely inhibits activation of prothrombin and the function of factor V may be to bind to the fragment 2 region of prothrombin and so prevent binding of the inhibiting fragment 1.

Thrombin formed as a result of these reactions is a two-chain serine protease with a trypsin-like specificity. The longer B chain exhibits a large degree of sequence homology with the pancreatic serine proteases, but differs from them in that it contains carbohydrate and only three intrachain disulphide bonds compared with the four usually found in pancreatic enzymes. The short chain is peculiar to thrombin and is likely to be responsible for the specificity of the enzyme.

Plasma thrombin activity is inhibited by heparin with which it forms a 1:1 complex.

c. Conversion of fibrinogen to fibrin

The final stages of the blood coagulation process involve the formation of fibrin monomer from fibrinogen, followed by polymerization and stabilization of the fibrin clot. The essential sequence of reactions is

In the first stage a peptide is split off from each of the A chains by the action of thrombin, whilst in the second stage, which requires Ca^{2+}, the tetramer dissociates into an active dimer of A chains and an inactive dimer of B chains.

Activated factor XIIIa is a transamidase enzyme that catalyses the formation of γ-glutamyl-lysine bonds between pairs of γ and α chains. The reaction between the γ chains is rapid, but it is believed that the α chains link more slowly.

The resultant cross-linked fibrin is

$$\text{Fibrinogen} \xrightarrow{\text{II}_a \text{ (Thrombin)}} \text{Fibrin monomer} + \text{2A peptides} \atop \text{2B peptides}$$

$$\text{Polymerization} \left\{ \begin{array}{l} \text{Fibrin monomer} \rightarrow \text{Fibrin polymer} \\ \text{Fibrin polymer} \xrightarrow{\text{XIII}_a} \text{Cross-linked fibrin} \end{array} \right.$$

Fibrinogen is a rod-shaped protein (mol. wt 340 000) composed of two monomeric units each containing Aα chains (mol. wt 63 000), Bβ chains (mol. wt 56 000) and γ chains (mol. wt 47 000) which are interconnected by disulphide bonds. The two halves of the molecule are joined by disulphide bonds between the N-terminal regions of pairs of α and γ chains (*Fig.* 26.17).

The first stage of the conversion of fibrinogen to fibrin is the hydrolysis of peptide A from the N-terminus of the Aα chains. This is believed to cause a conformational change in the molecule so allowing end-to-end association of the fibrin molecule and it also exposes the Bβ chain to the action of thrombin. Peptide B is then split off by thrombin and end-to-end association of the fibrin molecule occurs (*Fig.* 26.17). Final stabilization of the fibrin clot is achieved by factor XIII, a protein with a molecular weight of 320 000 and composed of two A chains (mol. wt 75 000) and two B chains (mol. wt 88 000). Factor XIII must first be activated to XIII$_a$ in two stages:

$$\text{XIII} \xrightarrow{\text{II}_a} \text{Intermediate} \xrightarrow{Ca^{2+}} \text{XIII}_a$$

$$\downarrow$$

$$\begin{array}{l}\text{Peptide} \\ M_r = 4000 \end{array} \qquad \left[\begin{array}{l} \text{Active dimer (A chains)} \\ + \\ \text{Inactive B chains} \end{array} \right]$$

insoluble and resistant to attack by proteolytic enzymes.

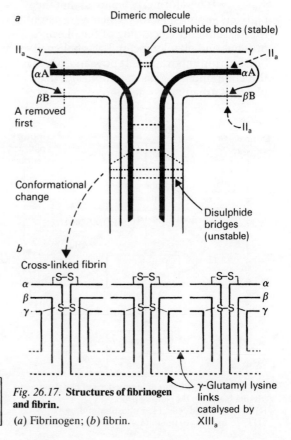

Fig. 26.17. **Structures of fibrinogen and fibrin.**

(*a*) Fibrinogen; (*b*) fibrin.

26.9 Interactions of platelets and plasma factors in the clotting process

Earlier (*see* Section 26.3) the importance of platelet aggregation in the clotting process was described and also the role of the plasma factors in producing the fibrin clot (*see* Sections 26.5–26.8). These processes do not, of course, proceed independently and there are several possible areas of interaction designed to make the process as effective as possible.

Possible sites of interactions are shown in *Fig.* 26.18 and of the greatest importance is, no doubt, the formation of the fibrin clot that provides a very effective network of fibres for enmeshing platelets and enhancing formation of the platelet plug. Thrombin and fibrinogen also play important roles in stimulating the aggregation of the platelets.

It was formerly believed that the platelets supplied factors which were essential for the sequence of clotting of the plasma proteins and platelet factor 3 was believed to be of major importance. It now appears that this component is a phospholipid and that platelets are not the sole source.

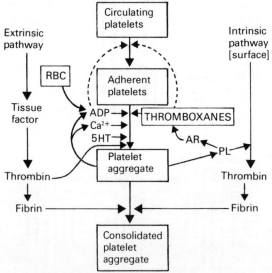

Fig. 26.18. Possible sites of interactions of platelets and plasma factors in the clotting process.
RBC = red blood cells; PL = phospholipid,
5HT = 5-hydroxytryptamine; AR = arachidonic acid.

26.10 Fibrinolysis

Although the fibrin clot is resistant to the action of proteolytic enzymes, it is attacked by a special protease called 'plasmin'. This enzyme does not occur in plasma but its precursor, plasminogen, occurs in the human and most animal plasmas at a total concentration of 200 μg/ml. The molecular weights of the plasminogens vary between 82 000 and 92 000. One group of plasminogens contains glutamate as its terminal amino group whilst the other group contains lysine in this position. Multiple forms of plasminogen occur in each group and can be separated by electrophoresis.

Activation of plasminogen is analogous to the activation of trypsin or thrombin and has been detected in the blood, in many tissues, in saliva, and in the vascular epithelium. Activating enzymes, such as urokinase, have been isolated from urine, and streptokinase has been isolated from several bacteria. The mechanism of activation of plasminogen may vary with the activator, but it has been studied in detail for urokinase:

$$\text{Plasminogen} \xrightarrow{\text{Urokinase}} \sim\!\!\sim\!\!\sim A \text{ (Heavy chain)}$$
$$+$$
$$\sim\!\!\sim B \text{ (Light chain)}$$

The split between Arg–Val residues in plasminogen causes the formation of a heavy A chain and a light B chain, the active site being located in the B chain.

In addition to a large group of plasminogen activators, many inhibitors are known. These fall into two groups: (*a*) those which inhibit plasminogen activation (anti-activators) and (*b*) those which inhibit plasmin (anti-plasmins). Anti-activators are less well defined than anti-plasmins but they occur naturally in the blood and are proteins of molecular weight 75 000–80 000 associated with the α_2 globulin fraction.

Five different plasma protease inhibitors acting as anti-plasmins have been purified. They form complexes with the enzyme proteins which are inactive. Platelets contain an anti-plasmin factor and vitamin E (α-tocopherol) in physiological concentrations also possesses this activity. As a consequence of the presence of this wide spectrum of inhibitors, plasmin is not detectable in the plasma.

Plasmin tested *in vitro* is a relatively non-specific protease and the question arises as to why, *in vivo*, it has a special affinity for the fibrin clot. Although this problem has not been resolved, three possible explanations have been advanced:

a.   There is preferential absorption of plasminogen to the fibrin clot to allow proteolysis to occur subsequently

b.   Plasmin is normally bound to anti-plasmins in plasma but that these dissociate when the complex is bound to the fibrin clot

c.   Activators are selectively bound to the fibrin clot which activates the plasma as it becomes attached.

Hypothesis (*a*) involving plasminogen absorption is the most widely accepted.

The stages of the plasma degradation of fibrin are shown in *Fig.* 26.19. The first stage of proteolysis involves the removal of peptides with a molecular weight of about 40 000 from the carboxyl ends of the Aα chains leaving the residual Aα chain remnants bonded to intact Bβ and γ chains by S–S links. The group of products, of molecular weight 260 000–300 000 is called 'fragment X'.

The next stage of cleavage involves the removal of peptides (mol. wt 6000) from the N-terminal ends of the Bβ chains, followed by the asymmetrical splitting of a large fragment D (mol. wt 94 000) leaving fragment Y. Proteolysis of the Aα, Bβ and γ chains of fragment Y yield an additional fragment D and a smaller fragment E (mol. wt 50 000).

Although this scheme describes the main process of fibrinolysis, it should be noted that there are many details remaining to be resolved. For example, fragment X is not a single well-defined protein but a heterogeneous mixture with a range of molecular weight of 240 000–300 000. Furthermore many smaller peptides, of molecular weights 20 000–30 000 can often be detected during the course of fibrinolysis. Proteolytic digestion of cross-linked fibrin by plasmin also yields a dimer of fragment D and fragment E (*Fig.* 26.19).

Fibrinolysis is an important defence mechanism of the body against thrombosis. Fibrin deposited in thrombi is cross linked and this linking is an important rate-limiting factor in lysis. In view of the chemical importance of formation of thrombi in blood vessels, more research is therefore being devoted to the investigation of the process, for example by measurements of the rate of formation of the fragment D–dimer complex in the plasma by immunological assays.

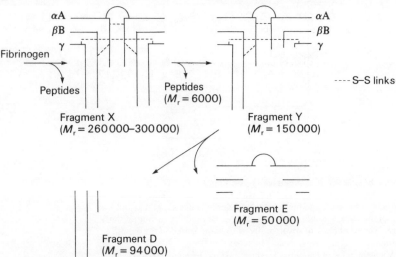

Fig. 26.19. **Degradation of fibrin by plasma.**

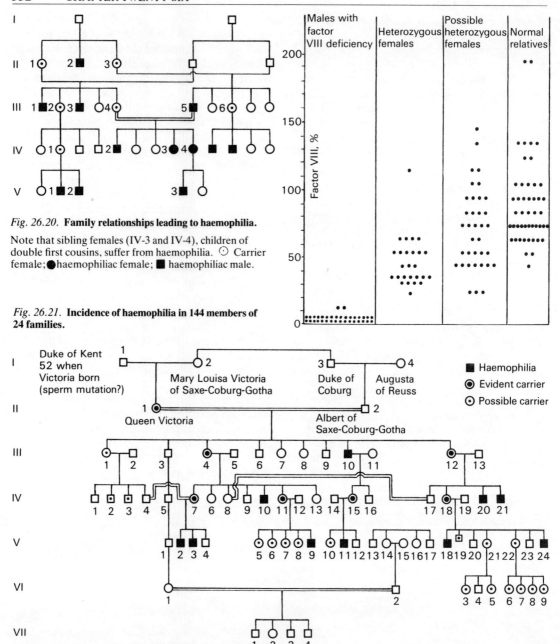

Fig. 26.20. Family relationships leading to haemophilia.

Note that sibling females (IV-3 and IV-4), children of double first cousins, suffer from haemophilia. ⊙ Carrier female; ●haemophiliac female; ■ haemophiliac male.

Fig. 26.21. Incidence of haemophilia in 144 members of 24 families.

Fig. 26.22. Haemophilia in the descendants of Queen Victoria.
Partial pedigree of descendants of Queen Victoria and resultant appearance of hemophilia A in one of her sons and his descendants and, via her daughters and granddaughters, in the royal families of Prussia, Hesse, Battenberg (Mountbatten), Russia and Spain. Note that present royal family of England is free of the gene despite inbreeding.
III–1, Princess Victoria, wife of Emperor Frederick III of Germany (III–2);
III–3, King Edward VII of England;

26.11 Genetic defects and abnormalities of blood clotting

In view of the large numbers of special proteins that must be synthesized for normal blood clotting to occur, it is perhaps, not surprising, that several genetic abnormalities have been observed in which defective synthesis of certain factors occurs.

The best known and well documented of these is deficiency of factor VIII, which gives rise to the classical symptoms of haemophilia: this is a haemorrhagic disorder characterized, in severe cases, by bleeding into the muscles and joints. Factor VIII is also deficient in Von Willebrand's disease, which causes mucous membrane bleeding. Combinations of factor V and factor VIII deficiencies are also known. It is interesting that factor VIII deficiency causes such severe symptoms, since it is not an enzyme but solely a cofactor for the activation of factor X_a (*see* Section 26.6).

Factor VIII deficiency is a classical example of a sex-linked or sex-related inheritance. The deficiency syndrome, haemophilia, is determined by an X-linked recessive gene. This gene is expressed by the male even though he has received it from one parent, his mother, whereas his female sibling must have received the gene from both parents in order to express it, which is clearly much less likely. The situation is further complicated by the poor life expectancy of haemophilians who, until recently, rarely survived childhood and did not reach marriageable age. Relationships which could lead to haemophilia are shown in *Fig.* 26.20.

It should be noted that in families displaying haemophilia, many heterozygous females have reduced concentrations of factor VIII in their plasma, but the concentration is usually sufficient to present clinical symptoms (*Fig.* 26.21).

Haemophilia has attracted a great deal of attention because of its presence in the British Royal family. Queen Victoria was a carrier of haemophilia and serious clinical symptoms were evident in many of her descendants (*Fig.* 26.22).

Fig. 26.22. (contd.)

III–4, Princess Alice, wife of Grand Duke Ludwig IV of Hesse-Darmstadt (III–3);
III–10, Prince Leopold, Duke of Albany, died, age 31, of hemorrhage after a fall;
III–12, Princess Beatrice, wife of Prince Henry Maurice of Battenberg (III–13);
IV–1, Kaiser Wilhelm II of Germany;
IV–2 and IV–3, Prince Sigismund (d. age 2) and Prince Waldemar (d. age 11) of Prussia;
IV–7, Princess Irene of Hesse, wife of Prince Henry of Prussia (IV–4);
IV–8, Princess Victoria of Hesse, wife of Prince Louis Alexander of Battenberg (IV–17), founder of English Mountbatten family;
IV–10, Prince Friedrich of Hesse, died as a child of hemorrhage after a fall;
IV–11, Princess Alix, later Queen Alexandra, wife of Tsar Nicholas II of Russia (IV–12);
IV–15, Princess Alice, wife of Alexander, Prince of Teck (IV–14);
IV–18, Princess Victoria Eugenié of Battenberg, wife of King Alfonso XIII of Spain (IV–19);
IV–20, Prince Leopold of Battenberg, died, age 33, presumably of hemorrhage, after surgery;
IV–21, Prince Maurice of Battenberg, died, age 23, in Battle of Ypres;
V–2, Prince Waldemar of Prussia, lived to be 56 but had no children;
V–3, Prince Henry of Prussia, died age 4;
V–1, King George V of England;
V–9, Tsarevitch Alexis of Russia;
V–11, Rupert, Lord Trematon, died of hemorrhage following auto accident;
V–14, Alice Mountbatten, married to Prince Andrew of Greece (V–15);
V–18, Alfonso Pio, Prince of Asturias, died age 31, of hemorrhage after auto accident;
V–24, Prince Gonzalo of Spain, died age 20, of hemorrhage after auto accident;
VI–1, Queen Elizabeth II of Gt. Britain, married to Prince Philip Mountbatten, Duke of Edinburgh (VI–2), slightly more closely related than third cousins.

Chapter 27 — Blood: catabolism of haemoglobin

27.1 Role of the reticuloendothelial system

The red blood cell has a limited lifespan in all mammals and in man it survives for about 100–120 days, a fact that can be demonstrated by administering ^{15}N-labelled glycine or [^{15}N]aminolaevulinate to an animal. Both of these molecules are precursors utilized in the synthesis of the porphyrin ring of haemoglobin and thus some of the ^{15}N will become incorporated into the porphyrin ring. After administration of the labelled compound, measurements are made of the amount of ^{15}N incorporated into the red cells by removing blood samples at intervals and determining the radioactivity; excreted ^{15}N is also measured by collection of the faeces. In fact, the faeces provide the main pathway for excretion of the bile pigments formed by degradation of haemoglobin (see Section 27.5).

A result of a typical experiment is shown in Fig. 27.1; after a very short delay, ^{15}N becomes rapidly incorporated into the haemoglobin of the red blood cells; for approximately the following 100 days in man the ^{15}N concentration remains nearly constant, since it is retained in the red blood cells as the porphyrin moiety of haemoglobin. After this period, however, the ^{15}N concentration starts to decline as the red blood cells begin to be destroyed. Simultaneously, there is also a sharp rise in the ^{15}N concentration of the faeces, demonstrating that the porphyrin is being extensively excreted in the faeces. It does not, of course, establish the chemical nature of the ^{15}N-labelled compound, but other experiments show that this is mainly in the form of 'stercobilin' (see Section 27.5). This experiment also shows that there is a sharp peak in the increase of ^{15}N excretion in the faeces very soon after administration of the [^{15}N]glycine. The reason for this is not entirely clear, but it is believed to be caused by a rapid destruction of some haemoglobin or immature red blood cells very soon after synthesis or formation.

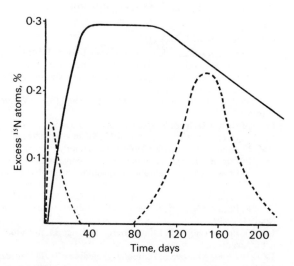

Fig. 27.1. **Incorporation of ^{15}N into haemoglobin and excretion as stercobilin.**
Glycine labelled with ^{15}N was injected into an animal at day zero. Samples of blood removed at regular intervals, the red cells separated and the ^{15}N determined. Samples of faeces were also collected regularly for the measurement of ^{15}N excretion in the form of stercobilin. —— Red-blood-cell haemoglobin; – – – faecal stercobilin.

Site of degradation of red blood cells

Red blood cells are normally described as being degraded in the 'reticuloendothelial system'. This is only a 'system' of the body in a very loose sense and it really describes the activity of special cells in several tissues, mainly macrophages with a high lysosome content, that all carry out similar functions. The reticuloendothelial system, besides degrading red blood cells, is also involved in the immune response, processing antigens (cf. Chapter 42) and in the body's defence mechanism, removing dead bacteria. Degradation of the red blood cells occurs mainly in the spleen, bone marrow, liver and lymph glands. Complete degradation of the haem of haemoglobin to bile pigments occurs in some tissues and always in the liver. A proportion of the haemoglobin liberated from red cells in other tissues can be transported to the liver bound to special plasma proteins or 'haptoglobins' that exist for this purpose. Complete degradation of the haemoglobin then occurs in the liver.

Although the tissues described above perform the main bulk of red blood cell catabolism, haemoglobin degradation is, in fact, possible in nearly all tissues. The pigmentation of bruises is primarily caused by conversion of haemoglobin to bile pigments.

Recognition of red cells suitable for degradation

Cells that are degraded by the reticuloendothelial system fall into one of two categories:

i. Senile cells
ii. Young cells that are structurally or functionally abnormal.

Clearly, old cells must possess some abnormality not possessed by younger normal cells that enables them to be recognized by the reticuloendothelial system. Many factors or components have been considered including:

a. Changes in membrane structure or rigidity
b. Loss of the activity of certain enzymes
c. Changes in haemoglobin conformation
d. Formation of abnormal metabolic intermediates
e. Changes in electrolyte concentrations.

Despite much research, the exact nature of the triggering process of red blood cell degradation has not yet been elucidated, but it appears likely that membrane flexibility must play a vital role, especially when it must be remembered that the red cells, 8 μm in diameter, must often pass through vessels only 3 μm in diameter. Membranes of red cells tend

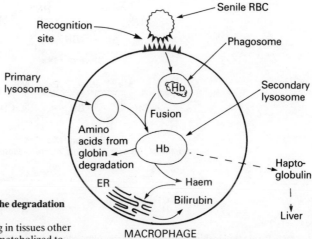

Fig. 27.2. **Probable sequence of events in the degradation of senile red blood cells by macrophages.**
Note that if degradation of cells is occurring in tissues other than the liver, haemoglobin may be either metabolized to amino acids and bilirubin *in situ* or transferred to the liver by means of the haptoglobulins.

to become more rigid as the cells age. Although this may be caused by changes in membrane structure, other factors, such as the cellular ATP concentration, may be very important because it has been demonstrated that there is a great increase in membrane rigidity when the ATP concentration falls. A very complex sequence of events may occur. Stagnation of older red cells will lead to an increase in deoxygenated haemoglobin which binds strongly to ATP, thus reducing the cellular ATP·concentration. When an aged red cell is recognized by a macrophage it is rapidly engulfed by phagocytosis to form a 'phagosome', and this in turn fuses with a primary lysosome, to form a secondary lysosome (cf. Chapter 3). Complete

degradation of the cellular proteins, including the globin of haemoglobin, to amino acids by lysosomal cathepsins then occurs (cf. Chapter 3). The haem is released from the tertiary lysosome subsequently formed and converted to bile pigments on the endoplasmic reticulum of the cell (*Fig. 27.2*).

27.2 Mechanisms of bile pigment formation

The existence of 'bile pigments', substances that give an intense yellow-green colour to bile, is a very old biochemical discovery and was made by Virchow in 1847. The bile pigments are formed from the haem group of

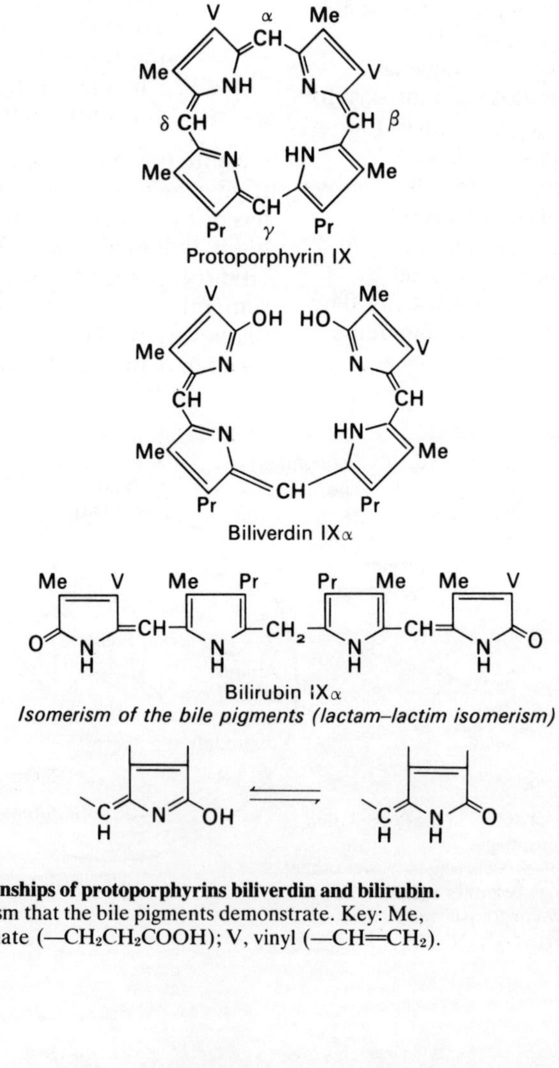

Protoporphyrin IX

Biliverdin IXα

Bilirubin IXα

Isomerism of the bile pigments (lactam–lactim isomerism)

Fig. 27.3. **Structural relationships of protoporphyrins biliverdin and bilirubin.** Note the tautomeric isomerism that the bile pigments demonstrate. Key: Me, methyl (—CH$_3$); Pr, propionate (—CH$_2$CH$_2$COOH); V, vinyl (—CH=CH$_2$).

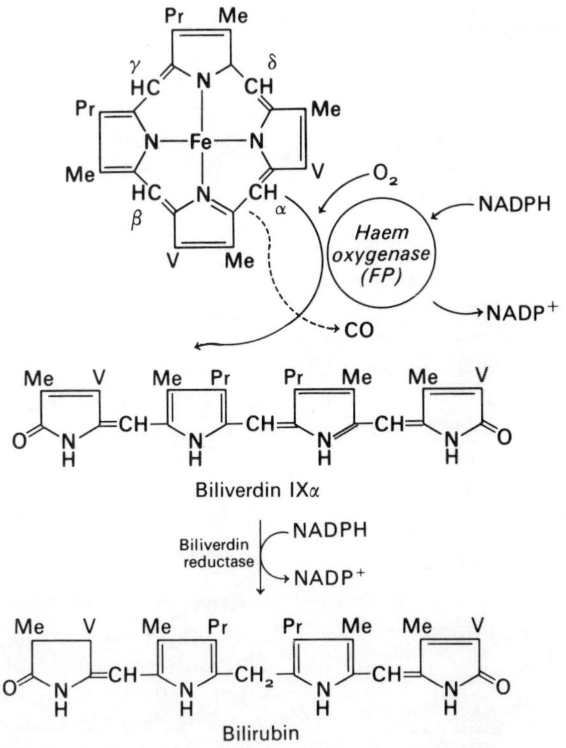

Fig. 27.4. **Mechanism of bilirubin formation from haem on the endoplasmic reticulum.**

haemoglobin by an opening of the protoporphyrin ring and removal of the iron.

The primary pigment formed is 'biliverdin' which is converted to 'bilirubin' by reduction; the structural relationships are shown in *Fig.* 27.3. Note that the bile pigments are described as 'IXα': the figure IX refers to the species of porphyrin that occurs in haemoglobin and the 'α' to the fact that the split of the porphyrin ring occurs at the α carbon.

The bile pigments are extremely lipid soluble and consequently (cf. Chapter 2) will readily pass through membranes. They are toxic substances with serious adverse effects on the brain and nervous system and, although the precise mechanism by which the bile pigments damage the brain is not yet established, it has been demonstrated that they cause uncoupling of oxidative phosphorylation in brain mitochondria, reduce respiratory control, and allow ingress of water and ions. Very young babies often suffer severely from the toxicity of bile pigments, a condition known as 'kernicterus' or 'jaundice of the newborn'.

The bile pigments are formed by oxidative metabolism of haem on the endoplasmic reticulum of the cell. The α carbon is oxidized to carbon monoxide by 'haem oxygenase' and this oxidation is NADPH dependent (*Fig.* 27.4). The fate of the carbon monoxide formed is uncertain but the majority is believed to be trapped by haemoglobin and eventually exhaled. Oxidation of the porphyrin ring of the haem releases iron which is transferred from the cells to the iron transport protein transferrin and made available for resynthesis into new haemoglobin (*see* Chapter 28).

Oxidation of the porphyrin to biliverdin is followed by reduction to bilirubin; the enzyme involved is biliverdin reductase and NADPH supplies the reducing equivalents (*Fig.* 27.4).

27.3 Bilirubin transport

If the red cell degradation occurs in the liver, further metabolism of bilirubin occurs *in situ* but bilirubin formed in other tissues is transported to the liver, bound to serum albumin; normally one albumin molecule out of sixty will be carrying a bilirubin molecule. Bilirubin is removed by the liver by a relatively slow process and of the 1200 ml blood flowing through the human liver each minute, only 44 ml is cleared of bilirubin every minute. Thus the half-life of a bilirubin molecule in the circulation is 17·7 min.

The liver traps bilirubin by means of a specific binding protein called 'ligandin' which is a basic protein composed of two subunits and with a total molecular weight of 44 000 (*Fig.* 27.5).

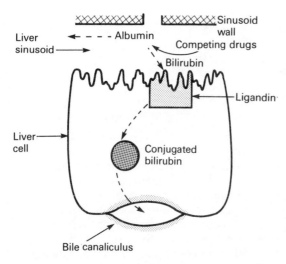

Fig. 27.5. **Mode of action of ligandin which traps plasma bilirubin and facilitates its transfer to liver cells.**

Conjugated bilirubin subsequently formed is excreted by the bile duct.

Many molecules including fatty acids, thyroxine and drugs, such as acetyl salicylate (aspirin) or sulphonamides, compete with bilirubin for binding sites on albumin. These molecules thus strongly inhibit uptake of bilirubin by the liver and, indirectly, enhance bilirubin uptake by other tissues, particularly by the nervous system with the attendant serious consequences described above. The situation is particularly acute in the newborn.

27.4 Conjugation of bilirubin: role of the liver

Bilirubin formed in the liver or transported to the liver from other tissues is conjugated with glucuronic acid to form a glucuronide. The enzyme, described as a glucuronyl transferase, is situated on membranes of the liver endoplasmic reticulum and utilizes UDP-glucuronic acid as cosubstrate. A mono- or diglucuronide can be formed (*Fig.* 27.6) and the process is identical to that used for conjugation of drugs (cf. Chapter 36). Although conjugation with glucuronic acid is the most important pathway, conjugation with xylosides and other hexuronic acids is also possible.

The conjugation of bilirubin renders the bilirubin molecule much more water soluble and, in this form, it is excreted via the bile duct together with the remainder of the bile constituents into the intestine. Conjugation makes the bilirubin molecule much more polar than bilirubin itself so that it does not cross any membrane readily and is much less toxic than free bilirubin. If any conjugated bilirubin were reabsorbed during passage through the intestine, it would thus be much less damaging to other tissues, including the nervous system, than bilirubin.

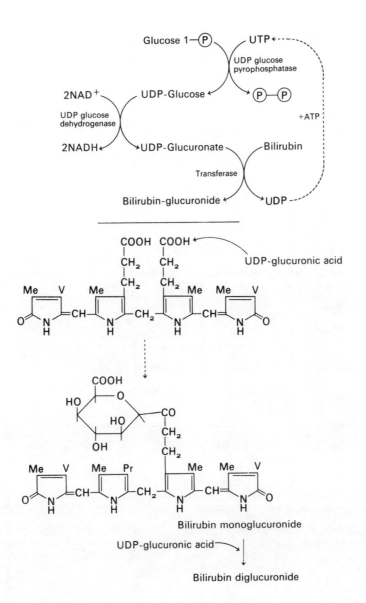

Fig. 27.6. **Conjugation of bilirubin in the liver.**

A monoglucuronide is formed first and this may be converted subsequently to a
diglucuronide.

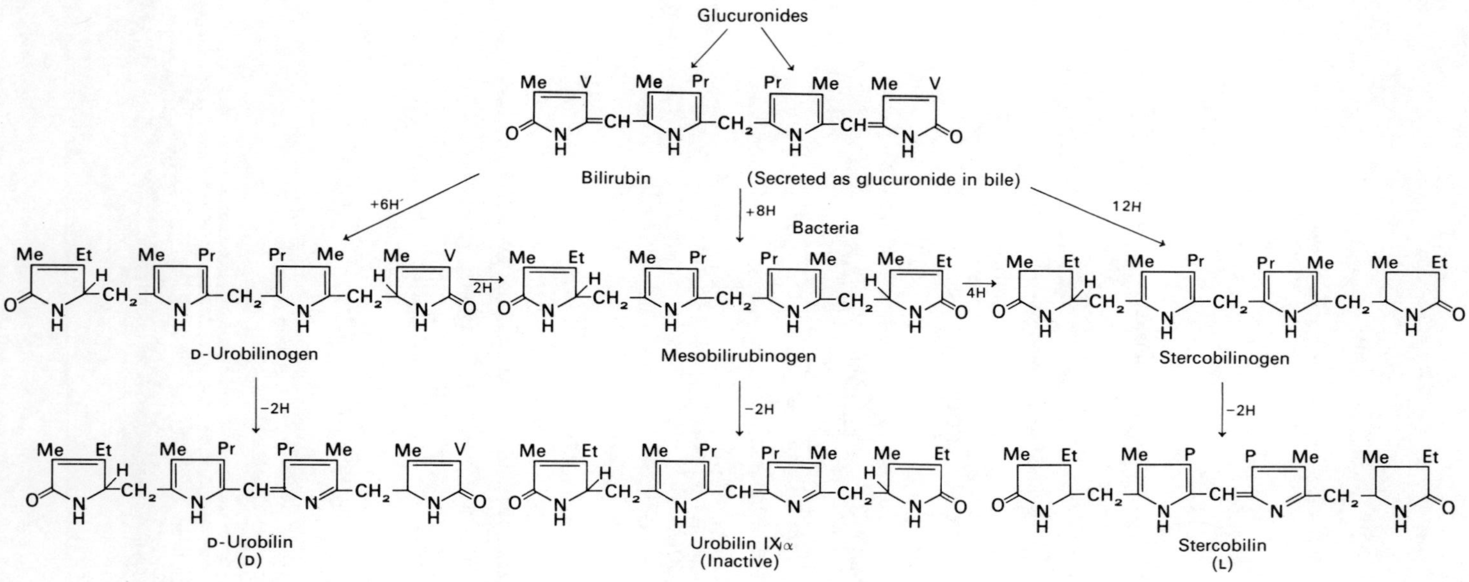

Fig. 27.7. **Bacterial degradation of the bile pigments in the intestine.**
Me, methyl; Et, ethyl; Pr, propionate; V, vinyl.

27.5 Excretion of bile pigments and bacterial metabolism

During their passage through the small and the large intestine, the bile pigments are attacked by bacteria that can reduce or oxidize the carbon atoms of the molecule. The result of the series of complex metabolic changes is that the bilirubin molecule excreted is substantially changed from that passed into the intestine. The main excretory product is stercobilin, but doubtless complex mixtures of this and several other pigments occur in the faeces (*Fig.* 27.7).

A proportion of the pigments is reabsorbed after bacterial degradation from the large intestine and excreted as urobilin or urobilinogen into the urine, and it is this which gives urine its characteristic colour.

27.6 Development of conjugating enzymes

In all mammals studied, including humans, the activity of the glucuronyl transferase enzyme on the endoplasmic reticulum is very low at birth but increases steadily after birth (*Fig.* 27.8). The activity of the enzyme producing UDP-glucuronic acid, UDP glucose dehydrogenase, may also be weak in very young animals and the quantity of UDP-glucuronic acid can increase over 30 times from fetal to adult life.

The reason for the low activity of glucuronyl transferase is not certain because many other enzymes of the endoplasmic reticulum are fully active at birth. It appears that initiation of synthesis of the enzyme may require some specific stimulus to the nuclear DNA genome involved in the coding or to the ribosomes synthesizing the enzyme. The lack of the transferase enzyme makes the newborn infant especially vulnerable to bilirubin, since most of the bilirubin formed from haemoglobin catabolism is not conjugated and therefore readily passes into the brain and nervous tissue. Severe and permanent damage may result, particularly in premature infants.

Attempts have been made to alleviate this condition by treatment of the mother in the last few days of pregnancy, or of the young child, with drugs such as phenobarbitone. Phenobarbitone is a powerful inducer of drug metabolizing enzymes, particularly those involved in oxidative drug metabolism (cf. Chapter 36). During the induction there is extensive proliferation of the membranes of the endoplasmic reticulum and, during this proliferation, synthesis of other enzymes

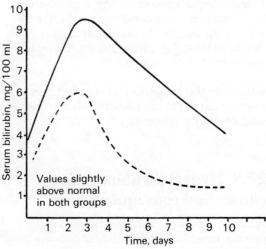

Fig. 27.9. **Effect of phenobarbitone injection on serum bilirubin levels in the newborn human infant.**

Phenobarbitone (10 doses of 5 mg at 8-h intervals) was injected into the mother just before birth. —— No treatment (placebo); – – – phenobarbitone treated.

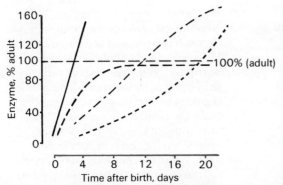

Fig. 27.8. **Development of glucuronyl transferase in the liver endoplasmic reticulum.**
—— Mouse; – – – guinea-pig; — — rabbit; - - - - - rat.

normally associated with the endoplasmic reticulum, such as glucuronyl transferase, is increased. Several trials have shown that the treatment is successful and the result of a typical trial is shown in *Fig*. 27.9. Phototherapy, using light at 450 nm, is also an effective treatment for jaundice of the newborn. Light at this wavelength causes destruction of the bile pigments.

27.7 Inherited defects of conjugation

Several inherited conditions have been described in which defective conjugation of bilirubin occurs.

Gilbert's disease. Hyperbilirubinaemia is observed in this condition which is autosomal dominant. The level of glucuronyl transferase activity is very low but may be induced by drugs, such as phenobarbitone, so that the capacity to synthesize the enzyme may not be impaired but the normal induction mechanism is faulty.

Crigler–Najjar syndrome. This is an inherited autosomal recessive condition in which there is complete absence of the conjugating enzyme. A special inbred strain of rats, the 'Gunn rat', also demonstrates this condition.

Arias type II syndrome. This is a condition intermediate between Gilbert's disease and the Crigler–Najjar syndrome.

27.8 Hyperbilirubinaemias: causes and consequences

A raised concentration of bilirubin in the plasma, described as hyperbilirubinaemia, may be caused by an excessively high concentration of either (*a*) free bilirubin or (*b*) conjugated bilirubin. The patient's skin, eyes and urine gradually become pigmented with a yellowish colouration and the condition is described as 'jaundice'.

The two distinct forms of bilirubin in plasma may be distinguished by a test devised by Van den Bergh in 1913. He showed that conjugated bilirubin reacted rapidly with a diazo reagent to give a purple 'azo-bilirubin': this reaction he described as 'direct'; however unconjugated bilirubin reacted very slowly unless an alcohol such as methanol was present and this reaction was described as 'indirect'. The test has been used for many years to distinguish the two forms of bilirubin in the plasma and is thus a valuable aid to diagnosis.

Conjugated bilirubin

Increased concentrations of conjugated bilirubin accumulate in the plasma as a result of intrahepatic choleostasis. The conjugated bile pigments then enter the sinusoidal plasma by reversed pinocytosis from the liver cells or from leakage of a damaged bile duct.

Choleostasis can be caused by:

a. A blockage of the bile duct—usually by gall stones
b. Liver damage caused by drugs or hormones
c. Liver damage caused by viruses—viral hepatitis.

Unconjugated bilirubin

Increased concentrations of unconjugated bilirubin in the plasma may be caused by the following factors.

a. *Increased pigment production.* The commonest cause of this condition is excess haemolysis of red blood cells. However, the healthy adult liver normally has a large reserve capacity to conjugate an increased load of bile pigment and conjugation increases to meet the demand. Increases of unconjugated bilirubin in plasma, unless production is in very large excess, will therefore be mainly due to transport from other tissues, such as the spleen, to the liver.
b. *Reduced liver uptake.* This, accompanied by normal production of pigment, is caused by a reduced

activity of the hepatic acceptor protein and has been observed in patients treated with extract of male fern for tapeworm infections. The extract contains flavaspidic acid which competes with the bilirubin for the binding sites on the liver ligandin.

c. *Defective conjugation.* This is common in newborn children, especially premature babies, since the glucuronyl transferase enzyme is in very low concentration at birth (*see* Section 27.6). The enzyme is also in a very low concentration in the inherited syndromes described in Section 27.7.

d. *Hepatocellular damage.* Several liver poisons, e.g. carbon tetrachloride or chloroform, or a disease such as viral hepatitis all cause damage to the conjugating enzymes and thus lead to a raised level of unconjugated bilirubin in the plasma. However, in conditions of this type the

concentrations of both conjugated and unconjugated bilirubin are usually raised (*Table* 27.1).

The relationships between the various conditions are summarized in *Fig.* 27.10.

Table 27.1 **The relationship between bilirubin, bilirubin monoglucuronide and bilirubin diglucuronide in various forms of jaundice**

Type	Bilirubin	Bilirubin monoglucuronide	Bilirubin diglucuronide
Haemolytic (adult)	+	±	±
Haemolytic (infant)	+ + +	−	−
Neonatal	+ +	−	−
Cirrhosis Hepatitis	+	+ + +	+ +
Biliary obstruction	+	+ +	+ + +

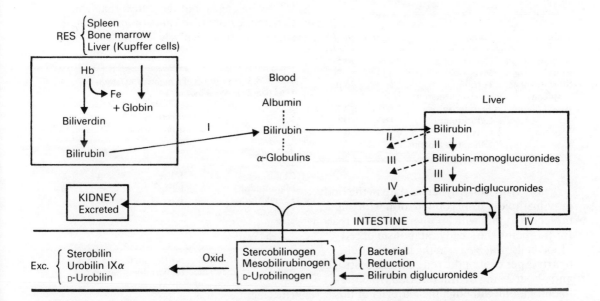

Fig. 27.10. **Schematic representation of bile pigment formation, organ inter-relationships and possible pathological conditions.**
RES, reticuloendothelial system. Pathological conditions: I, excess haemoglobin breakdown; II, III, inadequate conjugation; IV, biliary obstruction.

Chapter 28 Blood: iron and iron metabolism

28.1 Introduction

The total quantity of iron in the adult human body is not large: on average it amounts to about 4 g, the mass of a 3-inch nail. Most of the iron plays an important role as a component of haemoglobin and myoglobin, but the smaller amounts that occur in other iron-containing proteins, such as the cytochromes, are vital for normal cellular function (*see Table* 28.1).

Table 28.1 **Distribution of iron in a typical adult (weight 70 kg)**

Tissue	Form	Total weight, g	Iron, % total
Red blood cell	Haemoglobin	2·72	70·5
Muscle	Myoglobin	0·12	3·2
Store (liver & spleen)	Ferritin (Haemosiderin)	1·0	26·0
Blood plasma	Transferrin	0·003	0·1
	Cytochromes	0·0035	0·1
Various {	Catalase	0·0045	0·1
	Other iron proteins		

Iron is in a dynamic state in the body, moving rapidly from combination with one molecule to that with another, the main reason for this being the steady catabolism of red blood cells. Each day, in the human adult, nearly 1 per cent of the red cells are destroyed and about 25 mg of iron released from the haemoglobin, although the majority of this iron is conserved and reused. It is strange that, in an environment that contains iron in such abundance, the mammalian body conserves iron as though it were a rare metal. This suggests that at an earlier evolutionary epoch, iron may not have been so readily available.

Despite this efficient conservation, iron deficiency disease is widespread. There are various reasons for this, some of which are not yet fully understood and *Table* 28.2 illustrates the incidence of deficiency in a cross section of typical North American families. The significance of some of these measurements will become clearer in the following pages, but they show that iron deficiency states are very common in modern society, and particularly in females.

Table 28.2 **Iron status of a population (studies of 1564 people in Washington, USA)**

Individual	Incidence of iron-deficiency anaemia[†], %	Incidence of iron deficiency[*], %
Children	4·9	9·6
Males (adult)	1·2	3·3
Females (adult)	8·4	20·0
Males (>50 years old)	2·4	4·2
Females (>50 years old)	2·0	5·6

[*] Based on measurements of serum ferritin, saturation of transferrin and red blood cell protoporphyrin.
[†] Based on haemoglobin measurements.
From J. D. Cook et al., 1976.

28.2 Iron balance

For an adult to stay healthy, the amount of iron lost each day must be made up by an equivalent intake. Young, rapidly growing, children, on the other hand, will require a positive iron balance.

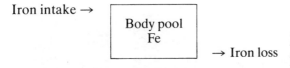

Iron intake → Body pool Fe → Iron loss

We must, therefore, examine, firstly, the factors affecting daily iron losses and, secondly, the daily iron intake and absorption from the gastrointestinal tract.

28.3 Daily iron losses

These may be divided into (*a*) normal loss, and (*b*) abnormal or pathological loss.

a. Normal losses

The adult male and non-menstruating female lose approximately 1·0 mg iron per day of which 50–60 per cent is lost from the gastrointestinal tract, and excreted in the bile or lost by shedding of mucosal cells. The remaining 40–50 per cent is lost from the skin, the hair, and in the urine.

Iron losses in menstruating females are much greater, and range from 1·4 to 3·2 mg/day. A recent survey in Sweden has shown that the losses of blood do not, in any one individual, vary from one period to another, but large differences exist between individuals. Fifty per cent of the women studied lost approximately 1·4 mg iron per day, 25 per cent lost 1·7 mg/day and 2·5 per cent lost 3·2 mg/day.

b. Abnormal losses

Many conditions can give rise to excessive blood loss and therefore to excessive iron loss, examples of which are listed below.

i. Excessive menstrual blood loss

ii. Haemorrhages from the gastrointestinal tract
iii. Postgastrectomy anaemia
iv. Pregnancy
v. Hook worm infections that cause internal haemorrhages.

28.4 Iron intake and dietary iron

Daily loss of iron therefore ranges from 1·0 mg to 3·2 mg for normal individuals. It can be much greater in pathological states and these losses must be replaced by intake of dietary iron. The iron in food is either in the form of a haem complex as part of haemoglobin or myoglobin, or in the form of non-haem iron. In the latter form, the iron is always bound (sometimes loosely, sometimes strongly) to the many complex molecules in the diet, such as proteins or polysaccharides. Although iron is widely distributed in foodstuffs, only a proportion, often less than 5 per cent, is finally absorbed from the diet. This is shown in *Fig.* 28.1, which also shows that the haemoglobin iron of meat is more efficiently absorbed than the iron contained in vegetables such as spinach or corn. There are three reasons for this. First, many organic molecules in vegetables and cereals bind strongly to non-haem iron, making it unavailable for absorption. Secondly, haem iron is already strongly bound to the haem complex and is unavailable to other complexing agents. Thirdly, there is probably a separate mechanism for the absorption of haem iron.

Although studies on the absorption of iron from pure foodstuffs have been valuable, more recent evidence using ordinary mixed meals shows that there is a complex interaction between the iron of different foodstuffs. Each food is capable of enhancing or inhibiting the absorption of iron from another food when the two are taken together in a normal meal. Experiments using isotopically labelled iron have shown, for example, that the absorption of iron from corn can be greatly increased by the addition of fish or meat.

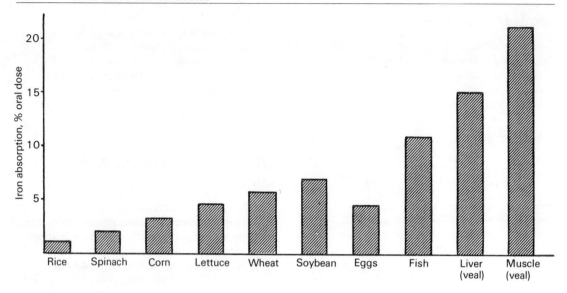

Fig. 28.1. **Absorption of iron from vegetable and meat foods in human subjects.**
11–137 subjects used in each test.

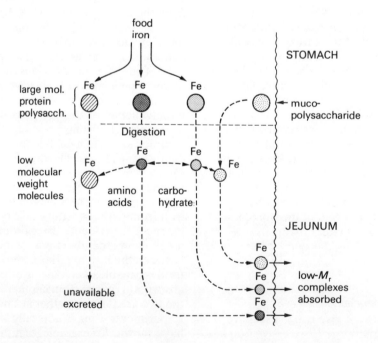

Fig. 28.2. **Role of complexes in iron absorption.**

The role of complexes in iron absorption is summarized in *Fig*. 28.2. The non-haem iron taken in from the diet is released, or partially released, from the food component that binds it. The iron then enters a pool accumulated from various dietary sources and is finally absorbed, probably complexed with the digested food components such as amino acids or carbohydrates (*see below*).

The low percentage of iron absorbed from many foodstuffs is an important factor in anaemia; another is the relatively low concentration of iron in most foods. As discussed in Chapter 11, humans, like most mammals, normally eat to satisfy their calorie requirements and may easily select a diet that leads to an iron deficiency anaemia and this is particularly true of females, as shown in *Table* 28.3. Note that the

Table 28.3 **Relation between iron intake and calorie intake**

Individual	Recommended calorie intake per day	Recommended	Probable
		Iron, mg/day	
Reference man	3200	5	10·5
Reference woman	2400	14	7·7
Young female	2500	12	8·8
Young male	2500	5	8·8
Child (18 months old)	1200	5	4·2

recommended iron intake is approximately five times the estimated loss, which is based on the assumption that only approximately 20 per cent of the dietary iron is absorbed. But, as explained in *Fig*. 28.1, the actual iron absorbed may be even less than this. As a consequence of the relatively low calorie intake in females, the dietary iron intake is often inadequate.

28.5 Factors affecting iron absorption

Due to the widespread nature of iron deficiency anaemia, there have been many investigations into the mechanism of iron absorption and the factors that regulate it.

Some of these are now firmly established, but others are not and further research is required to elucidate a complex problem. Below are listed those factors that have been firmly established.

I. The quantity of iron in the diet
Generally, iron absorption increases with intake, although the net intake is a result of the complex interactions between the various foods in the diet, and other factors discussed below.

II. The chemical form of the iron
Dietary iron may be in the inorganic ferrous (Fe^{2+}) or the ferric (Fe^{3+}) form, or it may be bound in a haem complex. Haem iron is generally well absorbed, because it is unavailable to other complexing agents, and because a separate absorption mechanism probably exists for haem iron as mentioned above. There is good evidence that ferrous iron is more effectively absorbed than ferric iron; inorganic iron for therapy is always supplied in the ferrous form. Ferric salts are much less soluble than ferrous salts and tend to precipitate out in the intestine.

III. Other components of the diet
These may be divided into two categories
  a. Inhibiting agents
      As mentioned above, iron in vegetables is often poorly absorbed, owing to the presence of organic complexing agents, such as oxalates. In cereals, the poor absorption of iron is usually ascribed to the presence of phytic acid, although it has recently been shown that the iron–phytate complex is well absorbed when pure, but when bound to other components, possibly fibre of the cereal, the iron becomes unavailable for absorption.
  b. Stimulating agents
      Amino acids, and especially cysteine, stimulate iron absorption very effectively in experiments that are carried out *in vitro*. It is thought that iron chelates with the amino acids which are then rapidly absorbed. It is

uncertain, however, whether iron–amino acid complexes are very effectively absorbed under normal physiological conditions, because many proteins, such as milk, cheese and eggs which would be digested to amino acids, do not enhance iron absorption. Meat protein, however, increases the absorption of non-haem iron and the porphyrin complexes may play an important role.

Ascorbic acid (vitamin C) strongly increases the rate of iron absorption both *in vitro* and *in vivo* and increases of up to four times the rate in the absence of the vitamin may occur. The reason for this powerful effect of vitamin C is not certain but the following points may be relevant.

i.   As a powerful reducing agent, it will reduce most of the inorganic iron to the ferrous state which is more efficiently absorbed than ferric

ii. It will release iron from the stored ferritin if this form of iron is taken in the diet (*see* Section 28.8)

iii. It may itself form complexes with iron that are efficiently absorbed.

IV. Iron status of the body

Iron stores in the body regulate the percentage of dietary ingested iron that is absorbed, absorption varying inversely with the quantity of iron in the body's stores. This point can be seen in *Fig*. 28.3 which depicts the relationship between iron absorption and circulating haemoglobin.

In addition to the above factors, other regulators may be involved in iron absorption, though evidence for their action is less well established.

a.   *Gastric juice: gastric factor*
     Postgastrectomy iron deficiency anaemia has often been reported. To account for this it has been proposed that the stomach secretes an iron-binding component, probably a proteoglycan, that aids iron absorption. The importance of this

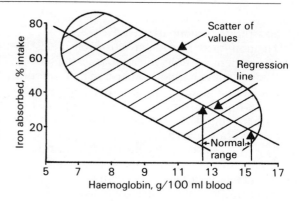

Fig. 28.3.   **Relation between haemoglobin and percentage iron absorption.**
Studies on 500 women in Glasgow.

component is not, however, firmly established. Further, it has been proposed that hydrochloric acid secreted by the gastric mucosa aids iron absorption by stimulating the pepsin digestion of protein, which then tends to release protein-bound iron. An acid medium also favours the stability of ferrous iron, which in turn is likely to enhance iron uptake.

b.   *Pancreatic juice*
     In certain pathological conditions, such as pancreatitis, that result in decreased production of pancreatic juice, the absorption of iron is often increased and, as a result, iron is deposited in the liver in excess. Pancreatic juice may, therefore, contain a factor that inhibits iron absorption, but its nature is unknown.

28.6 The mechanism and control of iron absorption

There are several preliminary basic questions to consider before we discuss the biochemical mechanism of iron transport.

a. In what form is the iron absorbed?
It would be possible for iron to be absorbed in either the free ionic form or bound in a

complex. The question is not yet fully resolved, but it appears likely that iron is absorbed as a natural haem complex or bound to another complexing agent, such as an amino acid, peptide or carbohydrate. The rate of absorption of different complexes is likely to vary from one to the other. Experiments on the interactive effects of different foodstuffs on iron absorption (*see* Section 28.4) support this concept.

b. In what form does iron occur in the mucosal cell?

Absorbed iron is transferred to a small soluble iron-binding protein in the mucosal cell. It is then transferred to the plasma, and bound to the iron transport protein *transferrin*, or to the iron storage protein *ferritin*. These two important proteins are described below.

The absorption process is illustrated in *Fig*. 28.4. If the iron absorbed greatly exceeds the demand, direct incorporation into ferritin is possible, but once so incorporated, it is unlikely that the iron can be made available for transfer to the soluble protein and thence to the plasma. The life-span of mucosal cells is relatively short; only 3–5 days in the human adult, and iron bound as ferritin is constantly lost with the degraded cells in the faeces. The incorporation of iron into the ferritin of the mucosal cell may therefore be regarded as a 'buffer' or protection against excess transfer of iron to the plasma. A mechanism of this type is essential, for it must be remembered that iron, although vital in small quantities, is toxic, even lethal, in high doses.

c. How is absorption regulated?

It has been known for many years that the iron status of the body (i.e. the quantity of iron in store) can regulate iron absorption from the intestine, but the precise mechanism remains obscure. It seems virtually certain that the iron transport protein *transferrin* must play a vital role, because this protein is moving rapidly

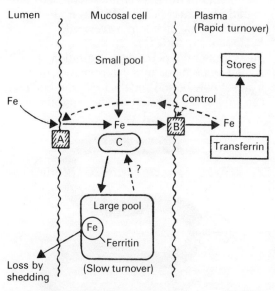

Fig. 28.4. **Mechanism of iron absorption.**
Iron is transported into the mucosal cell as a complex with a molecule such as an amino acid. It is then transferred to a soluble protein C. If the body's iron stores are depleted, it is then rapidly transferred to plasma transferrin, but if the iron stores are full then the iron is transferred from soluble protein C to *ferritin*. Ferritin iron is then lost as the mucosal cells are shed. Transfer of iron back to the soluble form C and then to transferrin is possible but considered unlikely. Iron absorption is controlled by the saturation of transferrin (*see also* Figs. 28.5 and 28.8). A and B are possible control sites (*see also* Fig. 28.6).

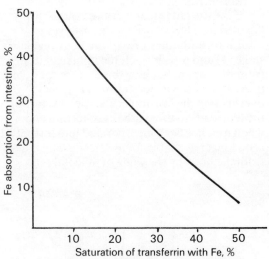

Fig. 28.5. **Relation of transferrin saturation to iron absorption.**
Saturation of transferrin with iron is an important regulator of iron absorption. It will be noted that absorption from the intestinal lumen is inversely proportional to the percentage saturation of transferrin.

between store and mucosa. Furthermore, in iron-deficiency states the quantity of iron carried on transferrin (the percentage 'iron saturation') is reduced (*Fig.* 28.5).

The transfer of iron from the intestinal lumen to transferrin in the plasma is likely to take place in a series of stages shown in *Fig.* 28.6. The iron, probably in the form of a Fe^{2+} complex, is taken up by a carrier protein $\boxed{C}$. In conditions of iron deficiency this protein may release the iron directly to transferrin but when adequate iron stores are available, transfer is likely to be to a low-molecular-weight storage protein $\boxed{S}$. This store may be located within organelles such as the mitochondria.

Uptake of iron by transferrin requires oxidation of Fe^{2+} to Fe^{3+} and this is believed to be accomplished by the action of the enzyme xanthine oxidase which produces hydrogen peroxide as one of its end products. As shown in *Fig.* 28.6, five possible sites of regulation of iron absorption are possible: (1) transfer of iron from the lumen, (2) uptake on to the carrier protein, (3) transfer to store, (4) transfer back to the carrier protein, and (5) transport across the membrane for uptake by transferrin.

Recent research using isotopically labelled iron has shown that transferrin saturation and iron stores of the body regulate stage (1) and possibly (2), the uptake of iron from the lumen, but how this is achieved is, at present, unknown. As shown in *Fig.* 28.4, overloading the low-molecular-weight storage protein leads to the synthesis of ferritin and when iron has been incorporated into ferritin it is no longer available for absorption and is lost from the body by shedding of mucosal cells.

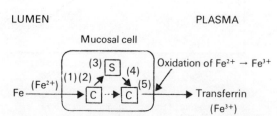

Fig. 28.6.   **Regulation of iron absorption.**
$\boxed{C}$ Carrier protein; $\boxed{S}$ low-molecular-weight storage protein; (1) → (5) possible regulation sites.

28.7   Iron transport: transferrin

Iron is transported in the plasma on the glycoprotein *transferrin*. On electrophoresis of plasma this protein migrates with the β-globulin fraction. It is composed of a single peptide chain whose molecular weight is between 74 000 and 93 000. The carbohydrate moiety, whose molecular weight is about 4500, is 5·3 per cent of the total. The carbohydrate unit is composed of two identical branches and there are four terminal *N*-acetylneuraminic acid residues; there are two iron-binding sites in the molecule, and it can be represented diagrammatically as in *Fig.* 28.7.

Iron is always transported in the ferric form and, although ferrous iron may be bound, it is rapidly oxidized to the ferric form. In the normal healthy human adult, approximately one-third of the iron-binding sites are occupied, but this may be altered dramatically in pathological conditions such as iron-deficiency anaemia or iron overload, haemochromatosis (*Fig.* 28.8). The measurement of percentage saturation of transferrin is made by a determination of the iron actually bound and the total iron-binding capacity of the transferrin. The saturation can,

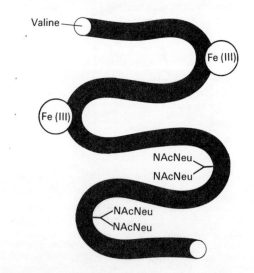

Fig. 28.7.   **Schematic representation of a transferrin molecule (a single polypeptide chain).**
Note that both iron-binding sites are unlikely to be occupied by iron atoms. NAcNeu = *N*-acetylneuraminic acid.

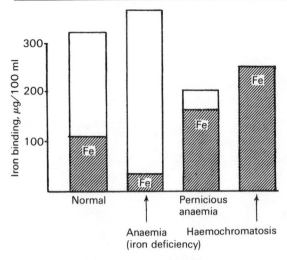

Fig. 28.8. **Iron-binding capacity of serum.**
The unshaded total block height represents the *total iron-binding capacity* of serum that is directly related to the transferrin concentration. The shaded block represents the proportion of iron-binding sites occupied in the transferrin under different conditions.

therefore, be used as a sensitive indicator of iron-deficiency anaemia (cf. *Table* 28.2).

Iron-binding experiments show that each Fe^{3+} ion is bound by the hydroxyl groups of specific tyrosine and histidine imidazole groups of residues in the polypeptide chain. Bicarbonate ions are also needed for a stable binding configuration. Many attempts have been made to demonstrate that the two iron atoms are bound with different affinities, and by different groupings, but it is now generally agreed that the sites are equivalent.

The interesting, probably unique, fact about transferrin is that a single asymmetric polypeptide chain arranges itself in three dimensions to form two very similar binding sites. In other similar examples of multisubstrate binding, more than one polypeptide subunit is involved.

Transferrin must, at least indirectly, regulate iron absorption from the intestine, and it has been proposed that the function of one of the iron-binding sites is to act as an

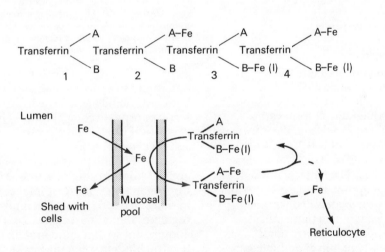

Fig. 28.9. **Mode of action of transferrin and possible forms of transferrin.**
A (the transporter) and B (the indicator) are iron-binding sites. In *Fig.* 28.5, the important role of iron saturation of transferrin in the regulation of iron absorption is shown. The mechanism of action of transferrin is not established, but a possible mechanism is shown above. It is postulated that only one of the two iron-binding sites in transferrin (B in the above scheme) acts as an indicator Fe(I) and that the other site (A) acts as the main transport site. When the body stores are loaded with iron, site B is occupied by an iron atom and this form of transferrin will reject iron absorbed from the intestine. When, however, stores are depleted, site B is not occupied by iron and then the transferrin molecule has a strong affinity for iron, which is carried on the A site.

indicator of body iron stores. If the site were saturated with iron this would indicate adequate stored iron, but if the sites bound little iron then the body iron stores would be low. The other site of the transferrin would then act as the main transport site (*see Fig.* 28.9). A hypothesis of this type appears to conflict with the equivalence of the two iron-binding groups, but the sites may be recognized by specific interaction of the receptor cell surface with the carbohydrate moiety or with specific amino acid residues of the transferrin molecule.

28.8 Iron storage: ferritin and haemosiderin

As mentioned previously, iron is incorporated into the iron storage protein, ferritin, in the mucosal cells, but the main stores of iron in the mammalian body are in the liver and spleen.

The storage proteins

a. Ferritin

The protein component of ferritin (i.e. apoferritin) is composed of 24 subunits arranged spherically. Two species of subunits have been described, the H form with a molecular weight of 21 000 and the L form with a molecular weight of 19 000. The 24 units may be composed entirely of H or L forms or a mixture of H and L. The molecular weight of H_{24} apoferritin is approximately 550 000 and that of L_{24} is approximately 460 000 and intermediates occur within these ranges. In heart apoferritin, the H form tends to predominate whereas liver and spleen apoferritins are composed mainly of the L form, and the L form is the main constituent of ferritin under conditions of iron overload. Each apoferritin can take up very large amounts of iron, but in a normal healthy adult about 2500 atoms of iron are bound in the whole ferritin complex. Iron will then form about 20 per cent of the total molecular weight of the complex protein, which increases from about 500 000 to about 750 000. The iron is

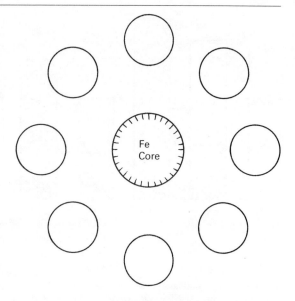

Fig. 28.10. **Ferritin and apoferritin: schematic representation of the ferritin molecule.**
The molecule is composed of 24 subunits arranged to form a sphere around the iron core of $(Fe-OOH)_8 \times (FeO-OPO_3H_2)$ and the subunits have a molecular weight of 19 000 or 21 000. The normal iron loading is 1500–2000 atoms/molecule. A small proportion of the iron is much more labile and enters the circulation more readily than the main bulk of the stored iron.

not bound to a specific site, as in transferrin, but exists mainly as a ferric hydroxide–ferric phosphate complex $(Fe-OOH)_8 - (FeO-OPO_3H_2)$ in the centre of the protein sphere (*Fig.* 28.10).

Ferritin plays a very important role as an iron storage protein, acting as both a 'sink' for iron in times of excess intake, and a 'source' in times of deprivation. The intake of iron also regulates the rate of synthesis of the apoferritin, so that more protein is available for iron binding when the intake is high.

b. Haemosiderin

In addition to storage as ferritin, iron can also be found in cells in the form described as 'haemosiderin'. Particles of haemosiderin are very large and can be detected under the light microscope; in fact it was described before the discovery of ferritin.

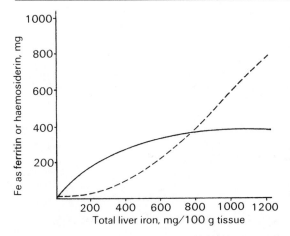

Fig. 28.11. **Relationship between ferritin and haemosiderin in iron overload.**
Animals were given regular iron injections as an iron–dextran complex for a period of 10 weeks. Note that when the store of iron in the liver is low, it is nearly all in the ferritin form, but progressively more is converted into haemosiderin. —— Ferritin; – – – haemosiderin.

Many years on, its nature is still obscure. It appears to be an ill-defined insoluble agglomerate of hydrated iron oxide with several organic constituents, and it may be formed by partial degradation of the protein of ferritin by proteolytic enzymes of the lysosomes followed by the release of iron oxide micelles to form insoluble aggregates trapped in tertiary lysosomes (cf. Chapter 3). In the normal healthy subject very little haemosiderin is detectable in cells but the quantity increases steadily during iron overload (*Fig.* 28.11). Conversely, however, iron can be made available to the body from haemosiderin during periods of deprivation, and exchanges of iron between ferritin and haemosiderin are possible.

28.9 Iron kinetics

We are now in a position to consider the movement of iron from one tissue to another, and from one complex to another. The inter-relationships of the tissues involved are summarized in *Fig.* 28.12.

Red blood cells are continually catabolized in the reticuloendothelial system to liberate 20–25 mg of iron per day. This is taken up by the transport protein transferrin, but some is stored as ferritin if the catabolism of the red blood cells is excessive. A small amount of iron, about 1 mg, is absorbed from the diet, but this is only about 10 per cent of the total dietary intake for the day (10 mg) (cf. Sections 28.4 and 28.5). The iron is then transported to the marrow by the transferrin where it is later taken up by the developing red blood cells, the reticulocytes. Mature red blood cells will not take up iron from transferrin—this property is specific to the reticulocytes and occurs essentially in a three-stage process.

 a. Absorption of transferrin onto the surface of the reticulocytes: this process is dependent on surface

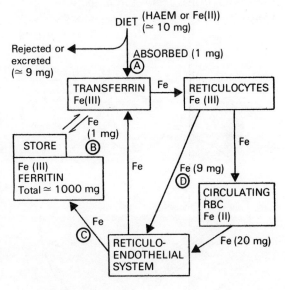

Fig. 28.12. **Iron kinetics.**
Catabolism of the red blood cells in the reticuloendothelial system (RES) releases iron that is transported by transferrin to developing erythrocytes. The small normal daily loss (1 mg) is replenished from the store (B) or from the diet (A). If breakdown of the red blood cells becomes excessive, the store becomes heavily loaded (C). Note that a relatively large amount of iron is circulated as a result of destruction of immature reticulocytes (D). Figures indicate transfer per day in normal man. In haemochromatosis, the store may reach 40 g.

charges, e.g. on the positive charge of NH_3^+ or negative charges of COO^- but not on energy.

b.   A progressive and slow transfer of iron (possibly still bound to transferrin) to the endoplasmic reticulum of the reticulocytes then occurs. This process is inhibited by metabolic inhibitors, and is energy dependent.

c.   The release of transferrin that is unchanged by the process.

The overall process is illustrated in *Fig.* 28.13.

During erythropoiesis approximately 9 mg iron are released each day, and returned to the plasma pool. Thus 30–35 mg iron passes through the pool each day.

It should also be noted from *Fig.* 28.12 that, under normal conditions, the quantity of iron in store remains constant, but in situations of dietary deficiency or excess blood loss, iron will move from the store. Conversely, in conditions of increased dietary absorption or excess degradation of red blood cells, iron will be moved into the store.

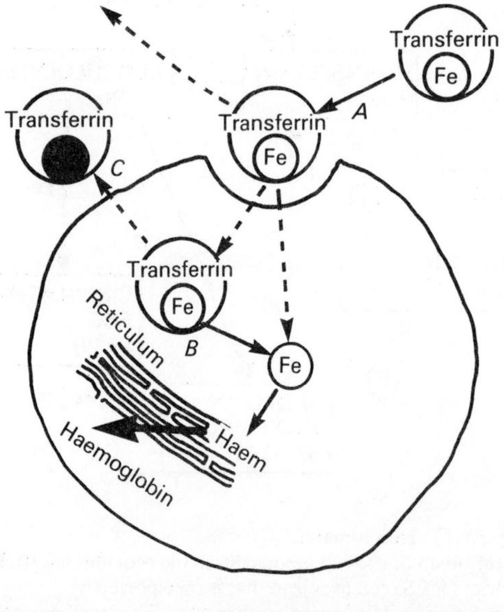

The rate of iron exchange (i.e. both uptake and release) between ferritin and transferrin is greatest when ferritin is already loaded with between 1000 and 1500 atoms of iron. The rate of uptake and release is much slower in the apoprotein and in ferritin heavily loaded with iron. This is an important property of the storage protein, and has implications in the treatment of iron overload (*see* Section 28.11).

28.10 Valency of iron during metabolism

It is well known that iron in its main physiological form, haemoglobin, is in the reduced or *ferrous* state and, as discussed above (*see* Section 28.5), iron in the *ferrous* state is much more effectively absorbed from the diet than *ferric* iron. Yet iron is transported in transferrin and bound in ferritin in the oxidized or *ferric* state.

The valency of the iron is physiologically very important. Haemoglobin will not function as an oxygen carrier if the iron is oxidized to the ferric state to form methaemoglobin. Conversely, iron, normally strongly bound to apoferritin in the ferric state, is released if the ferritin is treated *in vitro* with reducing agents such as ascorbic acid (vitamin C) or cysteine.

It is therefore clear that the tissues must possess elaborate mechanisms for reducing ferric to ferrous iron and for oxidizing ferrous to ferric iron. The nature of these processes is still not completely resolved but the current knowledge is summarized in *Fig.* 28.14. The release of ferric iron from both ferritin and transferrin requires reduction of the ferric iron by an NADH-dependent flavoprotein.

The oxidation of ferrous iron for incorporation into ferritin or transferrin requires one, and possibly two, copper-containing proteins, termed 'ferroxidase I' and 'ferroxidase II'. An interesting recent discovery is that ferroxidase I appears to be identical with the plasma copper-transporting

Fig. 28.13. **Transfer of transferrin iron to reticulocytes.**

The process involves three distinct stages: (*A*) an adsorption of transferrin on to specific sites on the reticulocytes; (*B*) transfer of iron to the site of haemoglobin synthesis in the reticulum (it is likely that the whole transferrin molecule migrates into the cell); (*C*) release of transferrin.

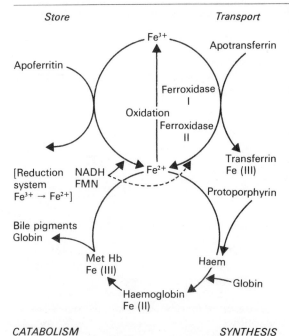

Store   Transport

Fe³⁺

Apotransferrin

Apoferritin

Ferroxidase I

Oxidation

Ferroxidase II

Transferrin Fe (III)

[Reduction NADH   Fe²⁺
system   FMN
Fe³⁺ → Fe²⁺]

Protoporphyrin

Bile pigments
Globin

Met Hb
Fe (III)   Haem

Globin

Haemoglobin
Fe (II)

CATABOLISM   SYNTHESIS

Fig. 28.14. Valency states of iron and ferrous ⇌ ferric cycles in iron metabolism.
Note that although iron is stored and transported as the ferric (III) form, it must be reduced to the ferrous (II) form before release from the store and then reoxidized to Fe(III) for uptake by transferrin. A similar process occurs when iron is transferred from transferrin to ferritin. FMN, flavin mononucleotide.

protein—ceruloplasmin. It is likely that the role of ferrous iron oxidation is one of the most important physiological roles of ceruloplasmin. A relationship between copper deficiency and anaemia has been known for nearly 50 years, and it now appears likely that deficient ceruloplasmin, caused by deficient copper intake, can result in the inability of transferrin to take up absorbed iron efficiently.

Despite this progress in the field, it is not yet known how many different mechanisms there are for iron oxidation and reduction, and how these vary from tissue to tissue.

28.11 Iron pathology

If the iron content of the tissues is too low, iron-deficiency anaemia will develop; if too high, a disease called 'haemochromatosis' develops.

I. Iron-deficiency anaemia

Anaemia is a condition which is characterized by a reduced number of circulating red blood cells, or a reduced amount of haemoglobin in the red blood cells, or both. Anaemias may be classified into three main types, iron deficiency not being the only cause.

i.   *Normocytic*. The red blood cells are unchanged in size and haemoglobin content, but the number in circulation is reduced. This type of anaemia is usually the result of internal or external haemorrhage.

ii. *Macrocytic*. The size of the red blood cells is normal, but the numbers are reduced. This type of anaemia is usually caused by deficiency of folic acid or vitamin B_{12} (cf. Chapter 24). This is the characteristic blood picture of *'pernicious anaemia'*.

iii. *Microcytic*. The cells are much smaller than normal and have a much reduced haemoglobin content. The number of cells may be reduced but to a smaller extent than in other forms of anaemia. Microcytic anaemia is caused by *iron deficiency*.

Causes of anaemias

Iron deficiency anaemias can result from (*a*) inadequate intake or absorption of iron or (*b*) abnormal losses of iron.

The development of anaemia can be traced through a number of sequential events:

Iron mobilization from store
until stores are exhausted
↓
Fall in percentage saturation
of transferrin
↓
Reduced supply of iron
to bone marrow
↓
Normocytic–microcytic
anaemias

Table 28.4 **Sequential stages in iron deficiency**

	Normal	*Iron depletion*	*Iron-deficient erythropolesis*	*Iron-deficient anaemia*
Iron stores	+ + + +	+	0	0
Erythron iron	+ + + +	+ + + +	+ + +	+ +
Transferrin, $\mu g/dl$	330	360	390	410
Plasma ferritin, $\mu g/l$	100	20	10	<10
Iron absorption	Normal	↑	↑	↑
Saturation of transferrin, %	35	30	<15	<10
Erythrocytes	Normal	Normal	Normal	Microcytic Hypochromic

☐ indicates significant deviation from normal values.

The sequence of events is illustrated in *Table* 28.4 and it is important to note that 'anaemia' is often a final end product of iron deficiency and that many earlier indicators of iron deficiency can be demonstrated.

Clearly, it is desirable to diagnose, and thus prevent, iron deficiency, before anaemia develops. This is usually accomplished by measuring the plasma iron concentration and the total iron-binding capacity of the transferrin in the plasma (*Fig.* 28.8). Inorganic iron is added to a sample of the serum and the percentage saturation of transferrin is measured. Normally the figure is 30 per cent, but if it is less than 15–16 per cent, the subject is regarded as suffering from an iron-deficiency condition (*Table* 28.2).

It would be more useful to measure the total body iron stores, but until recently this has proved impossible without taking biopsy specimens of tissue. Recently it has become possible, by the use of sensitive radioimmunoassay techniques, to measure the concentration of the circulating ferritin as distinct from transferrin. This has been found to be, on average, 140 ng/ml for males and 39 ng/ml for females. The method is thought to give a very good estimate of the stores of ferritin iron in the tissues.

a. Inadequate intake   As discussed in Section 28.4 inadequate intake usually results from an inadequate dietary intake together with the low percentage absorption of iron. Menstruating females are the main sufferers,

and it has been decided in some countries (e.g. Sweden) to recommend a regular iron supplement for this group of the population.

b. Abnormal losses   The main categories are as follows:

i.   Excess menstrual blood loss
ii. Losses from the gastrointestinal tract caused either by blood loss, or by excess sloughing off of mucosal cells (cf. Section 28.2)
iii. Postgastrectomy anaemia. Complete or partial gastrectomy frequently leads to iron-deficiency anaemia. The reason for this is not certain, but the defective iron absorption is thought to be caused by either the loss of acidity or lack of gastroferrin (cf. Section 28.5)
iv. Pregnancy. This makes considerable demands on the iron stores of the mother, although this is partially compensated for by the cessation of menstruation
v.   Hook worm infections. In many tropical and subtropical countries, hook worm is a major cause of iron-deficiency anaemia. As many as 500 million people are likely to be affected, and 250 ml of blood may be lost each day from a heavily infected subject.

Although the condition of iron deficiency is described as 'iron-deficiency anaemia', we now know that the anaemia is the end result of an 'iron-deficiency state'.

II. Haemochromatosis

At the opposite end of the spectrum, there are clinical conditions in which excessive deposits of iron are present in the tissues, particularly in the normal iron storage organs, the liver and spleen. In severe cases the total iron in these organs can increase from the normal value of about 1 g to 40–50 g. Many complex clinical symptoms can result from this overload, the chief of which are

  i.   Hepatomegaly
  ii. Skin pigmentation of a dusty brown to sunburnt colouration
  iii. Diabetes in 60–80 per cent of patients
  iv. Hypogonadism and especially atrophy of the testes
  v.   Joint disease
  vi. Heart disease.

It is not clear why pathological conditions such as diabetes develop when the tissues are overloaded with iron in large concentrations. But iron excess is very toxic, and many patients ultimately die from this condition. Clinical diagnosis is carried out most effectively by serum ferritin measurements. Concentration of the latter may be increased to between 3000 and 6000 ng/ml, from a normal value of 140 ng/ml.

Development

Haemochromatosis can arise from several different causes which can be subdivided into primary and secondary groups.

Primary causes

a. Genetic abnormality   The greater incidence of the disease amongst certain families indicates that a genetic disturbance is involved, but the genetics is not straightforward because there are many mild or intermediate forms of the disease. The exact nature of the genetic defect is unclear, but it appears that unregulated or poorly regulated synthesis of apoferritin in storage organs, such as the liver and spleen, may be responsible. The increased concentration of apoferritin then gradually traps more iron, until an iron overload condition develops.

b. The failure of iron absorption regulation   It was noted earlier that in adult males only 1 mg iron is lost per day. If the regulation of the intake of iron is impaired and an excess is absorbed, then increased amounts will be deposited. Once again, the precise nature of the defect is unknown, but several patients appear to absorb haemoglobin iron very effectively whilst maintaining normal absorption of inorganic iron.

c. Excess dietary intake   In certain groups iron intake is excessively high and haemo-chromatosis can develop as a result. This condition is prevalent amongst the Bantu people of South Africa, and has been extensively studied. The dietary iron appears to originate from the use of cooking utensils containing much rusty iron and of the consumption of beer with a high iron content. Consumption of large numbers of the attractive green iron pills by young children causes many cases of serious iron poisoning and even death every year in this country.

Secondary causes

a. Alcoholic cirrhosis   In alcoholics who have developed liver cirrhosis, iron overload is commonly found and it is likely that alcohol can increase iron absorption. Furthermore, some alcoholic drinks such as cider and some wines contain high concentrations of iron which is readily absorbed (*Table* 28.5).

Table 28.5 **Iron content of alcoholic beverages**

Beverage	Iron content, mg/l
US beer	0·1
Gin, whisky	0·6
US wines	2·3–2·6
Red and white French wines	6·2
Cider and wine from Rennes (France)	10–16

From McDonald R. A., 1963, *Arch. Intern. Med.*, Vol. 112, p. 184.

b. Portacaval shunt   Excess deposits of iron are frequently found in the livers of patients who have had a portacaval shunt, an operation performed to divert the flow of blood around the liver when the circulation is obstructed in cases of liver disease. This fact indicates that the liver is likely to play an important role in regulation of the body iron pool, and the absorption from the intestine. Diverting blood flow from the liver would impair the regulation.

c. Chronic pancreatitis   Experiments, both on man and animals, indicate that pancreatic juice may play an important role in the regulation of iron absorption because a reduced secretion of pancreatic juice leads to a marked increase in iron absorption and thence to excess iron deposits in the body. The mode of action of the pancreatic juice on iron absorption is unexplained. In experiments carried out *in vitro*, it has not been possible to demonstrate the presence of an iron-binding factor in pancreatic juice.

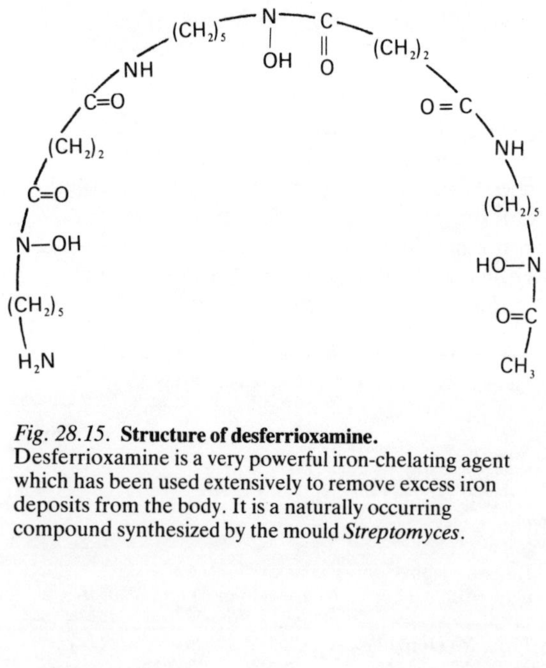

Treatment of haemochromatosis

The treatment is based on one of the two following principles.

a. Venesection therapy   Blood (usually 500 ml/week) is regularly removed from the patient. This stimulates the synthesis of new red blood cells and thus the demand for iron from the stores.

b. Chelation therapy   In this treatment attempts are made to remove iron from the stored form by means of a powerful iron-chelating agent. One very powerful iron-binding agent that is frequently used is desferrioxamine (*Fig.* 28.15). Removal of the stored iron is not easy since the binding in the ferritin complex is very strong, although it was noted that iron was readily released from ferritin if a reducing agent was added to reduce the bound ferric iron to the ferrous form. Treatment with desferrioxamine has been found to be successful *in vivo*, and deposits of iron are removed much more successfully, if the patient is also treated with large doses of ascorbic acid as a reducing agent (*Table* 28.6).

Fig. 28.15. **Structure of desferrioxamine.**
Desferrioxamine is a very powerful iron-chelating agent which has been used extensively to remove excess iron deposits from the body. It is a naturally occurring compound synthesized by the mould *Streptomyces*.

Table 28.6 **Treatment of iron overload with desferrioxamine and ascorbate**

Condition	Increase of iron excretion in urine produced by ascorbate, %
Large number of blood transfusions to combat anaemia	88
Haemochromatosis (genetic)	60
Haemochromatosis (excess dietary intake by Bantu subjects)	350

Patients are treated with desferrioxamine and then with desferrioxamine + 1·5 g ascorbate per day.
From Charlton R. W., Bothwell T. H. and Seftel H. C., 1973, *Clinics in Haematology*, Vol. 2, p. 383.

Chapter 29 Functions of the liver

Most of the important functions carried out by the liver are described in detail in other chapters, the objectives of this chapter being to summarize the known functions of the liver to help medical students comprehend the wide range of consequences that can arise as a result of liver damage or disease.

Some metabolic processes that have not been dealt with in detail, such as the synthesis of the bile salts, are discussed in this chapter.

29.1 Structure of the liver

Although the structure of the liver is described in detail in anatomy textbooks, there are certain aspects of the structure that bear an important relationship to its biochemical function and which are, therefore, emphasized in this chapter.

The liver is the largest organ in the human body and, in the adult, usually weighs approximately 1·5 kg. Despite the large size, the cells are remarkably uniform and parenchymal cells form 60 per cent of the total cell population. The other important cell type in the liver is the Kuppfer or Littoral cell and

these cells form 15–33 per cent of the total liver cell population in man.

The main blood supply to the liver, about 1·2 l/min, is received from the portal vein, the majority of the portal blood coming from the gut and, after a meal, containing high concentrations of digested food, such as glucose, amino acids and some short-chain fatty acids. A proportion, approximately 30 per cent, of the portal blood comes from the spleen.

The hepatic artery delivers much less blood to the liver than the portal vein, arterial blood delivered at a rate of approximately 0·3 l/min being the major source of oxygen for the liver. The blood provided by these two sources mixes in the liver sinusoids. The system of sinusoids (*Fig.* 29.1), differs from capillaries in having no collagen in their walls. Sinusoids are lined by the Kuppfer cells, and are ideally suited for the efficient diffusion of metabolites into liver cells.

The Kuppfer cells contain large numbers of lysosomes (*see* Chapter 3) that phagocytose very actively and can, therefore, swell to block the passage of blood through the sinusoids if ingested particles are not digested and dispersed.

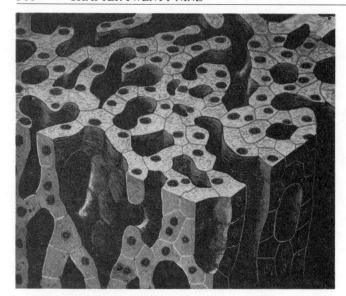

a

Fig. 29.1. Structure of human liver.
(*a*) Sinusoids and parenchymal cells;
(*b*) ultrastructure.

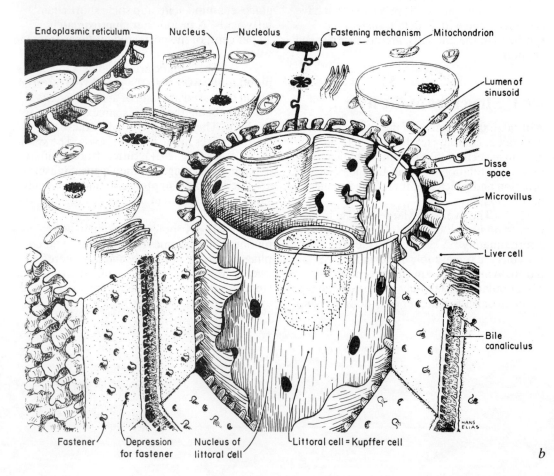

Endoplasmic reticulum — Nucleus — Nucleolus — Fastening mechanism — Mitochondrion

Lumen of sinusoid

Disse space

Microvillus

Liver cell

Bile canaliculus

Fastener   Depression for fastener   Nucleus of littoral cell   Littoral cell = Kupffer cell

b

29.2 Metabolic roles of the liver: general considerations

The liver plays a very important role in regulating the nutrition of the body tissues, which it does in both the fed and fasted states. In the fed state it will utilize the ingested glucose and amino acids, metabolizing them to form new amino acids. There is, therefore, extensive interconversion of amino acids by transamination and other pathways, so that the normal pattern of amino acids in the plasma is reached as soon as possible after a meal.

During the fasted state, the liver plays a similar role in processing nutriments. In this situation, however, nutriments are received from the storage tissues, for example amino acids from muscle and fatty acids from adipose tissues. The liver will convert the glucogenic amino acids, and particularly alanine (cf. Chapter 20), into blood glucose and glycogen, and the fatty acids into very-low-density lipoproteins that are transported to provide the tissues with fatty acids as fuel (cf. Chapter 19).

In summary, therefore, the liver can be described as an important 'nutriment regulating organ'.

29.3 Role of the liver in carbohydrate metabolism

The functions of the liver in carbohydrate metabolism can be illustrated diagrammatically by the triangle shown in *Fig.* 29.2. This represents the supply of monosaccharides, glucose, fructose and galactose received from the portal vein, and the supply of pyruvate and lactate from the other tissues, such as muscle, primarily via the hepatic artery. The metabolic pathways occurring in the liver are listed below.

A.   Interconverts monosaccharides:

Glucose ⇆ Fructose

Galactose ⇆ Glucose

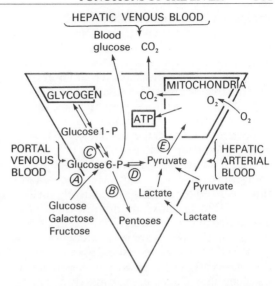

Fig. 29.2.   **Roles of the liver in carbohydrate metabolism.**

B.   Converts glucose to pentose by the pentose-phosphate pathway (Appendix 16).

C.   Synthesizes glycogen from glucose, stores the glycogen and reconverts the glycogen to blood glucose (Chapter 5).

D.   Converts glucose to pyruvate by the glycolytic pathway (Appendices 14, 15); synthesizes glucose from lactate or pyruvate by reversing the glycolytic pathway (Appendices 14, 15).

E.   Oxidizes pyruvate completely by the citrate cycle (Appendices 17, 18).

29.4 Role of the liver in fat metabolism

The fat, mainly triacylglycerol (triglyceride), content of the liver can vary widely depending on the state of nutrition. On an average mixed diet, high-carbohydrate or high-protein diet, fat will normally be relatively low and in the range 2·0–5·0 per cent of the liver weight. After feeding a high-fat diet, however, the triacylglycerol content of the liver can rise to very high values, and form between 25 and 50 per cent of the total weight. The fat content will also increase to 5–10 per cent of the liver weight during fasting.

These figures may give an impression that the liver synthesizes fats and fatty acids very efficiently but, for several reasons, fatty acid synthesis in the liver is unlikely to be important in men. In the normal well-fed adult, in Western societies, a large proportion of the total energy requirements (45–50 per cent) is provided in the diet as fatty acids and thus the need for synthesis is relatively small.

Furthermore, the adipose tissue is also active in the synthesis of fatty acids, but the rate in the liver is approximately twice as great as that in adipose tissue. However, total body fat synthesis over a 4-week period in mice has been shown to be approximately 50 per cent greater than in the liver. It will thus be seen that, contrary to many previous concepts, the liver does not play a unique role in the biosynthesis of fatty acids and the contribution of the adipose tissue is highly significant.

The liver does, however, play a very important role in the synthesis of triacylglycerols and phospholipids from preformed fatty acids, supplied from the diet or, more commonly, from the adipose tissue. As discussed in Chapter 19, the triacylglycerols and phospholipids are packaged into lipoproteins (very-low-density lipoproteins) for transport from the liver. If this process is overloaded, for example by the supply of very large quantities of fatty acids from the diet (high-fat diet) or of large quantities of fatty acids from the adipose tissues (starvation), the liver may not be able to synthesize the requisite phospholipids and protein at a rate sufficient to maintain lipoprotein synthesis. The supply of precursors, for example essential fatty acids for the phospholipids and amino acids for protein, will naturally be critical. Deficiencies of these precursors in relation to the supply of fatty acids will lead to the deposition of large amounts of triacylglycerols in the liver, as indicated above.

The liver can also oxidize fatty acids by β-oxidation to conserve energy as ATP, but under conditions of excess production and under-utilization, the end-product of

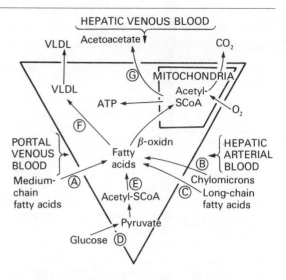

Fig. 29.3.   **Roles of the liver in fat metabolism.**

(A) Medium-chain length fatty acids received from the digestive tract; (B) chylomicrons not taken up by adipose or other tissues; (C) free fatty acids released from adipose tissue particularly during periods of fasting (cf. Chapter 19); (D) glycolysis; (E) fatty acid synthesis; (F) synthesis of lipoprotein, in particular very-low-density lipoproteins (VLDL); (G) production of acetoacetate.

β-oxidation, acetyl-SCoA, cannot be further utilized. Acetyl-SCoA molecules are condensed to form acetoacetyl-SCoA. Loss of the CoASH as a result of the action of the hydroxymethylglutaryl-SCoA synthetase causes the formation of acetoacetate which is liberated into the plasma. When produced in moderate quantities, this acid can be utilized by extrahepatic tissues, such as heart muscle, kidney or brain, but if extensive fatty acid oxidation occurs in the liver, as will happen during starvation or diabetes, large quantities of ketone bodies will be formed as a result of the inability of the liver to utilize acetoacetyl-SCoA. The sequence of the reactions is shown in Chapter 23. A summary of the role of the liver in fat metabolism is shown in Fig. 29.3.

29.5 The role of the liver in amino acid metabolism

The essential functions of the liver in amino acid metabolism are summarized in Fig. 29.4.

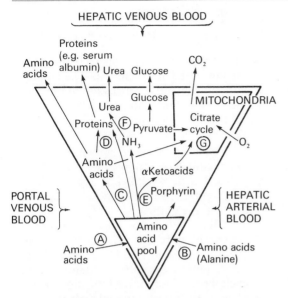

Fig. 29.4. **Roles of the liver in amino acid metabolism.**

(*A*) Amino acids received from the digestive tract after a meal; (*B*) amino acids (particularly alanine) produced by the muscle during a period of fasting; (*C*) interconversions of amino acids by transaminations and other reactions; (*D*) protein synthesis; (*E*) ketoacid formation; (*F*) urea formation; (*G*) oxidation of ketoacids by the citrate cycle.

Amino acids are received by the liver from two sources: during digestion and absorption of protein food via the portal blood (*A*) and during postabsorption and fasting conditions via the extrahepatic tissues, mainly the muscles (*B*). Alanine forms a major proportion of the amino acids released from muscle during fasting (Chapter 20).

Amino acids from either source enter the amino acid pool where extensive interconversion can occur as a result of the operation of the citrate cycle and by transamination (*C*; cf. Appendix 24). Depending on the requirements of the animal, at the time the amino acids are received from the portal blood, they may be recirculated following interconversion, incorporated into other molecules such as purines, pyrimidines or porphyrins or into liver or plasma proteins (*D*). During fasting conditions, most of the amino acids received from the hepatic artery will be converted to blood glucose, but a proportion may be used for protein synthesis

or be oxidized to produce energy after conversion to ketoacids (*E*).

It will be noted that an important function of the liver is to synthesize urea from the ammonia arising during deamination (*F*; cf. Appendix 25). The quantity of urea produced will, therefore, depend on the surplus amino acids from the diet not required for protein synthesis and that are being used for energy or, during fasting, to produce blood glucose.

28.6 The role of the liver in protein synthesis

As indicated in *Fig.* 29.4, the liver plays a vital role in the synthesis of two categories of proteins, those of the liver tissue itself and those that will be exported into the plasma or the plasma proteins. The liver is a very active organ in the biosynthesis of liver proteins and uses the conventional pathways of tissue protein synthesis (Appendix 32). It should be noted, however, that the total liver proteins are much more responsive to changes in the diet than those of other tissues in the body. Thus, in dogs, the liver weight given as a percentage of the total body weight, can be increased from 3·0 to 4·5 per cent by feeding protein, and the response of liver weight in

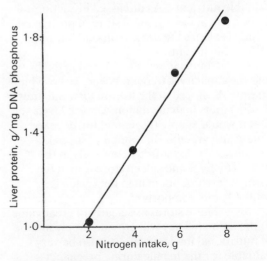

Fig. 29.5. **The effect of dietary protein intake on liver protein in young rats.**

growing rats to a high-protein diet is very marked (*Fig.* 29.5). No special storage protein appears to be laid down and the liver stores proteins by hypertrophy and hyperplasia.

Many of the major proteins of the plasma are synthesized in the liver, including albumin, α- and β-globulins, fibrinogen and, as discussed in Chapter 26, the blood clotting factors II(prothrombin), VII, IX and X, all of which are very unusual in requiring vitamin K for their synthesis. Many special transport proteins are also synthesized in the liver, including transferrin, ceruloplasmin and transport proteins for folate and for vitamin B_{12} (transcobalamins).

29.7 The role of the liver in storage

The liver's role in glycogen storage is well known and the glycogen can vary from about 0·3 per cent of the total liver weight in fasting conditions to about 10 per cent when a high carbohydrate diet is fed. As discussed in Chapter 23, the stored glycogen will normally supply the blood glucose for about 12 hours after a meal. In addition, the liver plays an important role in the storage of many other valuable nutrients. As discussed in Section 29.6, it can act as a general store of protein and it can also store the protein blood-clotting factors, such as prothrombin.

Many vitamins are stored in the liver and the capacity is frequently very large. Thus, vitamin A, stored in the liver in rats, can after feeding diets high in vitamin A reach levels so that it would supply the animal for 70 years! Folate and vitamin B_{12}, bound to special proteins, are also stored extensively in the liver. The liver and spleen store iron in the form of ferritin, but in man the spleen is probably more important.

These stores are clearly of great value to the animal. They will provide supply in time of nutritional inadequacy, and also be very valuable for the female during pregnancy.

29.8 The role of the liver in providing digestive secretions

The digestive secretion produced by the liver is the bile described in Chapter 16. The major components that aid the digestion of fats are the bile salts, but the phospholipids also aid fat emulsification and the bicarbonate helps to neutralize the gastric acidity.

Bile salts in the liver contain 30–40 per cent cholyl conjugates (*Figs.* 29.6 and 29.7). The deoxycholyl conjugates are secondary products formed by bacterial reduction in the gut.

Bile acids are synthesized by a complex series of reactions from cholesterol and the process is outlined in *Fig.* 29.8. Several oxidative stages are involved and the side chain of the cholesterol is removed, the double bond is saturated and two additional hydroxyl groups are introduced. The cholic acid that is formed is conjugated with glycine to form glycine conjugates (e.g. glycocholic acid) or with taurine to form taurine conjugates (e.g. taurocholic acid; *Fig.* 29.6).

The bile acids are secreted into the lumen of the gut where they are acted on by bacterial deconjugase enzymes that remove the glycine or taurine and then by a 7α-dehydroxylase that removes the 7-hydroxyl group. A complex mixture of different bile acids is then reabsorbed, the mixture containing unchanged conjugates and unconjugated bile acids, e.g. cholic and deoxycholic acid, that are returned to the liver where these unconjugated bile acids may be reconjugated with glycine or taurine to form new conjugates of, for example, deoxycholic acid. This process is part of the enterohepatic circulation and is illustrated in *Fig.* 29.9.

The synthesis of bile acids is under control at two points. The feedback of the bile acids into the liver dampens down their synthesis from cholesterol, but in addition control of cholesterol synthesis itself is important. This is also controlled by the dietary cholesterol carried on the chylomicrons and by the cholesterol on the low-density lipoprotein complex in the plasma (*Fig.* 29.10).

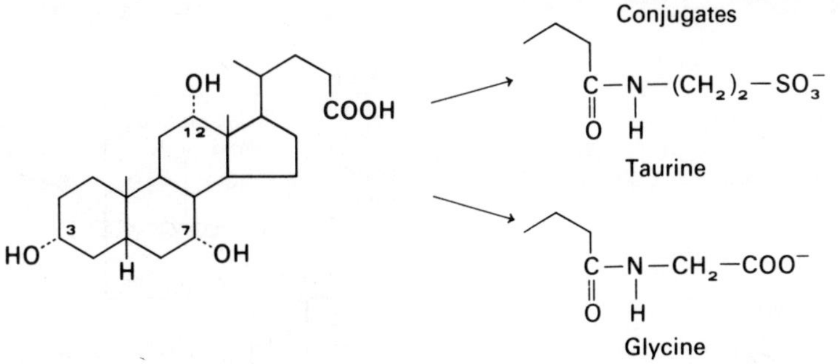

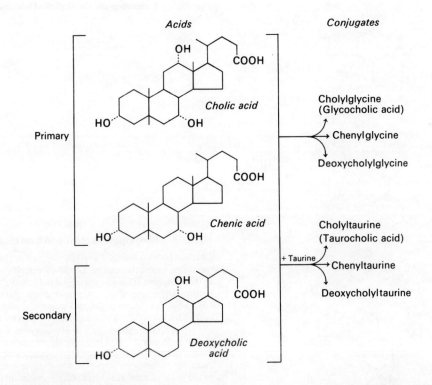

Fig. 29.6.   **Chemical structures of bile acids.**

	Hydroxyl groups	*Bile acid*
Trihydroxy	$3\alpha, 7\alpha, 12\alpha$	Cholic
Dihydroxy	$3\alpha, 7\alpha$	Chenocholic
	$3\alpha, 12\alpha$	Deoxycholic

Fig. 29.7.   **Bile acids of man.**
Normal composition of bile in man: 30–40 per cent cholyl conjugates; 30–40 per cent chenyl conjugates; 10–30 per cent deoxycholyl conjugates.

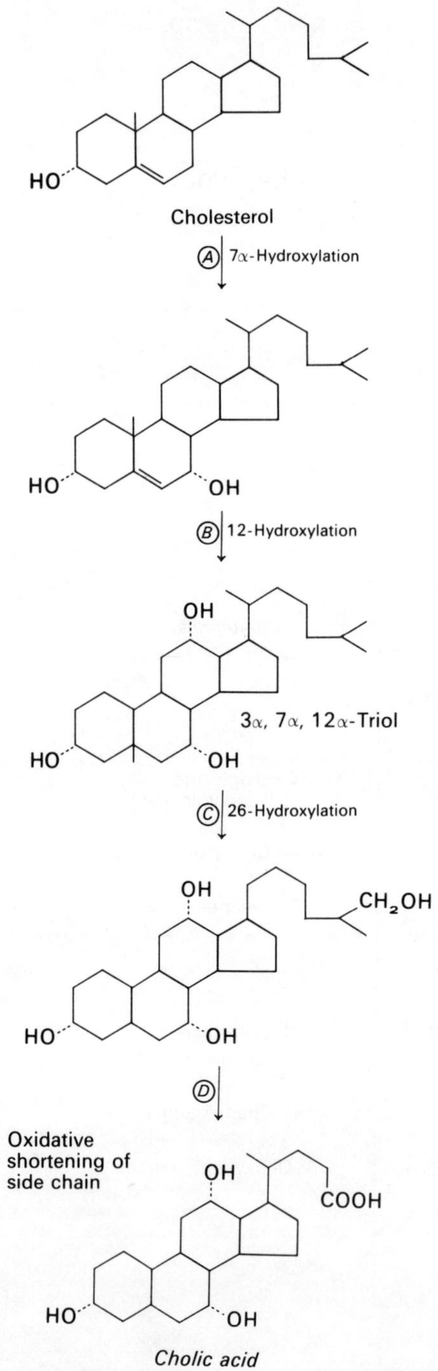

Cholesterol

Ⓐ 7α-Hydroxylation

3α, 7α, 12α-Triol

Ⓑ 12-Hydroxylation

Ⓒ 26-Hydroxylation

Ⓓ

Oxidative
shortening of
side chain

Cholic acid

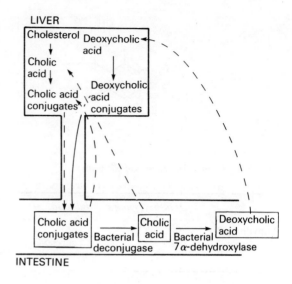

LIVER

Cholesterol	Deoxycholic acid
Cholic acid	
Cholic acid conjugates	Deoxycholic acid
	Deoxycholic acid conjugates

| Cholic acid conjugates | Bacterial deconjugase | Cholic acid | Bacterial 7α-dehydroxylase | Deoxycholic acid |

INTESTINE

Fig. 29.9.   **Enterohepatic circulation of bile acids.**

Fig. 29.8.   **Biosynthesis of cholic acid from cholesterol.**

(A) 7α-Hydroxylation is a regulatory step occurring in the endoplasmic reticulum. (B) 12-Hydroxylation occurs in three stages: (i) oxidation of C-3 to form the ketone, occurring in the endoplasmic reticulum; (ii) 12-hydroxylation in the endoplasmic reticulum; (iii) reduction of the ketone group at C-3 to hydroxyl, occurring in the cytosol. (C) 26-Hydroxylation occurring in the mitochondria. (D) Oxidative shortening of the side chain occurring in three stages: (i) oxidation of C-26 to carboxyl in the cytosol; (ii) C-24 hydroxylation, may occur in the endoplasmic reticulum; (iii) β-oxidation occurring in the mitochondria or cytosol.

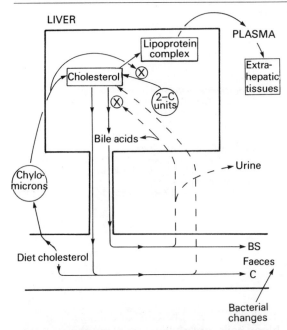

LIVER

PLASMA

Lipoprotein complex

Cholesterol ⊗

Extra-hepatic tissues

⊗

2–C units

Bile acids

Chylo-microns

Urine

Diet cholesterol

BS

Faeces

C

Bacterial changes

Fig. 29.10. **Control of cholesterol/bile acid synthesis in the liver.**
⊗ Points of feedback control of synthesis. Note that diet cholesterol is believed to be taken up into parenchymal cells and recycled to Kupffer cells which may eventually be saturated. BS = bile salts, C = cholesterol.

The metabolism of cholesterol to bile acid has many important medical implications. The plasma cholesterol plays a very important role in arterial disease (cf. Chapter 19), so that its excretion in the bile or conversion to bile acids is of major importance. The relation between the cholesterol and bile acid concentrations in the bile is also important in maintaining the cholesterol solubilized in bile acid micelles. Precipitated cholesterol is the major cause of 'gall stones'.

29.9 The excretory role of the liver in synthesizing or processing metabolites for excretion

The liver plays an important role in the processing of metabolites for excretion, using two routes. Water-soluble compounds are passed out into the blood for excretion by the kidney and insoluble or lipid-soluble materials are excreted through the bile and thence to the faeces.

a. Formation of water-soluble metabolites

In the former category, are included the formation of urea described in Section 29.5 and in Appendix 25, and the very important role of the liver in the metabolism of xenobiotics or the detoxication processes. As described in Chapter 36, the liver converts a vast range of toxic, often lipophilic, molecules into relatively non-toxic hydrophilic molecules that can easily be excreted by the kidney. The efficiency with which the liver metabolizes a drug will be a very important factor in determining its lifespan and its action in the body.

The liver is also the main site of the metabolism of many steroid hormones that are frequently oxidized, usually hydroxylated, by the cytochrome P450 system in a manner similar to that for drugs. Steroid hormones, such as cortisol, corticosterone, aldosterone, progesterone, oestradiol, oestrone, oestriol and testosterone, are all metabolized by the liver and it has recently been demonstrated that a special form of cytochrome P450 may be involved in testosterone metabolism. Examples of this metabolism are shown in *Fig*. 29.11. Subsequent to the formation of hydroxyl groups, many of the steroids are then conjugated with glucuronic acid or sulphate to render them water soluble, and the urine will normally contain a complex mixture of these metabolites. The reactions are, in fact, similar to those used for drug and xenobiotic metabolism and are described in detail in Chapter 36.

b. Formation of lipid-soluble and excretion of insoluble metabolites

The liver is the main organ of the body involved with the catabolism of haemoglobin and the excretion of the bile pigments. These lipophilic molecules are conjugated to give water-soluble glucuronides that are excreted in the bile as described in Chapter 27. Many other toxic molecules and particulate matter are

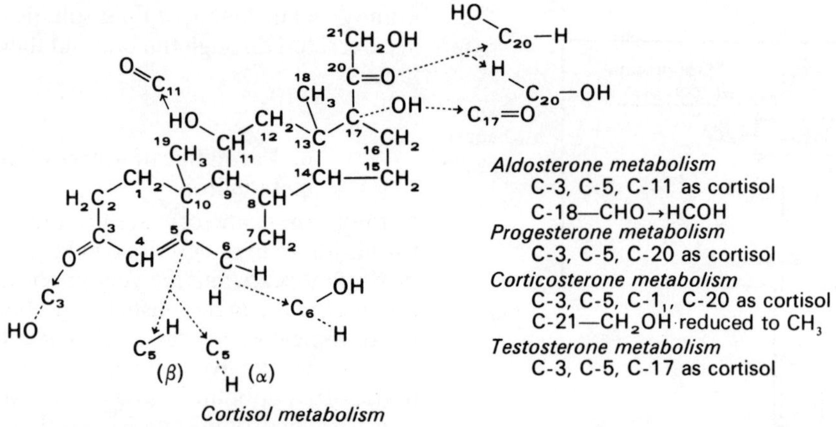

Fig. 29.11. **Metabolism of steroid hormones in the liver.**

→ represents metabolic pathways.

excreted by the liver through the bile and the faeces, and the Kuppfer or Littoral cells lining the sinusoids play a very active role in this process. These cells possess powerful phagocytic action and many active lysosomes, the mode of action of which is described in Chapter 3.

If any particulate material, e.g. carbon particles, gains access to the blood as a result of a wound, the particles are actively phagocytosed by the Kuppfer cells and, after ingestion into tertiary lysosomes, excreted into the bile. The Kuppfer cells can also recognize partially degraded proteins, such as serum albumin, and in fact their ability to recognize proteins of this type surpasses all conventional methods. These proteins, or damaged red blood cells, are phagocytosed but the proteins are degraded by the lysosomal cathepsins and the amino acids made available to the circulation or to the liver parenchymal cells. These cells will also phagocytose bacteria providing they have first been attacked by agglutinating antibodies.

Some particulate matter is, however, toxic after phagocytosis by the Kuppfer cells. Thus some inert material, such as silica or methylcellulose or lipopolysaccharides produced by Gram-negative bacteria are toxic. The reason why some particles or molecules are harmless and others toxic to these cells is at present unknown.

Chapter **30** The kidney

30.1 Major functions of the kidney

The kidney is a very important regulatory organ in the mammalian body and carries out the following functions:

- *a.* Regulation of the body water content
- *b.* Regulation of the plasma electrolytes
- *c.* Regulation of the plasma pH
- *d.* Excretion of toxic and waste products formed in the course of normal metabolism, of ingested toxic products and of the metabolites formed as a result of detoxication of toxic products in the liver.

30.2 The structure of the kidney

In longitudinal section, two distinct regions of the kidney, the cortex and the medulla, are shown (*see Fig.* 30.1).

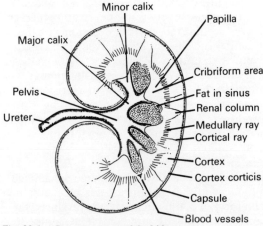

Fig. 30.1.   **Gross structure of the kidney.**

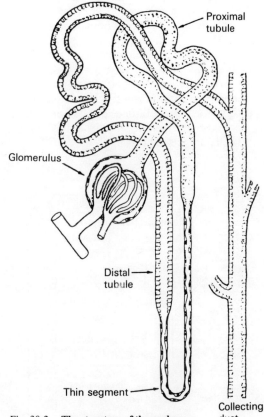

Fig. 30.2.   **The structure of the nephron.**

The main structural unit of the kidney is the nephron which is composed of a glomerulus in contact with a network of capillary vessels, the proximal tubule, the loop of Henle which descends deeply into the medulla, the distal tubule and the collecting duct that collects the urine before joining the ureter (*Fig.* 30.2). The human kidney contains

369

about one million nephrons, each of which is 20–50 μm wide and 50 mm long, so that the total length of tubules in two human kidneys has been calculated to be about 70 miles! The glomerulus filters the blood, retaining proteins, whilst the tubules are involved in absorption of water, small organic molecules and electrolytes, and in secretion.

30.3 Mechanism of action

A diagrammatic representation of the action of the kidney is shown in *Fig.* 30.3. Plasma is filtered under a pressure of about 70 mm mercury through a semipermeable membrane in the glomerulus, of which there are about 10^6 in the human kidney. Large protein molecules with a molecular weight of 60 000 or greater are retained. Albumin (molecular weight of 68 000) is normally retained, but haemoglobin (mol. wt = 66 700) would, if it leaked from red blood cells, be passed through.

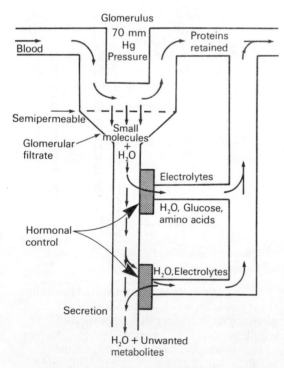

Fig. 30.3. **Diagrammatic representation of the action of the kidney.**

All electrolytes and small organic molecules, such as urea, glucose and amino acids, are passed through and these molecules are of two main types. Some are waste products, e.g. urea, that pass out into the urine, but others, e.g. glucose, are valuable metabolites and are reabsorbed. Electrolytes are reabsorbed, but the absorption may be regulated by hormones such as aldosterone. Water reabsorption also occurs but this is also under the control of a hormone, vasopressin. The pH of the urine is regulated mainly by the control of hydrogen ion secretion and also by excretion of hydrogen ions in the form of NH_4^+.

Approximately 180 litres of protein-free filtrate are produced by the kidney each day, and approximately 1 litre is finally produced so that the kidney clearly possesses enormous reabsorption capacity.

30.4 Composition of the urine

The composition and concentration of various end-products found in the urine can vary throughout the 24-hour period but, if samples collected throughout this period are pooled to give a '24-hour sample', a reasonable degree of constancy is achieved.

The volume, pH and composition of the urine can vary over wide limits depending mainly on dietary intake, including water and electrolytes, but urine normally contains the constituents shown in *Table* 30.1. In the UK, a typical urine volume is 1–2 litres, but under conditions of severe water deprivation this can be reduced to 150–900 ml, depending on the quantity of solute to be removed. The average solute excretion is 50 g per day, although it can vary from 30 to 70 g. Excessive fluid intake will normally cause the excretion of a large volume of urine unless there is extensive loss by sweating. *Table* 30.2 lists the major electrolytes lost in the urine and the minor inorganic constituents. A large number of organic acids are found in the urine, as listed in *Table* 30.3 and it should be noted that a large percentage of glucuronic and hippuric acids are excreted in the form of conjugates specially produced for excretion of toxic compounds (cf. Chapter 36).

Table 30.1 **Summary of the composition of the urine**

Components	Constituents
Water	
Inorganic substances	Trace constituents
	Electrolytes
Organic substances	Organic acids
	Carbohydrates
	Nitrogenous compounds
	Urea
	Uric acid (purines)
	Creatine
	Indoles (metabolites of
	tryptophan)
	Porphyrins
	NH_4^+
	Amino acids
	Hormones—vitamins
	Enzymes

Table 30.2 **Inorganic constituents of urine**

Cations	Cation concn, mg per 24 h	Anions	Anion concn, mg per 24 h
Na^+	3000–5000	Sulphate	
K^+	2000–4000	a. Inorganic	1000–2000
Ca^{2+}	100–450	b. Neutral	200–400
Mg^{2+}	100–300	c. Organic	150–300
Zn^{2+}	0·3–0·5	Phosphate	70–835
Fe^{2+}/Fe^{3+}	0·02–1·1	I^-	0·018–0·483
Ni^{2+}	0·15	Br^-	2
		Cl^-	5000–9000
$Cu^{2+}, Co^{2+},$ $Pb^{2+}, Mn^{2+},$ Sn^{2+} } microgramme quantities		HCO_3^-	Varies extensively with pH of urine

Table 30.3 **Organic acids and carbohydrates found in normal urine**

Constituent in urine	Concentration mg per 24 h	Average, mmol per 24 h
a. Organic acids		
Citric acid	200–1200	1·0–6·0
Glucuronic acid	100–1325 } conjugates	0·5–6·8
Hippuric acid	1000–2500	5·3–14·0
Lactic acid	73	0·8
Pyruvic acid	2·5–60	0·03–0·7
α-Ketoglutaric acid	13·5–59	0·1–0·4
b. Phenolic acids (at least 10 identified) e.g. m-Hydroxybenzoic acid	10–16	

	mmol/l	mg/l
c. Carbohydrates		
Glucose	0·24	43
Galactose	0·28	50 (Babies)
	0	0 (Adult)
Fructose	0·28	50
Lactose	1·1	370 (1–6 months)
	0·2	70 (Adult)
Ribose	1·0	187

Table 30.4 **Major nitrogenous constituents of urine after feeding on high or low-protein diet—typical values**

Nitrogen source	Nitrogen excretion, g/24 h, after feeding	
	High protein	Low protein
Total	16·8	3·60
Urea	14·7 (87·5)	2·20 (61·7)
Uric acid	0·18 (1·1)	0·09 (2·5)
Ammonia	0·50 (3·0)	0·42 (11·3)
Creatine	0·60 (3·6)	0·60 (17·2)

Values in parentheses are percentages of the total nitrogen.

As shown in this table, small quantities of carbohydrates are also excreted. The group of nitrogenous compounds listed in *Table* 30.4 are of major importance.

Total nitrogen excretion is closely related to the dietary intake of nitrogen and the quantity of urea, the major product of nitrogen metabolism, varies directly with the protein intake.

Small quantities of all the amino acids can be detected in the urine; total excretion of the majority of amino acids is less than 10 mg in a 24-hour period, but for some the excretion is much greater (*Table* 30.5). Uric acid is the

Table 30.5 **Amino acid, purine and porphyrin composition of normal urine**

Constituent	Concentration mg/24 h	Concentration mmol/24 h
a. Amino acids		
(except those below)	10	0·1
Exceptions: Glycine	132	1·77
Glutamine	100	0·80
Histidine	216	1·40
Serine	43	0·40
Taurine	156	1·25
Tyrosine	35	0·19

	Concentration, μg/24 h
b. Tryptophan metabolites	
e.g. Tryptamine	45–120
Serotonin	45–160
Indoles	4–30

	Average concentration mg/24 h	Average concentration mmol/24 h
c. Purines		
Uric acid	560	3·3
Adenine ⎱ Guanine ⎰	1·5	0·01
Xanthine	6·1	0·04
Hypoxanthine	9·7	0·07

	Concentration, μg/24 h
d. Porphyrins	
Porphyrins	120
Urobilinogen	500–2000

major purine excreted in the urine resulting from the catabolism of adenine and guanine, although small quantities of these purines and others can also be detected. A relatively large quantity of the porphyrin metabolite, urobilinogen, is present in urine and produces the pale-yellow colour, but small quantities of unchanged porphyrins also occur. Several enzymes, such as amylase, lipase, phosphatase, fibrinolysin and histaminase, also occur in the urine.

Urine tests can often be valuable as an indicator of pathological conditions, but they do not always indicate damage to the kidney. For example, large quantities of glucose in the urine may be caused by kidney disease or by diabetes which gives rise to a high level of circulating glucose.

A very large range of tests is available for urine examination and these tests are normally designed to search for raised concentrations of particular metabolites. Such values are very useful indicators in diagnosis and subsequent follow-up in the treatment and cure of pathological conditions. These analyses form a large section of the subject usually described as 'chemical pathology'.

30.5 Major metabolic processes in the kidney

The kidney carries out all the normal energy-providing reactions, e.g. glycolysis, reactions of the citrate cycle and β-oxidation of fatty acids. Many reactions involving amino acids also take place and transamination is very active in the kidney. The major part of the energy, produced as ATP, is used to provide energy for the many transport processes and particularly for reabsorption processes occurring in the kidney. Many of these have been studied, but the mechanism has been elucidated in detail in only a few instances. In the following sections, we shall discuss the absorption of (a) water, (b) electrolytes, (c) glucose and (d) amino acids.

The kidney is also an important regulator of pH and the mechanisms for achieving this will be discussed.

30.6 Energy provision in the kidney

During the last few years, much evidence has been accumulated to demonstrate that, despite its capacity to oxidize glucose, a large proportion of the ATP formed in, and oxygen utilized by, the kidney is a consequence of fatty acid oxidation. In fact, some believe that all the ATP used by the kidney is formed by fatty acid oxidation. The kidney is capable of oxidizing fatty acids that are released by adipose tissue and brought via the blood; it can additionally utilize fatty acids liberated from triacylglycerols that are stored within its own cells. Lipogenesis from glucose may also occur in the kidney.

The proportion of energy provided by fatty acid oxidation is still controversial, but recent reports indicate that the supply of energy from glucose oxidation may be confined to certain regions of the nephron, particularly the collecting duct and the distal tubule, whereas fatty acid oxidation occurs over the whole length of the nephron.

30.7 Water absorption

Water absorption is closely linked to that of electrolytes, particularly Na^+ and Cl^- reabsorption. The loop of Henle is believed to be the structure that is mainly involved, the hairpin-type structure being essential for its action. In its simplest form, the theory of water absorption by the loop considers that no active transport of water occurs and that movement is

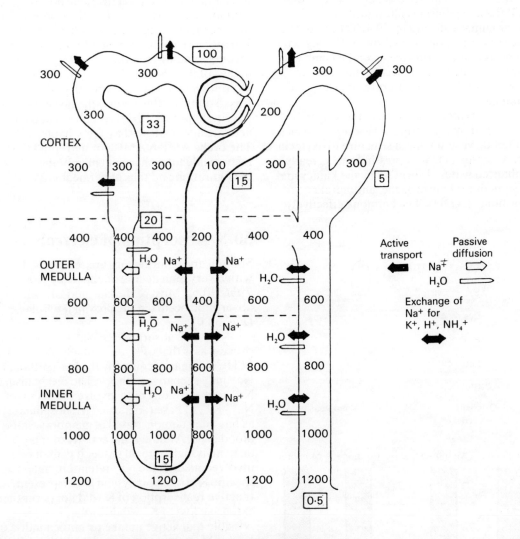

Fig. 30.4. The role of the loop of Henle in water absorption by the kidney.
The numbers in blocks are the percentage of glomerular filtrate remaining in the tubule. The other numbers indicate electrolyte concentrations in tubules and tissue.

under osmotic control. Active absorption of Na^+ from the ascending limb of the loop increases the osmotic pressure so attracting water from the descending loop; the system has been described as a countercurrent exchange. Cl^- is actively transported with the water in the ascending limb to increase the osmolarity of the intestinal cells and so increase water absorption. For this theory to explain absorption in this manner, the ascending limb of the nephron must be impermeable to water and the mechanism is illustrated diagrammatically in *Fig*. 30.4. Take note that the concentration of electrolytes leaving is the same as that entering the ascending limb and that concentration of electrolytes does not begin until the urine enters the collecting tubule.

Water absorption in the kidney is controlled by the hormone vasopressin which is a peptide very similar in structure to oxytocin (cf. Chapter 17). Vasopressin causes greatly enhanced water absorption by the kidney and is sometimes known as the 'antidiuretic hormone' (ADH). The hormone effectively increases the 'pore size' of membranes to water; consequently its action is not confined to the kidney, since it causes increased permeability of many membranes, including those of mitochondria; on treatment with vasopressin, mitochondria will therefore swell due to water accumulation. The effect of vasopressin is not confined to water: it appears to alter the structure of membranes resulting in increased permeability to molecules such as urea and in increased active transport of sodium. The mode of action of vasopressin in precise biochemical terms is still not clear. It may cause changes in membrane structure through binding membrane —SH groups by means of its —S—S— bridge (*Fig*. 30.5) or, alternatively, it may activate adenylate cyclase. Based on this latter theory, the hormone would have to bind to a special receptor that could activate adenylate cyclase (cf. Chapter 17). The cyclic AMP (cAMP) formed could then be responsible for a change in membrane configuration and thus of permeability.

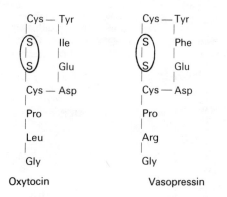

Oxytocin

Vasopressin

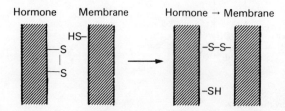

Fig. 30.5. **Structure of oxytocin and vasopressin and binding of hormones to membrane —SH groups.**

30.8 Absorption of electrolytes

Sodium and chloride ions are reabsorbed with a very high degree of efficiency (*see Table* 30.6). Most of the Na^+ and Cl^- absorption occurs in the proximal tubule (*Fig*. 30.6).

The absorption of Na^+ is an active process as is that of K^+; negatively charged Cl^- or HCO_3^- may accompany the Na^+, although Na^+ may exchange with K^+. The mechanism of Na^+ reabsorption could involve the action of a Na^+–K^+ ATPase (adenosine triphosphatase) as has been demonstrated in membranes of red blood cells. Although the activity of this enzyme is high in renal tissue, it cannot be involved in all of the Na^+ retention, unless a specific K^+ transport system is also present. If active reabsorption of K^+ did not occur then extensive loss of K^+ would ensue, but it is possible that active uptake by mitochondria of K^+ could be the mechanism by which K^+ is withdrawn from the lumen. Na^+ may enter the lumen cells driven by the Na^+–K^+ ATPase activity on the serosal side of the cells

Table 30.6 **Absorption efficiencies of the kidney for Na$^+$, K$^+$, Cl$^-$, HCO$_3^-$ and water**

Ion/ electrolyte	Plasma concn, mmol equiv./l	Rate of glomerular filtration, l/24 h	Quantity filtered, mmol equiv./24 h	Quantity excreted, mmol equiv./24 h	Quantity reabsorbed, mmol equiv./24 h	Percentage reabsorbed, %
Na$^+$	140	180	23 940	103	23 837	99·6
Cl$^-$	105	180	19 845	103	19 742	99·5
HCO$_3^-$	27	180	5103	2	5101	99·9 +
K$^+$	4	180	684	51	633	92·6
Water	0·94 l/l	180	169·2 1/24 h	1·5 1/24 h	167·7 1/24 h	99·1

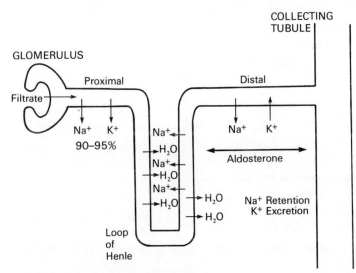

Fig. 30.6. **Sites of Na$^+$ and K$^+$ absorption by the kidney tubules.**
Cl$^-$ normally accompanies Na$^+$ or K$^+$ unless an exchange occurs.

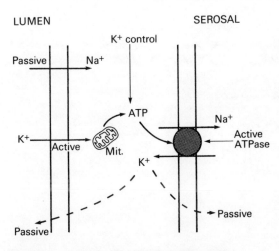

Fig. 30.7. **Possible role of Na$^+$–K$^+$ ATPase in Na$^+$ transport.**

(*Fig.* 30.7). Sodium absorption in the distal tubule is under direct control of the hormone aldosterone, the release of which is controlled initially by the release of the hormone renin. The latter is released from the kidney under conditions of reduced arterial pressure, reduced plasma volume, depletion of Na$^+$, oedema or by activation of the renal nerve (*Fig.* 30.8).

Renin catalyses the conversion of angiotensinogen (hypertensinogen), which is synthesized in the liver and circulates in the plasma, to angiotensin I (hypertensin) which is then hydrolysed by a second enzyme (converting enzyme) to the active form 'angiotensin II' (*Fig.* 30.9). This hormone stimulates the adrenal to produce the steroid hormone aldosterone which, in turn,

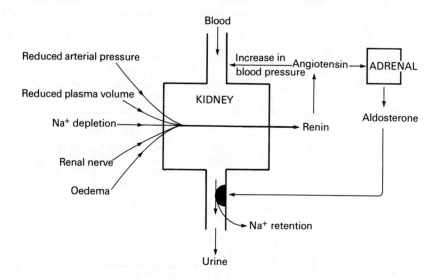

Fig. 30.8. **Factors that are involved in the release of renin.**

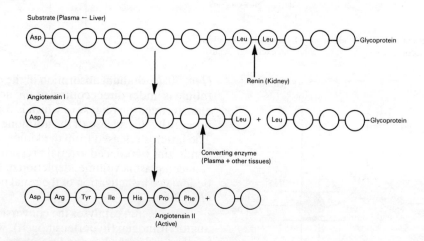

Fig. 30.9. **Formation of angiotensin II.**

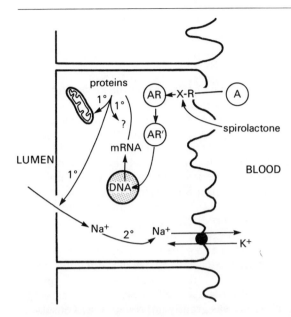

Fig. 30.10. **Mode of action of aldosterone.**
A, aldosterone; R, receptor; AR, aldosterone–receptor complex; 1°, primary action; 2°, secondary effect on Na^+ transport.

There is now little doubt that aldosterone, attached to its protein carrier, does regulate the synthesis of a specific protein, although the nature of this protein still remains obscure. It is possible that the synthesis of a vital protein in the Na^+ transport system is regulated, since control of ATP synthesis is likely to affect many metabolic energy-requiring processes and control by ATP is not sufficiently specific for Na^+ transport to explain the actions of aldosterone.

stimulates Na^+ reabsorption in the distal and collecting tubules. Na^+ reabsorption in the proximal tubules is not believed to be under aldosterone control and, although the mechanism by which aldosterone stimulates Na^+ retention is not yet elucidated, it is likely to involve a fundamental control in the cell nucleus. The steroid is believed to bind to a cytosolic protein receptor which binds to DNA and then allows synthesis of an mRNA that codes for the synthesis of a specific protein (*Fig.* 30.10). Although Na^+ transport, and therefore Na^+–K^+ ATPase, is probably regulated eventually, it may be indirectly controlled by the provision of ATP. Thus, in adrenalectomized animals, the Na^+–K^+ ATPase activity falls significantly, but so does the citrate synthase activity. On treatment with aldosterone the activity of citrate synthase rises over a short period of time, but that of Na^+–K^+ ATPase requires much more time to reach normal values. Experiments of this type suggest that ATP synthesis, which is indirectly controlled by citrate synthase, may be vital in the regulation of Na^+–K^+ ATPase.

30.9 Absorption of glucose

Under normal conditions, virtually no glucose is excreted by the kidney and all that is filtered is reabsorbed. However, as the plasma concentration is increased above a critical value, glucose appears in the urine and, in general, the higher the plasma concentration, the greater the quantity excreted. It should be appreciated, however, that in addition to the plasma concentration of glucose, the rate of blood flow through the kidney and the blood pressure can affect the excretion of glucose in the urine.

The mechanism of glucose transport across the tubule membrane is still not understood in detail: the process is known to be energy dependent, requiring ATP. It is generally believed that a specific carrier molecule is involved and that the process is electrogenic and likely to be associated with the transport of Na^+. Na^+–K^+ ATPase may be involved as described for the absorption of glucose through the intestinal mucosa (*see* Chapter 16). Phloridzin (cf. Chapter 13).

is a powerful inhibitor of glucose absorption and is believed to bind to glucose receptors on the membrane so preventing binding of glucose to the cell.

30.10 Absorption of amino acids

The glomerular filtrate contains amino acids in concentrations very close to their individual plasma concentrations. The rate of reabsorption of different amino acids is not equal and glycine, for example, is reabsorbed much more rapidly than lysine or arginine (*see Fig.* 30.11). Experiments using different amino acids in competition show that at least four separate types of site are available for reabsorption of groups of amino acids:

a.   Basic amino acids, i.e. arginine, histidine, lysine, ornithine and cystine

b.   Neutral amino acids, e.g. leucine and isoleucine

c.   Acidic amino acids, e.g. glutamic acid

d.   Glycine.

The existence, in the kidney tubules, of absorption sites specific for groups of amino acids was first indicated many years ago by the discovery of an absorption abnormality. In this condition, large amounts of cystine were excreted in the urine which, being relatively insoluble, deposited from the urine samples. This condition was described as 'cystinuria'. The original discovery was made at the beginning of the century by Garrod who observed that cystinuria was genetically controlled and appeared to be present in certain close relatives. He described this and some other similar conditions as 'inborn errors of metabolism'. Much later, when amino acid analysis was greatly improved, it was observed that cystinuric urine also contained relatively large quantities of lysine, arginine and ornithine, in addition to cystine, but these amino acids were not detected due to their greater solubility. These amino acids possess similar structures (*see Fig.* 30.12) and it was therefore apparent that they were all absorbed by a common transport mechanism. The site in the kidney tubule appeared to require the existence of two positively charged nitrogen atoms separated by 4–6 carbon atoms. Subsequently, it was shown that similar absorption sites existed in the intestine and,

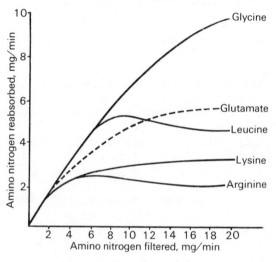

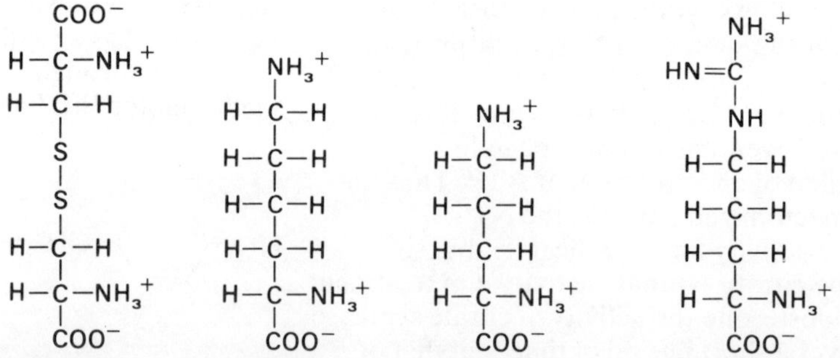

Fig. 30.11. **Rates of amino acid reabsorption.**

Excretion, μmol/day	Cystine	Lysine	Ornithine	Arginine
Normal	50	100	0	0
'Cystinuria'	3800	9500	2000	7000

Fig. 30.12. **Comparison of structures of cystine, lysine, arginine and ornithine and their excretion in 'cystinuria'.**

furthermore, the individuals suffering from this condition usually possessed inadequate absorption sites for these amino acids in both the liver and the kidney. The synthesis of these absorption receptors is, therefore, likely to be controlled by a single gene.

As for many other absorption processes in the kidney, the mechanism of amino acid absorption is not understood in detail, but Meister suggested that glutathione played an important role in the kidney and took part in a transport system involved in the reabsorption of amino acids. The trapping of the amino acid is considered to involve a transpeptidation that utilizes the glutamyl residue of glutathione to convert the amino acid into a dipeptide. Glutamate is eventually reformed and combines with cysteine and glycine to form glutathione, the cysteine and glycine being released during the transpeptidation (see Fig. 30.13).

Although it is well established that a glutathione transpeptidase is active in the brush border of the kidney lumen, it is not universally accepted that this enzyme plays an important role in amino acid transport. It has been demonstrated that the γ-glutamyl transpeptidase is extracellular and not available to intracellular glutathione. The enzyme's main role may involve a stage in the conversion of the glutathione conjugate to the mercapturic acid conjugate in the kidney lumen.

30.11 Regulation of pH

The kidney plays a very important role in pH regulation of the plasma and of the whole body. It is capable of extracting large quantities of HCO_3^- if the plasma becomes alkaline and consequently produces alkaline urine (Fig. 30.14). Under conditions of acidity, which are more common because many metabolic processes produce acid, HCO_3^- is completely reabsorbed and H^+ is excreted, as such, associated with phosphate or as NH_4^+. The urine pH can fall to values as low as 4.8, whereas it is unusual for the plasma pH to fall below 7.2 so that H^+ can be excreted directly. A comparison between the total compositions of acid and alkaline urines is shown in Fig. 30.15. The mechanism of reabsorption of HCO_3^- is shown in Fig. 30.16; H^+ is excreted and is normally exchanged for Na^+. Association of the H^+ with HCO_3^- takes place in the lumen to form H_2CO_3 that dissociates into H_2O and CO_2. Absorption of CO_2 allows reassociation with H_2O to form H_2CO_3, a reaction that is catalysed by carbonic anhydrase. Dissociation yields H^+ and HCO_3^- and the latter is transported back to the plasma. The process causes very efficient retention of HCO_3^- in the plasma, thus tending to buffer the plasma against increase of acidity.

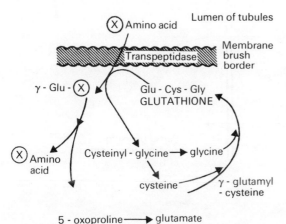

Fig. 30.13. **Role of glutathione in amino acid transport: the γ-glutamyl cycle.**

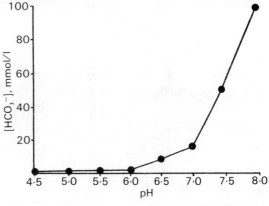

Fig. 30.14. **The relation between the urine pH and the bicarbonate concentration.**

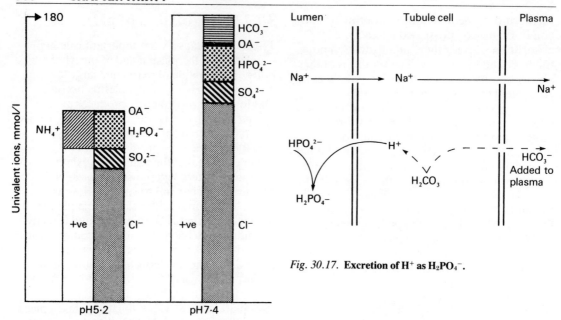

Fig. 30.15. **Comparison of the electrolyte compositions of acid and alkaline urines.**

Na$^+$ is the major positive ion, but as indicated in *Table* 30.2, significant quantities of K$^+$, Ca^{2+} and Mg^{2+} are also excreted. OA—organic acid.

Fig. 30.17. **Excretion of H$^+$ as H$_2$PO$_4^-$.**

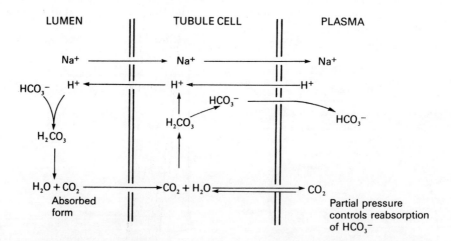

Fig. 30.16. **Mechanism of reabsorption of HCO$_3^-$.**

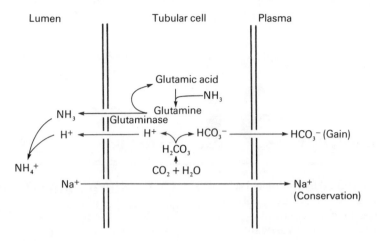

Fig.30.18. **Excretion of NH₄⁺.**

A second important process is shown in *Fig*. 30.17; hydrogen ions once again exchange with Na^+, but are accepted by HPO_4^{2-} to form $H_2PO_4^-$ which is excreted. The excretion of phosphate in this form therefore permits direct excretion of H^+.

A third and important mechanism involved in the secretion of H^+ is the production of NH_4^+: the process is outlined in *Fig*. 30.18. Ammonia (NH_3) is produced by the action of glutaminase on glutamine and is then liberated in the lumen where it associates with H^+. This may be formed by dissociation of H_2CO_3 as shown, or from an acid such as lactic or acetoacetic acid. NH_4^+ is produced in the lumen and HCO_3^- and Na^+ returned to the plasma. As discussed in Chapter 7, glutaminase, which is phosphate dependent, is located inside the mitochondria so that glutamine must be transported into the mitochondria before hydrolysis takes place and the ammonia released is transported across the mitochondrial membrane. Reformation of NH_4^+ can occur in the cytosol, but it is more probable that NH_3 is transported through the tubular cell membrane to reassociate with H^+ in the lumen.

Under normal acid–base conditions, the amount of ammonia added to the tubular fluid by the last portion of the proximal tubule is equivalent to that excreted in the urine. Ammonia delivery to the bend of the loop of Henle is approximately twice the amount present at the end of the proximal tubule, but ammonia delivered to the distal convoluted tubule is less than 50 per cent of the ammonia present in the proximal tubule. Ammonia is, therefore, removed probably between the tip of the loop of Henle and the distal convoluted tubule, but it is added back to the tubular fluid between the distal convoluted tubule and the medullary collecting duct.

During metabolic acidosis, the patterns of ammonia addition to, and subtraction from, the tubular fluid is qualitatively similar but the majority of ammonia excretion is accounted for by enhanced addition to the tubular fluid. The mode of regulation of the production of ammonia from glutamine is still unresolved, but it appears likely that the glutamine transported from the cytosol into the mitochondria and/or the phosphate-dependent glutaminase could be important sites of regulation.

Chapter 31 Muscle

Muscle has been studied by biological scientists for many hundreds of years. The ability of living tissue to perform mechanical work has always fascinated scientific investigators, probably because the end result of muscle action was always so clearly apparent and important. The vital importance of muscle, coupled with the relative ease of its study, has ensured that early investigators paid great attention to the tissue. As a consequence, muscle has been studied extensively, by physiological methods, by light microscopy, by electron microscopy and, more recently, by biochemical methods. For many years these different lines of investigation tended to proceed independently, but more recently, greater efforts have been made to inter-relate the experimental findings using the different approaches so that a much better understanding of the structure and functions of muscle is gradually evolving.

31.1 Microscopic structure of muscle

Muscle fibres viewed under the high-powered light microscope are seen to be composed of a number of longitudinal units called 'sarcomeres'. Each sarcomere is defined by a 'Z' line and contains a dark 'A' band and a lighter 'I' band extending across the Z lines (*Fig.* 31.1).

Closer examination of the structure by electron microscopy shows that the dark 'A' band is formed of a large number of thick filaments, whereas the light 'I' band is composed of a comparable number of thin filaments. Careful study of the electron micrographs demonstrates that the thin filaments are interspaced between the thick filaments (*Fig.* 31.2).

During contraction, the 'I' band becomes much reduced in size and this is believed to be caused by extensive overlapping of the thick and thin filaments (*Fig.* 31.3).

Fig. 31.1.   **Muscle fibres viewed under a low-power electron microscope.**
Myofibrils of rat skeletal muscle cell. A = A band; Z = Z line; I = I band; Mi = mitochondria.

Fig. 31.2. **Muscle fibres viewed under a high-powered electron microscope, showing thick and thin filaments.** Longitudinal section of a skeletal muscle (baboon). A = A band, Z = Z line, N = nucleus.

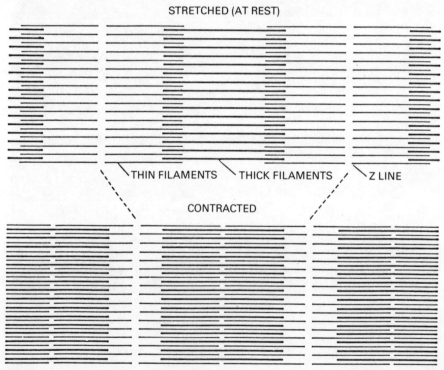

Fig. 31.3. **Diagrammatic representation of a muscle fibre showing thick and thin filaments and their overlap resulting from contraction.**

31.2 The proteins of muscle

Muscle contains four main proteins, myosin, actin, tropomyosin and troponin, and a number of minor proteins, such as actin cross-linking proteins.

Myosin

All muscle myosin molecules contain two heavy chains of approximate molecular weight 2×10^5 and two pairs of light chains of approximate molecular weight 2×10^4. In diagrammatic form, the myosin can be represented by a type of head and tail structure (*Fig.* 31.4).

In several theories of muscle action, the light chains are considered to play an important active role. This activity may,

however, be confined to one pair of chains possessing ATPase activity and described as essential, whilst the other pair are described as 'non-essential', because their removal does not affect the ATPase activity associated with myosin.

The subject is currently controversial, but some modern research tends to support the view that the activity of myosin is resident entirely in the heavy chains and that the light chains represent vestigial proteins involved in the regulation of muscle activity in an earlier phase of evolution. The fact that one pair of light chains can undergo phosphorylation supports the possibility of a regulatory role.

Actin

Actin is basically a monomeric protein with a molecular weight of 42 000 called 'G' actin. The 'G' actin monomers readily form large elongated polymers containing 1000 units that are several micrometres in length and may be visualized as a string of beads. For assembly of the polymers, Mg^{2+} is essential and the 'G' actin units are orientated with a so-called 'pointed' region and 'barbed' region. Assembly of the monomeric units occurs at the barbed

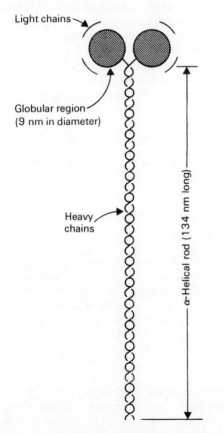

Fig. 31.4. **Diagrammatic representation of a myosin molecule.**

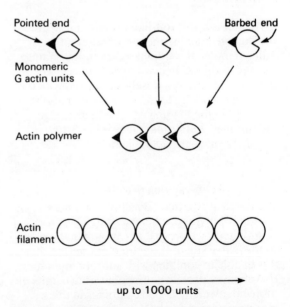

Fig. 31.5. **Diagrammatic representation of assembly of G actin monomers.**

end and disassembly at the pointed end (*Fig.* 31.5).

Several proteins can regulate the assembly of actin monomers into polymers. One group of proteins, called 'profilins', binds to 'G' actin, stabilizes the monomer and prevents polymerization. Other proteins cross-link actin strands.

Although, in this chapter, we are concentrating our attention on muscle, it should be appreciated that actin occurs in many different types of eukaryotic cells and the contraction of actin is responsible for a wide variety of motile processes, including cell locomotion, cytoplasmic streaming and transport, secretion, phagocytosis and cytokinesis.

Actin cross-linking proteins

A number of proteins have been described that cross-link, and stabilize, actin chains.

Filamin is a high-molecular-weight flexible protein that usually exists as a dimer with a molecular weight of 54 000. It binds to actin at 40-nm intervals giving a ratio of 1 filamin:15 actin monomers.

α-Actinin is a stiff rod-like protein, quite unlike the flexible filamin, and is composed of two units of molecular weight 105 000. It is mainly located in the 'Z' band of muscle fibres. The fact that it is usually closely associated with the membrane has led to the suggestion that it is involved in the binding of actin filaments to membranes, but its main role is likely to be simply to space the actin filaments.

Actin-capping protein

We noted above that 'G' actin monomers normally polymerize head to tail. This process is inhibited by a process of capping that binds to the active site of the actin monomer so that it is unable to combine with another monomer. *β-Actinin*, a protein component of two units of molecular weight 35 000 combines in this manner, stabilizing the 'G' actin units and preventing them from polymerizing.

Tropomyosin

Tropomyosins are long, thin molecules of molecular weight 70 000, composed of double-stranded α-helices, that attach end to end, forming a long thin filament on each strand of actin. Each strand of actin carries its own filament of tropomyosin that lies in the groove between strands of actin and one filament of tropomyosin extends over seven actin strands. Each filament of skeletal muscle is approximately 1 μm in length containing 300–400 actin molecules and 40–60 tropomyosin molecules.

Troponin

Troponins are globular proteins composed of three polypeptide chains: TpC (mol. wt 18 000), TpI (mol. wt 24 000), and TpT (mol. wt 37 000). One troponin molecule sits astride each tropomyosin molecule, a short distance from one end of the molecule. One of the troponin peptides, troponin C, plays an important role in the binding of Ca^{2+}. The nerve impulse causes a release of Ca^{2+} from the sarcoplasmic reticulum that then binds to troponin C causing a configurational change in the protein. This initiates a series of events eventually leading to ATP hydrolysis and muscle contraction (*see* Sections 31.4, 31.5 and 31.6).

31.3 Assembly of proteins into filaments

The thick filaments are composed of myosin molecules placed end to end with the light chains sticking out in the form of double heads (*Fig.* 31.6). The assembly of thick myosin filaments is interspersed with thin filaments consisting of chains of actin molecules with associated tropomyosin and troponin molecules (*Fig.* 31.7). In the relaxed state of muscle, gaps exist between the ends of chains of actin molecules, the thin filaments giving a clear zone, which decreases in width during contraction (*Fig.* 31.8). The 'Z' line has a flat protein structure to which thin filaments are attached.

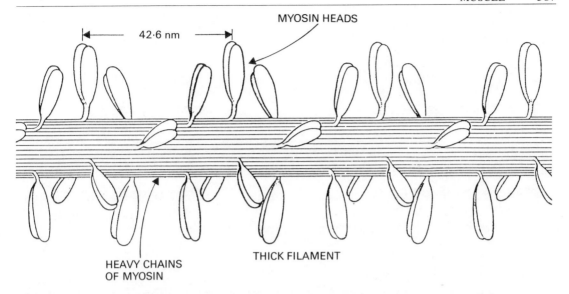

Fig. 31.6.   **The structure of the thick filaments of muscle showing the myosin heads sticking out from the filament.**

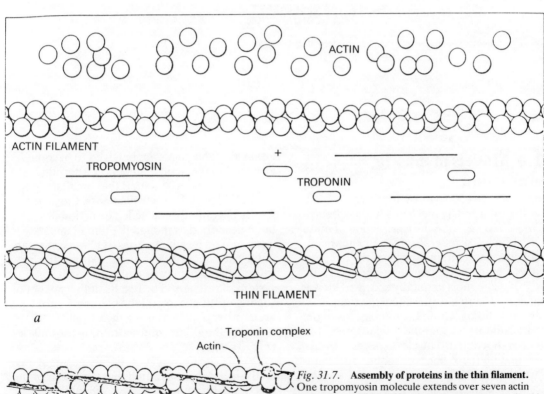

Fig. 31.7.   **Assembly of proteins in the thin filament.**
One tropomyosin molecule extends over seven actin molecules and 300–400 actin molecules are contained in the filaments that are about 1 μm in length.
(*a*) Mode of assembly; (*b*) detail of structure of thin filament.

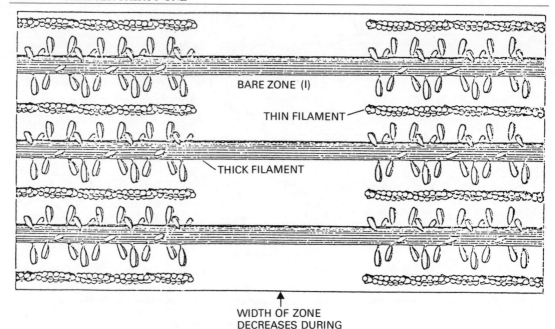

WIDTH OF ZONE
DECREASES DURING
CONTRACTION

Fig. 31.8. **Assembly of thick and thin filaments into fibres.**

31.4 Models of muscle contraction

During the last 50 years of intensive investigation, many mechanisms have been proposed to describe the contraction of muscle and, although a general agreement has been reached about certain aspects of the overall process, many details remain to be resolved.

The most generally accepted view is that contraction is achieved by the thin filaments sliding within the interspaces of the thick filaments, so that the total length of the fibres is shortened. The clear spaces surrounding the Z lines consequently diminish in width (*Fig.* 31.3).

A mechanism that has received much support, originally proposed by Huxley, is the 'cross bridge cycle'; this theory assumes that the small heads of the myosin molecules play a vital role. The heads are believed to become attached to the actin filaments at a certain angle and then twist so that the thin filaments are pulled past the thick filaments, the heads finally becoming detached. The action has been described as though the thin filaments were moved by a long series of rowers, the oars being the heads of the myosin. Alternatively, this mechanism may be compared with a system of ratchets. The results of this ratchet or rowing motion is to pull the filaments into greater overlap decreasing the distance between the Z lines and shortening the muscle (*Fig.* 31.9).

The most recent research using X-ray analysis has demonstrated that the heads of the myosin (S1 units) are 'comma'-shaped molecules. The broad head of the comma interacts with F actin near the groove between the two strands of actin monomers. Individual

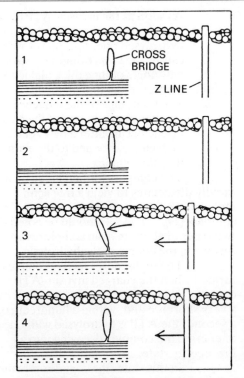

a

CROSS BRIDGE

Z LINE

1

2

3

4

Fig. 31.9. The cross bridge cycle.
(*a*) Movement of the thin filament as a result of binding of the myosin head to the actin thin filament, followed by a swivelling or twisting movement. (*b*) Movement of thin filaments closer together as a result of multiple action of cross bridge cycles. The closer proximity of the thin filaments should be compared with the gap shown in *Fig.* 31.8, which represents the situation before contraction of the filaments began.

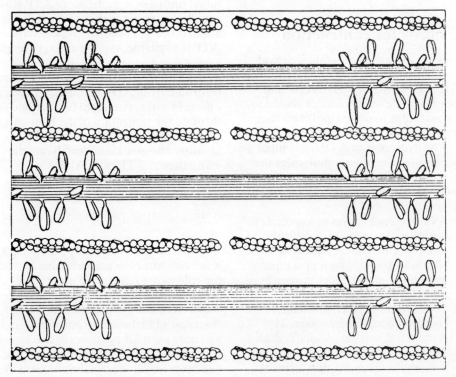

b

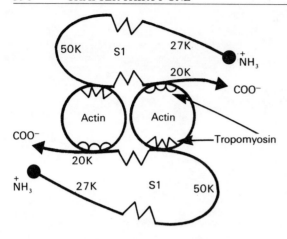

Fig. 31.10. **S1 (head) units of myosin straddling two actin strands and the regulatory protein tropomyosin.** The *K* values are $10^3 \times$ the molecular weight in daltons.

S1 units lie between, and interact with both strands of a filament, thus straddling the regulatory protein tropomyosin in the groove as shown in *Fig.* 31.10.

31.5 Energy for contraction

The energy for muscle contraction is provided by ATP and the essential sequence is shown in *Fig.* 31.11. ATP is hydrolysed to ADP and inorganic phosphate and its energy is used to cause a cross bridge to swivel, pulling a thin filament past a thick one to shorten the muscle.

The hydrolysis of ATP takes place on the head region of the myosin molecules that form the cross bridges projecting from the thick filaments. The splitting of ATP proceeds in the distinct steps listed below.

a.   ATP becomes bound to a special receptor site on the surface of the protein. The tendency for ATP to bind is so great that under normal conditions almost every myosin head is bound to ATP.

b.   The myosin–ATP complex alters the conformation of the protein. This protein is described as 'charged' and shows a great tendency to bind to an actin molecule in the thin filament.

c.   As soon as the myosin–ATP complex binds to actin, the ATP is rapidly hydrolysed releasing energy which causes the cross bridges to swivel to a new angle pulling thick filaments with respect to the thin filaments.

d.   The final stage of the sequence involving detachment of the cross bridge occurs after a new ATP molecule has bound to the actin–myosin complex.

The complex formed in the final step rapidly dissociates to yield a free actin molecule and an uncharged myosin–ATP complex (*Fig.* 31.11). It should be noted that two types of actin–myosin complexes are formed in the course of the hydrolysis.

The high-energy 'active complex' is formed when the charged myosin–ATP complex links with an actin molecule. It has a very short life and, within a hundredth of a second, the ATP is hydrolysed with release of energy, the complex decaying into a low-energy state.

In this state, the complex remains intact until a new molecule of ATP is available, usually occurring within a very short time, about 0·001 s, in normal muscle. When no ATP is available, the low-energy complex is very stable and is, in fact, the form occurring in rigor mortis. After hydrolysis ATP can be reformed from ADP in muscle in several different ways. An important method is through the utilization of the high-energy compound creatine phosphate, the enzyme creatine phosphokinase catalysing the formation of ATP from ADP

$$\text{ADP} + \text{Creatine phosphate} \overset{\text{Creatine phosphokinase}}{\rightleftharpoons} \text{Creatine} + \text{ADP}$$

The reaction was originally called the 'Lohman reaction', after the man who discovered it and the role of the reaction in muscle has posed a puzzle for many years. It was originally believed that it provided an emergency reservoir of high-energy phosphate and that it was only used for purposes of this type when other sources of ATP were unavailable, for example during the first few seconds of

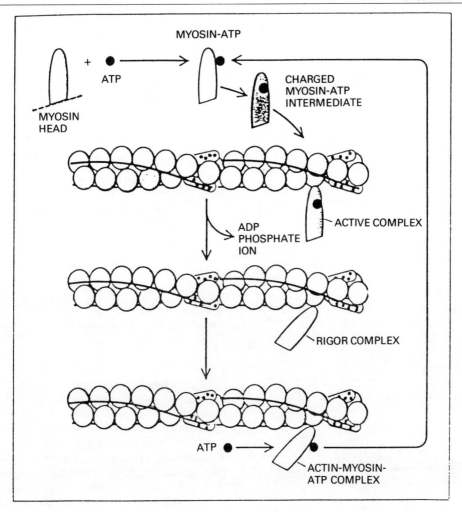

MYOSIN-ATP

ATP

MYOSIN
HEAD

CHARGED
MYOSIN-ATP
INTERMEDIATE

ACTIVE COMPLEX

ADP
PHOSPHATE
ION

RIGOR COMPLEX

ATP

ACTIN-MYOSIN-
ATP COMPLEX

Fig. 31.11. **The role of ATP in muscle contraction.**

strenuous exercise. More recently, however, experiments have shown that a large proportion of the energy may be made available directly from creatine phosphate. If this is the case then two forms of ATP may effectively exist, one form being used directly in muscle contraction and resynthesized by the Lohman reaction, and another form that could be synthesized during the course of metabolism and used to synthesize creatine phosphate, but not to provide energy for muscle directly. The two forms of ATP do not, of course, differ chemically but their different localizations within the muscle could have an important bearing on their function.

Muscle possesses the capacity to generate ATP under anaerobic conditions, utilizing its store of glycogen, and metabolizing this by the process of glycolysis. If no, or a limited, supply of oxygen is present then lactic acid will be produced (Appendix 15). A short period of strenuous exercise, for example running up a long flight of stairs, will cause a

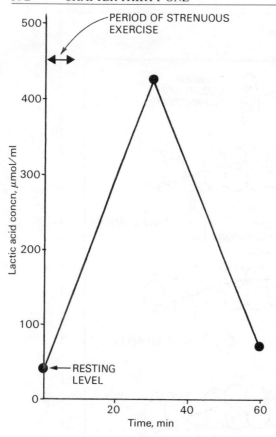

Fig. 31.12. **The effect of a short period of strenuous exercise on the concentration of lactic acid in urine.**

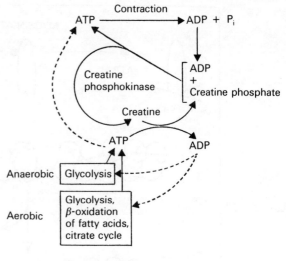

Fig. 31.13. **Synthesis of ATP in muscle.**
As described in the text, all the ATP produced metabolically may be used to regenerate creatine phosphate, but it is possible that some may be used directly.

large increase in the concentration of lactic acid in blood and urine (*Fig.* 31.12). During prolonged exercise the oxygenation of muscle gradually increases to a maximum and then

ATP may be produced by several different metabolic processes, glycolysis that will become aerobic, β-oxidation of fatty acids and by the citric acid cycle (Appendices 17–20).

The importance of fatty acid oxidation in the muscle is often overlooked, but it should be noted that muscle will oxidize fatty acids very efficiently. Measurements on human limbs have demonstrated that over 50 per cent of the energy for contraction can be supplied by plasma free fatty acids.

The different modes of ATP synthesis are shown in *Fig.* 31.13.

31.6 The role of calcium in muscle contraction

The important role that calcium plays in muscle contraction was one of the classic discoveries in physiology, made by Sydney Ringer over 100 years ago. This important discovery was, in fact, made by accident since Ringer used tap water for his initial experiments. His results were discussed in the *Journal of Physiology* (1882) from which comes the following extract: 'Concerning the influence extended by the constituents of blood on the contraction of the ventricle.
I discovered that the saline solution which I had been using had not been prepared with distilled water but with pipe water supplied by the New River Water Company. I tested the activity of the saline solution made with distilled water and did not get the effects described. It is obvious that the effects are due to some inorganic constituents of the pipe water.'

Extensive research during the ensuing 100 years has demonstrated that Ca^{2+} plays a very important role in the regulation of muscle contraction. The main action of the nerve impulse to muscle is to cause a release of Ca^{2+} from the T-tubules and sarcoplasmic reticulum into the sarcoplasm of the muscle fibres. Released Ca^{2+} rapidly becomes bound to troponin (*Fig.* 31.14).

The sequence of the cross bridge cycle sensitive to calcium is the linking of the charged form of the myosin–ATP complex to actin, that is the formation of the active complexes. As Ca^{2+} binds to troponin, the question arises as to how the binding of Ca^{2+} to troponin can regulate the binding of myosin–ATP complexes to actin. The problem is complicated since each troponin molecule is associated with several actin subunits and all subunits can interact with myosin heads.

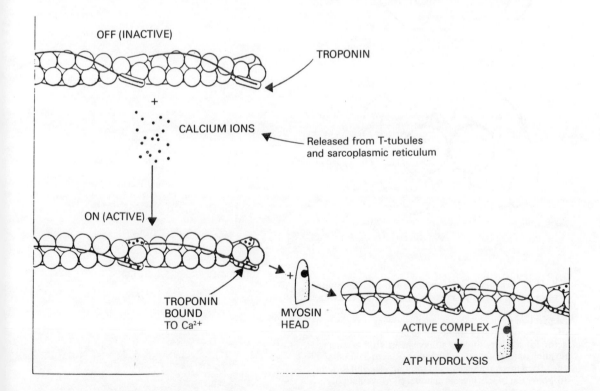

Fig. 31.14. **Binding of Ca^{2+} to troponin and the conversion of the muscle filament to an active state.**

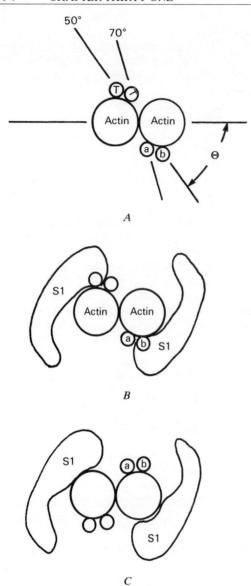

To solve this problem the 'site-blocking model' has been proposed and is illustrated in *Fig.* 31.15. This theory proposes that tropomyosin physically blocks the attachment of myosin to actin. Binding of Ca^{2+} to troponin molecules causes the tropomyosin to role into the groove between the peptide chains out of the way so that actin can gain access to myosin and contraction occurs.

Several modifications of this theory have been suggested but most research tends to support the concept that Ca^{2+} regulates, by means of troponin and tropomyosin, the access of myosin to actin.

Fig. 31.15. **Schematic representation of the relationship of myosin heads, tropomyosin and actin: the steric blocking model.**
(*A*) Section through a thin filament showing two positions of the tropomyosin (T) molecule (*a* and *b*) on strands of actin: (a) active position of tropomyosin; (b) blocking position of tropomyosin. (*B*) Attachment of S1 units (heads) of myosin molecules to tropomyosin and to actin: tropomyosin blocks myosin action. (*C*) Movement of tropomyosin to expose active '*a* site' and release of myosin heads from blocking '*b* site'.

Chapter 32   Bone and collagen: calcification

Soft tissues of vertebrates probably started to become hardened with minerals about 400 million years ago. Prior to that period, animals were formed with perishable backbones consisting of cartilage and muscle and such backbones are still found in sharks and lampreys today.

Calcium phosphate, the main mineral of the bones of higher animals, is however not universal throughout the animal kingdom and many invertebrates construct hard tissues of calcium carbonate; indeed many millions of generations of early invertebrates have formed the white cliffs clearly visible along many coastlines.

Although the concept of mineralization of bone and the consequent strengthening of the tissues is extremely important, it should not be overlooked that bone is a living tissue like liver and brain and that the calcium and phosphate of bone are in dynamic equilibrium with the calcium and phosphate of plasma. The very large body mass of calcium phosphate in bone makes the tissue a very important reservoir of both calcium and phosphate. It should also be appreciated that bone salt is not a random precipitate, but part of complex subcellular structures. The formation of bone salt, or calcification, is clearly of great importance during growth and during repair after accidental damage to the skeleton. Under certain pathological conditions, calcification can also occur in several other tissues, such as the kidneys or arteries, where it can cause serious damage. An understanding of the process of calcification is, therefore, of great value in medicine, but details of the mechanisms of the process are still obscure, despite a great deal of research spanning many years.

32.1 Bone structure

Bone consists of inorganic salts, mainly calcium phosphate, proteins of which collagen is the most important, proteoglycans, and a smaller fraction of lipids and water.

The dense mineral particles are generally of the same width, about 5 nm in the form of long cylinders overlaid in the direction of the collagen fibres. Some mineral cylinders may aggregate to form much thicker cylinders, with a width of approximately 20 nm. In sections of bone, lamellar, trabecular and Haversian structures can be distinguished in different parts of the bone (*see Fig.* 32.1).

Three types of cell have been described in bone: osteoblasts, osteoclasts, and osteocytes. The osteoblasts are involved in the formation of the matrix of new bone; they are migratory cells that synthesize webs of collagen fibres and proteoglycans that form the matrix to be eventually calcified. This new matrix is termed the 'osteoid tissue'. It is gradually permeated by cells, described as 'osteocytes' which possess long filamentous projections and which interlace the osteoid tissue. They are initially formed from osteoblasts (*Fig.* 32.2). The osteoclasts which are large multinucleated cells, are mainly concentrated on the surface of the bone. Their main function appears to be the resorption of bone and the release of calcium and phosphate into the plasma.

A diagrammatic representation of the cells making up the structure of bone is shown in *Fig.* 32.3.

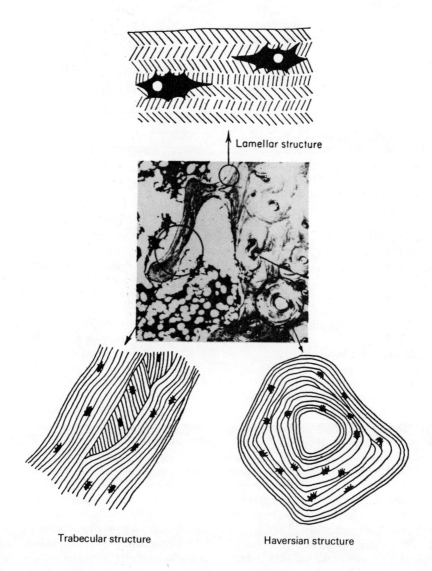

Lamellar structure

Trabecular structure

Haversian structure

Fig. 32.1. **Structures of bone.**

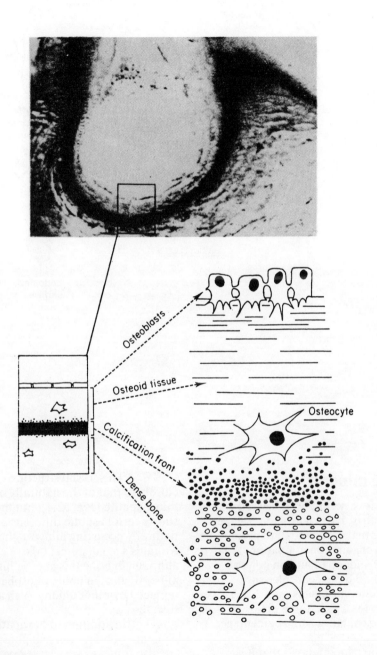

Fig. 32.2.   **Bone formation.**

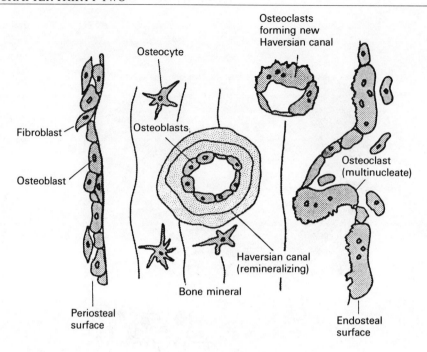

Fig. 32.3. **Cells in bone structure.**

32.2 Bone mineral

Bone mineral is very closely associated with the organic matrix, particularly the collagen fibres, and this makes it difficult to prepare the mineral free from any contamination. Analysis shows that the mineral of human bone contains: 19·3–25·8 per cent calcium, 9·0–11·7 per cent phosphates, and 2·3–3·7 per cent carbonate (as CO_2). Some bones are reported to contain a small percentage (0·22–0·34 per cent) of magnesium. X-ray analysis of bone mineral shows that it is composed of narrow crystallites about 5 nm in diameter but of indeterminate length, probably about 60–70 nm, arranged parallel to the collagen fibre axes.

The crystal structure corresponds closely to that of the naturally occurring fluoroapatite (*Fig.* 32.4), but the bone salt shows certain subtle differences from the naturally occurring mineral: the mineral contains a small percentage of carbonate, although in bone this may be higher than in the mineral; also, in bone, phosphate may be replaced by other anions, such as sulphate or silicates.

In addition to the apatite crystal structure of bone, calcium phosphate deposits are also found as very large crystals on external surfaces, such as dentine. Non-crystalline deposits occur additionally within other tissues, often within certain subcellular organelles, such as the mitochondria.

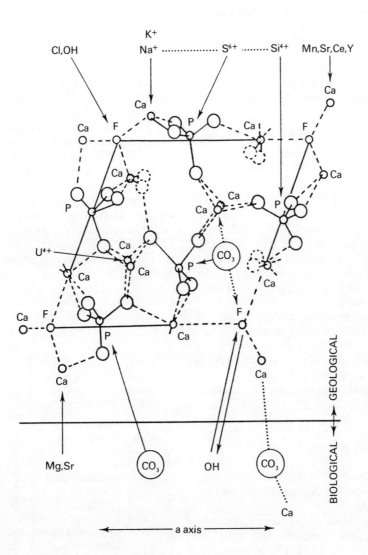

Fig. 32.4. **Comparison of geological and biological apatites.**
Note that the carbonate concentration is increased in bone and that Ca^{2+} may be
replaced by other metals such as Mg^{2+} or Sr^{2+}.

32.3 Precipitation of calcium phosphate

Calcium phosphate, the main constituent of the bone mineral, is relatively insoluble in water. The precipitation of calcium phosphate, like other relatively insoluble salts, is regulated by the law of mass action. A constant may be derived which is described as the 'solubility product' and this may be represented by a general equation for a typical relatively insoluble salt, MA:

$$M^+ + A^- \leftrightharpoons MA$$
$$\text{Insoluble}$$
$$\text{precipitate}$$

$$\frac{[M^+][A^-]}{[MA]} = K$$

The concentration of the solid, [MA], is constant and thus a new constant may be defined $K_{sp} = [M^+][A^-]$. K_{sp} is the solubility product and, if the product $[M^+][A^-]$ exceeds K_{sp}, we may predict that precipitation of the solid MA will occur.

An important application of the principle is that the solubility product can be exceeded, and precipitation occur, if the concentration of *either* of the ions M^+ or A^- is increased, i.e. increase in $[M^+]$ or $[A^-]$.

For calcium phosphate the equation is represented by

$$3Ca^{2+} + 2PO_4^{3-} \leftrightharpoons Ca_3(PO_4)_2$$

and the solubility product is

$$K_{sp} = [Ca^{2+}]^3[PO_4^{3-}]^2$$

The actual value is pH dependent and, at pH 7·0, is 25.

Precipitation of calcium phosphate can, therefore, occur spontaneously if the solubility product is exceeded by increases in the concentration of calcium ions, phosphate ions or both and many earlier theories of calcification were based on this concept. Some measurements of calcium and phosphate concentrations under which calcium phosphate deposits do not, however, relate closely to the solubility product of calcium phosphate $[Ca_3(PO_4)_2]$, but more closely to the K_{sp} of calcium hydrogen phosphate $CaHPO_4$.

$$Ca^{2+} + HPO_4^{2-} \leftrightharpoons CaHPO_4$$

For calcium hydrogen phosphate:

$$K_{sp} = [Ca^{2+}][HPO_4^{2-}] = 5·3–6·6.$$

Although, in the regulation of bone and tissue mineralization, the solubility product principle may be important, it must be remembered that application of even the simpler chemical theories is not always easy or accurate in biological systems.

Calcium and phosphate concentrations in the aqueous medium may be altered by binding to proteins, and complex localized concentration changes may occur as a result of metabolic processes. Furthermore, the 'micro-environmental' concentrations of calcium and phosphate may not be identical to those in the bulk of the solution. Some of the effects of the biological environment on precipitation of calcium phosphate are shown in *Fig.* 32.5.

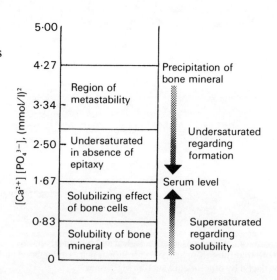

Fig. 32.5. **Precipitation of calcium phosphate and factors affecting it.**

32.4 Collagen

Collagen is the most abundant protein in the mammalian body and also forms the extracellular framework for all multicellular animals. The strong fibres of collagen form the main structure of bone and cartilage, but collagen also forms vital parts of the structure of woven sheets like those that occur in skin and in filtration membranes of the glomerulus.

All collagen molecules are formed from a triple helix: this structure is unique to collagen and formed from a coiled coil of thick polypeptide subunits and chains (*Fig.* 32.6). Each α chain twists in a left-handed helix with three amino acid residues per turn and the three chains are wound in a right-handed superhelix to form a rod-like molecule almost 1·4 nm in diameter (*Fig.* 32.6).

At least five different types of collagen are known: the type forming about 90 per cent of collagen in the human body is Type I, the chains of which are described as $[\alpha 1(I)]_2\alpha 2$. In this form of collagen, the chains each contain about 1050 amino acid residues and the molecule is 300 nm in length.

The amino acid sequence of collagen is extremely unusual, being composed of triplets of glycine + two other amino acids:

$$—(Gly—X—Y)—(Gly—X—Y)—$$

The large proportion of glycine is essential to the structure, because it is the only amino acid small enough to fit in the central core of the molecule. Ends of the collagen molecule are not, however, in the triple helix form and they lack glycine. Proline and hydroxyproline together account for about one-third of the X–Y positions so that they form 25 per cent of the total amino acid residues.

The high concentration of hydroxyproline is very unusual in proteins in general and collagen is unique in this respect. The hydroxyl group of proline is now known to be essential for stabilization of the triple helix, probably by intrachain hydrogen bonds bridging through water molecules. About 90 hydroxyproline residues are needed per chain to preserve the triple helix at body temperature, 37 °C and, if the biosynthesis of hydroxyproline is impaired, unstable collagen is formed in which the chains tend to unwind.

Most of the hydroxyproline is 4-hydroxyproline, although a small proportion of 3-hydroxyproline residues are also found in the collagen molecules. Type I collagen contains only one or two residues of 3-hydroxyproline near the carboxyl end of the chain, but other types of collagens may contain 10 per cent of the total hydroxyproline content as the C-3 derivative.

Lysyl residues are also hydroxylated and, subsequently, chains of glucose or glucose–galactose are coupled to the hydroxyl groups. The side chains of lysyl groups containing free —NH_3^+ groups are also oxidized to aldehydes, which form Schiff bases with adjacent chains producing stable cross-links.

The function of all these side-chain modifications are unknown, but 4-hydroxyproline stabilizes the helix of the molecule and hydroxylysine is essential for sugar attachment; the carbohydrate chain may

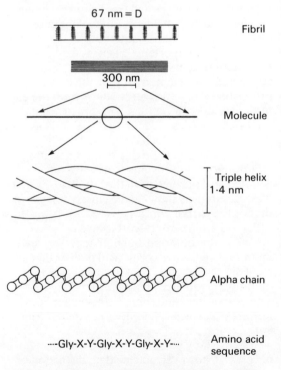

67 nm = D

Fibril

300 nm

Molecule

Triple helix 1·4 nm

Alpha chain

····Gly-X-Y-Gly-X-Y-Gly-X-Y-····   Amino acid sequence

Fig. 32.6. **Basic features of the collagen molecule and its relation to the structure of the fibril.**

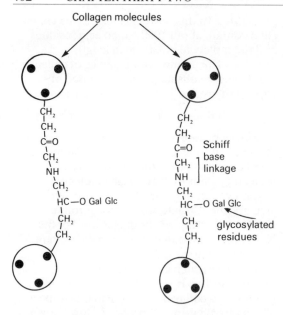

Collagen molecules

Schiff base linkage

glycosylated residues

Fig. 32.7. **Cross-linking of collagen chains.**
Formation of Schiff base type linkages between aldehyde groups of one chain and the lysyl amino groups of another chain.

be effective in water retention and also in cross-link formation. A summary of the modifications and their roles is indicated in *Fig.* 32.7.

32.5 Biosynthesis of collagen and the role of ascorbic acid (vitamin C)

The disease described as 'scurvy' became prominent amongst seafarers during the fifteenth and sixteenth centuries. During this period large numbers of ships set sail from Europe on long voyages of exploration and many of the sailors on them were, after a period at sea, struck down with this serious disease, 'scurvy', characterized by a large range of symptoms and ultimately fatal (*see Table* 32.1). The seriousness of the disease can be judged from reports of the voyages during this period. In Vasco de Gama's voyage around the Cape of Good Hope in 1498, 100 of his 160 crew died from scurvy and nearly 100 years later in 1593, Sir Richard Hawkins

Table 32.1 **Symptoms of scurvy**

Tendency to bleed
Gums become soft and spongy—teeth become loose
Bleeding into joints and under skin with formation of petechia
Bones become weak
Wound healing is poor
Anaemia
Infections develop
DEATH

reported records of 10 000 seamen dying from scurvy. In addition to the diseases of seamen, records showed that land-dwellers often suffered, particularly in winter.

The reason for the widespread incidence of the disease was the poor rations carried on these voyages and, in particular, the lack of fresh fruit and vegetables; but it was not until 1747 that this fact was clearly recognized by a British Naval Surgeon, James Lind. He showed that it was possible to cure scurvy by the incredibly simple treatment of feeding patients two oranges and one lemon each day, and he reported an almost miraculous cure after one week of this treatment.

The importance of citrus fruit for the treatment of scurvy was soon widely appreciated and lemon, or lime juice, was prescribed daily for British sailors from the end of the eighteenth century. This practice led the Americans to describe the British sailors as 'Limeys', a term which is still used today.

The active principle in orange and lemon juice that prevents scurvy was not isolated and purified until the late 1920s, when its structure was shown to be closely related to that of the carbohydrates (*Fig.* 32.8). It was called 'vitamin C' or 'ascorbic acid'.

It is of interest to note that very few mammals require ascorbic acid in their diet and most synthesize their own from glucose, the metabolic sequence being shown in *Fig.* 32.8. The guinea-pig, and some of the higher apes, require ascorbic acid, as does the Indian fruit bat.

Many years of research were necessary, however, to elucidate the action of ascorbic acid; when guinea-pigs deficient in ascorbic acid were studied it became clear that

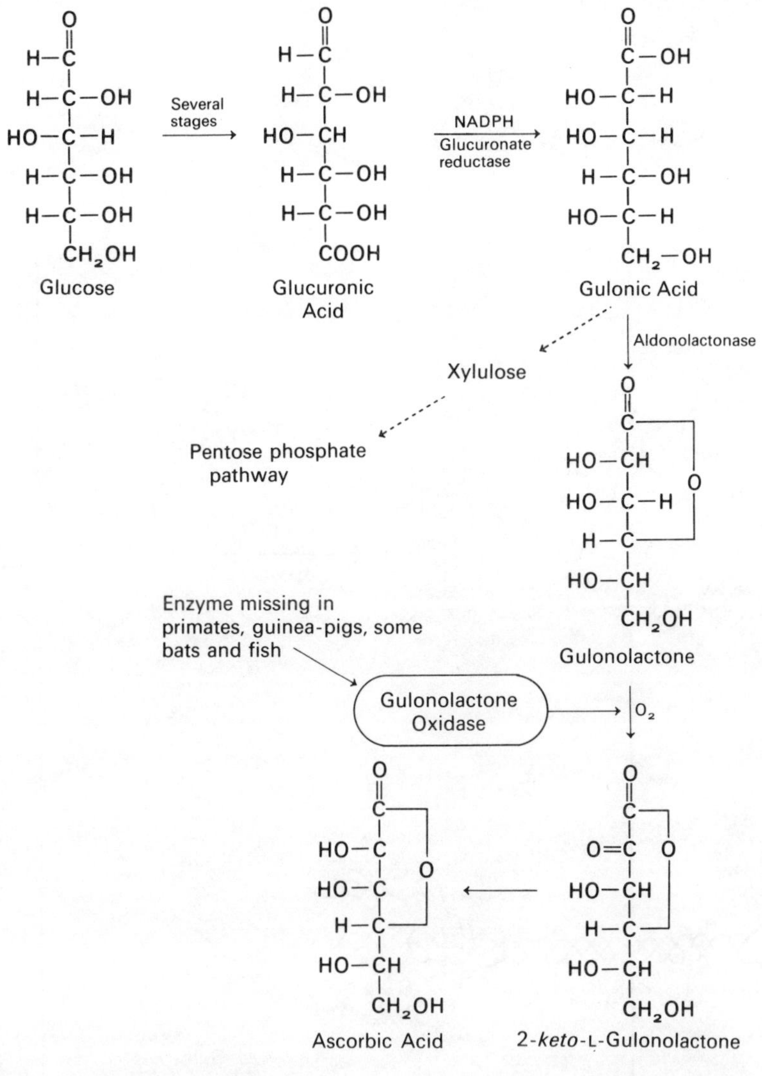

Fig. 32.8.   **Genetic defect in ascorbic acid synthesis.**

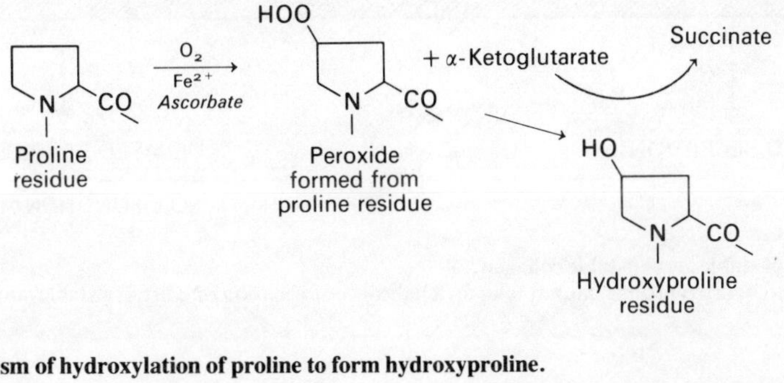

Fig. 32.9.   **Mechanism of hydroxylation of proline to form hydroxyproline.**

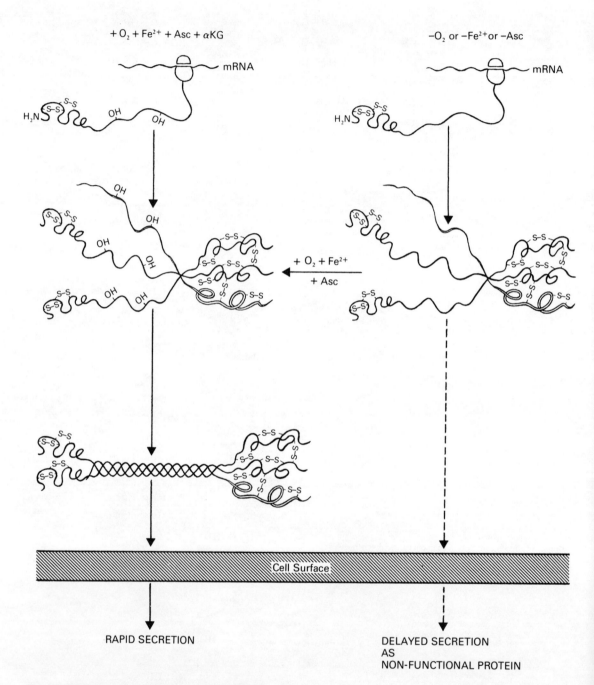

Fig. 32.10. **Formation of stable and unstable collagen.**
'Collagen' containing no hydroxyproline cannot take up a helical conformation or exist in a stable state.

they were suffering from an inability to synthesize collagen efficiently. Furthermore, many of the symptoms of scurvy in man, such as the weak blood vessels and poor wound healing, appear to be a result of ineffective synthesis of collagen. It, therefore, became apparent that ascorbic acid played a very important role in collagen biosynthesis.

This role has, in recent years, been finally established and ascorbic acid has been shown to be intimately involved in the biosynthesis of hydroxyproline. The chains of procollagen are initially synthesized on the ribosome and these chains contain glycine, proline and all the other amino acids in the correct sequence, but no hydroxyproline residues. In fact there is no tRNA molecule which can present the hydroxyproline residue to the ribosomes. As the peptide is formed, certain of the proline residues, but not all, are hydroxylated by a proline hydroxylase enzyme which also requires iron, oxygen and ascorbic

acid. During the process, α-ketoglutarate is oxidized to succinate (*see Fig.* 32.9). If oxygen, iron or ascorbate is absent, a type of collagen is formed that cannot take up a helical conformation and is unstable (*see Fig.* 32.10). The role of ascorbic acid is that of a reducing agent, supplying electrons for conversion of oxygen to a form active in the hydroxylation process. In parenthesis, it is important to note that although the function of ascorbic acid in the synthesis of hydroxyproline is important, it is not its only role. Ascorbic acid is likely to be involved in other hydroxylation reactions requiring active oxygen.

Three enzymes are involved in the process of hydroxylation, forming 4-hydroxyproline, 3-hydroxyproline or hydroxylysine. The enzymes are specific for proline or lysine residues located in a peptide chain and occupying correct positions in the amino acid sequence. Furthermore, the protein must be in a non-helical conformation

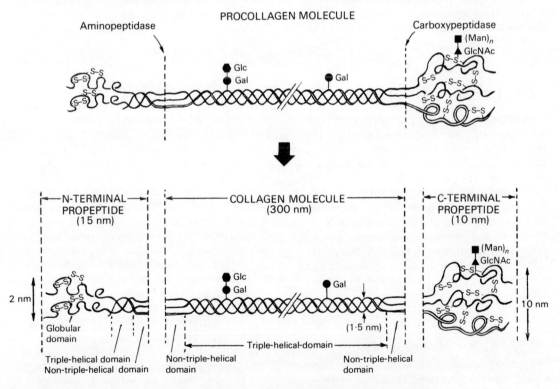

Fig. 32.11. **The conversion of procollagen to collagen by carboxypeptidase and aminopeptidase.**
Glc, glucose; Gal, galactose; Man, mannose; GlcNAc, *N*-acetylglucosamine.

for the enzyme to be effective. The 4-hydroxylase attacks only those proline residues which are in the 'Y' position, whereas the 3-hydroxylase hydroxylates the proline residues in the 'X' position, in the —(Gly—X—Y)$_n$— sequence.

As the lysyl residues are hydroxylated, sugar residues are added to the residues. Glycosylation of lysine is catalysed by a glucosyl and a galactosyl transferase and the protein must be in a non-linear form for the addition of carbohydrate residues to occur. This protein, described as 'procollagen', is assembled in the rough endoplasmic reticulum and passes through the Golgi apparatus before leaving the cell.

After the procollagen has been formed, it must be attacked by two enzymes outside the cells: an aminopeptidase which splits a peptide from the amino-terminal part of the molecule and a carboxypeptidase which splits a peptide from the carboxyl-terminal end of the molecule. When these peptides have been split, spontaneous assembly into the collagen occurs (*Fig.* 32.11).

32.6 Calcification of bone

Calcification of bone is preceded by the formation of a web of collagen fibres and polysaccharides by the osteoblast cells. These cells are at first migratory, but gradually they become differentiated to form sheets of cells producing the collagen, and polysaccharide calcification then proceeds within the matrix; this may, at first, be initiated in the cells themselves. The precise mechanism is still not completely understood, and it is, therefore, desirable to review the various theories and experimental evidence on which they are based.

a. Robison ester theory

As long ago as 1923, Robison discovered that calcifying bone contained relatively high concentrations of a phosphatase. This phosphatase, unlike the phosphatase of lysosomes, was most active at a very alkaline pH and was, therefore, described as 'alkaline phosphatase'. Robison proposed that this enzyme would hydrolyse phosphate from phosphate esters, such as those of glucose, to increase the concentration of phosphate in the calcifying zone. Increased phosphate concentration would, as shown in Section 32.3, cause the solubility product of calcium phosphate to be exceeded and the deposition of calcium phosphate to occur. This was an attractive theory which held sway for many years but, unfortunately, several serious objections made the simple theory untenable. Firstly, the optimum pH of the enzyme, pH 9·5–10·0, is far greater than that likely to occur in any tissue fluids; secondly, the concentrations of phosphate esters available for hydrolysis are too low to permit appreciable calcification; thirdly, the alkaline phosphatase is present in the mesenchyme prior to differentiation of the osteoblasts. Furthermore, alkaline phosphates may be demonstrated in many cells which do not calcify.

b. Epitaxic mechanisms

Bone mineral is composed of small crystallites of regular size oriented with respect to collagen. It has, therefore, been suggested that crystals of apatite can form only where a site is available which has the correct geometry for initiation of the formation of a lattice of fixed dimensions. In support of this theory, it can be demonstrated that decalcified bone can initiate crystallization of calcium phosphate at concentrations below those required for spontaneous precipitation. This theory is consistent with the equilibrium conditions present between serum fluids and bone salt and with the regular arrays of crystals on collagen fibres.

Criticism of the role of epitactical mechanisms is based on the fact that precipitation of calcium phosphate can occur in other fibres, such as keratin, and on the fact that the orientation of crystals is also shown when calcium phosphate deposition is studied *in vitro*.

c. The role of globules and vesicles

It has also been observed that calcifying bone contains large numbers of extracellular membrane-bound 'globules' or 'vesicles'. Calcification appears to occur within these vesicles and several enzymes, including alkaline phosphatase, are present; the phosphatase is believed to be a glycoprotein and to bind calcium. The role of the vesicles is as yet unclear but they may be a means of concentrating calcium phosphate outside the cells before deposition on the epitactic sites of the collagen fibres.

d. Calcification within the cells: the role of the mitochondria

Many of the original investigations of bone mineral deposition tended to consider that calcification occurred outside the cells, in the extracellular spaces. More recently, however, it has been suggested that the initial deposition

of the bone salt may occur within the cells. If this is so, then extrusion of the calcium phosphate, in an encapsulated form, into the extracellular space could subsequently occur.

Evidence that the cells may play an important role in calcium phosphate deposition has been provided by the work of Lehninger on calcium uptake by mitochondria of many tissues. When suspensions of mitochondria from several different tissues are incubated in a medium containing Ca^{2+} in which they can actively respire, they accumulate large amounts of Ca^{2+} from the medium, up to several hundred times the initial calcium content of the mitochondria. If phosphate and ATP are present, the calcium and phosphate accumulate in the mitochondria in a ratio of $1\cdot7-2\cdot0$ atoms of calcium for each phosphate ion. This can lead to a massive loading, up to 3000 nmol Ca^{2+} per mg protein, of granules of calcium phosphate in the mitochondria. The granules can be seen in the electron

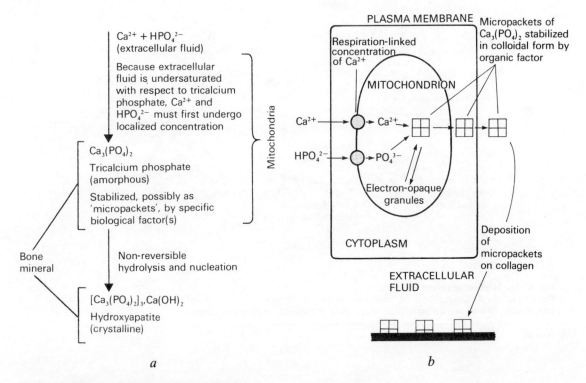

Fig. 32.12. **The role of mitochondria in calcification.**
(*a*) The formation of crystalline hydroxyapatite; (*b*) the role of the mitochondria.

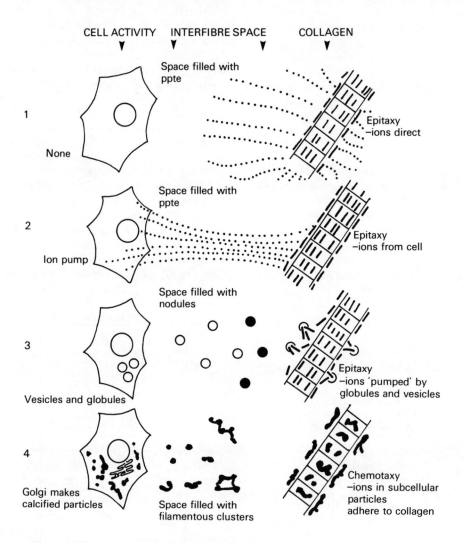

Fig. 32.13. **Mechanisms of calcification.**
Possible inter-relationships of cells, interfibre spaces and collagen fibres. 1, 2, 3 and
4 represent possible mechanisms. Note that if the mitochondria play an important
role, as proposed in *Fig.* 32.12, the mechanism 3 is the most likely to represent the
system.

microscope and the calcium phosphate content can increase the dry weight of the mitochondria by 25 per cent. The calcium phosphate initially laid down is amorphous and for bone salt formation it must be subsequently converted to the crystalline hydroxyapatite.

The concentrations of Ca^{2+} and phosphate in blood plasma and intestinal fluid are normally undersaturated with respect to amorphous calcium phosphate, but supersaturated with respect to hydroxyapatite.

Lehninger proposed that the role of the mitochondria was to concentrate the calcium and phosphate ions from the extracellular fluid into the mitochondria (forming 'micropackets'); the solubility product of calcium phosphate was then exceeded and deposition of amorphous calcium phosphate occurred. These micropackets of calcium phosphate could then be transported out of the mitochondria, leave the cell, possibly by reversed pinocytosis, and form the vesicles or granules described above. The vesicles would then diffuse to the collagen where the epitactic sites would cause deposits of the granules, eventually forming crystalline hydroxyapatites.

This theory that mitochondria could play an important role in calcification of bone is not completely and finally verified, but it does help to link a large number of biological and biochemical observations and the essential points are summarized in *Fig.* 32.12. A summary of the possible theories of bone calcification showing the relationship of the cells to the collagen fibres is shown in *Fig.* 32.13.

32.7 Resorption of bone

Resorption of the bone mineral is necessary for two quite separate reasons: it is required during new formation, or remodelling, of bones to ensure that the correct structure and shape is maintained and, secondly, bone resorption is essential to supply calcium to the plasma wherever this may be depleted. Typical conditions under which this may occur are in diets deficient in calcium and vitamin D or in pregnancy. As discussed in Chapter 22, the maintenance of the blood Ca^{2+} concentration is of vital importance.

Mature bone is demineralized by destruction of the hard matrix by special multinucleated cells, the osteoblasts. These cells possess powerful proteolytic enzymes within their lysosomes, such as collagenase, which hydrolyse the collagen fibres releasing the bone mineral.

The osteoblasts are not the only cells involved in bone resorption; osteocytes are also involved. Histological examination of resorbing bone has shown that three distinct processes may occur: these are described as 'feathering', i.e. the reduction of mineral in regions of the bone without destruction of the organic matrix, 'osteocytic' osteolysis, i.e. the enlargement of the lacunae in which the osteocytes live by demineralization of the cavities, and 'autoclasis', i.e. the fragmentation and dispersion of specific regions of the bone and possibly of the bone matrix itself as a result of osteocytic activity. The three processes are shown diagrammatically in *Fig.* 32.14.

The role of the bone as a reservoir of calcium and phosphate may be mediated by a surface membrane of cells surrounding the bone. Although not universally accepted, the concept is termed the 'membrane hypothesis' and is supported by much direct and circumstantial evidence.

This theory is illustrated in *Figs.* 32.15 and 32.16. It will be noted that the bone is believed to be covered with a sheath of external cells connected by fluid channels to the osteocyte cells within the mineralized tissue. Calcium and phosphate can be transported, possibly as amorphous calcium phosphate or as calcium and phosphate ions, to the oestocytes within the bone for mineralization to occur. Alternatively, the calcium and phosphate of the lining cells is available for transport into the plasma when the need arises (*Fig.* 32.16).

Strong support for a mechanism of this type is provided by the action of the parathyroid hormone which is known to release calcium and phosphate into the plasma from the bone (cf. Chapter 22). This hormone stimulates the activity of oesteoblast cells which increase in size, divide and show greatly

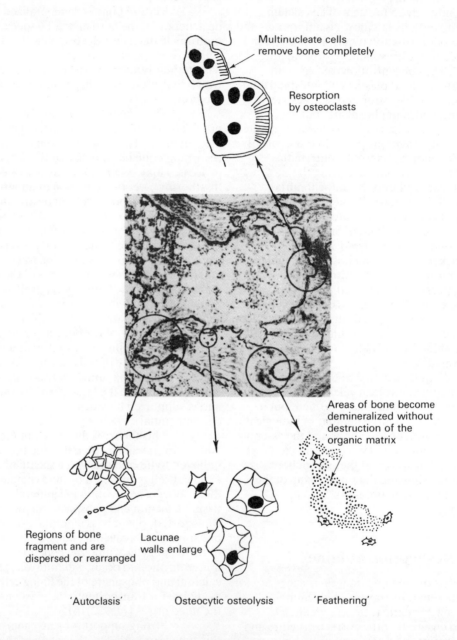

Multinucleate cells
remove bone completely

Resorption
by osteoclasts

Areas of bone become
demineralized without
destruction of the
organic matrix

Regions of bone
fragment and are
dispersed or rearranged

Lacunae
walls enlarge

'Autoclasis'   Osteocytic osteolysis   'Feathering'

Fig. 32.14. **Histological observations of the processes involved in bone resorption.**

enhanced hydrolytic enzyme activity. These cells cause bone resorption but the process is slow, requiring 2–3 hours to become fully active. Parathyroid hormone does, however, cause a rapid release of calcium into the plasma which can be detected a few minutes after injection of the hormone and, therefore, osteoblast activation must involve a different mechanism. It is likely that, initially, parathyroid hormone causes a release of Ca^{2+} from the bone membrane cells, probably by stimulating its release from the mitochondria and by increasing the permeability of the cell membrane to Ca^{2+} (*Fig.* 32.16).

Parathyroid hormone, therefore, acts by a two stage process: firstly it stimulates release of calcium from the surface membrane cells and secondly, if more Ca^{2+} is still urgently required, it stimulates the osteoclasts to degrade the bone mineral to release both Ca^{2+} and phosphate into the plasma.

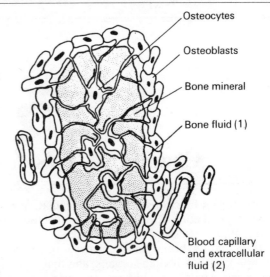

Fig. 32.15. **Diagrammatic representation of bone with its surrounding cell membrane.**

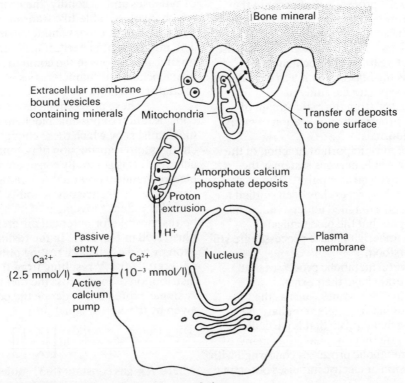

Fig. 32.16. **The role of bone membrane cells in calcium regulation.**
Note that Ca^{2+} may be released rapidly from the cell cytoplasm and then from storage in the mitochondria. Release from bone mineral is a slower subsequent process. During Ca^{2+} deposition the processes would proceed in the reverse direction.

Chapter 33 The brain and the central nervous system

The complex functions of the brain regulating a vast range of control mechanisms, intellectual thought processes and emotions make it certain that the brain's biochemistry is the most complex of that of all the organs, and that many years will elapse before a complete understanding is reached.

Progress has, however, been made in the knowledge of brain biochemistry and many of the metabolic processes that occur in the brain have now been studied and accurately described: the subject is usually called 'neurochemistry'. Exactly how this term arose as an alternative to 'neurobiochemistry' is obscure, as is the reason why it was considered necessary to separate the study of 'neurochemistry' from the main stream of biochemical progress by publication of many specialized journals.

The most important function of the brain and the whole nervous system is the transmission of electric impulses. The physiology of this process has been studied for many years and a detailed understanding has been developed, but the biochemical mechanisms underlying these processes are still poorly understood.

Several metabolic processes that occur in the brain have their exact counterpart in other tissues, such as the liver or kidney, but in this chapter emphasis is placed on the metabolism that is particularly relevant to brain function, including many specialized metabolic processes concerned with the transmission of electric impulses occurring in the brain.

33.1 Excitation and conduction

Excitation is defined as the ability of a cell or organelle to respond to stimuli: the response to a physicochemical stimulus is itself manifested by a reversible change in the membrane. Two stages of excitation are recognized: firstly, the initiation which takes place within the neurone and involves the summation of large numbers of synapses and, secondly, the propagation which involves cable-like transmission along a nerve axon or other conductive medium. The initial event in excitation is very likely to involve alterations in the configuration of physicochemical characteristics of the excitatory membrane.

These phenomena naturally lead to consideration of possible biochemical mechanisms by which these configurational changes and transmission of potential might be achieved. It is generally considered that the electrochemical potential is maintained by variation in selective permeability for K^+ influx and Na^+ efflux. Movement of ions would then be responsible for potential differences as illustrated in *Fig.* 33.1. In the resting state, the axon or nerve cell has a resting potential (E_R) of about -90 mV resulting from the unequal distribution of ions across the membrane. A single expression to define the potential is given by the Nernst equation:

$$E_R = \frac{RT}{F} \ln \left[\frac{[K^+]_i}{[K^+]_o} \right]$$

where R = gas constant (8·31 joules per kelvin mole or 8·31 JK^{-1} mol^{-1})

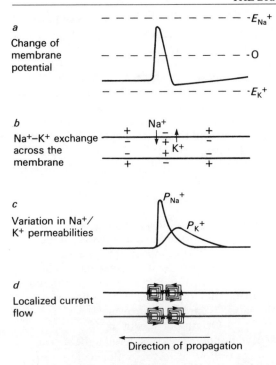

a

Change of membrane potential

b

Na⁺–K⁺ exchange across the membrane

c

Variation in Na⁺/K⁺ permeabilities

d

Localized current flow

Direction of propagation

Fig. 33.1. **Propagation of the nerve impulse along an axon.**

T = temperature in kelvins (K)
F = Faradays constant (96 500 coulombs per mole or 96 500 C mol⁻¹)

This expression gives a reasonable approximation of the potential, but to obtain an accurate value a much more complex expression incorporating Na^+ and Cl^- concentrations must be employed.

Although these ion movements can explain the potential changes and transmission of the impulse, the biochemical problem is much more difficult. It is necessary to try and explain exactly how the membrane is altered in order that differential permeability to Na^+ and K^+ is achieved. Many theories have been proposed which can be based on the following ideas.

a. Alterations in conformation of the structural proteins of the membranes, possibly by combination of the proteins with divalent cations, such as Ca^{2+}

b. Modification of the state or charge of membrane lipids

c. Changes in the pore size of the membranes

d. Changes in the charge or configuration of membrane glycoproteins

e. Initiation of a chemical reaction within the membranes, probably enzyme catalysed

f. Initiation of an electrochemical process controlling ion conductance.

Despite a great deal of investigation, the mechanisms of excitation and transmission are still obscure. The ionic hypothesis appears to explain the electrical measurements during nerve excitation, but it does not provide any insight into the biochemical mechanisms involved. In consideration of possible mechanisms, however, certain points should be noted: calcium ions play an important role in the maintenance of excitability of nerves and Ca^{2+} is likely to regulate the Na^+ and K^+ conductance during excitation; ATP may also play an important role because Na^+–K^+ exchanges through many different membranes in cells are known to be dependent on membrane ATPases (cf. Chapters 2, 25).

Some ingenious theories have linked Ca^{2+}, ATP and phospholipids; it has been suggested that Ca^{2+}–ATP–phospholipid complexes could form important membrane constituents. The initial chemical event in excitation is then believed to be the displacement of Ca^{2+}, followed by release of ATP which, in turn, is hydrolysed by ATPase to provide the energy needed for ion displacement. However, experimental attempts to demonstrate the requirement for ATP during nervous concentration have, so far, given equivocal results.

33.2 Chemical transmission and transmitters

It is now considered likely that the majority of communications that take place between receptor and neurone, neurone and neurone, and neurone and effector probably involve

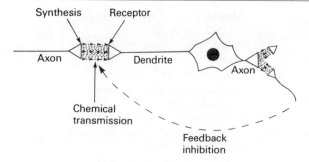

Fig. 33.2. **Chemical transmission.**

presynaptic liberation of chemicals that either stimulate or inhibit specialized regions of membranes (*Fig*. 33.2). It should be noted that these chemicals may have either a stimulatory or inhibitory effect on the receptors so that feedback control is possible as illustrated in *Fig*. 33.2.

A large group of metabolites are now known that can act in either a stimulatory or inhibitory manner: one of the best known is acetylcholine, but, in addition, catecholamines, serotonin and several amino acids are also involved. All transmitter molecules normally proceed through a set sequence: they are synthesized, released, taken-up by receptors to effect their action, and then destroyed.

a. Acetylcholine

Acetylcholine is synthesized by the enzyme 'choline acetylase' from choline and acetyl-SCoA. Acetyl-SCoA can be formed from pyruvate or from acetate in the presence of ATP:

$$
\begin{cases}
\text{ATP} + \text{Acetate} \rightarrow \text{Adenylacetate} + \text{Pyrophosphate} \\
\text{Adenylacetate} + \text{CoASH} \rightarrow \text{Acetyl-SCoA} + \text{AMP}
\end{cases}
$$

$$\text{Acetyl-SCoA} + \text{Choline} \xrightarrow[\text{acetylase}]{\text{Choline}} \text{Acetylcholine} + \text{CoASH}$$

The enzyme involved has been shown to be located in neurofilaments, synaptic vesicles and on the neuronal membrane. At the nerve endings, acetylcholine is released almost explosively so that storage, probably in synaptic vesicles after synthesis, is likely.

Immediately following interaction with the receptor, acetylcholine has to be destroyed and this is achieved by the enzyme acetylcholinesterase:

$$\text{Acetylcholine} + \text{H}_2\text{O} \xrightarrow{\text{Acetylcholinesterase}} \text{Choline} + \text{Acetate}$$

Cholinesterases of different degrees of specificity are found in many tissues of the body, in addition to those of the nervous system, and include the plasma and red blood cells. This function is believed to be protective, ensuring that any acetylcholine released into the blood is rapidly destroyed.

Originally 'receptors' for acetylcholine were considered to be special 'patches' on the surface of the receptor cell, but it is now clear that receptors are likely to be protein molecules built, in an ordered fashion, into the excitable membranes.

An overall view of the synthesis and utilization of acetylcholine at the nerve endings is shown in *Fig*. 33.3.

b. Catecholamines and serotonin

Using histochemical fluorescence methods, it has been possible to localize tracts in the brain that specifically utilize serotonin (5-hydroxytryptamine), dopamine (3,4-dihydroxyphenylethylamine) or noradrenaline as transmitters (*Fig*. 33.4).

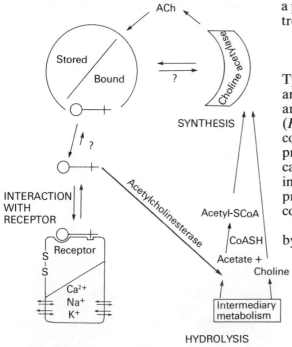

Fig. 33.3.   **Synthesis and hydrolysis of acetylcholine (ACh).**

The cell bodies of the noradrenaline neurones occur in ten or more clusters in the medulla oblongata, pons and mid-brain and they are mostly packed in the locus, the coerulus. Axons from those cells in the medulla oblongata descend in the lateral columns of the spinal cord, while others ascend into the medial forebrain bundle.

The cell bodies of the serotonin-containing neurones are localized in a series of nuclei in the lower membrane and upper pons and are called 'raphe' nuclei. Their axons also ascend in the medial forebrain bundle.

The dopaminergic tracts are more circumscribed than those of the two other types. The largest tract originates in the zona compacta of the substantia nigra, while other tracts arise from cell bodies in the mid-brain and in the arcuate and anterior perventricular nuclei of the hypothalamus. In Parkinson's disease, the nigrostriatal tract degenerates with depletion of brain dopamine.

L-3,4-Dihydroxyphenylalanine (dopa),

a precursor of dopamine, is effective in treatment.

Catecholamines

Tyrosine is the precursor of both dopamine and noradrenaline, two hydroxylase enzymes and one decarboxylase enzyme being involved (*Fig.* 33.5). Tyrosine hydroxylase primarily controls the synthetic role of both of these precursors and is confined mainly to the catecholamine nerve terminals. The enzyme is inhibited by catecholamine, indicating that the products can normally exert a feedback control.

Dopa is decarboxylated to dopamine by 'dopa decarboxylase', but this enzyme can

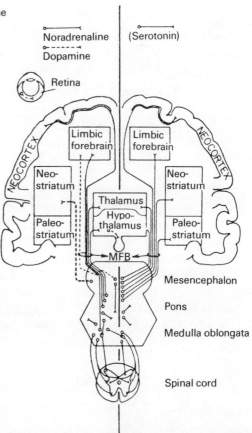

Fig. 33.4.   **Tracts in the brain utilizing noradrenaline, dopamine or serotonin as transmitters.**
MFB, medial forebrain bundle.

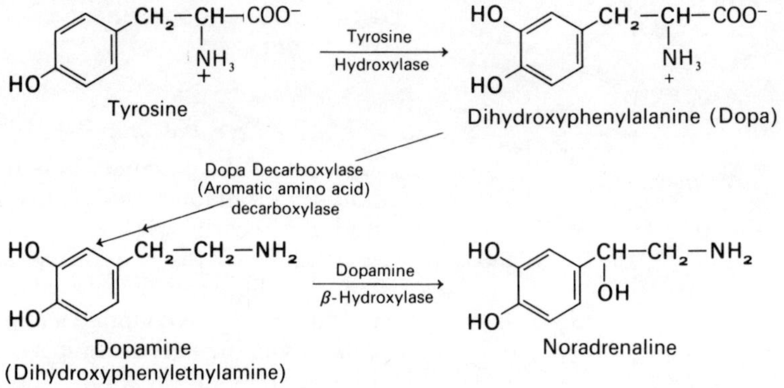

Fig. 33.5.   **Synthesis of dopamine and noradrenaline.**

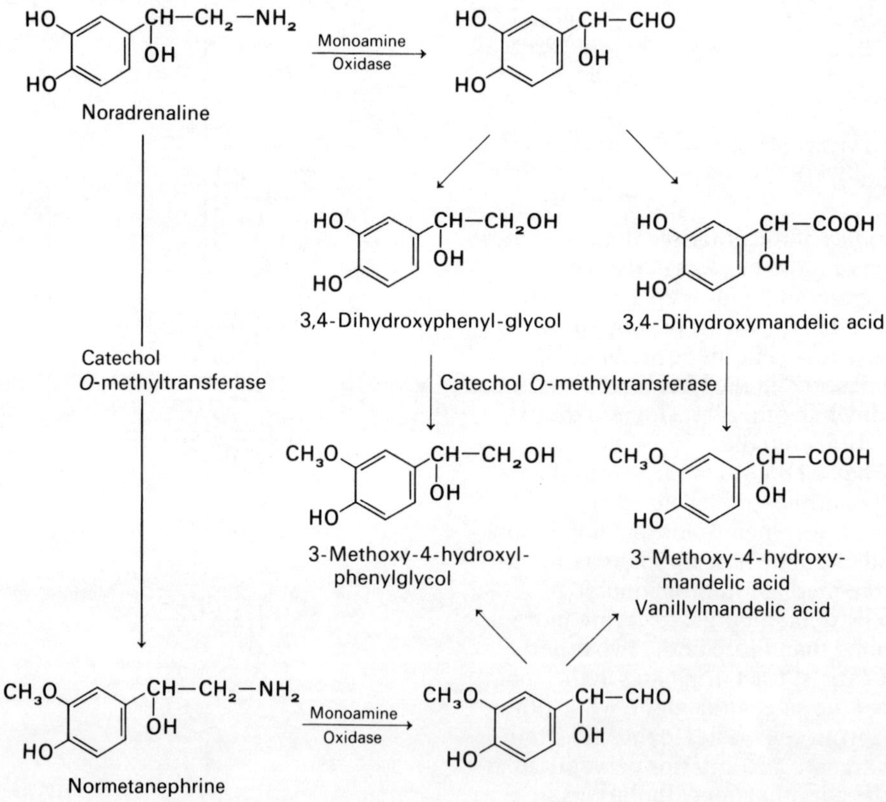

Fig. 33.6.   **Degradation of catecholamines.**

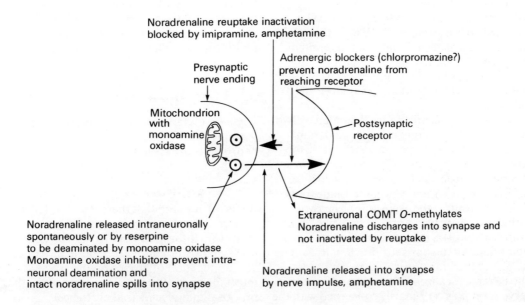

Noradrenaline reuptake inactivation
blocked by imipramine, amphetamine

Adrenergic blockers (chlorpromazine?)
prevent noradrenaline from
reaching receptor

Presynaptic
nerve ending

Mitochondrion
with
monoamine
oxidase

Postsynaptic
receptor

Noradrenaline released intraneuronally
spontaneously or by reserpine
to be deaminated by monoamine oxidase
Monoamine oxidase inhibitors prevent intra-
neuronal deamination and
intact noradrenaline spills into synapse

Extraneuronal COMT O-methylates
Noradrenaline discharges into synapse and
not inactivated by reuptake

Noradrenaline released into synapse
by nerve impulse, amphetamine

Fig. 33.7. **Events at catecholamine synapses showing roles of reuptake and metabolic degradation.**

COMT = catechol O-methyltransferase.

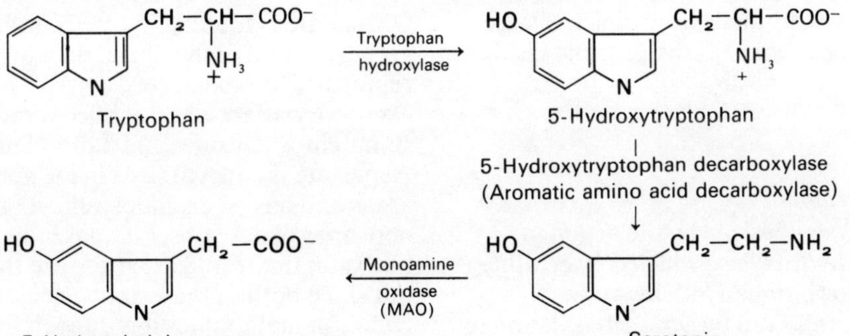

Tryptophan

Tryptophan
hydroxylase

5-Hydroxytryptophan

5-Hydroxytryptophan decarboxylase
(Aromatic amino acid decarboxylase)

5-Hydroxyindole acetate

Monoamine
oxidase
(MAO)

Serotonin
(5-Hydroxytryptamine)

Fig. 33.8. **Biosynthesis of serotonin and the action of monoamine oxidase.**

decarboxylate many amino acids including 5-hydroxytryptophan, so that it is often described as an 'aromatic amino acid decarboxylase'. As for tyrosine hydroxylase, this enzyme is also contained within the soluble portion of the catecholamine nerve terminals. It appears to be greatly in excess of requirements since drugs inhibiting 95 per cent of its activity do not reduce brain levels of catecholamines.

Catecholamines are degraded by two different pathways: one involves oxidation to the aldehyde by monoamine oxidase and the other is by methylation. The enzyme, catechol *O*-methyltransferase (COMT), transfers the methyl group of *S*-adenosylmethionine to the hydroxyl group in the 3-position. The two pathways are shown in *Fig*. 33.6. Monoamine oxidase (MAO) is present in the outer membranes of most tissue mitochondria and in the glia of the brain, as well as in neurones. The enzyme localized on mitochondria of the nerve terminals is primarily concerned with the deamination of surplus catecholamine that leaks out of synaptic vesicles. The catechol *O*-methyltransferase appears to be located outside the catecholamine nerve terminal.

Although these two elaborate metabolic processes are available for degradation of catecholamines, it appears more likely that the primary mode of inactivation is the reuptake of catecholamines into the nerve terminals from which release occurred. A summary of the processes is shown in *Fig*. 33.7.

Serotonin

Serotonin is synthesized by the action of tryptophan hydroxylase which converts the amino acid tryptophan to 5-hydroxy-tryptophan, followed by a decarboxylation to form 5-hydroxytryptamine (serotonin). Monoamine oxidase metabolizes serotonin to the corresponding inactive amino acid (*Fig*. 33.8). Serotonin, once synthesized, can be stored in vesicles at the nerve endings; it is inactivated in similar fashion to the catecholamines, mainly by a process of reuptake.

The raphe nuclei of the cell bodies are the main sites of the serotoninergic neurones and there is good evidence that these are concerned with sleep since inhibition of serotonin synthesis with drugs, such as *p*-chlorophenylalanine, causes insomnia.

33.3 Amino acid transmitters

Glycine, glutamate, aspartate and γ-aminobutyric acid are known to act as neurotransmitters in the brain.

a. Glycine

It has been found that there is a correlation between the glycine content and the number of inhibitory interneurones involved in mediating polysynaptic inhibitory reflexes in the cord; also application of glycine inhibits spinal interneurones. These observations led to the proposal that glycine is likely to be a transmitter for a population of interneurones in the spinal cord.

Studies with the drug strychnine have indicated that glycine is the transmitter normally liberated at strychnine-sensitive spinal synapses.

b. Glutamate and aspartate

Glutamate is found to be concentrated to a greater extent in the dorsal than in the ventral regions of the spinal cord and to have an excitatory effect when applied directly to the spinal interneurones and motor neurones. Aspartate is concentrated in the ventral cord areas and has an excitatory effect on spinal neurones. These observations indicate that both glutamate and aspartate are transmitters, but since both of these amino acids are common metabolites it is difficult to unequivocally establish this fact.

For verification of their role, it is desirable to establish that the transmitters are synthesized, stored in granules, released, have a postsynaptic action and are inactivated. Many experiments have been carried out to attempt to establish the role of glutamate and aspartate as transmitters and the main bulk of evidence supports this role; the most convincing evidence has been obtained by the study of neuromuscular junctions in arthropods.

c. γ-Aminobutyric acid

In contrast to much of the tenuous evidence concerning the possible roles of glycine, aspartate and glutamate as transmitters, it is very well established that γ-aminobutyric acid is an important inhibitory transmitter in the mammalian central nervous system.

γ-Aminobutyrate is formed by decarboxylation of glutamate

is believed to be mediated by special γ-aminobutyrate binding sites on the membranes which appear to need a high Na^+ concentration for maximum activity. After binding, γ-aminobutyrate equilibrates rapidly with the intracellular cell contents.

Several drugs bind specifically with γ-aminobutyrate binding sites. Of these picrotoxin has been known for a considerable time, but a more recently discovered alkaloid, bicucrilline, is even more specific. γ-Aminobutyrate plays an important inhibitory role in the cerebellum and its concentration and that of the decarboxylase are particularly

$$^-OOC-CH_2-CH_2-CH\underset{\textstyle COO^-}{\overset{\textstyle NH_3^+}{\big<}} \xrightarrow[\text{decarboxylase}]{\text{Glutamate}} \quad ^-OOC-CH_2-CH_2-CH_2-NH_3^+$$

Glutamate   γ-Aminobutyrate (GABA)

and the activity of glutamate decarboxylase regulates the rate of γ-aminobutyrate formation. γ-Aminobutyrate can undergo reversible transamination with α-ketoglutarate to form glutamate and succinate semialdehyde and these reactions link with the reactions of the citrate cycle (*see Fig*. 33.9). This metabolic sequence is sometimes termed the 'γ-aminobutyrate shunt'. Stimulation of axons of several nerves that inhibit the action of lobster muscle, has been shown to release γ-aminobutyrate, but evidence from vertebrate muscle experiments is, in general, indirect. γ-Aminobutyrate has been shown to be released from preparations of rat brain synaptosomes.

γ-Aminobutyrate controls the permeability of membranes to the chloride ion, increasing the permeability so that the membrane potential stays close to the resting level. It is unlikely that γ-aminobutyrate is catabolized by transamination since this enzyme, the only known enzyme to be involved in the catabolism of γ-aminobutyrate, is localized in the mitochondria that are intracellular and not associated with neuronal membrane components. It is, therefore, virtually certain that γ-aminobutyrate is inactivated by reuptake. This process

high in the Purkinje cells. γ-Aminobutyrate is probably released onto membranes of these cells. On the other hand, the cerebellar white matter has very low concentrations of γ-aminobutyrate and glutamate decarboxylase.

γ-Aminobutyrate is an extremely important transmitter, acting as an inhibitor within the cerebellum and mediating virtually all the signals leaving the cerebellum. It also takes part in information processing beyond the Purkinje cell synapses.

d. Interactions of amino acid transmitters in the spinal cord

A schematic, and somewhat speculative, scheme for the role of amino acid transmitters in the spinal cord is shown in *Fig*. 33.10.

Glutamate could be the main postsynaptic excitatory transmitter liberated by presynaptic endings of dorsal root fibres. Most of these fibres terminate in the dorsal region of the cord at the beginning of polysynaptic pathways that end on motor neurones (2) but some continue to form monosynaptic contacts with motor neurones (1). Most of the small dorsal interneurones are inhibitory (3), possibly γ-aminobutyrate neurones, that play an important role in the processing of

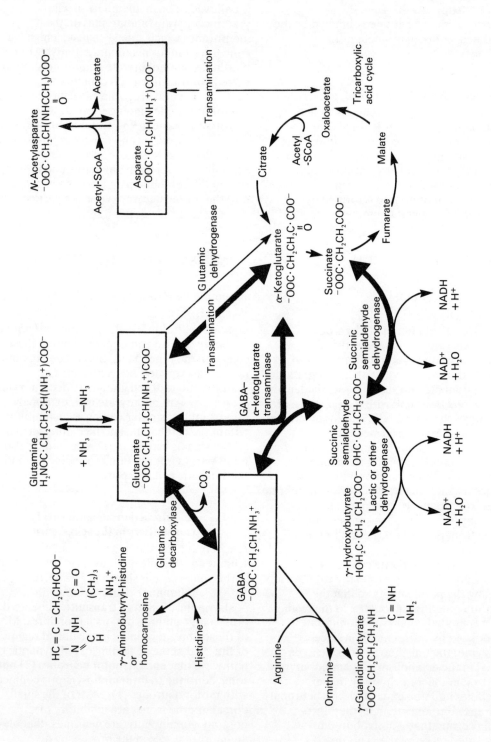

Fig. 33.9. **Metabolism of glutamate in the brain.** GABA, γ-aminobutyric acid.

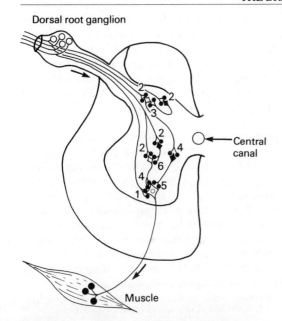

Fig. 33.10. **Interactions of amino acid transmitters in the spinal cord.**
($\ominus$) Excitatory interneurone; ($\blacklozenge$) inhibitory interneurone; ($\ominus$) motor neurone. (1, 2) glutamic acid; (3) γ-aminobutyric acid; (4) aspartic acid; (5) glycine; (6) glycine or γ-aminobutyric acid; (7) acetylcholine.

information and the point of entry into the cord. Certain interneurones depolarize motor neurones by releasing the excitatory transmitter aspartate (4), whilst others inhibit motor neurones by release of glycine (5). Glycine and γ-aminobutyrate are present at all levels of the cord's grey matter so that γ-aminobutyrate and glycine interneurones may participate in modulating activities in all regions of the cord (6). Acetylcholine is the excitator transmitter at the neuromuscular junction (7).

33.4 Myelin

Myelin is a very important constituent of the white matter of the brain and the myelin sheath forms 50 per cent of the total dry weight. Myelin is mainly responsible for the gross chemical difference between white and grey matter, and accounts for the glistening white appearance and high lipid content of white matter.

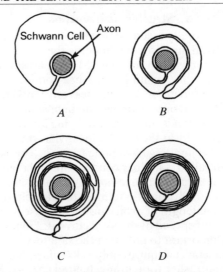

Fig. 33.11. **The formation of myelin in the central nervous system.**
(A–D) Stages in the formation of layers of myelin around the axon by the action of the Schwann cells. The cytoplasm forms a ring both inside and outside the sheath.

The myelin sheath is greatly extended and modified plasma membrane that is wrapped around the nerve axon in a spiral fashion (*Figs*. 33.11 and 33.12). The sheath is not continuous because each myelin-generating cell synthesizing myelin forms only a segment of the axon; the short portions of uncovered axons are called the 'nodes of Ranvier'.

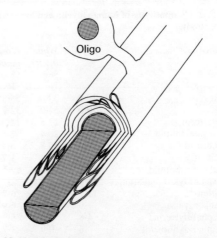

Fig. 33.12. **Myelin layers surrounding an axon in the central nervous system.**

a. Structure and function

It is generally assumed that myelin acts as an electrical insulator surrounding the axon, the 'wire' carrying the current. Myelin, however, also facilitates conduction, since non-myelinated nerves conduct less efficiently than myelinated ones. Conduction in myelinated fibres depends on the sheath being interrupted periodically at the nodes of Ranvier, the impulse jumping from node to node. This is often called 'saltatory conduction' and is about six times faster than in non-myelinated fibres.

Myelin also reduces the capacitance per unit length of axon causing increase in the speed of spread of the local current. Finally, it is important to note that the saltatory conductance requires only 1/300 of the Na^+ flux necessary for non-myelinated nerve and this is economical in energy requirement.

b. Composition of myelin

Myelin contains a relatively low percentage of water, but a high lipid content. Compositions of human myelin, white and grey matter are shown in *Table* 33.1. The major lipids of myelin are cholesterol, cerebrosides and phosphatidylethanolamines, mainly in the form of plasmalogens.

The long-chain fatty acid residues of myelin are characterized by a very high

proportion of fatty aldehydes, occurring primarily as components of phosphatidylethanolamine and phosphatidylserine; they constitute about 12 per cent of the total fatty chain of the myelin lipids. Oleic acid makes up a large proportion of the total fatty acids; the polyunsaturated content is relatively low and α-hydroxy acids occur in sphingolipids, cerebrosides and sulphatides.

Myelin contains a number of proteins some of which have been studied in detail. The major protein is a lipoprotein or proteolipid of molecular weight 20 000–30 000 containing about 40 per cent polar amino acids. The basic protein in myelin has been studied extensively since on injection into animals it produces an antibody response causing a disease of the brain that involves focal areas of inflammation and demyelination resembling multiple sclerosis. This protein has a molecular weight of 18 000, is highly basic with an isoelectric point greater than pH 12 and contains 54 per cent polar amino acids. It is an unfolded molecule with no tertiary structure.

A third group of myelin proteins are acid-soluble proteolipids called 'Wolfgram proteins' after their discoverer.

c. Development and metabolism of myelin

During development, rapid myelinization overlaps cellular proliferation, although the myelin first laid down is of a different composition from that isolated in the adult. As the animal matures, myelin galactolipids increase by about 50 per cent and phosphatidylcholine decreases by a similar amount. Polysialogangliosides decrease and monosialogangliosides increase to form 90 per cent of the total. Complex changes in the protein ratios also occur.

Myelin is synthesized and turned over rapidly during development, but is relatively stable in the adult. Some components of myelin, for example phosphatidylinositols, have a very high turnover rate in the adult, while other lipid components turnover at different rates. For example, the half-lives of phosphatidylinositol, phosphatidylcholine and phosphatidylserine

Table 33.1 **Compositions of human myelin and brain white and grey matter**

Substance	Myelin, %	White matter, %	Grey matter, %
Protein	30·0	39·0	55·3
Lipids	70·0	54·9	32·7
Cholesterol	27·7	27·5	22·0
Total galactolipid	27·5	26·4	7·3
Cerebroside	22·7	19·8	5·4
Sulphatide	3·8	5·4	1·7
Total phospholipid	43·1	45·9	69·5
Phosphatidylethanolamines	15·6	14·9	22·7
Lecithin	11·2	12·8	26·7
Sphingomyelin	7·9	7·7	6·9
Phosphatidylserine	4·8	7·9	8·7
Phosphatidylinositol	0·6	0·9	2·7
Plasmalogens	12·3	11·2	8·8

are 5 weeks, 2 months and 4 months, respectively, half-lives which are much longer than the same constituents of liver membranes.

33.5 Metabolism in the brain

a. Carbohydrate metabolism

Under normal conditions the most important substrate for brain metabolism is glucose and the brain depends on glucose as a major source of energy and as a source of carbon for synthesis into other molecules. Even a small decline in the oxidative metabolism of glucose leads to a disruption of brain function.

Glucose is metabolized by glycolysis, the pentose-shunt pathway and, eventually, as in other tissues such as liver, by the citrate cycle following pyruvate formation. The brain can also synthesize glycogen, which is stored in granules.

The activity of the pentose–phosphate shunt is high in the developing brain and reaches a peak during myelination. This is believed to be associated with the provision of NADPH for lipid synthesis. Even in the adult brain, the pathway has been estimated to be 3–8 per cent of that of glycolysis.

b. Lipid metabolism

Of the main organs of the body, the brain is particularly rich in lipids and an important aspect of lipid metabolism is the synthesis of lipid components. The brain synthesizes cholesterol, glycerophospholipids and sphingolipids. This latter group is particularly important in the brain and nervous system in general: sphingolipids are built up from sphingosine acylated to form ceramide and then linked to short chains of carbohydrate residues in the form of gangliosides, i.e. with one or more N-acetylneuraminic acid (sialic acid) residues attached (*Fig*. 33.13). The total lipid compositions of the human brain and major gangliosides are shown in *Tables* 33.2 and 33.3 and in *Fig*. 33.14.

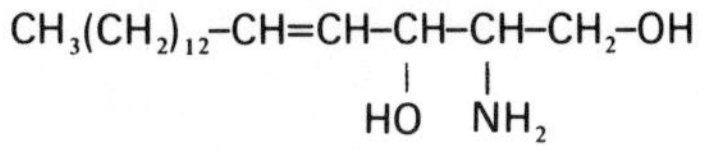

It will be noted that white matter is much richer in galactolipids and poorer in phospholipids than the grey matter. Also,

$$CH_3(CH_2)_{12}-CH=CH-CH-CH-CH_2-OH$$

with HO and NH_2 substituents

Sphingosine

$$CH_3(CH_2)_{12}-CH=CH-CH-CH-CH_2-OH$$

with HO and NH substituents, NH bearing $C=O$ and R

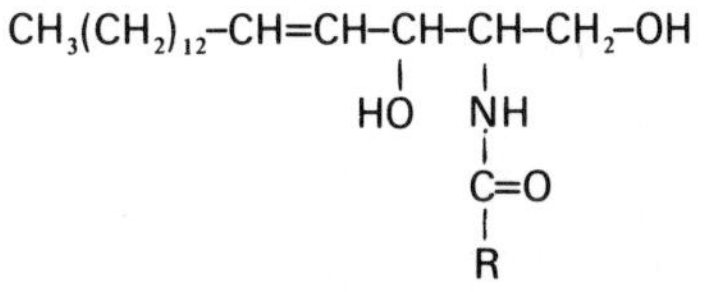

$$R= -(CH_2)_nCH_3$$

Ceramide (N-acylsphingosine)

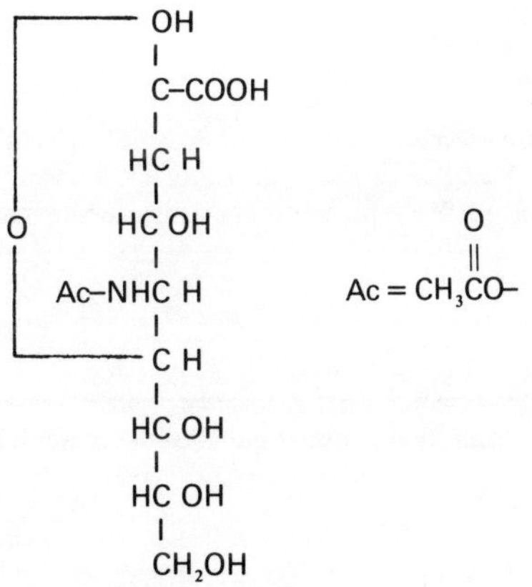

N-acetylneuraminic acid (NAcNeu) structure:

OH
|
C–COOH
|
HC H
|
HC OH
|
Ac–NHC H
|
C H
|
HC OH
|
HC OH
|
CH$_2$OH

$Ac = CH_3CO-$

N-acetylneuraminic acid (NAcNeu)

Fig. 33.13. **Molecular units used in the formation of gangliosides.**

plasmalogen constitutes 75–80 per cent of the ethanolamine phospholipids present in white matter but less than half of that present in grey matter.

The biosynthetic and catabolic pathways in brain lipid metabolism are generally similar to those found in other organs. An important group of diseases, the sphingolipidoses, is known to be caused by deficiencies in specific catabolic enzymes; their deficiency tends to cause accumulation of certain sphingolipids. This type of disease was first recorded in 1881 by Tay in England

Table 33.2 **Lipid composition of normal adult human brain**

Constituents	Grey matter		White matter	
	Dry wt, %	Lipid, %	Dry wt, %	Lipid, %
Proteolipid	2·7	—	8·4	—
Total lipid	32·7	100	54·9	100
Cholesterol	7·2	22·0	15·1	27·5
Phospholipid, total	22·7	69·5	25·2	45·9
Ethanolamine phospholipid	7·2	22·7	8·2	14·9
Lecithin	8·7	26·7	7·0	12·8
Sphingomyelin	2·3	6·9	4·2	7·7
Monophosphatidylinositol	0·9	2·7	0·5	0·9
Serine phospholipid	2·8	8·7	4·3	7·9
Plasmalogen	4·1	8·8	6·4	11·2
Galactolipid, total	2·4	7·3	14·5	26·4
Cerebroside	1·8	5·4	10·9	19·8
Sulphatide	0·6	1·7	3·0	5·4
Ganglioside, total	1·7	—	0·18	—

Water content: 81.9% (grey matter); 71.6% (white matter).

Table 33.3 **Composition of major gangliosides in adult human brain**

Constituents	Grey matter†		White matter†	
	Average	Range	Average	Range
Total N-acetylneuraminic acid*	812	744–918	110	80–180
G_0	3·9	3·2–4·8	4·8	2·8–6·1
G_{T1}	19·7	15·8–25·7	19·1	14·1–21·2
G_{D1b}	16·7	14·3–19·9	14·8	12·2–18·1
G_{D2}	3·0	1·2–4·2	1·6	1·2–3·1
G_{D1a}	38·0	29·1–43·7	36·2	30·0–38·2
G_{D3}	2·0	1·0–2·8	3·2	1·2–5·0
G_{M1}	14·2	13·0–15·6	18·8	14·6–21·2
G_{M2}	1·7	1·5–2·0	1·0	0·6–2·0
G_{M3}	<1	—	<1	—

* The total N-acetylneuraminic acid is given in $\mu g/g$ wet weight.
† Values are expressed as a percentage of total N-acetylneuraminic acid (sialic acid) in each ganglioside.
G_0 represents all sialic acid that has mobility slower than G_{T1}.
G_{D2} is a Tay–Sachs ganglioside (G_{M2}) with an additional N-acetylneuraminic acid.
G_{D3} is a haematoside (G_{M3}) with an additional N-acetylneuraminic acid.

Table 33.4 **Metabolic diseases characterized by inabilities to degrade sphingolipids**

Disease	Major sphingolipid accumulated	Enzyme defect
Niemann–Pick	Cer ┤┆├ (PChol) Sphingomyelin	Sphingomyelinase
Gaucher	Cer ┤┆β├ ⟨Glc⟩ Ceramide glucoside (glucocerebroside)	β-Glucosidase
Krabbe	Cer ─β β─ ⟨Gal⟩ Ceramide galactoside (galactocerebroside)	β-Galactosidase
Metachromatic leucodystrophy	Cer ─β─ ⟨Gal⟩ OSO₃⁻ Ceramide galactose-3-sulphate (sulphatide)	Sulphatidase
Ceramide lactoside lipidosis	Cer ─β─ ⟨Glc⟩ ─β┆─ ⟨Gal⟩ Ceramide lactoside	β-Galactosidase
Fabry	Cer ─β─ ⟨Glc⟩ ─β─ ⟨Gal⟩ ─α┆─ ⟨Gal⟩ Ceramide trihexoside	α-Galactosidase
Tay–Sachs	Cer ─β─ ⟨Glc⟩ ─β─ ⟨Gal⟩ ─β┆─ ⟨NAc-Gal⟩ NAcNeu Ganglioside GM₂	Hexosaminidase A
Tay–Sachs variant	Cer ─β─ ⟨Glc⟩ ─β─ ⟨Gal⟩ ─α─ ⟨Gal⟩ ─β┆─ ⟨NAc-Gal⟩ Globoside (plus ganglioside GM₂)	Total hexosaminidase
Generalized gangliosidosis	Cer ─β─ ⟨Glc⟩ ─β─ ⟨Gal⟩ ─β─ ⟨NAc Gal⟩ ─β┆─ ⟨Gal⟩ NAcNeu Ganglioside GM₂	β-Galactosidase

Cer = N-acylsphingosine, or ceramide.	Glc = glucose.
NAcNeu = N-acetylneuraminic acid.	PChol = phosphatidylcholine.
Gal = galactose.	NAcGal = N-acetylgalactosamine.

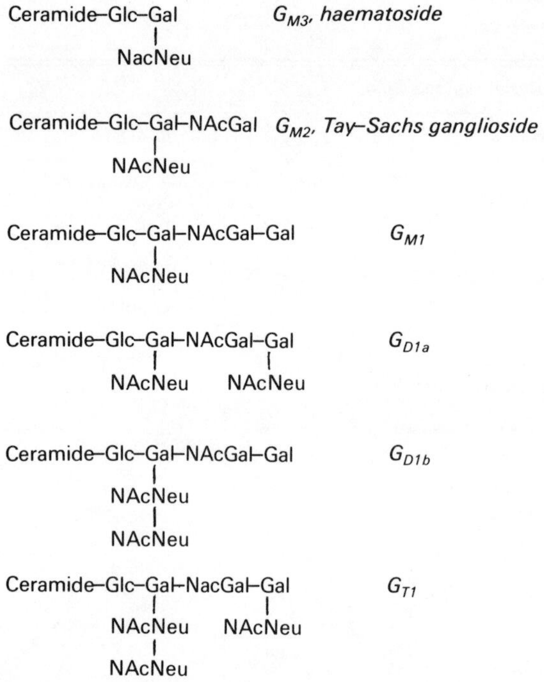

Ceramide–Glc–Gal G_{M3}, haematoside
|
NacNeu

Ceramide–Glc–Gal–NAcGal G_{M2}, Tay–Sachs ganglioside
|
NAcNeu

Ceramide–Glc–Gal–NAcGal–Gal G_{M1}
|
NAcNeu

Ceramide–Glc–Gal–NAcGal–Gal G_{D1a}
| |
NAcNeu   NAcNeu

Ceramide–Glc–Gal–NAcGal–Gal G_{D1b}
|
NAcNeu
|
NAcNeu

Ceramide–Glc–Gal–NacGal–Gal G_{T1}
| |
NAcNeu   NAcNeu
|
NAcNeu

Fig. 33.14. **Structures of the major gangliosides of human brain and nervous tissues.**
Glc, glucose; Gal, galactose; NAcGal, *N*-acetylgalactosamine; NaNeu, *N*-acetylneuraminic acid.

and, in 1887, by Sachs in the USA and the condition has since been described as 'Tay–Sachs' disease. The reason for the accumulation of sphingolipid was not understood for many years until, in 1965, it was shown that a related condition, termed 'Gaucher's disease', was caused by the deficiency of a hydrolytic enzyme involved in the catabolism and accumulation of a sphingolipid of specific striatum. It gradually became apparent that a large group of sphingolipidoses was caused by genetic conditions manifested by the inability of the patient to produce specific enzymes involved in the hydrolysis of sphingolipids. The conditions are summarized in *Tables* 33.4 and 33.5.

c. Amino acid/protein metabolism

The major part of amino acid and protein metabolism occurring in the brain is similar to

Table 33.5 **Clinical aspects of lipodystrophies**

Disease	Clinical symptoms
Tay–Sachs disease (classic and variant forms)	Mental retardation; amaurosis; cherry-red spot in macula; neuronal cells distended with 'membranous cytoplasmic bodies'
Gaucher's disease	Mental retardation (infantile form only); hepatosplenomegaly; hip and long-bone involvement; oil red and periodic acid–Schiff positive lipid-laden (Gaucher) cells in bone marrow
Niemann–Pick disease	Generally similar to Gaucher's disease; 30% with cherry-red spot in macula; marrow cells (foam cells) stain for both lipid and phosphorus
Fabry's disease	Reddish-purple maculopapular rash in umbilical, inguinal, and scrotal areas; renal impairment; corneal opacities; peripheral neuralgias and abnormalities of ECG
Globoid leucodystrophy	Mental retardation; 'globoid bodies' in brain tissue sections
Metachromatic leucodystrophy	Mental retardation; psychological disturbances (adult form); decreased nerve-conduction time; nerve biopsy shows yellow-brown droplets when stained with cresyl violet (metachromasia)
Generalized gangliosidosis	Mental retardation, cherry-red spot in macula; hepatomegaly; bone marrow involvement
Ceramide lactoside lipidosis	Slowly progressing CNS impairment; organomegaly; macrocytic anaemia, leucopenia and thrombocytopenia due to involvement of bone marrow and spleen

that in other tissues, but there are some important distinctions discussed below.

If the concentrations of free amino acids in the brain and plasma are compared, it is immediately apparent that certain amino acids, such as glutamate, its derivatives and aspartate, are specifically concentrated in the brain (*Table* 33.6). The high concentration and

Table 33.6 **Comparisons of free amino acids of human brain and plasma**

Amino acid	Concentration, μmol per g or ml	
	Brain	Plasma
Glutamic acid	10·6	0·05
N-Acetylaspartic acid	5·7	—
Glutamine	4·3	0·7
γ-Aminobutyric acid	2·3	—
Aspartic acid	2·2	0·01
Cystathionine	1·9	—
Taurine	1·9	0·1
Glycine	1·3	0·4
Alanine	0·9	0·4
Glutathione	0·9	—
Serine	0·7	0·1
Threonine	0·2	0·15
Valine	0·2	0·25
Lysine	0·1	0·12
Leucine	0·1	0·15
Proline	0·1	0·1
Asparagine	0·1	0·07
Methionine	0·1	0·02
Isoleucine	0·1	0·1
Arginine	0·1	0·1
Cysteine	0·1	0·1
Phenylalanine	0·1	0·1
Tyrosine	0·1	0·1
Histidine	0·1	0·1
Tryptophan	0·05	0·05

form of amino acids, glutamate, glutamine, aspartate and γ-aminobutyrate, the latter being an important inhibitory transmitter discussed earlier in this chapter in Section 33.3.

It is clear that specific transport systems exist for amino acids from the plasma to the brain and five specific systems are probably involved: these sites are believed to be specific for (*a*) small neutral amino acids, (*b*) large neutral amino acids and aromatic amino acids, (*c*) small basic amino acids, (*d*) larger basic amino acids and (*e*) acidic amino acids. Other sites may exist for γ-aminobutyrate and imino acids. Each site has a high affinity for its particular group of amino acids but complex interactions with other sites are likely to occur. Such competitive interactions may play an important role in the pathology of several disease states, an example being shown in *Fig.* 33.15.

The mechanisms of brain protein synthesis are identical to synthesis in other

extensive metabolism of glutamate and its derivatives are amongst the most significant aspects of brain metabolism and this group of amino acids forms more than two-thirds of the free amino nitrogen in the brain.

Acetylaspartate, the concentration of which is often two–three times that of aspartic acid, occurs only in brain and is generally at a low concentration at birth, rising to adult levels during development. It is synthesized from aspartate and acetyl-SCoA, but its primary function is as yet uncertain: it may form part of the intracellular fixed anion pool, a reservoir of acetyl groups or a source of N-blocked end groups for synthesis into special proteins.

The metabolism of the glutamate family of amino acids is quantitatively the most significant aspect of amino acid metabolism in the brain. After injection of labelled glucose into an animal, 70 per cent of the isotope present in the soluble brain fraction is in the

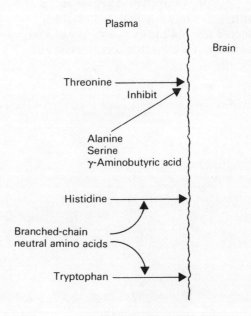

Fig. 33.15. **Competition of amino acids for uptake into the brain.**
High concentrations of branched-chain amino acids in the plasma can inhibit the uptake of histidine and tryptophan and thus the synthesis of serotonin in the brain.

tissues, although several problems arise in the consideration of details of the system. For example: are the proteins of the axons and nerve endings active when synthesized and where do they originate? Transport of proteins from neurones to axons is most likely to occur after synthesis. Studies of labelled proteins show that they move from the nerve cell through the axon in two major waves referred to as 'rapid flow' and 'slow flow'. Rapidly transported protein moves at a rate of several hundred millimetres per day, but most of the protein moves much more slowly at a rate of several millimetres per day. The mechanism of transport is not known; microtubules may be involved and energy is required.

Nerve endings contain two types of protein, soluble and particulate. Most of the soluble protein is believed to be transported from the nerve cell body, but the particulate protein is likely to be synthesized in the synaptosome.

d. Utilization of ketone bodies

Although it was originally believed that glucose was obligatory as a supply of energy for the brain, it is clear that acetoacetate and 3-hydroxybutyrate can also be sources of energy. If patients are starved, for example in the treatment of obesity, it can be calculated that more than 50 per cent of their total energy supply can be accounted for by the uptake of ketone bodies into the brain. The uptake of 3-hydroxybutyrate is several times greater than that of acetoacetate, since its blood concentration is much higher.

Under normal circumstances, when ample glucose is available and the levels of ketone bodies in the plasma are very low, the brain does not use ketone bodies; however, in starvation when carbohydrate stores are exhausted and gluconeogenesis cannot supply glucose sufficiently rapidly, ketone bodies are not used by the brain, presumably on account used. In diabetic acidosis, when ketosis is just as severe as in starvation, ketone bodies are not used by the brain, presumably on account of the presence of adequate supplies of glucose in the blood.

It should be noted, however, that in the absence of blood glucose although 3-hydroxybutyrate can partially replace glucose, it cannot fully satisfy cerebral energy needs or maintain cerebral function in the absence of glucose.

Part **4**

Environmental hazards—detoxication

Chapter 34 Toxicology: general aspects

In this part of the book, some general aspects of biochemical toxicology are dealt with and illustrated with descriptions of effects of toxic metals, drugs and organic chemicals, and some carcinogens.

Man can normally be regarded as being in equilibrium with three types of environment, described as physical, biological and chemical (*Fig.* 34.1). The biological environment includes his interactions with all other living organisms, such as his fellow human beings, plants and micro-organisms. The body's reaction to changes in temperature, pressure or humidity is included in the study of the effects of the physical environment and is normally discussed under the subject heading 'physiology'. The study of man's reaction to his chemical environment comes under the general title of 'biochemistry' and under this heading the effects of some toxic substances to which man is frequently exposed are discussed,

Table 34.1 **Classification of chemicals of the environment**

Group	
I	Desirable, e.g. foodstuffs
II	Harmless and inert
III	Desirable in small quantities
	Toxic in large quantities
IV	Toxic in any quantity

a study frequently called 'biochemical toxicology'.

Environmental chemicals may be classified into four categories as shown in *Table* 34.1. In group I are included most natural foodstuffs, such as glucose and amino acids. The possibility does exist that no chemical is completely harmless to man if taken in very large quantities, for example a large and continuous intake of glucose could eventually lead to obesity and arterial disease. However, when the intake is in reasonable dietary quantities, chemicals in group I can be classified as beneficial. In group II inert components of the diet and environment are classified, i.e. components with no nutritional value and also no toxicity; the group consists of substances not normally absorbed or absorbed only to a very limited extent, for example cellulose and sulphate ions. In group III several metals, such as copper and zinc, that are essential in small trace quantities but toxic in larger amounts, are placed.

The major proportion of toxic substances is placed in group IV since these are believed to be toxic in any quantity, however small. An important consideration in the assessment of the health risk of any organic

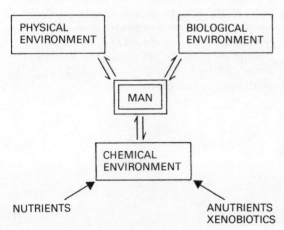

Fig. 34.1. **Relation of man to his environment.**

431

or inorganic compound is the threshold dose, i.e. the tolerance of the human body to repeated exposure over a prolonged period to a specified small dose of a toxic substance. This problem is often very difficult to resolve since it involves testing over very long periods, perhaps for many years, and such experiments which are carried out on animals are very costly and often yield results that are difficult to relate to the human situation.

The modern industrial environment of man has exposed him to many more toxic substances, in greater concentrations, than man's primitive ancestors were exposed to. However, even in earlier periods, it cannot be assumed that man avoided toxic substances altogether. Plants, fungi and micro-organisms produce many very toxic substances and bitter experience has taught man to recognize and avoid sources of dangerous chemicals in the fight for survival.

34.1 Biochemical damage caused by toxic substances

Tissues of the animal body can be damaged by toxic substances involving many different mechanisms and following a wide range of time intervals.

Firstly, it is important to appreciate that potentially toxic substances fall into two fairly well defined categories: those that are toxic without chemical modification and those that require metabolism for activation. The first group includes substances such as the mercuric salts; the second includes compounds such as cyanogen glycosides that, on hydrolysis, form HCN, and the polycyclic hydrocarbons that are oxidized to epoxides in the tissues, the latter being active carcinogens (Chapter 38).

Secondly, there is a wide range of time during which toxicants may be effective. Thus cyanide or carbon monoxide cause rapid death, but mutagens and carcinogens cause subtle changes in DNA and may take many years to exert their effects.

Many substances are toxic because they bind to proteins or to DNA: if the protein is a vital enzyme then rapid inactivation of an important metabolic pathway could result. Binding to DNA is likely to lead to mutation and will only manifest itself after treated cells have passed through many divisions. Toxicants may also be effective by acting as antagonists, there being sufficient structural similarity for the natural metabolite to interfere seriously with the action of the normal metabolite. Thus, dicoumarol, which occurs in sweet clover, is similar in structure to vitamin K and it will effectively block the action of the vitamin so diminishing the efficiency of the blood clotting process (cf. Chapter 26). Similarly, goitrogens contained in many plants produce thiocyanate (CNS^-) ions that compete with the I^- ion for uptake into the thyroid.

Several special proteins produced by plants, snake venoms and bacteria are toxic: lectins that agglutinate red cells are produced by plants; snake venoms contain many enzymes, the most important being phospholipases that cause haemolysis of red blood cells by splitting fatty acids from the phospholipid constituents of the membrane; several proteins, particularly those produced by bacteria, are potent neurotoxins. The botulinum toxin produced by *Clostridium botulinum* is one of the most toxic substances known and is believed to act by combining with postsynaptic receptors.

Chapter 35 Toxic metals

In this chapter, copper, mercury, lead and the radionuclides are considered as examples of toxic metals, whilst the toxic effects of a large iron intake are discussed in Chapter 28. Copper is an example of a metal that is required in low concentrations, but is toxic at high concentrations, whereas mercury and lead are toxic at any concentrations.

     For understanding of the toxic effects of metals, it is very desirable to become acquainted with the properties of metal chelate complexes and, in the first section, these will be discussed in an elementary style.

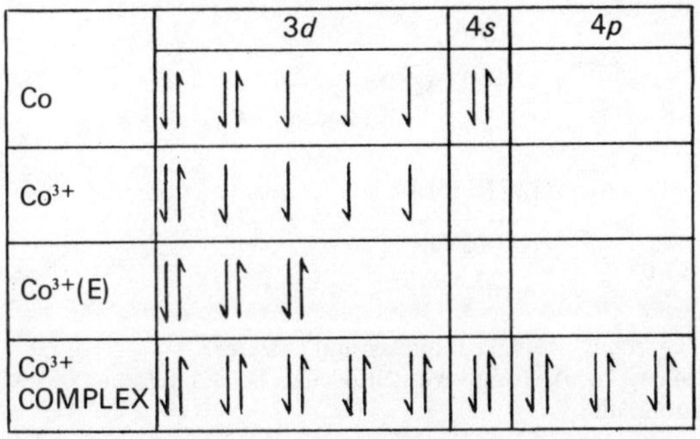

Fig. 35.1. **Electronic configurations of different forms of cobalt.**

35.1 Metal complexes and chelates

Complexes of metals with ammonium ions and with other ions or molecules were first studied in the early years of this century by the German chemist Werner, who observed that the transition elements, iron (Fe), cobalt (Co), chromium (Cr), nickel (Ni) and elements that showed transition element behaviour, such as copper (Cu), silver (Ag) and zinc (Zn), readily formed complexes with ammonia or amines. These complexes were called 'coordination complexes', a good example being the formation of the bright blue 'cuprammonium complex' resulting from the addition of ammonium hydroxide to a copper salt solution, a well-known experiment in elementary inorganic chemistry. Werner pointed out that the metals that formed stable complexes possessed unfilled inner electron shells. He proposed that the ammonia, as a donor, provided electrons to fill these shells, so forming a stable complex closely resembling that of the inert gases. This concept, although useful, is an oversimplification and consideration must be given to the electron spin as shown for the formation of the cobalt–oxalate complex (*see Fig.* 35.1).

     As shown in *Fig.* 35.1, the cobalt atom has three unpaired electrons in the $3d$ shell and can become a Co^{3+} ion by losing one pair of electrons from the $4s$ shell and the other electron from an electron pair in the $3d$ shell. The outer two electrons of the four unpaired electrons in the $3d$ shell can then reverse their spins and pair with the inner electrons to form $Co^{3+}(E)$. This is an excited species and highly unstable due to the tendency of electrons to move to higher orbitals and reverse their spins to reform unpaired electrons. If, however, the electron shells are filled by a donor molecule, such as oxalate, a very stable configuration results, in which all shells are filled with paired electrons. The formation of such a structure is typical of that found in all metal complexes.

433

Complex $M + 4A \rightleftharpoons MA_4$

Chelate (bidentate)

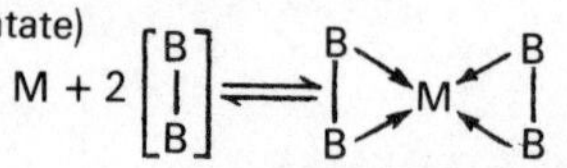

$$M + 2\begin{bmatrix} B \\ | \\ B \end{bmatrix} \rightleftharpoons \begin{matrix} B \\ B \end{matrix} M \begin{matrix} B \\ B \end{matrix}$$

Stability constants

$$K_{MA_4} = \frac{[MA_4]}{[M][A]^4}$$

Generally $K_{MB_4} >> K_{MA_4}$

$$K_{M_4} = \frac{[M(B-B)_2]}{[M][B-B]^2}$$

Fig. 35.2.   **Metal complexes and chelates.**
M, metal; A, complexing molecule; B–B, chelating molecule.

Ammonia complexes, e.g.

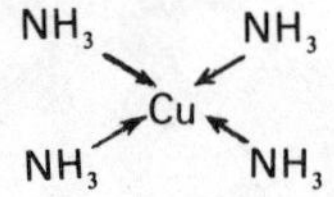

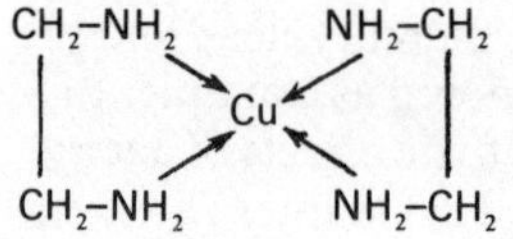

	Stability constants, log k		
	Co	Cu	Zn
	3·68	7·68	4·69

Ethylene diamine chelates, e.g.

	5·99	10·73	5·92

Fig. 35.3.   **Stability constants of complexes and chelates.**

Note that the coordination number of cobalt is 6, whereas that of copper and zinc is 4.

Amino acid

Sulphydryl group (cysteine)

Hydroxyl groups (salicylate)

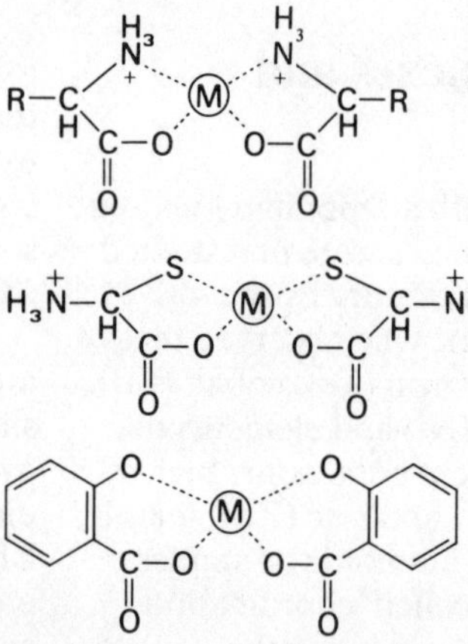

Fig. 35.4.   **Typical chelating agents of biological importance.**

Note that in the examples shown, the metal has a coordination number of 4. For metals with a coordination number of 6, 3 molecules of the ligand are required.

Further study of these metal complexes by others during the 1920s demonstrated another very important aspect: if binding by the donor molecule could take place at two sites so that a ring incorporating the metal was formed, a very stable complex would result. Complexes of this type, illustrated in *Fig.* 35.2, were termed 'chelates', the donor atoms being described as 'ligands'. Comparison of the stability constants for the ammonia and ethylene diamine complexes, shows that ethylene diamine which forms a ring and, therefore, a chelate, is 30–1000 times more stable than the corresponding ammonia complex. A diagrammatic representation of a chelate complex is shown in *Fig.* 35.3.

Typical chelating molecules of biological importance are shown in *Fig.* 35.4: the ligands are N and O for typical amino acids, S and O for cysteine and two O atoms for salicylates. In these examples, the metal is shown with only two donor ligands, but to form stable complexes either four or six are normally used. This is also true for simple complexes and was described by Werner as the 'coordination number'. For divalent copper, this number is four so that Cu^{2+} will form stable complexes with two amino acids or two salicylate molecules. For iron and chromium, the number is six so that three chelating molecules are needed to provide the six ligands.

Since the original discovery of these chelates, many hundreds of compounds have been studied in great detail and some useful general facts about the stability of the complexes have been elucidated. They are summarized below.

a. Nature of the ligand

Nitrogen, oxygen and sulphur are used almost exclusively, but they are not equally preferred by all metals. Thus copper, mercury and cobalt have a preference for N over O and copper and mercury for S over O. The alkaline earth metals form complexes when O is the ligand.

b. Nature of the metal

For many chelate complexes, the metals can be placed in order of descending chelate stability

Table 35.1 **Relative stability of metal chelate complexes**

Metal	Glycine complex log K	Metal	Malonate complex relative to Ca = 1
Pd	9·12	Cu	1100
Cu	8·51	Ni	32
Ni	6·12	Co	17
Pb	5·53	Mn	6·2
Zn	4·95	Cd	5·6
Co	4·95	Mg	2·3
Cd	4·74	Ca	1·0
Mg	3·45	Ba	0·16

as shown in *Table* 35.1. For most metal complexes, the order of stability follows the order shown in this table which, in turn, is dependent on the stability constants.

c. The size of the ring

The size of the ring formed by the metal and the chelate has an important effect on the stability and the most stable rings are those containing five or six atoms.

d. The pH

The pH of the medium has a significant effect on the stability of complexes: generally,

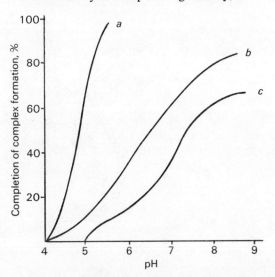

Fig. 35.5.   **Effect of pH on stability of cobalt complexes.**

(*a*) Histidine complex; (*b*) aspartate complex;
(*c*) glutamate complex.

stability increases as the pH is raised and vice versa, as shown for cobalt complexes in *Fig.* 35.5. This effect is caused by competition between the hydrogen ions and the metal ions for the chelator, so decrease in pH (or increase in acidity and therefore in H⁺ concentration) will bring about displacement of the metal. This effect could be of major importance during digestion: as the food reaches the acid conditions of the stomach, metals will be displaced from the complexes that are ingested in the diet; then, on transfer of the food to the more alkaline medium of the intestine, new complexes of the metals with other chelates will be formed.

35.2 Protein complexes

Many of the amino acid side chains of proteins will clearly be very effective chelating or ligand groups. Of particular importance are the —SH (thiol) of cysteine residues, the COO⁻ of glutamate and aspartate residues, the —OH of tyrosine and the imidazole nitrogen of histidine. Many different types of metal–protein interaction or binding are clearly possible, dependent on the peptide chain

conformation; *Fig.* 35.6 illustrates some of these diagrammatically.

If the metal has a high affinity for the ligands, then formation of the metal–protein complex can lead to the distortion of the protein structure in order to form the stable chelate. An example of this would be shown in *Fig.* 35.6, if the sections of peptide chain labelled A, B, C and D moved closer together. Distortion of configuration of this nature may lead to a loss of biological activity and ultimately to denaturation of the protein. If one of the groups bound, e.g. —SH or imidazole ═NH, is part of the active site of an enzyme, then inactivation of the enzyme will occur rapidly on exposure to the metal.

From knowledge of metal binding with simpler molecules, various predictions about the interaction of different metals with proteins can be made: thus copper, mercury, lead and zinc will bind very strongly to proteins by many different ligand groups. Imidazole ═NH and —SH are particularly important in binding of these metals. On the other hand, binding of calcium, magnesium and barium is much weaker and usually by the COO⁻ groups. Sodium and potassium do not form complexes because their inner electron shells are already filled.

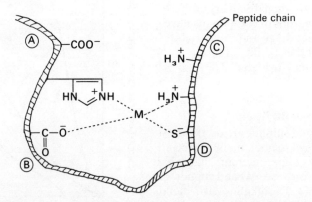

Fig. 35.6. Metal binding to proteins.
Metal-binding groups in proteins are: —COOH in glutamic and aspartic acids; —OH in serine and tyrosine; —NH₂ in lysine and arginine; ═NH in arginine and histidine; —SH in cysteine; —Ⓟ in phosphate.

35.3 Copper

Copper deficiency

Copper deficiency conditions are virtually unknown in man but are well described in animals and, particularly, in sheep grazing on pastures deficient in copper. The disease in lambs produced by copper-deficient ewes has been called 'sway back', since this describes some symptoms of the condition. It is characterized by cerebral demyelinization, degeneration of the motor tracts and necrosis of the large neurones. The pathology is of considerable interest since it closely resembles that of multiple sclerosis in man. Anaemia and defective bone development also occur.

Biochemical explanations for the development of all of these conditions have not yet been advanced, but it is known that copper is an essential component of some important enzymes, e.g. lysyl oxidase which oxidizes lysine to the corresponding aldehyde, contains copper and is essential for the cross-linking of collagen in the formation of bone.

The important respiratory enzyme, cytochrome oxidase, also contains copper and it has been suggested that the reduced activity of this enzyme in the neurones may be responsible for the neurological symptoms that result, but this has not yet been established as the primary cause.

Kinetics

The main aspects of copper kinetics are illustrated in *Fig.* 35.7. About 2 mg of copper is ingested per day and a similar quantity is excreted in the faeces; about 0·6 mg is absorbed, about 0·1 mg being excreted in the bile, giving the true net absorption as only a very small percentage of the ingested copper.

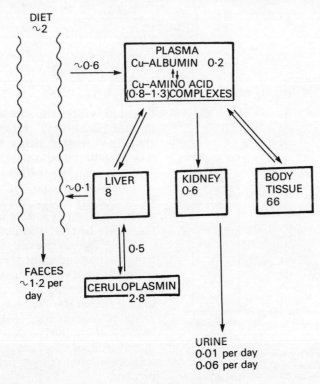

Fig. 35.7. **Copper kinetics.**
Figures in the blocks show normal copper stores; figures along the arrows show the average transfer of copper in 24 h; all figures are in milligrammes.

The absorbed copper is taken into tissues to balance loss in the urine and loss from the skin and hair. Copper is transported in the body complexed with albumin or amino acids and is then taken up by the body tissues. In the liver, copper is incorporated into a special protein, ceruloplasmin, which is a glycoprotein of molecular weight 150 000, containing 0·34 per cent copper. Ceruloplasmin is often described as a copper transport protein, but this is not true since, although it circulates in the plasma, it only deposits copper in the liver. Ceruloplasmin also performs an important function in catalysing the oxidation of Fe^{2+} to Fe^{3+} (cf. Chapter 28) and acts as a copper reserve.

Toxicity

When taken in large amounts, copper can act as a poison by binding with many proteins so inactivating enzymes. However, when the intake of copper is still within the range described in *Fig*. 35.8, a serious chemical

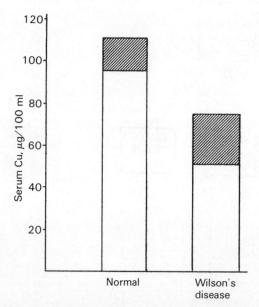

Fig. 35.8. **Serum copper in normal subjects and in those suffering from Wilson's disease.**
▨ Non-ceruloplasmin copper; ☐ ceruloplasmin copper.

condition caused by excess copper in the tissue can occasionally develop. It occurs in about 1 in 200 000 of the population, and was first described by Wilson in 1912; consequently it is known as 'Wilson's disease'. It is characterized by muscular rigidity, tremor, incoordination, damage to the basal ganglia, pigmentation in the liver and psychiatric problems.

Early observations indicated that it is an inherited (autosomal recessive) disease and that as many as 1 in 200 may be heterozygous for the condition. In an individual suffering from Wilson's disease, many tissues of the body contain much larger deposits of copper than present in a normal individual's tissues (*Table* 35.2). The genetic

Table 35.2 **Copper content of tissues in a normal individual and in a sufferer from Wilson's disease**

Tissue	Copper, mg/100 g dry tissue	
	Normal	Wilson's disease
Liver	5–17	30–160
Kidney	1	4–28
Brain (white cortex)	1–8	4–45
Spinal cord	1	10

defect in Wilson's disease may be an inability to synthesize normal ceruloplasmin, since study has revealed, firstly, that the concentration is below normal levels and, secondly, that the ability to carry copper is considerably reduced (*Fig*. 35.8). Consequently more copper would be carried by serum albumin and amino acids, and would, therefore, be readily available to the tissues.

The precise cellular damage resulting from the accumulation of excess copper in the tissues has not yet been resolved, but a knowledge of copper–protein interaction has indicated that it will bind with many vital proteins, particularly —SH proteins. This binding would thus be expected to cause inactivation of many enzymes and conformational changes in many other proteins.

35.4 Toxicity of mercury

It should be appreciated that mercury, like several other toxic metals, can exist in three distinct forms: as metallic mercury, as inorganic mercuric or mercurous salts or as part of an organic molecule. Quite often all these forms are described as 'mercury', but it should be noted that they possess many different biochemical properties leading to very different biological effects.

Organomercury compounds are frequently used as fungicides, particularly in industries such as paper making; examples are shown in *Fig.* 35.9.

Alkylmercury

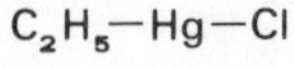

$$C_2H_5-Hg-Cl$$

Ethylmercuric chloride

Arylmercury

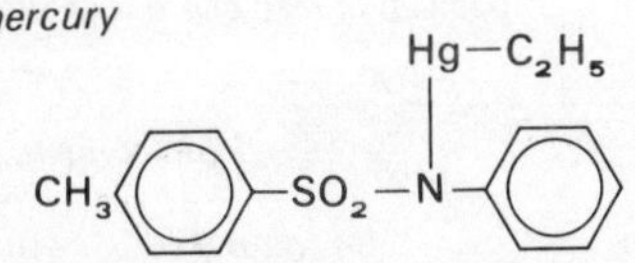

$$CH_3-\langle\bigcirc\rangle-SO_2-N\langle\bigcirc\rangle \quad Hg-C_2H_5$$

Ethylmercuri-toluene sulphonamide

Fig. 35.9. **Organomercury compounds.**

Inorganic mercury

The toxicity of inorganic mercury clearly results from the formation of stable mercury–protein complexes. Of the metals, mercury possesses one of the strongest affinities for —SH groups and will, therefore, bind extremely readily, inactivating any enzyme requiring an —SH group in its active centre.

Organic mercury

The first serious outbreak of mercury poisoning that has been adequately documented, occurred in Minamata Bay, Japan during 1953–60; 121 people were poisoned and 46 died as a result of the consumption of fish containing methylmercury. The cause was traced to the release of inorganic mecury by a plastic's factory into the effluent water and it was later discovered that the methylmercury had been formed by bacteria, probably in the gut of the fish. Symptoms of the poisoning were severe damage to the nervous system, including paraesthesia (abnormal prickling sensations) ataxia (lack of coordination, dysorthria), nervous stammering, loss of vision, hearing and, eventually, death.

A much more serious outbreak occurred in Iraq in 1972 when 6530 individuals were poisoned and 459 died as a result of eating bread made from wheat grain treated with methylmercuric chloride that should only have been used for planting. When the cause of poisoning was established, it was discovered that the treated wheat contained $3\cdot7$–$14\cdot9$ μg Hg per g grain.

Analyses of the tissue fluids of the victims showed that 80 per cent of the mercury was transported in the red blood cells and only 20 per cent in the plasma. The analyses showed that ingested organomercury compound tends to remain in the organic form in the tissues and that only a relatively small proportion is

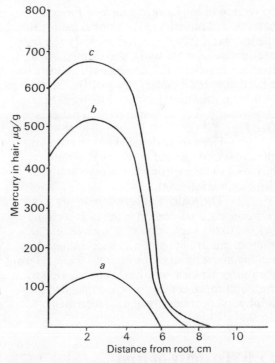

Fig. 35.10. **Mercury in hair after methylmercury poisoning.**
(*a*) Inorganic mercury; (*b*) methylmercury; (*c*) total mercury.

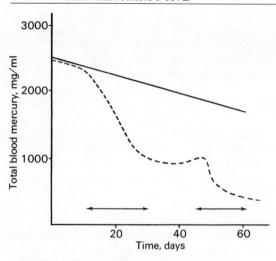

Fig. 35.11. **Clearance of mercury from blood after methylmercury poisoning.**
(——) Untreated patients; (– – –) patients treated with *N*-acetylpenicillamine (40 mg per kg per day) during periods indicated.

converted to the inorganic form, although excretion of mercury in the urine is mainly in the inorganic form. Hair has proved to be a useful tissue for the study of toxic metals and it frequently reflects the loading of the body with a particular toxic metal (*Fig.* 35.10). The rate of mercury removal from the blood has been shown to be very slow in untreated patients (*see Fig.* 35.11).

Even these detailed observations of clinical symptoms and tissue concentrations have not so far explained mercury's action at the biochemical level.

The toxicity of organomercury compounds, as discussed in the next section, has led to the suggestion that organometallic compounds are lipophilic and accumulate readily in the lipids of the nervous tissue. Here the inorganic form may be slowly released by metabolism, so causing severe damage to those sulphydryl proteins previously described.

35.5 Toxicity of lead

The problem of lead toxicity has been a major public issue in many industrial countries over the last decade. As in the case of mercury, lead exists in both inorganic and organic forms: inorganic lead, in the form of lead salts, has been widely used in making paint for many years and lead alloys are also widely used; the organic compound of lead causing major concern is lead tetraethyl $[Pb(C_2H_5)_4]$ which is added to petrol as an antiknock.

There is no doubt that the lead content in the human body has increased through the years and the dramatic increases observed from the analysis of teeth of people who died between 200 and 600 AD and the analysis of modern teeth samples is shown in *Table* 35.3. It is clear that the lead content of modern man is much greater than that of ancient man and that the lead content increases steadily throughout life.

Table 35.3 **Comparison of lead content of teeth of modern and ancient man**

	Lead content, ppm, in	
Age group, years	Modern teeth (USA, 1975)	Ancient teeth (200–600 AD)
0–9	11	5·3
10–19	17	3·0
20–29	19	3·7
30–39	24	1·3
40–49	27	2·0
50–59	50	—
60–69	54	—
70–79	55	—

Inorganic lead

Inorganic lead enters the body mainly through ingestion: children can obtain toxic quantities of lead by chewing toys or furniture painted with lead paints, but people of all ages also ingest some lead as a constituent of plant foodstuffs. The lead content of plants can be relatively high in some areas, particularly when grown in land close to lead smelting works. Vegetables grown in the vicinity of such a factory will contain much higher lead concentrations than vegetables grown in more distant areas.

Lead has a strong affinity for complex formation with proteins and, like mercury, binds strongly with —SH groups; it is through this binding that lead exerts its toxic effects.

The main symptoms of lead poisoning are, as with mercury, manifested by changes in the nervous system. Damage of the peripheral nerves, of the autonomic system and of the central nervous system are all involved. Children are generally considered to be much more sensitive than adults and some research has demonstrated that chronic lead poisoning can result in lower intelligence ratings and disturbed personalities. Children in many British cities have a blood lead concentration of 0·4 part per million (ppm) and some children with levels of 0·8 ppm develop clinical symptoms. Blood concentrations of 0·3–0·5 ppm are considered by some authorities to be undesirable.

In addition to the toxic effects on the nervous system, lead can also become incorporated into bones and teeth, in similar proportions to calcium. Although this may lead, in severe cases, to a weakening of the bones, it is of less immediate importance than the toxic effects on the nervous system. Lead is also known to cause anaemia: the major proportion of the lead in the blood is in the red blood cells and lead has been shown to inhibit all steps in haem synthesis. Particularly important are the inhibition of both amino acid synthetase and the incorporation of iron into the porphyrin ring catalysed by ferrochelatase. The action of lead on this process is shown in *Fig.* 35.12.

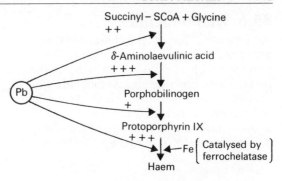

Fig. 35.12. **The effect of lead on haem synthesis.** The number of plus signs indicates the efficiency of the inhibition.

Organic lead

The organic form of lead in petrol, lead tetraethyl [$Pb(C_2H_5)_4$], is the form of organic lead in the human environment that is of most concern. Much controversy has developed over two issues: firstly, whether inhalation of exhaust fumes containing lead is a major human source of the metal and, secondly, whether lead-free petrols should be introduced. A large percentage of the petrol sold in the United States is now lead free, but some European countries have, in general, been more reluctant to remove lead from petrol and are less convinced by the arguments put forward for its removal. The main question in this issue is whether this form of lead provides a significant proportion of the lead entering the human body.

Like organic forms of mercury, lead tetraethyl has a very high affinity for lipids in the cells and, therefore, becomes incorporated into lipids of the nervous system so exerting its effects in this tissue. It has recently been discovered that lead tetraethyl can undergo oxidative metabolism in the liver endoplasmic reticulum, to the free radical $\overset{+}{Pb}(C_2H_5)_3$ and it has been suggested that this very reactive species is the main agent causing cellular damage. However, although this metabolism may explain damage in tissues, such as the liver, it has not been demonstrated to occur extensively in nerve tissue and transport of this reactive radical to the brain would be very unlikely. The exact mechanism of brain damage by lead, therefore, remains unclear.

35.6 Radionuclides

An entirely different type of metal toxicity arises when any of a group of metals, described as bone-seeking isotopes, gain access to the body. These bone-seeking isotopes are all divalent metals sufficiently similar to calcium to become incorporated into the hydroxyapatite-type structure of the bone. Typical members of this group are shown in *Table* 35.4.

Table 35.4 **Typical bone-seeking isotopes**

Element	Isotope	Type of radiation	Half-life
Radium	^{224}Ra	α, β, γ	3·6 days
Radium	^{226}Ra	α, β, γ	1602 years
Thorium	^{232}Th	α, β, γ	$1·4 \times 10^{10}$ years
Plutonium	^{239}Pu	α, β, γ	Stable
Strontium	^{90}Sr	β	28 years

Of these elements, an isotope of plutonium, ^{239}Pu, that is widely used in nuclear reactors of power stations is a potential hazard. Nuclear explosions, now vigorously controlled by international agreement, can produce over 200 radioactive isotopes of 30 elements. Most of these are harmless, firstly because they are relatively insoluble and not absorbed into the body, and secondly because they have short half-lives. One of the most serious hazards is provided by the strontium isotope, ^{90}Sr. Fission of uranium and plutonium yields 3·5 per cent krypton which, in turn, forms ^{90}Sr by the pathway shown in *Fig.* 35.13. Strontium possesses a very similar size and electronic configuration to calcium and is readily taken up

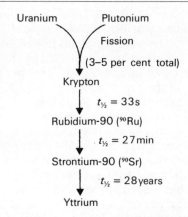

Fig. 35.13. **Formation of strontium-90 (^{90}Sr) from uranium or plutonium fission.**

into bone hydroxyapatite, and since it possesses a half-life of 28 years, it is a relatively stable isotope.

If the quantity of radioactive isotope incorporated into the bone is small, no serious consequences are likely to ensue and, in 1970, it was calculated that human bones, on average contained a radioactive component equivalent to $0·5–1·0$ pCi/g Ca^{2+}. One pCi (picocurie) delivers a dose to the bone of $0·003$ rad/year, a dose well below the maximum safe dose agreed by experts.

If, however, the quantity of radioactive isotope, such as ^{90}Sr, becoming incorporated into the bone is large, then serious consequences may follow. Ionizing radiations, such as β rays or γ rays can initiate production of tumours (sarcomas) by the bone cells or the transformation of blood-forming cells in the bone marrow into leukaemic cells.

30.7 Removal of toxic metals: chelation therapy

It would clearly be very desirable if excess deposits of toxic metals, such as iron or copper, could be removed from the tissues and excreted. This is a difficult problem, but through understanding of the chemistry of metal chelates, a number of very useful compounds has been developed.

Earlier in this chapter, the very high affinity of copper for the —SH group was described, and this knowledge led to the development of a very effective drug for removing copper from the tissues, penicillamine (*Fig.* 35.14). This drug can be used effectively for the treatment of Wilson's disease and mercury poisoning, since it provides two ligands, —SH and —NH$_3^+$ that bind powerfully to copper.

The removal of radionuclides from bone presents a much more difficult problem. As discussed earlier, metals such as Sr^{2+} do not bind to —SH or —NH$_3^+$ ligands, but do bind to —COO$^-$. A chelator strongly binding Sr^{2+} must, therefore, possess several —COO$^-$ groups. Examples of molecules that bind metals via —COO$^-$ ligands are ethylenediamine tetraacetic acid (EDTA) and diethylenetriamine pentaacetic acid (DTPA) as shown in *Fig.* 35.15.

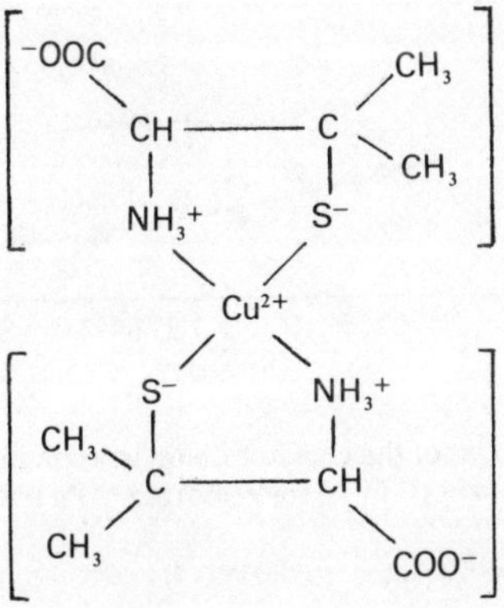

Fig. 35.14. **Binding of copper by penicillamine.**
Two molecules of penicillamine are shown in brackets.

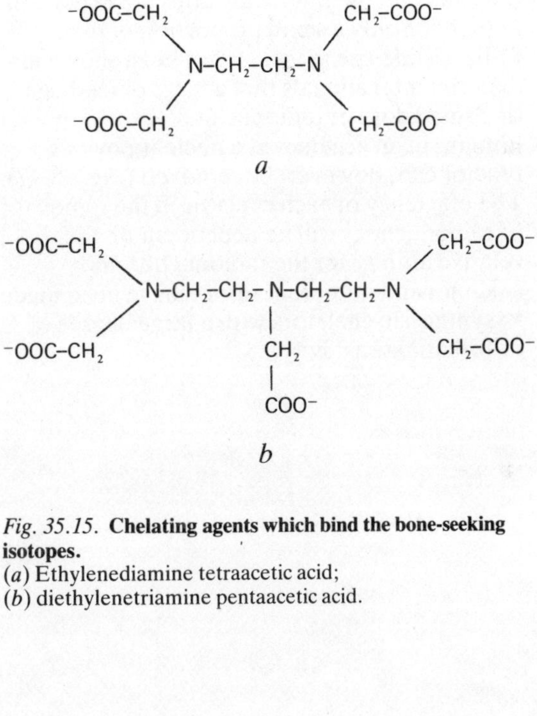

Fig. 35.15. **Chelating agents which bind the bone-seeking isotopes.**
(*a*) Ethylenediamine tetraacetic acid;
(*b*) diethylenetriamine pentaacetic acid.

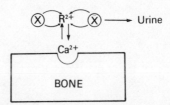

Fig. 35.16. **Use of chelators for the removal of isotopes (R^{2+}) from bone.**
Competition of two metals for the chelator.

Chelators of this type are, however, very toxic since they also bind calcium very strongly and drastically reduce calcium concentrations in the blood, giving rise to the well-known serious consequences of a fall in plasma calcium concentration. The problem is circumvented by treatment with Ca^{2+} chelator complexes. The uptake of another metal, for example ^{239}Pu by EDTA, will depend on the relative affinity of ^{239}Pu for the chelator as compared with that of calcium. If the affinity is only slightly greater than that of calcium, a slow excretion of the radionuclide as a chelate complex can be achieved (*Fig.* 35.16). It should be noted that complete exchange of the radionuclide, R^{2+}, for Ca^{2+} does not occur and in practice only a small proportion of the radionuclide exchanges. It has been shown in experimental animals that a large percentage of a small dose of radionuclide received by humans in an accident at a nuclear power reactor can, however, be removed (*Fig.* 35.17). The efficiency of each chelator in the removal of radionuclides will be dependent on the relative affinity for the radionuclide and calcium, and numerous efforts have been made to synthesize chelators with a large degree of differential sensitivity.

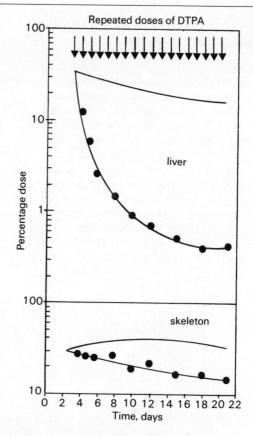

Fig. 35.17. **Use of the chelator diethylenetriamine pentaacetic acid (DTPA) for removing ^{239}Pu from the tissues.**
(———) Untreated; (–·–·–) treated.

Chapter 36

Metabolism of xenobiotics: xenobiochemistry

The metabolism discussed in this chapter has been described under many titles. The term 'xenobiotic' means a foreign compound. It, therefore, is usually of 'no biological value' and includes a vast range of chemicals, many organic, which are distinct from those of proven biological value, such as glucose, amino acids or vitamins. Recently the subject has greatly expanded and the term 'xenobiochemistry' is now frequently used to describe this large field of study.

During past years many other titles were used to describe this subject. The terms 'foreign substances' or 'anutrients' could replace 'xenobiotics' and xenobiochemistry was often described as 'drug metabolism', since many of the 'xenobiotics' were drugs used in clinical treatment. The first descriptive phrase used for xenobiotic metabolism was 'detoxication processes' or 'detoxication mechanisms' and, since many xenobiotics are toxic, this title must have arisen from the fact that many of the metabolic processes lead to a reduction in toxicity of the ingested xenobiotic. Indeed, the first authoritative book on the subject by R. T. Williams was entitled *Detoxication Mechanisms*. Metabolism may, however, lead to the formation of more toxic products and so this useful, but misleading, phrase is much less frequently used in current literature.

36.1 Origins of xenobiotics and mode of entry into the body

Xenobiotics, of which there are many tens of thousands, can enter the body by various

Table 36.1 **Sources of foreign compounds**

1. *Pharmaceuticals*
   e.g. Aspirin
   Tranquilizers (e.g. chlorpromazine, imipramine)
   Oral contraceptives

2. *Industrial chemicals*
   Solvents–benzene, CCl₄, trichloroethylene
   Dyestuffs and precursors (β-naphthylamine)
   Detergents
   Bleaching agents

3. *Cosmetics*
   Lipsticks
   Hair sprays
   Hair dyes (*p*-phenylene diamine)

4. *Food additives (industrial processing)*
   Colour: butter yellow
                   (*p*-dimethylamino-azobenzene)
                   azo dyes (sudan I, tartrazine)
   Sweeteners: cyclamates
                   saccharin
   Antioxidants: e.g. BHT*
                   propyl gallate

5. *Pesticides*
   Insecticides: DDT
                   Aldrin
                   Dieldrin
                   Parathion
   Herbicides

6. *Food anutrients (normal constituents)*
   Pigments—anthocyanins
   Oils—terpenes
   Caffeine

7. *Bacterial metabolites—toxins*
   Formed in gastrointestinal tract
   (e.g. amines—putrescine, cadaverine)

* BHT = 2,6-Di-*tert*-butyl-4-methylphenol.

445

routes. These may be classified as accidental, including, for example, the inhalation of toxic chemicals from polluted atmospheres or ingestion of contaminated food, or as deliberate. The latter term is usually used in drug treatment when drugs enter the body by ingestion or injection. A summary of the major sources of these compounds is shown in *Table* 36.1.

The following points are of particular interest. Although exposure to industrial chemicals is primarily a hazard for workers in various industries, frequent exposure to some chemicals, e.g. detergents or cosmetics, can occur during normal domestic use.

Food colouring is a well-established practice used to make food more attractive to the purchaser, but colouring additives have to be used with caution. In fact one such additive for margarine, 'butter yellow', was found to be carcinogenic and is no longer in use.

Similarly, many crops grown for food are treated with pesticides and the surrounding earth treated with herbicides. A proportion of chemicals used for treatment are then incorporated into the plants and, after eating, are incorporated into the tissues of the body. Insecticides which are lipophilic, e.g. DDT, are taken up by the fat depots and may be stored for long periods in the body.

Most of the chemicals listed in groups 1–5 in *Table* 36.1 are man-made and it could be thought that avoidance of such chemicals would completely eliminate the problem of exposure of the body to toxic chemicals. This is not the case, however, since toxic chemicals can be introduced into the body by eating plants uncontaminated by man-made chemicals. Toxic substances can arise from two sources: the plants themselves may contain toxic chemicals or toxic substances may be produced by the action of bacteria on completely harmless natural compounds in the gastrointestinal tract. For example, toxic amines can be produced by decarboxylation of unabsorbed amino acids. There is currently much public enthusiasm for the consumption of 'natural foods' and, although there is some justification for this view, it should be appreciated that consumption of 'natural foods' is not without its hazards.

36.2 General properties of xenobiotic metabolites

The ability of the mammalian body to convert ingested toxic materials into less toxic metabolites was first demonstrated just over a hundred years ago. It had been observed that ingestion of phenol by animals led to the excretion of phenyl sulphate in the urine.

Phenyl sulphate is much less toxic than phenol and from this and the discovery of other metabolites of reduced toxicity, the concept of 'detoxication' was gradually developed. Studies of the excretory products of large numbers of xenobiotics demonstrated, however, that although most of them were metabolized to less toxic products, several exceptions were found, one being the conversion of pyridine to the much more toxic methylpyridine (*Table* 36.2).

Two additional general principles concerning xenobiotic metabolites have come to light: most of the metabolites are more water soluble than the xenobiotic (*see Table* 36.3) and the majority are more acidic (*see Table* 36.4).

Table 36.2 **Toxicity of xenobiotics and their metabolites**

Toxicity or LD_{50} of xenobiotic, g/kg mice	Drug	Metabolite	Toxicity or LD_{50} of metabolite, g/kg mice
2·0	Benzoic acid	→ Hippuric acid	4·15
2·85	p-Aminobenzoic acid	→ p-Aminohippuric acid	4·93
1·8	Sulphadiazine	→ Acetylsulphadiazine	0·6
1·2	Pyridine	→ Methylpyridine chloride	0·22

LD_{50} is the dose giving 50 per cent deaths.

Table 36.3 **Solubility of xenobiotics and their metabolites**

Xenobiotic solubility, mg/100 ml H_2O	Xenobiotic	Metabolite	Metabolite solubility, mg/100 ml H_2O
184	Benzoic acid	→ Hippuric acid	463
915	Phenylacetic acid	→ Penaceturic acid	1145
		Phenylacetylglutamine	117
1138	Picric acid	→ Picramic acid	65
1480	Sulphanilamide	→ N-Acetylsulphanilamide	534

Table 36.4 **Acidity of xenobiotic metabolites**

Xenobiotic, pK_a		Metabolite		Metabolite, pK_a
10·0	Phenol	→ Phenylglucuronide		3·4
4·2	Benzoic acid	→ Hippuric acid		3·7
	Benzene	→ Phenylmercapturic acid		3·7
		Phenaceturic acid		3·7
2·9	o-Chlorobenzoic acid	→ o-Chlorohippuric acid		3·8

36.3 Role of the liver in xenobiotic metabolism

Although several tissues of the body can metabolize xenobiotics to a limited extent, the liver is the main site of metabolism and, following absorption into the body, most xenobiotics undergo metabolism in the liver (*Fig.* 36.1).

The metabolites produced by the liver are generally relatively water soluble and are passed into the blood for excretion by the kidneys. A limited proportion of metabolites is

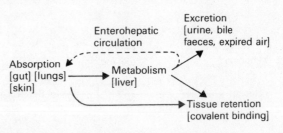

Fig. 36.1.   **Tissue inter-relationships in metabolism of xenobiotics.**

excreted via the faeces or expired air. Although this is the commonest method of processing xenobiotics, two other pathways for metabolite disposal may occur. Metabolites may be returned to the gut by the enterohepatic circulation or they may bind to a cellular macromolecule and be retained. This binding can cause serious consequences for the cell, such as a DNA mutation which is an important consequence of the metabolism of procarcinogens and is discussed fully in Chapter 38.

36.4 Phase I and phase II reactions

Several chemical transformations are used by the liver in the metabolism of xenobiotics.

These include conjugation, hydrolysis, reduction and oxidation, of which conjugation and oxidation are the most important. Occasionally a single transformation occurs, but more frequently two processes are involved, e.g. oxidation followed by conjugation. It is, therefore, useful to classify xenobiotic metabolism into phase I and phase II metabolism (*Fig.* 36.2). Phase I metabolites may be excreted after the xenobiotic has undergone oxidation, reduction or hydrolysis, but if the product is subsequently conjugated a phase II metabolite is produced. Direct conjugation may also occur so that a conjugate is formed and excreted without preliminary metabolism. These processes will be considered in detail, typical examples of phase I and phase II metabolism being shown in *Fig.* 36.3.

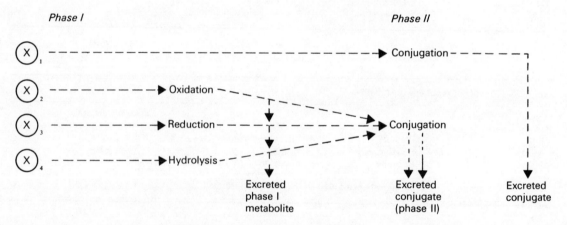

Fig. 36.2. **Phase I and phase II metabolism. X = xenobiotics.**
The pathway shown for X_2, i.e. oxidation followed by conjugation, is the most common pathway.

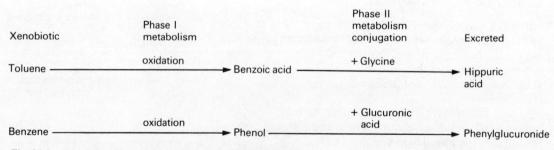

Fig. 36.3. **Examples of phase I and phase II xenobiotic metabolism.**

36.5 Conjugation reactions of xenobiotics

Xenobiotics can be conjugated directly, but as mentioned in Section 36.4 conjugation occurs more frequently as a phase II reaction subsequent to a preliminary modification of the molecule by, for example, oxidation. The purpose of conjugation is to make the xenobiotic more polar, more water soluble and, therefore, more easily excreted in the urine. Most conjugates are much less toxic than the parent xenobiotic but, in a few cases noted in the introduction, increased toxicity can result from conjugation.

Table 36.5 **Molecules used for conjugation**

Source	Conjugating molecule	Products
Carbohydrate	Glucuronic acid	→ Glucuronide
	Glucose	→ Glucoside
	Ribose	→ Riboside
	Xylose	→ Xyloside
Amino acids	Glycine	—
	(Glutathione	→ Mercapturic acids)
	Orithine	—
	Glutamine	—
Derived from amino acids	(Cysteine)	→ Sulphate
	(Methionine)	→ Methylated derivative
Various	Acetyl-SCoA	→ Acetylated derivative

Where no name is specified for the product this varies with the molecule with which conjugation occurs.

The wide range of molecules used for conjugation is shown in *Table 36.5*, although they are not all used to the same extent, e.g. conjugation with glucuronic acid is much more common than with glucose. Also the use of different conjugating molecules often varies with the species, so that conjugation with ornithine is common in birds but rare in mammals, and phenylacetic acid is conjugated with glycine in mammals but with glutamine in man and chimpanzee. Multiple forms of conjugation occur quite frequently; for example phenol will, in most mammals, be converted to both glucuronide and sulphate.

Conjugation reactions are synthetic processes and, therefore, energy requiring. The conjugation of xenobiotic X with molecule A can take place via two different pathways in which either the xenobiotic or the conjugating molecule is activated

i. A + High-energy molecule → Ⓐ (Active)

$$\boxed{X} \; + \; Ⓐ \;(\text{Active}) \; \rightarrow \; \boxed{X}\text{—}Ⓐ$$

ii. X + High-energy molecule → $\boxed{X}$ (Active)

$$\boxed{X} \; (\text{Active}) \; + \; Ⓐ \; \rightarrow \; \boxed{X}\text{—}Ⓐ$$

The main conjugating molecules can be divided into eight groups for discussion.

a. Glucuronides

In the formation of glucuronides the conjugating molecule is activated, the initial stage being identical to that used in glycogen synthesis (*see* Chapter 5) but differing in that the UDP-glucose is oxidized to the glucuronic acid form before transfer occurs

$$\text{UTP + Glucose 1–}Ⓟ \longrightarrow \text{UDP–Glucose} + Ⓟ\text{–}Ⓟ$$

$$\text{UDP-glucose + NAD}^+ \xrightarrow[\text{dehydrogenase}]{\text{UDP-glucose}} \text{UDP-glucuronic acid + NADH}$$

$$\text{UDP-glucuronic acid + X–OH} \xrightarrow[\text{transferase}]{\text{Glucuronyl}} \text{XO-glucuronate + UDP}$$

+ ATP

Conjugation with glucuronide is an important process and many drugs are excreted as glucuronides, examples being shown in *Table* 36.6.

Table 36.6 Examples of groups and drugs which form glucuronides

Site of binding of glucuronide	Example of drug
Hydroxyl group	
Phenolic	Morphine
Alcoholic	Chloramphenicol
Carboxyl	
Aromatic	Salicylic acid
Aliphatic	Indomethacin
Sulphydryl	Mercaptobenzothiazole
Amino	Dapsome
Imide	Sulphathiazole

b. Glucosides

Conjugation directly with sugars, such as glucose or xylose, was originally believed to occur only in the lower species of the animal kingdom. However, it has been demonstrated, more recently, that small amounts of glucosides and xylosides are excreted in the urine of mammals including man, although they are of much less importance than the glucuronides.

c. Glycine

An important conjugate of glycine with benzoic acid, called 'hippuric acid', is one of the oldest xenobiotic conjugates known, having been described by Liebig in 1829. It was found in the urine of cows and horses and takes its name from the Greek for horse—hippos.

Glycine conjugates are formed from many aromatic carboxylic acids, e.g. benzoic or phenylacetic acid, and their formation involves activation of the xenobiotic:

d. Glutathione

The existence of sulphur-containing detoxication products in the urine was known over a hundred years ago. Sulphur appears in several different forms in the urine, one important group of sulphur compounds being described as the 'mercapturic acids'. These were isolated from the urine of animals that had been fed small quantities of halogenated aromatic compounds. They were unusual in that the sulphur they contained was bound in an organic molecule rather than in the inorganic form which is also present in relatively large quantities in the urine (Chapter 30).

Only recently has the role of the tripeptide, glutathione, been established as a necessary factor in their formation. A very wide range of organic compounds are now known to be excreted after conjugation with glutathione, including aromatic halogenated compounds and nitro compounds, aliphatic halides, sulphate and nitro compounds, alkene halides and the carcinogenic epoxides formed from polycyclic hydrocarbons (cf. Chapter 38). The formation of mercapturic acids (from a typical xenobiotic) is shown in *Fig.* 36.4.

The initial reaction is sometimes spontaneous, the neutrophilic glutathione reacting readily with an electrophilic centre on the xenobiotic, although an enzyme glutathione transferase is often required. These transferase enzymes are clearly important in cellular metabolism and make up nearly 10 per cent of the total cytosolic protein in rat liver and 2 per cent in human liver. The glutathione conjugate can be excreted as such, but other reactions frequently take place leading to the formation of mercapturic acids. In this reaction, the glutamate residue and then the glycine residue is split off, leaving only the cysteinyl residue bound, which is acetylated before excretion (*Fig.* 36.4).

$$X—COOH + ATP \rightarrow X—CO—AMP + \text{\textcircled{P}}—\text{\textcircled{P}}$$

$$X—CO—AMP + CoASH \xrightarrow{\text{Acylthiokinase}} X—CO—SCoA + AMP$$

$$X—CO—SCoA + H_2N—CH_2—COOH \xrightarrow{\text{N-Acyltransferase}} X—CO—NH—CH_2—COOH + CoASH$$

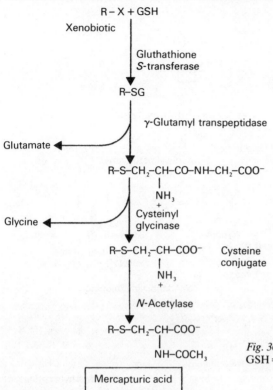

Fig. 36.4. **Formation of mercapturic acids.** GSH = glutathione.

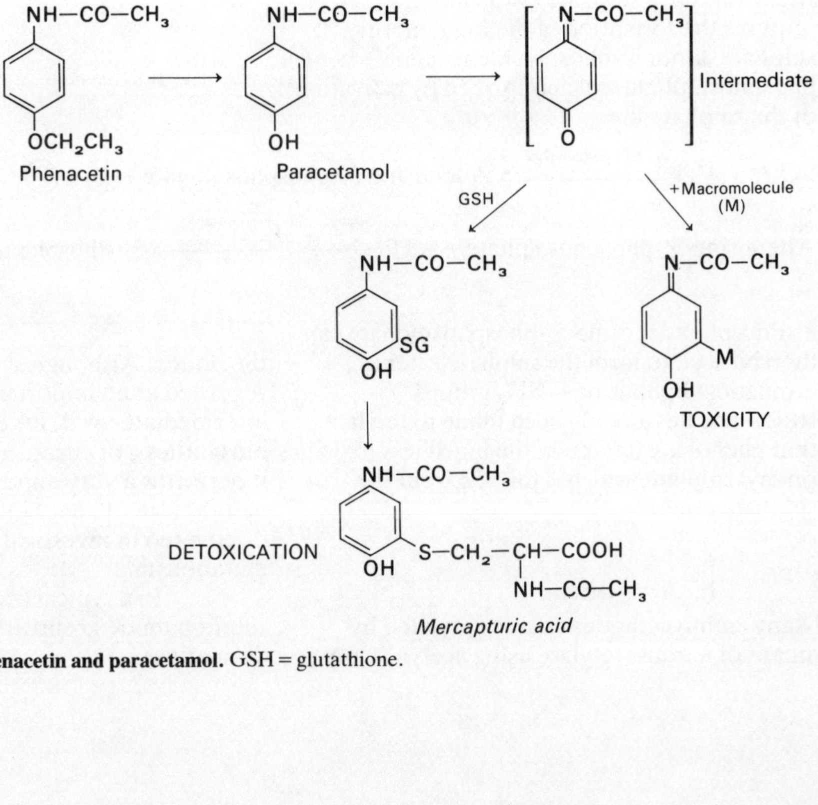

Intake of a large dose of a drug, such as phenacetin, can lead to extensive depletion of glutathione in the liver. If this occurs, then the drug metabolite, e.g. an epoxide formed by oxidation, is free to bind to cell macromolecules such as protein or nucleic acids, and severe cellular damage can result (*Fig.* 36.5).

e. Glutamine

Conjugation with glutamine occurs in the human liver, but is of relatively minor importance.

f. Methylation

The methyl group of *S*-adenosylmethionine is frequently used to methylate xenobiotics

Fig. 36.5. **Metabolism of phenacetin and paracetamol.** GSH = glutathione.

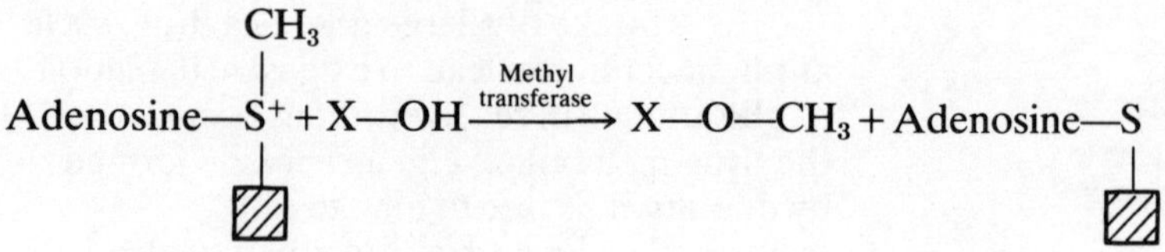

$$\text{Adenosine}\!-\!\overset{\overset{\displaystyle CH_3}{|}}{S^+}\!+\!X\!-\!OH \xrightarrow[\text{transferase}]{\text{Methyl}} X\!-\!O\!-\!CH_3 + \text{Adenosine}\!-\!S$$

S-Adenosylmethionine   *S*-Adenosylhomocysteine

Methylation can occur at oxygen, sulphur and nitrogen atoms, including the nitrogen of a heterocyclic ring, and these are referred to as *O*-methylation, e.g. R—OH → R—O—CH$_3$, *N*-methylation, e.g. R—NH$_2$ → R—NH—CH$_3$ and '*S*'-methylation, e.g. R—SH→R—S—CH$_3$. Many of the methylated products are less toxic than the parent xenobiotic although some are more toxic and, as with xenobiotic metabolism, in general, the methylation bears no relation to xenobiotic toxicity, e.g. an excess dose of the non-toxic B vitamin nicotinamide will be methylated and excreted as the more toxic methylnicotinamde (*see Fig.* 36.6).

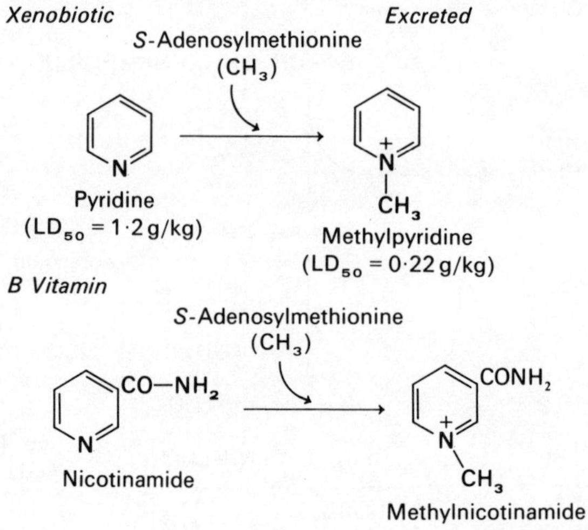

Fig. 36.6. **Examples of methylation.**

g. Sulphate esters

Biosynthesis of sulphate esters initially requires the formation of the high-energy sulphate donor 3'-phosphoadenosine-5'-phosphosulphate which is formed by activation of the sulphate ion

$$SO_4{}^{2-} + ATP \underset{}{\overset{\text{ATP-sulphurylase}}{\rightleftharpoons}} \text{Adenosine-5'-phosphosulphate} + \textcircled{P}\!-\!\textcircled{P}$$

$$\text{Adenosine-5'-phosphosulphate} + ATP \xrightarrow[\text{phosphokinase}]{\text{Adenosine-5'-phosphosulphate}} \text{3'-Phosphoadenosine-5'-phosphosulphate} + ATP$$

3'-Phosphoadenosine-5'-phosphosulphate can then be used to form the sulphate ester of aromatic, aliphatic or —NH$_2$ groups. Reference has already been made to the fact that phenol is excreted as the much less toxic phenyl sulphate which is formed from the sulphate donor.

h. Acetylation

Many amino compounds are acetylated by means of a transacetylase using acetyl-SCoA as the donor. Although acetyl-SCoA is commonly regarded as an important metabolic intermediate used, for example, in the biosynthesis of citrate and fatty acids, it performs a very important role in the metabolism of xenobiotics and was first discovered in investigations of drug metabolism.

In a typical acetylation, drugs of the sulphonamide group are converted to acetyl derivatives

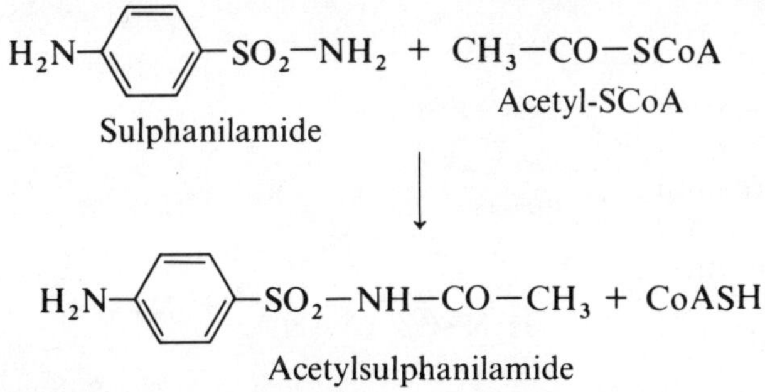

$$H_2N-\langle\text{benzene}\rangle-SO_2-NH_2 \;+\; CH_3-CO-SCoA$$

Sulphanilamide   Acetyl-SCoA

$$H_2N-\langle\text{benzene}\rangle-SO_2-NH-CO-CH_3 \;+\; CoASH$$

Acetylsulphanilamide

Indeed this is an example of a drug being made more toxic by metabolism, since the acetyl derivative, acetyl sulphanilamide produced is less soluble than the drug and tends to be deposited in the kidney, causing renal failure in severe cases.

36.6 Reduction in xenobiotic metabolism

Although oxidation is of much greater importance and is described in detail later (*see* Section 36.8), a limited number of reactions involving reduction do occur and these are illustrated below.

An interesting example of this metabolism is shown by the azo red dye prontosil which on arbitrary testing in the 1930s was found to be remarkably effective in curing mice of septicaemia caused by staphylococcal or streptococcal infections. The reason behind this cure was that prontosil was reduced in the liver producing the effective drug sulphanilamide, although, at the time of observation, the sulphonamide series of drugs had not been discovered

a. Nitro reduction

$$R-NO_2 \rightarrow R-NO- \rightarrow R-NHOH- \rightarrow R-NH_2$$

Aromatic nitro compounds are frequently reduced to the corresponding amine:

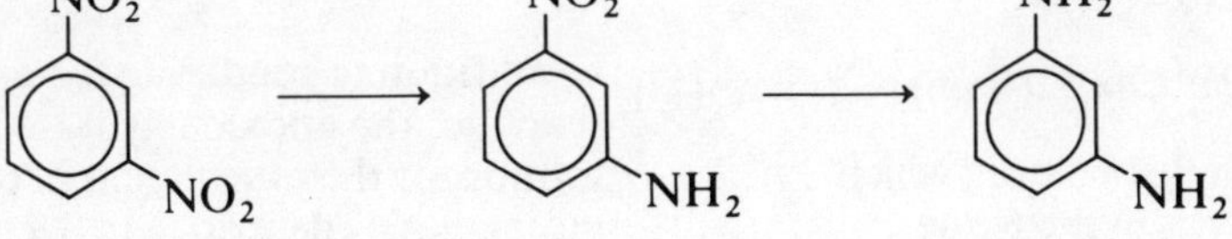

Subsequent metabolism of the amine by, for example, conjugation may then take place.

b. Azo reduction

Amines are also produced by azo reduction:

$$R_1-N{=}N-R_2 \rightarrow R_1-NH-NH-R_2 \rightarrow R_1-NH_2 + H_2-R_2$$

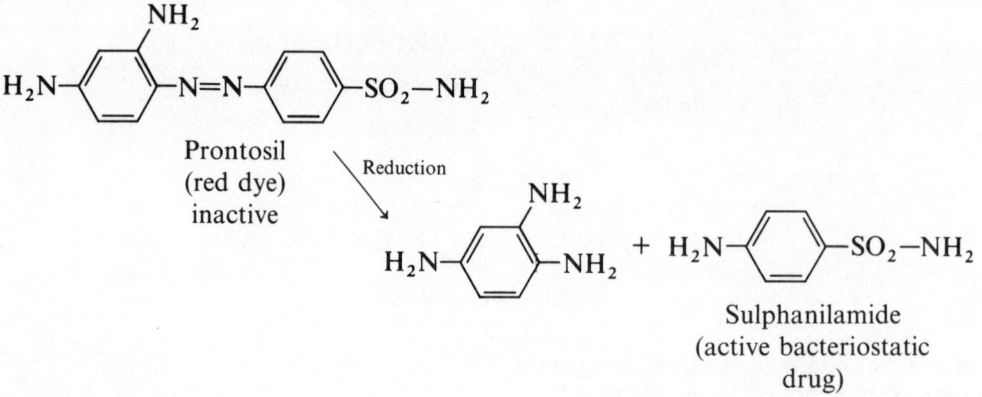

Prontosil
(red dye)
inactive

Reduction

Sulphanilamide
(active bacteriostatic
drug)

c. Reduction of halogenated compound

Some halogenated compounds undergo the following reduction:

$$R—CCl_3 \rightarrow R—CH—Cl_2$$

The hypnotic chloral is, in fact, metabolized by this means:

$$CCl_3—CHO \xrightarrow{2H} CCl_3—CH_2OH$$

36.7 Hydrolysis of xenobiotics

Some xenobiotics undergo hydrolysis and this can occur before other metabolism or as a phase II reaction subsequent to oxidation. The types of reactions include ester hydrolysis, amide hydrolysis and epoxide hydration.

a. Ester hydrolysis

$$R_1—CO—OR_2 \xrightarrow{H_2O} R_1—COOH + R_2—OH$$

Aspirin is an example of a drug which undergoes hydrolysis in the tissues.

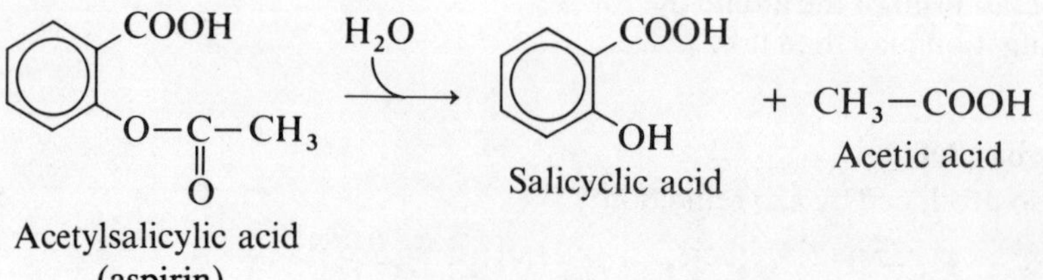

Acetylsalicylic acid
(aspirin)

Salicyclic acid

+ $CH_3—COOH$
Acetic acid

b. Amide hydrolysis

$$R—CO—NH_2 \xrightarrow{H_2O} R—COOH + NH_3$$

c. Epoxide hydration

Epoxides are frequently formed during oxidative metabolism of aromatic and some aliphatic xenobiotics and this is discussed in detail later. For example:

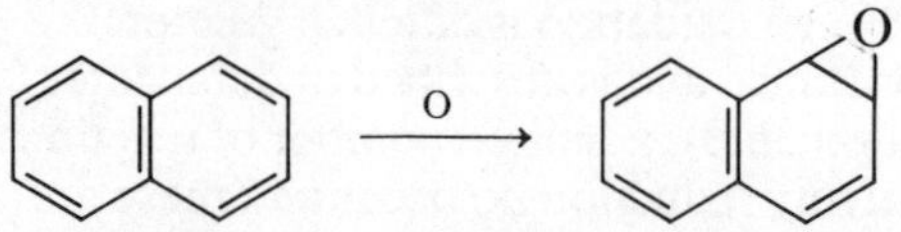

These epoxides are electrophiles and very reactive, but they are converted to relatively inactive hydroxyl compounds by an enzyme known as epoxide hydrase or hydratase.

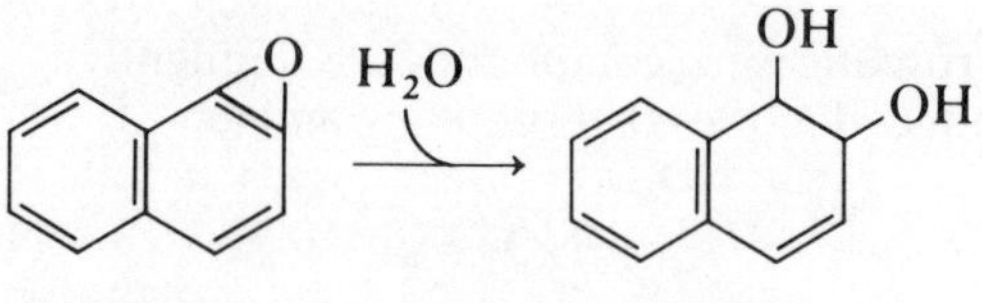

In addition to rendering the metabolites less reactive, the epoxide hydrase provides —OH groups for the attachment of conjugating molecules as described in Section 36.5.

36.8 Oxidative metabolism of xenobiotics

a. Introduction

Earlier investigations of xenobiotic metabolism demonstrated that xenobiotics underwent many different chemical transformations that appeared to be unrelated and therefore dependent on many different enzymes. For example, some compounds were converted to hydroxyl derivatives:

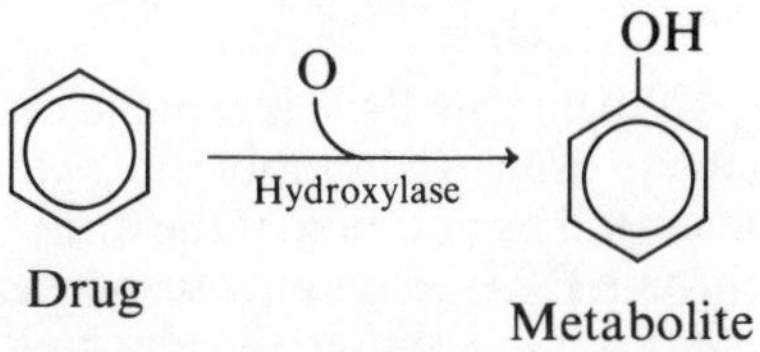

Drug

Metabolite

The enzymes carrying out catalysis of this type were described as 'hydroxylases' whilst the demethylation of other compounds was apparently catalysed by 'demethylases'.

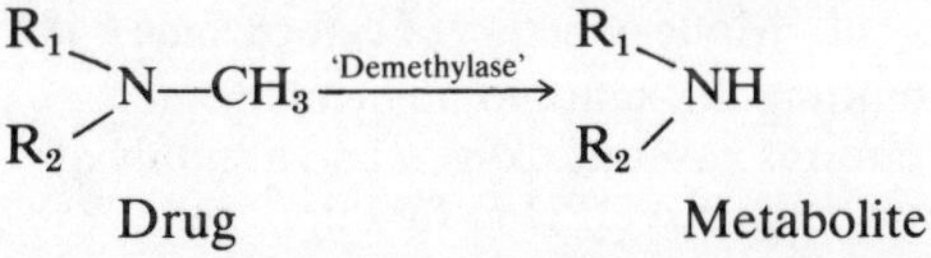

$$R_1, R_2 N-CH_3 \xrightarrow{\text{'Demethylase'}} R_1, R_2 NH$$

Drug

Metabolite

However, more careful study of the metabolism of large numbers of drugs and xenobiotics revealed that oxidation played an important role and that many reactions, such as demethylation, were oxygen dependent. The many and varied types of oxidations are summarized in *Fig.* 36.7.

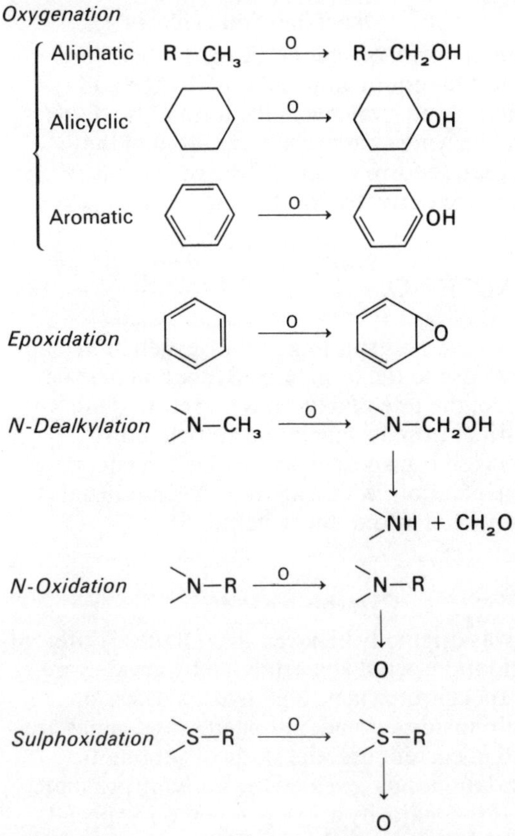

Oxygenation

Aliphatic $R-CH_3 \xrightarrow{O} R-CH_2OH$

Alicyclic

Aromatic

Epoxidation

N-Dealkylation $>N-CH_3 \xrightarrow{O} >N-CH_2OH$

$>NH + CH_2O$

N-Oxidation $>N-R \xrightarrow{O} >N-R$

Sulphoxidation $>S-R \xrightarrow{O} >S-R$

Fig. 36.7.   **Role of oxidation in the metabolism of xenobiotics.**

These oxidations constitute an important form of 'phase I' metabolism. This is particularly relevant to the introduction of an —OH group into an aromatic ring, since this group then provides an important attachment site for a conjugating group such as a glucuronide or sulphate (*see Fig.* 36.8).

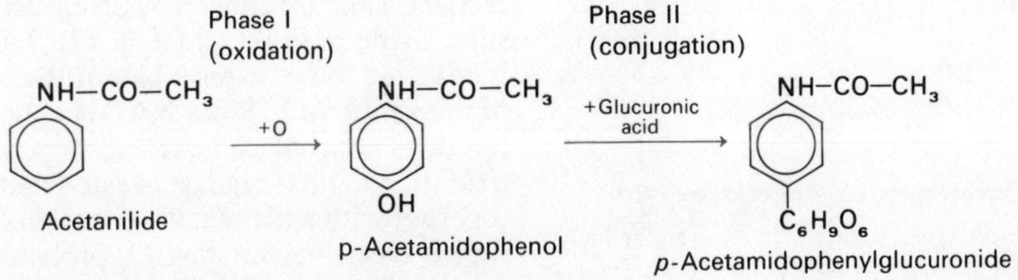

Phase I (oxidation)

Phase II (conjugation)

Acetanilide

p-Acetamidophenol

$C_6H_9O_6$

p-Acetamidophenylglucuronide

Fig. 36.8.   **Oxidation (phase I) followed by conjugation (phase II).**

b. 'Mixed function oxidases'

The majority of xenobiotic oxidations were found to require atmospheric oxygen and reducing equivalents in the form of NADPH. The enzyme system used one atom of the oxygen to form water and the other to form the oxidized derivative of the xenobiotic.

$$\text{NADPH} + \text{O}_2 + \text{X} + \text{H}^+ \xrightarrow{\text{Oxidative enzyme}} \text{X}-\text{OH} + \text{H}_2\text{O} + \text{NADP}^+$$

This splitting of the oxygen molecule gave rise to the term 'mixed function oxidase'. Here, the fate of oxygen is therefore quite distinct from its fate in the mitochondrial electron transport chain involved in energy conservation, where all the oxygen utilized is converted into water (Chapter 6).

c. Cytochrome P450

It was originally believed that all these different oxidations would be catalysed by a wide variety of specific enzymes, e.g. 'hydroxylases' or 'sulphoxidases' and, although these names are still in current use, the study of xenobiotic oxidations has revealed the unifying principle that the majority are dependent on a special cytochrome, cytochrome P450.

This cytochrome occurs in the liver endoplasmic reticulum, but in other tissues,

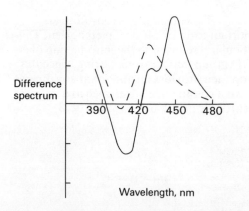

Fig. 36.9.   Difference spectrum of cytochrome P450.

Samples of reduced cytochrome P450 (– – –) and reduced cytochrome P450 treated with carbon monoxide (——) are placed in tubes in a double-beam spectrometer, and the difference spectrum is measured.

such as the kidney or adrenal, it is mainly located in the mitochondria. When a suspension of tissue containing a high concentration of the cytochrome in the reduced form is treated with carbon monoxide, the shift in the absorption spectra leads to the production of a difference spectrum with

a clearly defined maximum at 450 nm (*see Fig.* 36.9), and hence the term 'cytochrome P450'.

Cytochrome P450 is capable of binding many drugs and xenobiotics, which is demonstrated by addition of the drug to cytochrome P450-containing suspensions and which result in a shift in the absorption spectra. Most drugs fall into one of two categories which cause either a type I or type II spectral shift (*see Fig.* 36.10).

d. Mode of action of cytochrome P450

After binding the xenobiotic, cytochrome P450 requires gaseous oxygen and a supply of electrons from NADPH to function (*see Fig.* 36.11 for a simplified scheme). The detailed mechanism of action of the cytochrome is still not completely resolved but it is believed to be that shown in *Fig.* 36.12.

Cytochrome P450 containing iron in the Fe(III) state binds the xenobiotic X (stage A), followed by reduction of the iron to the Fe(II) state by means of electron transfer from NADPH via a flavoprotein (stage B). Oxygen is taken up by the cytochrome (stage C) and is reduced by an electron to the superoxide radical O_2^- (stage D). This radical is believed to be responsible for the oxidation of the xenobiotic. Since NADH enhances but does not replace NADPH-dependent oxidations, it has been suggested that NADH, together with a distinct flavoprotein, may supply electrons for stage D, probably via an intermediate cytochrome, cytochrome b_5 (*see Fig.* 36.12).

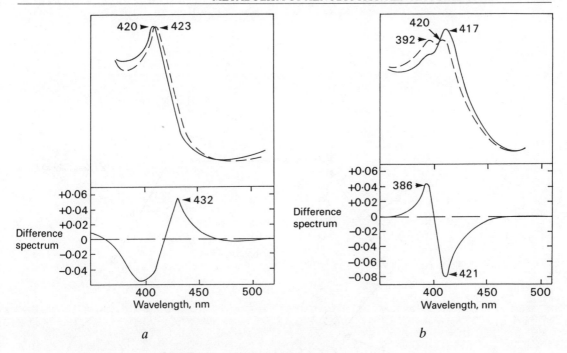

Fig. 36.10. **Spectral shifts caused by binding of drugs to cytochrome P450.**
(*a*) Type I group: (——) untreated; (– – –) + aniline. (*b*) Type II group: (——) untreated; (– – –) + hexobarbital.

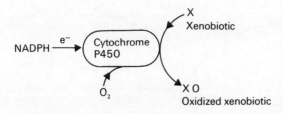

Fig. 36.11. **Simplified concept of function of cytochrome P450.**

e. Binding of cytochrome P450 system to membranes

Cytochrome P450 and its associated flavoprotein is bound, in the liver, to the endoplasmic reticulum membrane. The membrane is essential for maximum activity of the system, the activity being greatly reduced by disruption of the membrane with, for example, detergents or by treatment with phospholipase. Activity of purified components may be partially restored by addition of preparations of phospholipid micelles, demonstrating the vital role played by phospholipids in the system. The cytochrome and flavoprotein molecules are believed to be located in different positions in the membrane as shown in *Fig.* 36.13.

f. Role of cytochrome P450

Cytochrome P450 clearly plays a vital role in the metabolism of xenobiotics including many drugs, examples being the demethylation of morphine and codeine and the hydroxylation of amphetamines. Some drugs undergo multiple forms of oxidative metabolism, e.g. chlorpromazine (*see Fig*. 36.14).

In addition to its importance in the oxidation of xenobiotics, cytochrome P450 is

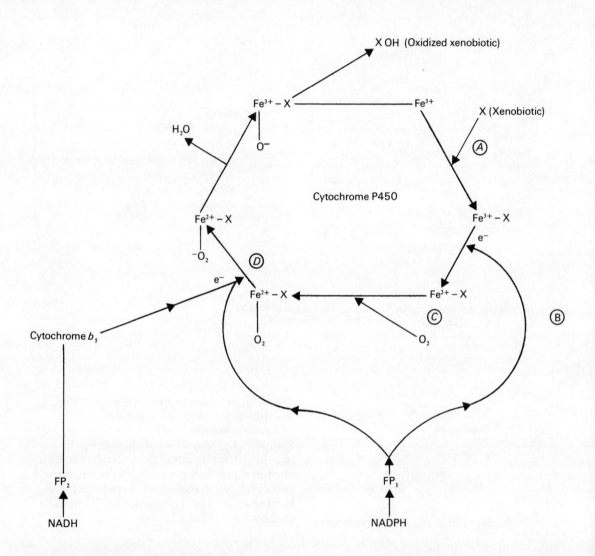

Fig. 36.12. **Mode of action of cytochrome P450.**
FP_1 and FP_2 are flavoproteins.

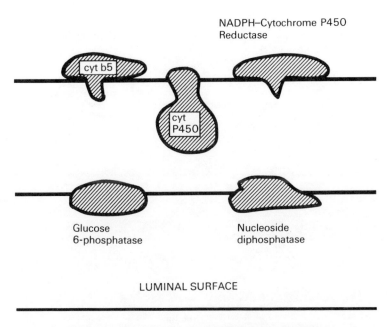

CYTOPLASMIC SURFACE

NADPH–Cytochrome P450
Reductase

cyt b5

cyt
P450

Glucose
6-phosphatase

Nucleoside
diphosphatase

LUMINAL SURFACE

Distribution of enzymes on the endoplasmic reticulum

Cytoplasmic surface   *Luminal surface*

Cytochrome b_5 Nucleoside diphosphatase
NADH–cytochrome b_5 reductase   Glucose 6-phosphatase
NADPH–cytochrome c reductase   Cytochrome P450
Cytochrome P450 β-Glucuronidase
ATPase
5′-Nucleotidase
Nucleoside pyrophosphatase
GDP–mannosyl transferase

Fig. 36.13. **Location of enzymes in the membranes of the
endoplasmic reticulum.**

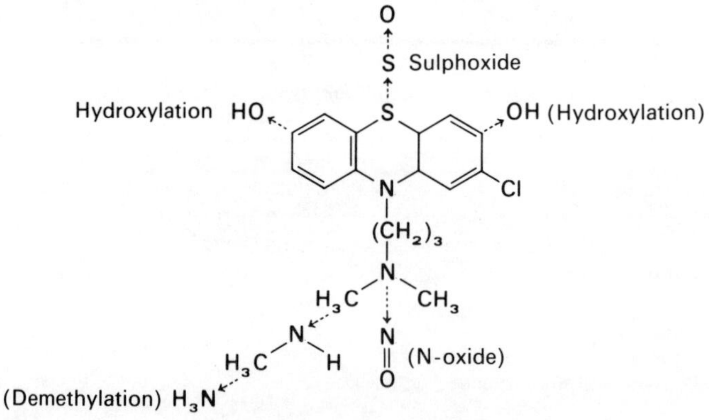

Fig. 36.14. **Oxidative metabolism of typical drugs.**

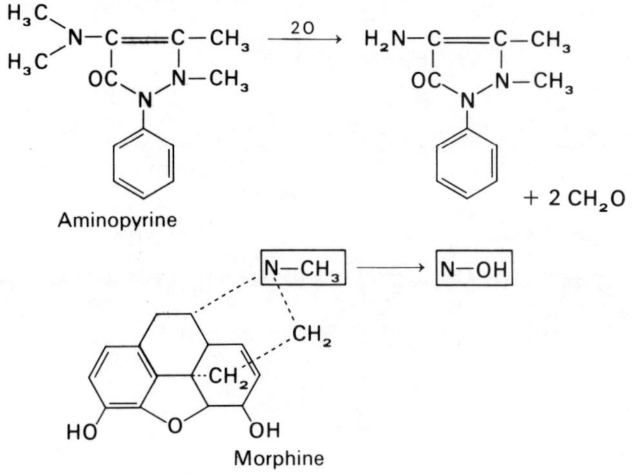

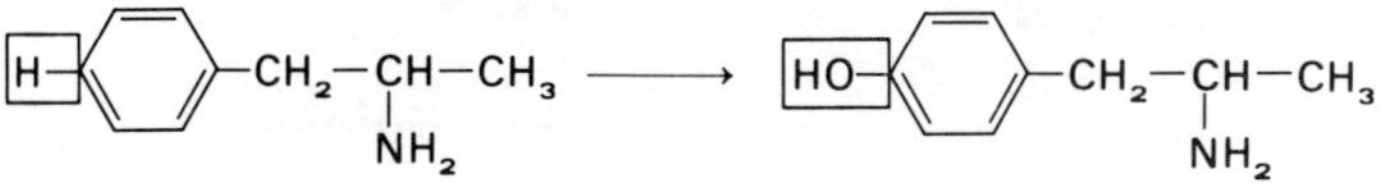

(*a*) Oxidative demethylation, e.g. aminopyrine or morphine; (*b*) hydroxylation,
e.g. amphetamine; (*c*) multiple oxidation of chlorpromazine.

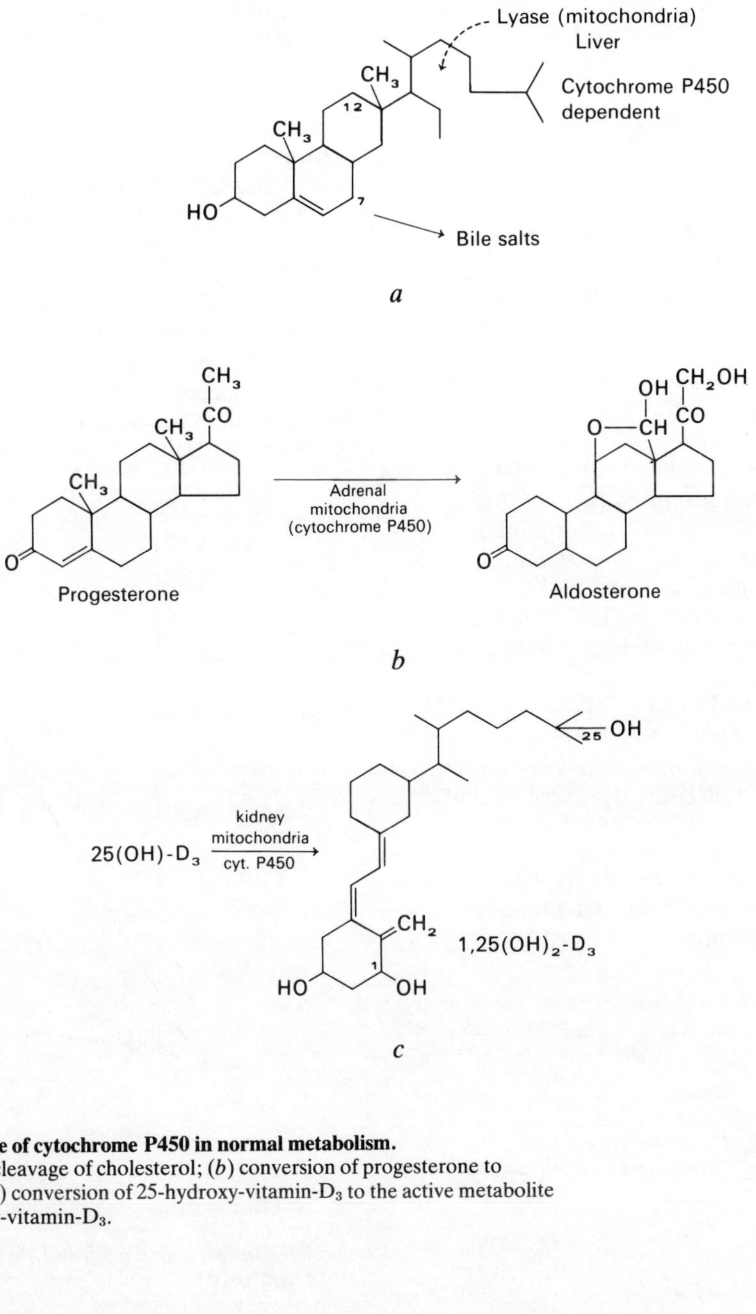

Fig. 36.15. **Role of cytochrome P450 in normal metabolism.**
(*a*) Side chain cleavage of cholesterol; (*b*) conversion of progesterone to
aldosterone; (*c*) conversion of 25-hydroxy-vitamin-D₃ to the active metabolite
1,25-dihydroxy-vitamin-D₃.

also essential for the metabolism of many steroids. It is known to be involved in the side-chain cleavage of cholesterol, an essential stage in the biosynthesis of bile salts. In the adrenal, cytochrome P450 is involved in the oxidation of progesterone to aldosterone and, in the kidney, the 1-hydroxylation of 25-hydroxy-vitamin-D_3 is cytochrome P450 dependent (cf. Chapter 22). A summary of these reactions is shown in *Fig.* 36.15.

g. Specificity and multiple forms of cytochrome P450

The discovery of cytochrome P450 has raised the problem of how a single cytochrome can possibly carry out such a vast range of substrate oxidations, particularly when these are often highly specific. Many different forms of cytochrome P450 are believed, by some, to exist but experimental effort to date has only defined three different forms. Two of these, extracted from rabbit liver, are sometimes referred to as LM_2 and LM_4 and the three forms from rat liver as a, b and c. The three cytochromes show specificity for oxidative demethylation of a drug such as aminopyrine, hydroxylation of the carcinogen benzo(a)pyrene and hydroxylation of testosterone (*see Table* 36.7). Forms of cytochrome P450 with similar specificities have been found in human liver. It is possible that many more forms of cytochrome P450 exist, although an alternative would be the regulation, by other proteins or the lipid membrane, of the specificity of a limited number of cytochromes.

Table 36.7 **Multiple forms of cytochrome P450 in rat liver**

Substrate	Catalytic activity, mmol per min per nmol P450		
	a	*b*	*c*
Aminopyrine (Demethylation)	0·5	7·2	1·6
Benzo(a)pyrene (Hydroxylation)	0·1	0·5	19·2
Testosterone (7α-Hydroxylation)	4·4	<0·1	<0·1

i. The effects of sex and species on oxidative drug metabolism

The rate of oxidative metabolism of a drug such as aminopyrine is much faster in males than in females. This has been clearly demonstrated in the rat where the rate of formation of the oxidative demethylation product by the liver endoplasmic reticulum is 5·75 units in the male and 1·25 units in the female.

This difference is apparently caused by the high concentration of circulating testosterone in the male, since the rate in castrated males is reduced to a value close to that of the females. It can be restored to the original rate by injection of testosterone, and this is shown clearly by the Lineweaver–Burke plots of $1/v$ against $1/[S]$ for oxidative demethylation in the male and female animals (*Fig.* 36.16). Rates of oxidative metabolism of

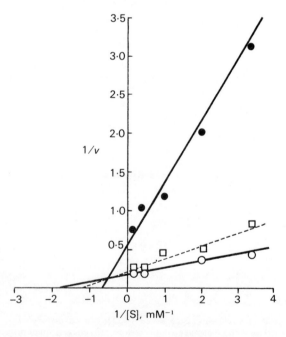

Fig. 36.16. **Effect of testosterone on oxidative demethylation.**
Lineweaver–Burke plot of $1/v$ versus $1/[S]$ for the oxidative demethylation of aminopyrine by liver endoplasmic reticulum, prepared from (○—○) normal rats, (●—●) castrated rats and (—□—□—) castrated rats injected with testosterone.

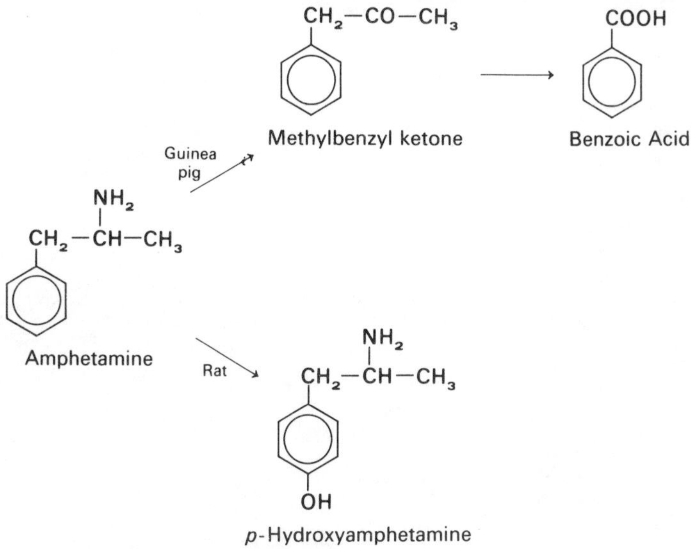

Fig. 36.17. **Species differences in oxidative metabolism.**

drugs are also greater in human males than in human females, but the differences are not as clearly marked as for rats.

There are also wide differences in the pattern of oxidative metabolism of drugs in different species. For example, in the guinea-pig, the drug amphetamine undergoes side-chain oxidation to methylbenzyl ketone and then to benzoic acid. In the rat, however, the drug undergoes hydroxylation in the ring to form p-hydroxyamphetamine (*Fig.* 36.17).

36.9 Induction of oxidative xenobiotic metabolism

Clinicians have observed, over many years, that patients who receive prolonged drug treatment usually develop a resistance and, consequently, require increased drug doses to elicit the same effect as the initial dose. The assumption was that the site of drug activity became more resistant, but this explanation became inadequate on demonstration of cross resistance, i.e. patients treated with drug A for a period, showed resistance to drug B soon after the initial administration.

Many drugs are capable of inducing the activity of the total system of oxidative xenobiotic metabolism, i.e. cytochrome P450, the flavoprotein and associated membranes, and this has been shown, by extensive biochemical investigations, to be a primary cause of the apparent cross resistance. The total system is non-specific and, consequently, any increase in enzymic activity will lead to more rapid drug oxidation. The effect of one drug on the rate of oxidative demethylation of another drug is shown in *Fig.* 36.18. After phenobarbitone is administered to rats, the concentration of cytochrome P450 and the rate of oxidative demethylation of aminopyrine in preparations of liver endoplasmic reticulum both increase rapidly for 3–4 days, followed by a sharp decline to normal values. If the phenobarbitone administration is continued, the metabolism rate is maintained at the elevated level. This results, *in vivo*, in the rapid metabolism and excretion of a wide range of drugs oxidized by the cytochrome P450 system. This induction of the system therefore explains the original observation of the low efficiency of the second drug B used to treat a patient receiving drug A.

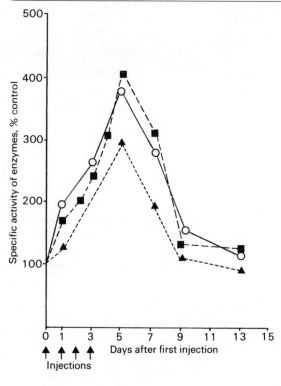

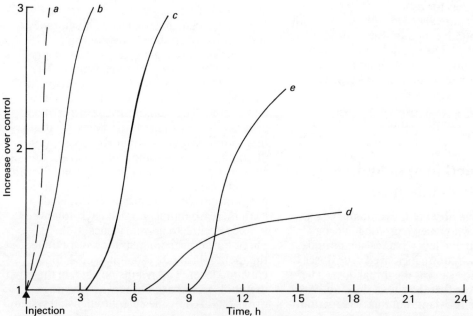

Fig. 36.18. **Effect of phenobarbitone on oxidative demethylation.**
Effect of phenobarbitone injection on the rates of oxidative demethylation of aminopyrine, on the concentrations of cytochrome P450 and flavoprotein in liver endoplasmic reticulum. (▲ – – – ▲) Rate of oxidative demethylation; (□—□) cytochrome P450 concentration; (○—○) flavoprotein concentration.

Fig. 36.19. **Changes in the liver endoplasmic reticulum after injection of the inducing drug phenobarbitone.**

(*a*) Drug binding; (*b*) phospholipid metabolism (turnover); (*c*) enzymes of rough membranes; (*d*) enzymes of smooth membranes; (*e*) phospholipid content of membrane.

Many drugs and other xenobiotics, such as insecticides, have been tested for their capacity to induce the oxidation system and their efficiency varies widely. It does not appear to be closely related to the pharmacological action or the structure of the drug (*see Table* 36.8) and the reason that some drugs are powerful inducers, whilst others are relatively weak inducers, remains obscure.

Some of the most effective inducers are insecticides such as DDT. Substances of this type are taken in with the food and become incorporated into the fat depots from which they are slowly released. This slow release leads to a prolonged and slightly elevated concentration of cytochrome P450 and the related flavoprotein and this, in turn, causes an increased rate of metabolism of many drugs. The sequence of events in the complex induction mechanism is illustrated in *Fig.* 36.19. The inducing drug binds very rapidly to the endoplasmic reticulum after injection, followed by a rapid turnover of phospholipid and proliferation of the rough membranes containing the ribosomes. Proliferation of the smooth membranes, which contain the additional cytochrome P450 and flavoprotein, follows.

Many carcinogens are also inducers of a special form of cytochrome P450 which is discussed in Chapter 38. This induction is known to be genetically controlled and carcinogen inducers are consequently believed to control the derepression of DNA.

Table 36.8 **Relative inductive effects of different classes of drugs**

Pharmacological action	Drug	Induction effect
Anaesthetics	Nitrous oxide	+
	Ether	+
	Halothane	0
Hypnotics Sedatives	Barbiturates	+ + +
	Thalidomide	0
	Ethanol	+/−
	Ethinanate (Valamid)	0
Anticonvulsants	Paramethadione	+ +
	Trimethadione	0
Tranquilizers	Chlorpromazine	+ +
	Promazine	0
Analgesics	Phenylbutazone	+ + +
	Aminopyrine	+ +
	Pethidine	0
Antihistamines	Diphenhydramine	+ +
Psychomotor stimulants	Imipramine	+ +
Hypoglycaemic drugs	Tolbutamide	+ + +
	Sulphanilamide	0
Insecticides	Chlordane	+ + + +
	DDT	+ + + +
	Dieldrin	+ + +
	Aldrin	+ + +
	Pyrethium	0
Carcinogenic hydrocarbons*	Benzo(a)pyrene	+ + +
	Methylcholanthene	+ + +

* The type of cytochrome P450 induced is not identical to that induced by most drugs and is discussed fully in Chapter 38.

Chapter 37 — Alcohol: effects on metabolism

The term 'alcohol' is widely understood by the general public to mean 'ethanol' (C_2H_5OH), but it should be remembered that the term 'alcohol' is a general name for a very large number of chemicals that contain the 'alcohol' hydroxyl group —OH.

All alcohols have slightly different pharmacological actions. Most alcoholic drinks contains small quantities of other alcohols,

particularly the longer-chain aliphatic alcohols, but the alcohol that causes most concern is 'methanol' (CH_3OH), since it is much more toxic than ethanol and causes blindness and rapid death. In spite of this, in many parts of the world 'meths drinkers' inflict serious problems on themselves.

In this chapter, the discussion of alcohols is confined to ethanol which has been much more carefully studied than any other alcohol. Ethanol consumption is now very large in many parts of the world and it has been calculated that ethanol currently forms 6 per cent of the total calorie input in the UK. Taking into account distribution in consumption in the population, this figure implies that there must be a significant number of people taking a large proportion of their energy (15–25 per cent) as ethanol.

The effects of alcohol on the central nervous system are well documented in pharmacological textbooks; this chapter is devoted to a discussion on the metabolism of alcohol and its effects on other metabolic processes.

37.1 Metabolism of ethanol

The major proportion of ingested ethanol, about 95 per cent, is metabolized in the liver. It is oxidized, in two stages, to acetate:

$$C_2H_5OH \xrightarrow[\text{Alcohol dehydrogenase}]{NAD^+ \quad NADH} CH_3CHO \xrightarrow[\text{Aldehyde dehydrogenase}]{NAD^+ \quad NADH} CH_3COOH$$

Ethanol   Acetaldehyde H_2O Acetic acid

A significant aspect of the metabolism is the large consumption of NAD^+ as a result of its conversion to NADH.

NADH can be reoxidized to NAD^+ by a variety of reactions, including reduction of pyruvate to lactate, acetoacetate to 3-hydroxybutyrate, dihydroxyacetone phosphate to glycerol 3-phosphate and the reduction step in glucose synthesis, i.e. the conversion of 1,3-diphosphoglycerate to glyceraldehyde 3-phosphate. The major route for the reoxidation of NADH is the electron transport system in the mitochondria. It was noted in Chapter 6, that the mitochondrial membrane is impermeable to NADH, but transport of metabolites, such as malate, through the membrane, enables NADH to be generated inside the mitochondria.

Even though all these potential pathways for the reoxidation of NADH exist, a very characteristic effect of ethanol is to increase the NADH/NAD^+ ratio in the cells by as much as four times; the cytosolic ratio may

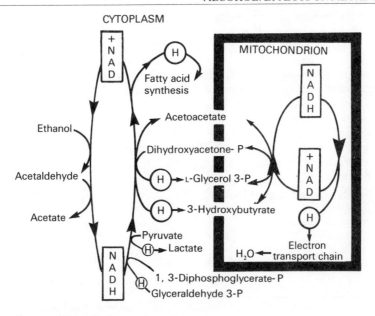

Fig. 37.1.   **Production of NADH during the metabolism of ethanol.**

be even higher. As a corollary to this, the lactate/pyruvate ratio and 3-hydroxybutyrate/ acetoacetate ratios increase substantially in venous blood leaving the liver. The metabolic stages are summarized in *Fig.* 37.1.

37.2 Hypoglycaemic effects of alcohol

The fact that alcohol causes pronounced hypoglycaemia has been known for over 40 years. It was originally considered to be associated only with malnourished chronic alcoholics, but subsequently it became clear that it can also occur, quite readily, after occasional ingestion of alcohol. Hypoglycaemia usually results in an inadequate supply of glucose to the brain which can then cause derangement of cerebral function and death. The effect of ethanol on the blood glucose levels under experimental conditions is shown in *Fig.* 37.2, which shows additionally that the hepatic glucose supply is also suppressed by alcohol.

It has become clear in recent years that a high value for the NADH/NAD$^+$ ratio in

the liver is probably the major factor in the suppression of gluconeogenesis, the metabolic steps of which are illustrated in *Fig.* 37.3, emphasizing the stages utilizing NAD$^+$ or NADH. The pathway from fructose 1,6-diphosphate is not blocked by ethanol and fructose rapidly overcomes hypoglycaemia induced by alcohol (*Fig.* 37.4).

The increase in the NADH/NAD$^+$ ratio, characteristic of the oxidation of ethanol, may affect the gluconeogenic sequences in several different ways.:

a.   The availability of oxaloacetate for conversion to phospho*enol*pyruvate following the slow-down of the citrate cycle occurring during ethanol oxidation is reduced. This is because increases in NADH/NAD$^+$ ratios tend to suppress several oxidative reactions of the cycle, such as the oxidation of isocitrate to α-ketoglutarate, of α-ketoglutarate to succinate and of malate to oxaloacetate.

b.   The availability of pyruvate for conversion to oxaloacetate and then to phospho*enol*pyruvate, is decreased by the rapid reduction of pyruvate to lactate under the impact of high concentrations of NADH formed during ethanol oxidation.

c. Glycerol produced by adipose tissues during lipolysis is a known source of glucose, especially during fasting. This conversion requires the formation of dihydroxyacetone phosphate from glycerol 3-phosphate and is also depressed by high concentrations of NADH.

The hypothesis that ethanol-induced inhibition of gluconeogenesis results from a marked increase in the NADH/NAD⁺ ratio can also explain many other observed effects, such as increase in fatty acid oxidation, inhibition of the galactose to glucose conversion and diminished transformation of serotonin to 5-hydroxyindole acetic acid.

From the above, it can be concluded that an increase in the NADH/NAD⁺ ratio is an important unifying hypothesis in the explanation of ethanol's metabolic effects.

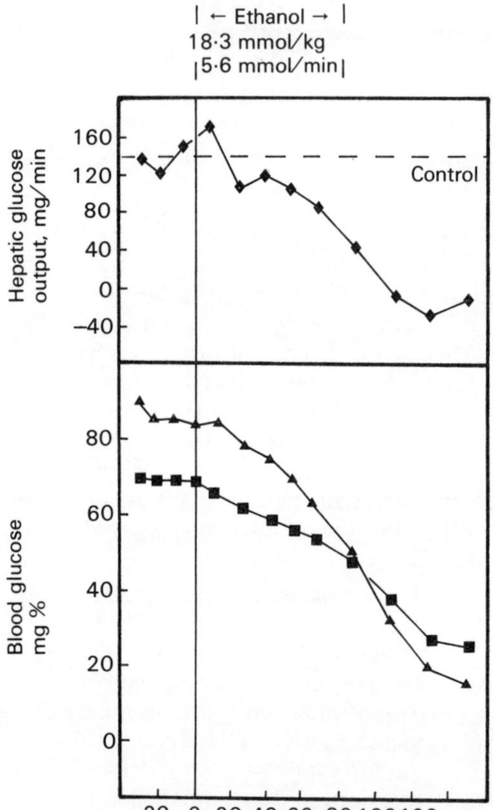

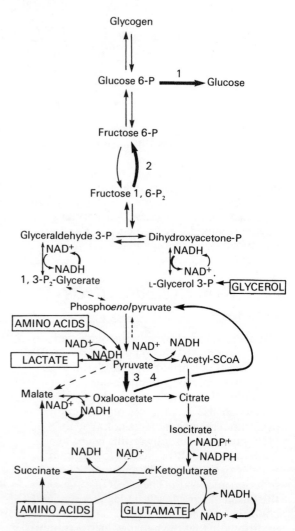

Fig. 37.2. **Effect of ethanol on blood glucose levels and liver glucose output.**
(▲—▲) Hepatic venous; (○—○) arterial. (– – –) Control, hepatic glucose output (133 mg/min).

Fig. 37.3. **The utilization of NAD⁺ and NADH in gluconeogenesis.**

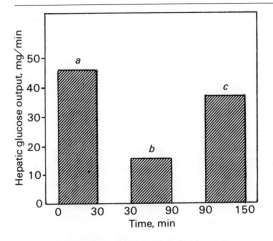

Fig. 37.4. **The effect of fructose on ethanol inhibition of hepatic glucose production.**
(*a*) Control; (*b*) ethanol alone, 162 mmol/h;
(*c*) ethanol plus fructose, 32 mmol/h.

37.3 Effects of alcohol on plasma lipids

In marked contrast to effects on plasma glucose, alcohol ingestion causes severe hyperlipidaemia. Furthermore, hyperlipidaemia can occur in both poorly nourished and well nourished people. In one study of men in the age range of 30 to 70 years, daily drinkers had double the mean plasma triacylglycerol (triglyceride) level and a median level 50 per cent greater than that of abstainers. These differences cannot be ascribed to obesity, diet differences or impaired glucose tolerance.

The effect of alcohol on plasma triacylglycerols arises from the fact that alcohol is a food. The capacity of the liver to oxidize alcohol to acetate is readily saturated, so that alcohol metabolism is maintained at a maximum rate for 1 hour after ingestion of 20 ml spirits, 33 ml wine or 124 ml beer. In the liver, alcohol is oxidized in preference to all other substrates; consequently oxidation of other substrates, mainly fatty acids and glucose, decreases during alcohol consumption often to 50 per cent of the norm. This depression of oxidation of free fatty acids entering the liver causes proportional increases in their conversion to triacylglycerols.

The majority of triacylglycerols produced are exported in the form of very-low-density lipoproteins (VLDL) (cf. Chapter 19), thereby producing a rise in total plasma triacylglycerols. Some are, however, retained in the liver and their accumulation is markedly increased when low-calorie diets are fed giving rise to the 'fatty liver' condition (cf. Chapter 19).

The effects of alcohol on fat metabolism are also manifested during the fed state. Absorption of triacylglycerols causes uptake of fatty acids by the liver. Inefficient oxidation in this organ results in the fatty acid being released as components of very-low-density lipoproteins. Furthermore, the rate of glucose oxidation is reduced so that a larger proportion is metabolized to triacylglycerols.

Although intake of alcohol always leads to an increase of very-low-density lipoprotein production by the liver, their high concentrations may not be maintained in regular drinkers of alcohol. This is believed to result from an increase in the activity of lipoprotein lipase in the tissues which is responsible for the hydrolysis of the triacylglycerols of the very-low-density lipoproteins (cf. Chapter 19). As an important consequence of this, high-density lipoproteins, which supply the important cofactors for lipoprotein lipase activity (CI and CII apoproteins) also increases in concentration. Epidemiological studies have indicated that there is a correlation between increased levels of high-density lipoproteins and reduced incidence of coronary heart disease (cf. Chapter 19), so that, in moderation, alcohol consumption may give some advantages to prolonged health!

Chapter 38 Chemical carcinogenesis

38.1 Introduction: what is cancer?

Cancer is probably the most serious disease affecting man. It is not, however, confined to man since most mammals and many other lower animals are affected. The seriousness of the disease is mainly a consequence of the difficulties of early diagnosis and treatment; even though these have been greatly improved during the last few years, cancer still presents formidable and often intractable problems.

A cancer may be defined as a 'malignant neoplasm' which defined literally is a strong new type of cellular growth that usually invades and takes over normal cellular tissue. The term 'cancer' is sometimes only used to describe neoplasms of epithelial cells, such as the skin, and other malignant neoplasms of mesodermal cells, for example those of the bone, are described as 'sarcomas'.

38.2 Cancer-causing agents

The fact that certain external physical agents cause cancer was clearly recognized when work was begun on radioactive elements and X-rays at the end of the ninteenth century. Many early workers suffered from serious cancers on their hands and fingers through working with either X-rays or radioactive substances. In the early 1920s, it was also proved that certain chemicals caused cancer.

It is now clear that cancer can be caused by the following:
- *a.* Radiation either ultraviolet (u.v.) or ionizing radiation, such as X-rays and gamma-rays (γ-rays)
- *b.* Viruses
- *c.* Chemicals.

Initially, it was considered that the role of these factors in the generation of cancer was of limited and possibly academic interest only, but during the last decade it has become clear that they are real and effective agents that cause cancer in man in his everyday life. The role of certain viruses in causing some animal cancers has been decisively proved, but their role in the initiation of human tumours is still much disputed. Radiation and viruses, both specialized topics, will not be discussed in this chapter which will be devoted to ways in which chemicals cause cancer, i.e. chemical carcinogenesis. It is now clear that many tumours in man are directly or indirectly caused by certain chemicals and much research is, therefore, concerned with establishing the nature of carcinogenic chemicals, their likely sources and how their effects are exerted.

38.3 Chemical nature of carcinogens—their occurrence in the environment and their origins

The fact that chemicals could cause cancer arose from the observations of Percival Pott of St Bartholomew's Hospital in London in 1775. He drew attention to the fact that there was a high incidence of cancer of the scrotum among London chimney sweeps who often had to climb up inside the large chimneys to sweep the soot down. He correctly deduced that the soot was responsible for the cancer, although at that

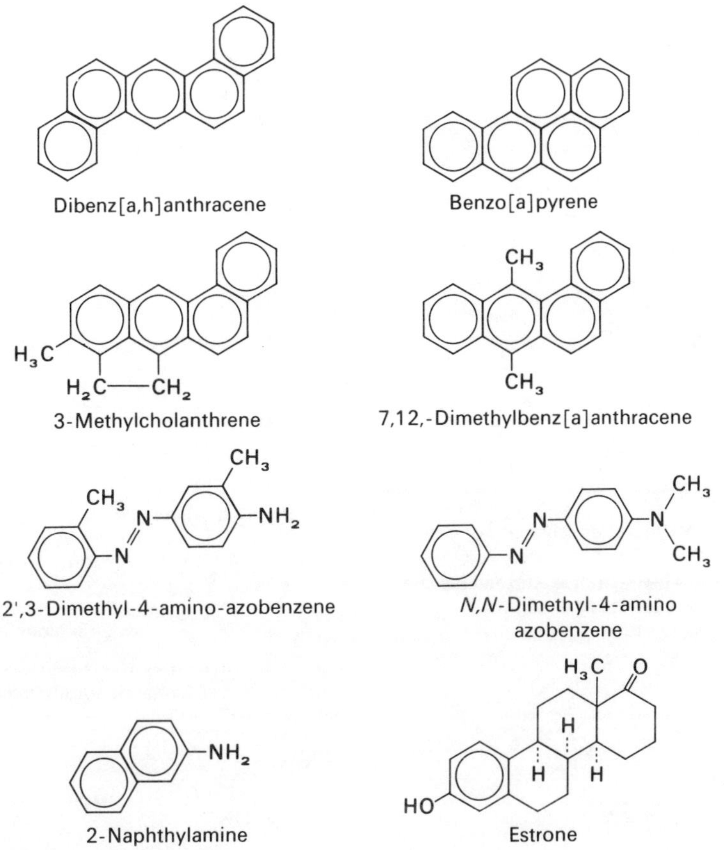

Dibenz[a,h]anthracene

Benzo[a]pyrene

3-Methylcholanthrene

7,12,-Dimethylbenz[a]anthracene

2',3-Dimethyl-4-amino-azobenzene

N,N-Dimethyl-4-amino
azobenzene

2-Naphthylamine

Estrone

Fig. 38.1.   **Structures of chemical carcinogens I.**
The structures of the principal chemical carcinogens discovered prior to 1940.

time it was difficult to visualize how such an apparently inert material as soot could cause serious tissue damage.

The explanation of Percival Pott's observation was not provided until 150 years later, in the 1920s, when it was discovered by Japanese workers that painting extracts of soot onto the skin of mice produced tumours of the skin.

The work was confirmed in London, and, in 1929, Kennaway working at the Chester Beatty Research Institute isolated, from soot extract, the first pure chemical carcinogen, dibenz-anthracene (*Fig*. 38.1).

The finding was quickly taken up by leading organic chemists of that time, notably by Cook in Scotland and Fieser in the USA,

who synthesized numerous compounds related to dibenzanthracene which were tested for their carcinogenicity. Typical compounds are listed in *Fig*. 38.1. It was hoped that the study of these structures would provide a vital clue as to the chemical structures which caused cancer but no easy solution appeared. The carcinogenicity varied from one molecule to another in a rather unpredictable manner.

Since chimney sweeping had become better organized and forms of heating other than coal fires came into vogue the carcinogenic effects of the polycyclic hydrocarbons, as this group of compounds was known, tended to be regarded of only academic interest for several years subsequent to their discovery. However, in 1953 Doll,

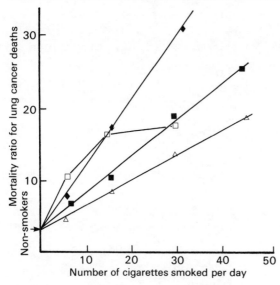

Fig. 38.2.   **Relationships between cigarette smoking and
lung cancer deaths in the UK, Canada and USA.**
(♦ — ♦) British doctors; (□– – –□) Canadian
ex-servicemen; (■– – –■) US ex-servicemen;
(△ — △) survey of men in 25 states in USA.

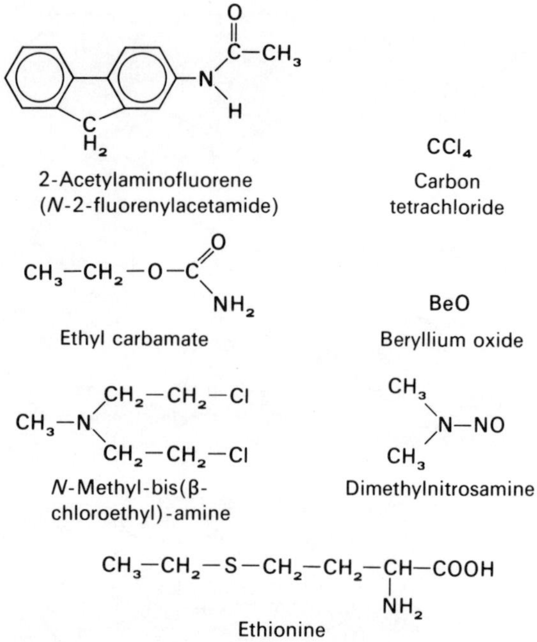

2-Acetylaminofluorene
(*N*-2-fluorenylacetamide)

CCl_4
Carbon
tetrachloride

Ethyl carbamate

BeO
Beryllium oxide

N-Methyl-bis(β-
chloroethyl)-amine

Dimethylnitrosamine

$CH_3-CH_2-S-CH_2-CH_2-CH-COOH$
                                         |
                                        NH_2

Ethionine

Fig. 38.3.   **Structures of chemical carcinogens II.**
The structures of some chemical carcinogens identified
between 1940 and 1960.

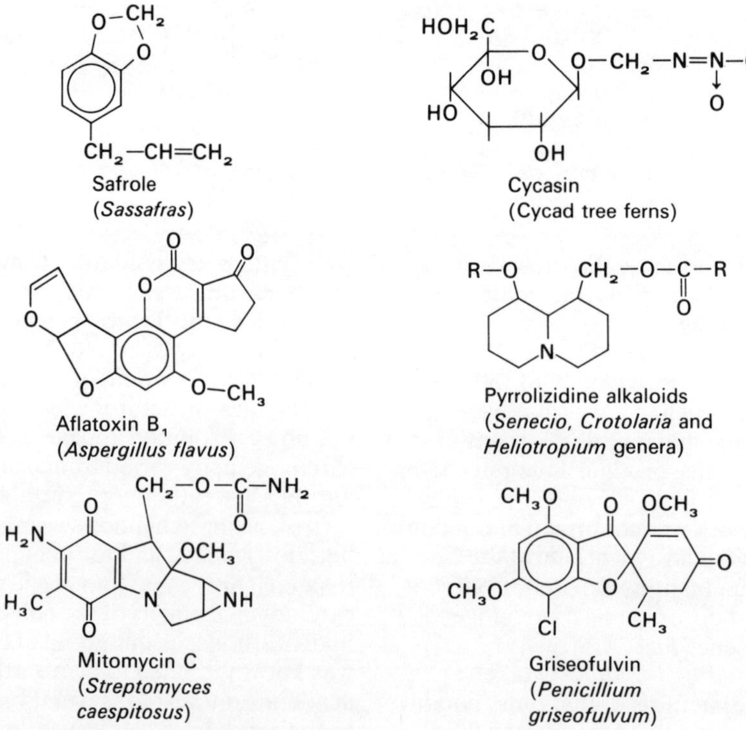

Safrole
(*Sassafras*)

Cycasin
(Cycad tree ferns)

Aflatoxin B$_1$
(*Aspergillus flavus*)

Pyrrolizidine alkaloids
(*Senecio, Crotolaria* and
Heliotropium genera)

Mitomycin C
(*Streptomyces
caespitosus*)

Griseofulvin
(*Penicillium
griseofulvum*)

Fig. 38.4.   **Structures of chemical carcinogens III.**
The structures of some chemical carcinogens that are products of plants and micro-organisms.

following intensive epidemiological and statistical analysis, proved clearly that cigarette smoking was a prime cause of lung cancer in man (*Fig*. 38.2). Careful analysis of the smoke and tar obtained from cigarettes demonstrated that it contained many carcinogenic hydrocarbons, the most important being benzo[a]pyrene. Thus, for the first time, a close link was established between academic research on cancer in animals and a very common environmental hazard. Consequently, the medical world began to take chemical carcinogenesis very seriously.

Following the discovery of a group of carcinogenic polycyclic hydrocarbons before 1940, further research has established two new aspects: (*a*) relatively simple aliphatic molecules, such as vinyl chloride, could be carcinogenic (*Fig*. 38.3); (*b*) many carcinogens are produced naturally by plants and micro-organisms (*Fig*. 38.4). This finding which established that naturally produced chemicals could be carcinogenic was of great importance.

38.4 How do carcinogens gain access to the body?

Carcinogens may gain access to the body by several routes, as described below.

a. Skin application

In the course of the investigation of carcinogenic effects of chemicals, it was observed that painting the carcinogen on the skin of mice could produce tumours, although repeated applications were usually necessary. Polycyclic hydrocarbons and products containing them, such as tars and soot, will also produce skin tumours in man. The risk, although greatest for those working in certain industries, is likely to be very small with modern knowledge and protective clothing for workers.

b. Inhalation

Many carcinogens may be inhaled and examples of typical chemicals gaining entry to

the body via this route are: soot, polycyclic hydrocarbons, asbestos, nickel, vinyl chlorides and naphthylamine. Workers in certain industrial processes, including those using asbestos or vinyl chloride, are exposed to the greatest risk and stringent precautions are now taken with these chemicals. However, the majority of the population must, to some extent, be exposed to carcinogenic chemicals. For example, smoke from most fires and exhaust fumes from vehicles contain low concentrations of polycyclic hydrocarbons. The intake of carcinogens from these sources is, however, generally regarded as very small compared with the relatively large intake resulting from smoking.

c. Through the digestive tract

It is now generally considered that carcinogens, ingested with the diet or produced during the course of digestion, present a serious hazard to man and recent epidemiological studies have focused attention on those carcinogens associated with food.

Carcinogens can originate in the digestive tract from several sources.:

i.   They can be formed during cooking
ii. They can be natural constituents of the food or fungal organisms contaminating it
iii. They can be formed by interactions of foodstuffs
iv. They can be formed by bacterial action on partially digested foods or on the bile acids.

These points can be illustrated by the examples given below.

i.   Many polycyclic hydrocarbons are formed as a consequence of certain methods of cooking, and are particularly prevalent in smoked fish and meat cooked in smoky flames, i.e. 'barbecued meat'. The current enthusiasm for 'barbecued food' conflicts with scientific progress in cancer prevention.
ii. The existence of potential carcinogens in foodstuffs was strikingly demonstrated in quite a

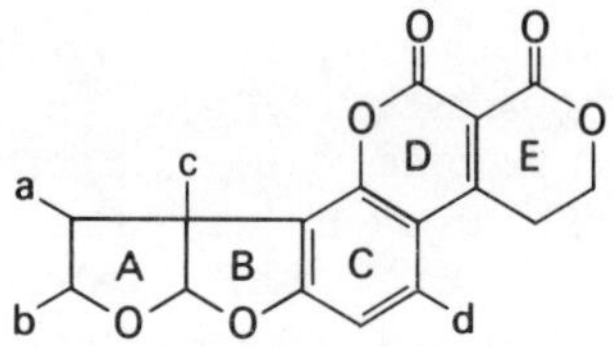

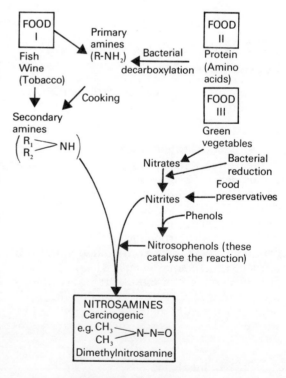

Aflatoxin I (G series)

	a	b	c	d
G1	H	H	H	H
G2	H_2	H_2	H	H
G2a	H_2	(H(OH))	H	OCH_3

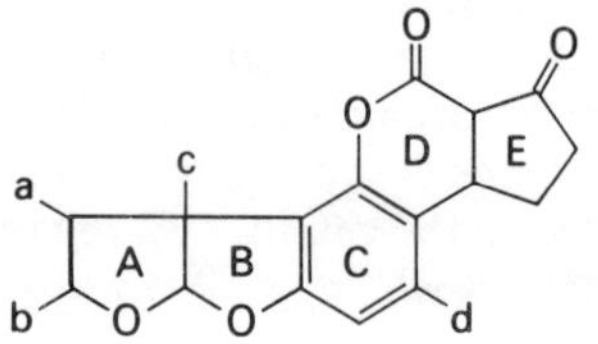

Aflatoxin II (B series)

	a	b	c	d
B1	H	H	H	OCH_3
B2	H_2	H_2	H	OCH_3
B2a	H_2	H(OH)	H	OCH_3
M1	H	H	OH	OCH_3
M2	H_2	H_2	OH	OCH_3

Fig. 38.5. **The aflatoxins.**
There are two series of aflatoxins termed I and II or G and B. Different substituents are found in positions a, b, c and d, and these are shown above for eight different compounds. Note that the 'E' ring is six membered in the I series, but five membered in the II series.

had become contaminated with *Aspergillus flavus*, an organism producing a group of compounds called 'aflatoxins' that are very potent inducers of liver tumours (*Fig.* 38.5).

iii. About 30 years ago, it was observed that small amounts of dimethyl-nitrosamine

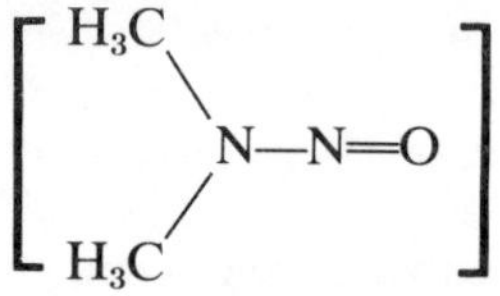

$$\left[\begin{array}{c} H_3C \\ \\ H_3C \end{array} \!\!\! N\!\!-\!\!N\!\!=\!\!O \right]$$

fed to rats caused a high incidence of liver tumours. Tumours were also observed in the kidney, oesophagus and lung. At the time the observations were regarded as of purely academic interest and of little relevance to cancer incidence in man.

remarkable way during the early 1960s. In Idaho in the USA, many thousands of trout reared in hatcheries died and subsequent pathological examination showed that the majority were suffering from serious hepatomas. During the following years, a serious outbreak of a disease amongst turkeys in England, killing 100 000 birds, was also characterized by acute hepatic necrosis. This disease was originally of unknown origin and termed 'turkey X disease'. Intensive research during the next few years established that the outbreaks, in both the trout and turkeys, were caused by their diet. Both groups had been fed on peanuts which, on account of damp storage conditions in Africa,

Fig. 38.6. **Possible mechanism of the production of carcinogenic nitrosamines in the gut.**

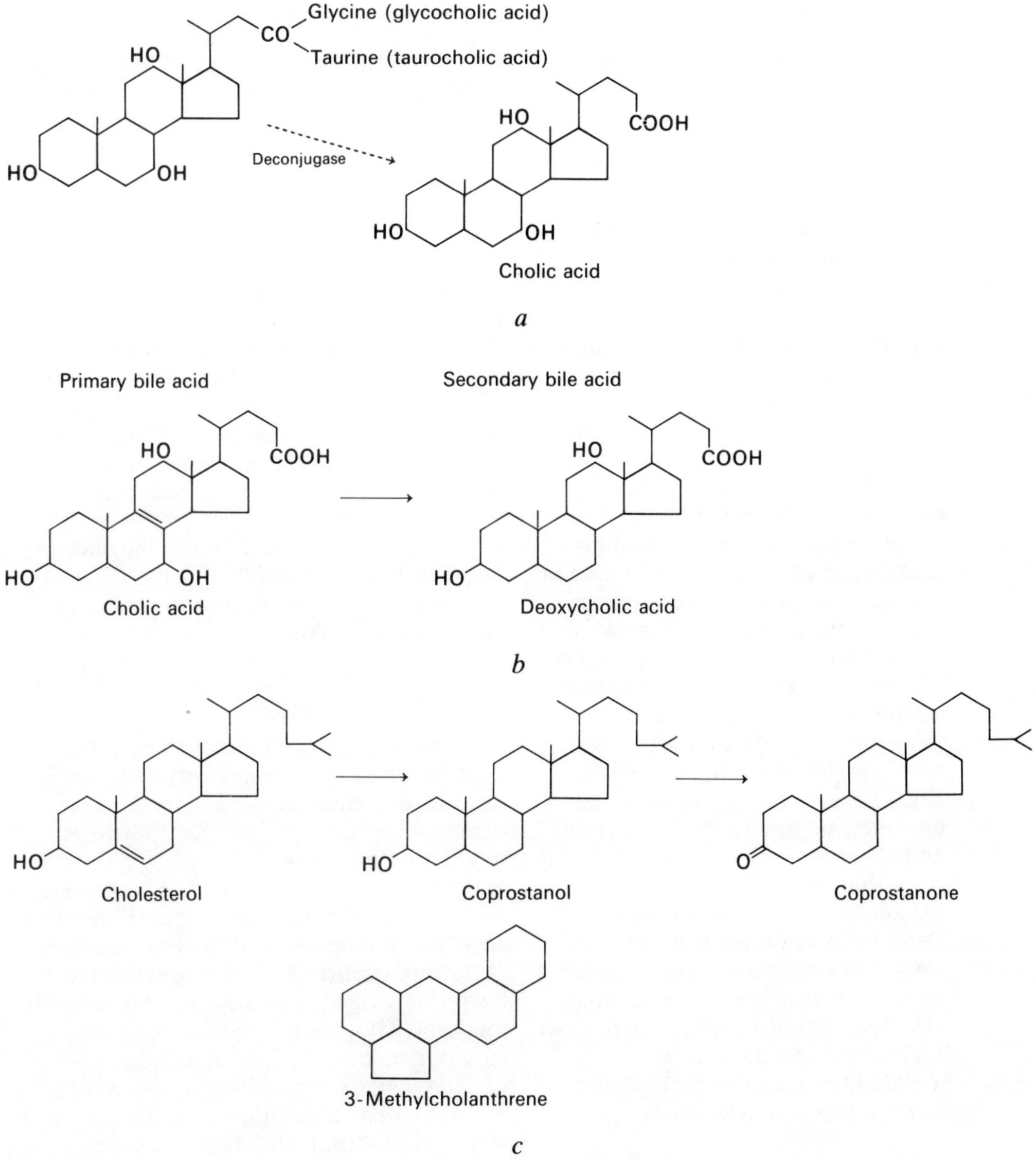

Fig. 38.7. **Microbial degradation of acid and neutral sterols.**
(*a*) Microbial deconjugation of bile acids; (*b*) microbial 7-dehydroxylase reaction; (*c*) microbial cholesterol dehydrogenase reaction.

About 15 years later, however, it was shown that, in acidic conditions, secondary aliphatic amines would react with nitrites to form nitrosamines. Furthermore, it was proved that lung and stomach tumours in animals resulted from feeding them with secondary amines mixed with nitrite. It is known that in the human diet, secondary amines can be present as components of fish products and nitrites can also be present either formed by plants or as added preservative. It is possible that the carcinogenic nitrosamines might be produced in the human digestive system by a scheme outlined in *Fig. 38.6*. It is known that stomach cancer is widespread in Japan where a lot of dried and smoked fish is eaten.

iv. For many years epidemiological evidence has accumulated to show that diets high in meat and fat tend to cause a greater incidence of cancer in the gastrointestinal tract than do vegetarian diets. The precise reason for this is not known, but many have surmised that bile salts may be important since their secretion is increased by ingestion of excess fat. Bile salts are attacked by bacterial enzymes, such as 7α-dehydroxylase, and cholesterol by enzymes, such as a dehydrogenase (*Fig. 38.7*). Deconjugation of bile acids can also occur yielding products structurally similar to known carcinogens, e.g. methylcholanthrene. Despite many efforts to establish a carcinogenic role for bile salt and cholesterol metabolites, unequivocal evidence has not been obtained yet.

38.5 Methods of testing for carcinogens

The recognition that many chemicals, both natural and synthetic, that come into contact with man may be carcinogenic has stimulated the development of new methods for testing carcinogens. At present, no method is 100 per cent reliable and it is important, generally, to test each chemical by as many different methods as possible. Animal tests are extremely time consuming and expensive and much effort has been directed to improving tests using cultured eukaryotic cells or bacteria. Typical tests can be classified as below.

a. Animal tests

Animal tests are frequently carried out by a method similar to the original tests for carcinogens, i.e. skin painting onto mice. The skin dose is repeated several times and the tumour incidence noted, but very large numbers of animals have to be used for statistically valid results to be obtained and the experiments are very prolonged. Since the tumours are, however, produced under natural conditions, this test is often considered to be the most reliable test for carcinogenicity. Some tests involve implanting the chemical subcutaneously or in the bladder, rather than painting on the skin.

b. Ames test

This test, named after its developer, is dependent on the mutagenicity of carcinogens. Strains of bacteria (e.g. *Salmonella*) were developed which are *His⁻*, i.e. they were unable to synthesize the amino acid histidine and would only grow if histidine was added to the medium. On addition of mutagenic agents, such as carcinogens, to the culture medium, mutations occurred which resulted in the bacteria being able to grow on a histidine-free medium. This test is a good indication of carcinogenicity, but, like most tests for carcinogenicity, is not 100 per cent reliable.

In a recent study on 175 carcinogens, 90 per cent caused mutation, i.e. were positive, and 10 per cent negative; and a study on 108 non-carcinogenic chemicals showed 13 per cent to be positive and 87 per cent negative. Note that many carcinogens require metabolism before they become active mutagens, as described in Section 38.8, i.e. a preliminary 'activation' stage has to be carried out before testing.

Fig. 38.8.   **Sister chromatid exchange.**

a. Mammalian cells are grown in presence of bromodeoxyuridine (BrdUrd).
b. During cell division, the BrdUrd becomes incorporated into the strands of DNA in place of thymidine.
c. The cells are allowed to undergo a second division in presence of BrdUrd.
d. The cells are treated with a drug colcemid to block the cell cycle in metaphase (M-phase).
e. The chromosomes are stained with a fluorescent dye.
f. The fluorescence is quenched by BrdUrd so that newly synthesized strands of DNA appear dark by comparison with normal DNA

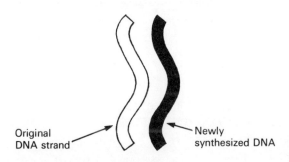

Original DNA strand — — Newly synthesized DNA

g. The process is repeated but the carcinogen or chemical under test is added to the culture medium.
Carcinogenic or mutagenic chemicals cause 'sister chromatid exchanges', that is switching of DNA synthesis containing BrdUrd and normal DNA between strands. This shows a distinctive pattern:

'Sister chromatid exchanges' caused by a carcinogen. The number of sister chromatid exchanges is generally proportional to the carcinogenicity of the chemical.

c. Sister chromatid exchange

Recently, it was discovered that most mutagens caused chromosomes of dividing eukaryotic cells to exhibit a phenomenon known as 'sister chromatid exchange', in which exchange between pairs of chromosomes could

be demonstrated, as illustrated in *Fig.* 38.8. The number of the exchanges is generally proportional to the mutagenicity of the chemial being tested.

38.6 Factors affecting the carcinogenicity of chemicals

a. Incidence

The incidence of tumours produced by any chemical is dependent on the total dose and the frequency of application, although simple quantitative predictions are seldom possible (*Fig.* 38.9).

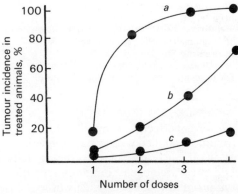

Fig. 38.9.   **Relation between tumour incidence, dose and frequency.**
Bladder tumours in mice treated with *N*-methyl-*N*-nitrosourea (mg): (*a*) 1·5; (*b*) 0·5; (*c*) 0·1.

b. Synergism

Two chemicals may frequently exhibit synergistic effects, i.e. the effect of the two chemicals applied together may be much greater than either applied singly (*Fig.* 38.10). Although it is possible that one of the chemicals may be acting as a promotor (*see below*), this is by no means definite.

c. Promotors

When carcinogenic chemicals were first tested on the skin of mice, it was frequently observed

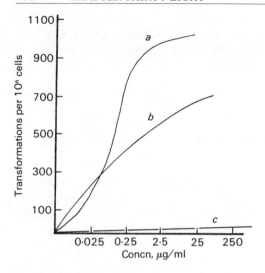

Fig. 38.10. **Synergistic effects of two chemicals used simultaneously.**
Mutagenic effects measured by studying 'transformations' in cultured mammalian cells. 'Transformed' cells are identified in culture morphologically.
(*a*) Biphenyl + aniline; (*b*) biphenyl alone; (*c*) aniline alone. The concentration is measured on a log scale.

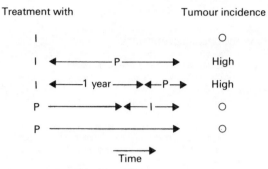

Fig. 38.11. **Relation between initiators and promotors.**
Mouse skin tests using initiator I, e.g. polycyclic hydrocarbon and promotor P, e.g. croton oil.

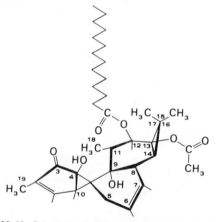

Fig. 38.12. **Schematic structure of a phorbol derivative found in croton seeds which is a powerful promotor.**

that they were ineffective or only weakly effective unless applied with certain seed oils, e.g. croton oil. Such substances were originally described as cocarcinogens, but more recently the term 'promotor' has been widely used.

The primary carcinogen is now usually termed 'initiator' and it has been found that the sequence of treatment is of importance, since addition of promotor after initiator will always induce tumours, whereas when promotor precedes initiator no tumours develop (*Fig.* 38.11). The pure promotor has now been isolated from croton oil and been found to be an ester of phorbol—phorbol-12-myristate-13-acetate (*Fig.* 38.12).

The phenomenon of promotion was originally associated with skin tests of carcinogens, but more recent research has demonstrated that it also occurs in other tissues, such as the liver or the bladder. Cyclamates, saccharin and even tryptophan can act as effective promotors for tumours of the bladder (*Table* 38.1). Metabolites of tyrosine, e.g. phenol or *p*-cresol, and tryptophan, e.g. indole or indole acetate, can

Table 38.1 **Promotion of bladder tumours in rats by dietary saccharin, cyclamates or tryptophan**

Initiator N-*methyl*-N-*nitrosourea* (*single injection*)	Dietary 5% cyclamate (C) or 5% saccharin (S)	Tumour incidence, %
+	−	0
−	+ (S)	2
+	+ (S)	52
−	+ (C)	1
+	+ (C)	58
FANFT* (*0·2% in diet*)	*Dietary tryptophan (2%)*	
+	−	20
−	+	0
+	+	53

* FANFT = *N*-[4-(5-nitro-2-furyl)-2-thiazolyl]formamide.

act as the promotors in the gastrointestinal tract; it is also possible that the carcinogenic effects of asbestos may be connected with promotion rather than with initiation and this view is supported by the significant fact that the incidence of cancer in asbestos workers is much greater in smokers than in non-smokers (*Table* 38.2).

Table 38.2 **Effect of exposure to asbestos and smoking on human lung cancer incidence**

Exposed to asbestos	Smokers	Cancer incidence, %
−	−	1
+	−	5
+	+	53

The mode of action of promotors is, as yet, only poorly understood. Phorbol esters are, however, known to stimulate the synthesis of many cellular macromolecules including DNA, RNA, protein and phospholipids, and they also cause a large increase in the activity of lysosomal proteinases although the

relationship between this activity and synthesis is uncertain. Promotors may also inhibit differentiation and cause production by the tissues of cells which, like typical tumour cells, are immature and not so well differentiated.

38.7 Multistage concept of carcinogenesis

Recent studies on the relationship between initiators and promotors have led to the development of a multistage concept of carcinogenesis as illustrated in *Fig*. 38.13. The initiator, e.g. a polycyclic hydrocarbon, is believed to cause a small number of cells to mutate to potential tumour cells. Since their proportion in the total tissue cells is so small, they are virtually undetectable. Promotors will then, by a mechanism as yet not fully understood, stimulate the rapid proliferation of the mutated cells to ultimately produce a tumour. In certain circumstances, a benign or very slow growing tumour may be produced, but it is not known whether these tumours slowly progress to carcinoma and, if so, how or why.

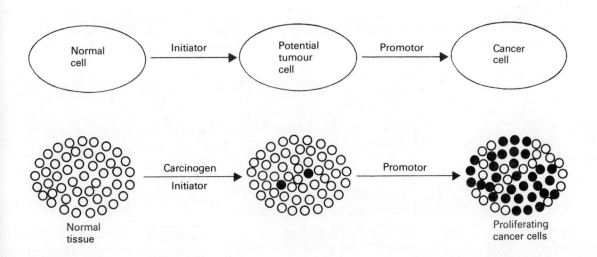

Fig. 38.13. **Possible roles of initiators and promotors in tumour development.**

38.8 Metabolism of carcinogens

For 40 years following the discovery of the polycyclic hydrocarbons, many attempts were made to elucidate their mode of action, but all ended in failure, the main reason being that chemical theory predicted that they would be very unreactive. Furthermore they did not, under normal experimental conditions, readily react with DNA or other important constituents of the cell. Thus the problem arose of how these inert molecules cause cancer.

The answer was surprisingly simple: they did not cause cancer themselves, their *metabolites* did. The breakthrough in the elucidation of the mechanism came in the late 1960s resulting from a simple experiment. It was discovered that if ³H-labelled polycyclic hydrocarbons were first incubated with a liver microsomal fraction in the presence of

NADPH and oxygen, strong binding to protein and DNA could be demonstrated, and that a relationship existed between the known carcinogenic effects and the DNA binding (*Table* 38.3). The metabolic mechanism and the nature of the metabolites were both unresolved by these experiments, and only intensive research has helped in the considerable progress made towards elucidation of both these aspects.

The lipophilic carcinogens are metabolized on the endoplasmic reticulum by means of an electron transport chain very similar to that described for drug oxidation (*see* Chapter 36). The cytochrome involved is a type of cytochrome P450 not identical to that normally involved in oxidative drug metabolism and which has a high degree of specificity for hydrocarbons. The absorption maximum is slightly different from cytochrome P450's and it is sometimes called 'cytochrome

Table 38.3 **Binding of metabolites of polycyclic hydrocarbons to DNA**

	Binding to protein, nmol/mol		Binding to DNA, μmol/g-atom	
Polycyclic hydrocarbon	Before metabolism	After metabolism	Before metabolism	After metabolism
Benzo[a]pyrene	0	0·78	0	1·41
Methylcholanthracene	0	0·73	0	0·78
Phenanthracene	0	0·72	0	0·05
Dibenzanthracene	0	0·95	0	0·44

From Grover P. L. and Sims P., 1968, *Biochem. J.*, Vol. 110, p. 159.

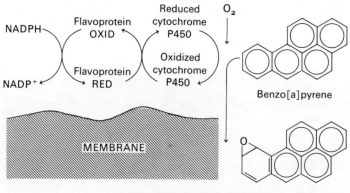

Fig. 38.14. **Oxidative metabolism of polycyclic hydrocarbons on endoplasmic reticulum.**

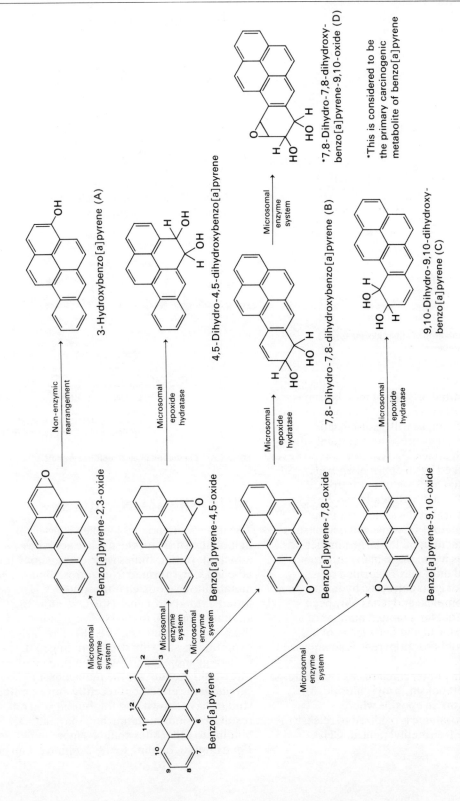

Fig. 38.15. **Major pathways of metabolism of benzo[a]pyrene.**
* Considered to be the primary carcinogenic metabolite of benzo[a]pyrene.

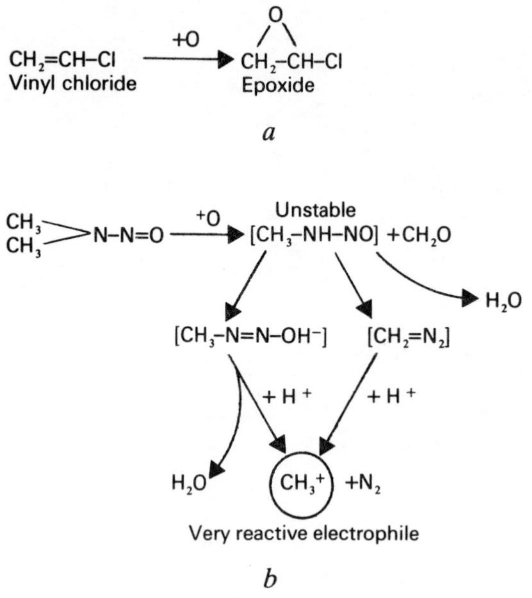

$$CH_2=CH-Cl \xrightarrow{+O} \overset{\displaystyle O}{\overset{\displaystyle /\backslash}{CH_2-CH-Cl}}$$

Vinyl chloride   Epoxide

a

$$\underset{CH_3}{\overset{CH_3}{>}}N-N=O \xrightarrow{+O} \underset{\text{Unstable}}{[CH_3-NH-NO]} +CH_2O$$

$\longrightarrow H_2O$

$[CH_3-N=N-OH^-]$ $[CH_2=N_2]$

$+ H^+$ $+ H^+$

H_2O $\left(CH_3^+\right)$ $+N_2$

Very reactive electrophile

b

Fig. 38.16. **Metabolism of vinyl chloride (*a*) and dimethylnitrosamine (*b*).**

P448'. The initial product of metabolism is always an epoxide (*Fig.* 38.14). Epoxides are very reactive and, as will be discussed in Section 38.10, they are almost certainly the species which attack DNA. They are, however, readily attacked by an enzyme, epoxide hydratase, that adds water to the epoxide converting it to a dihydrodiol (*Fig.* 38.15).

Perusal of the structure of the molecule of benzo[a]pyrene will demonstrate that, consequent upon the large number of double bonds in the rings, many different epoxides and diols may be formed so metabolism of each polycyclic hydrocarbon is extremely complex, and it has only been studied in detail for a limited number of hydrocarbons. Current knowledge of the metabolism of benzo[a]pyrene is summarized in *Fig.* 38.15.

Many other carcinogens also undergo oxidative metabolism. Vinyl chloride, for example, forms an epoxide whereas dimethylnitrosamine is oxidized to release the very reactive free methyl radical, CH_3^+ (*Fig.* 38.16).

38.9 Formation of electrophilic reagents

Once an understanding of the metabolism of more carcinogens had developed, it became clear that a unifying hypothesis could be applied, i.e. carcinogens are metabolized to electrophiles with a strong affinity for the nucleophilic sites that abound on nucleic acids (*Fig.* 38.17). Both the CH_3^+ radical and epoxides have electron deficiencies and fall into this category of electrophiles, and carcinogenic metabolites would, therefore, be bound to DNA by this mechanism.

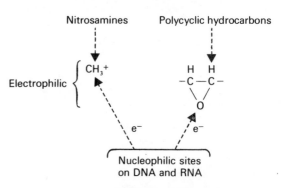

Fig. 38.17. **Formation of electrophilic reagents.**

38.10 Site of DNA attack

The use of carcinogens labelled isotopically has helped the study for the nucleic acid sites attacked. Initial studies were carried out with labelled CH_3^+ formed from nitrosamines. The metabolite of oxidized nitrosamine, CH_3^+, was found to bind strongly to the O–6 atom of guanine, and also to the N–7 of guanine and the N–3 of adenine (*Fig.* 38.18). The early experiments showing strong binding to the O–6 of guanine led to the assumption that this would account for the mutagenic or carcinogenic effect, but recently more careful studies have shown that this binding did not result in serious mutation and that, in fact, binding to other sites could be more important. For example, binding to the 2-amino group of

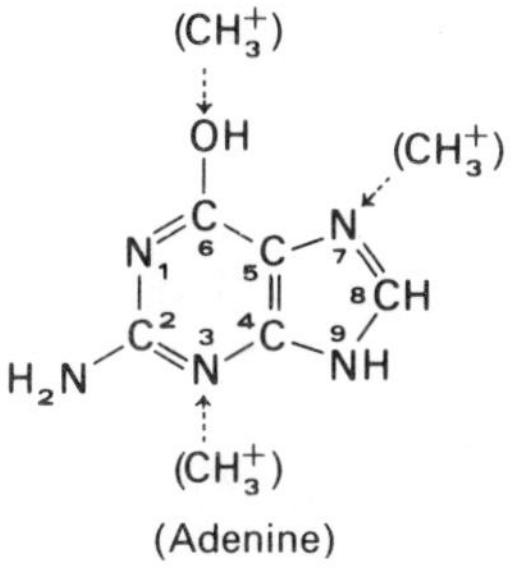

Fig. 38.18. **Sites of nucleophilic attack on guanine and adenine residues in DNA by methyl radicals.**

guanine by an epoxide of benzo[a]pyrene is possibly of major importance (*Fig.* 38.19). It was pointed out in *Fig.* 38.15 that only the 9,10-epoxide-7,8-diol metabolite is believed to be important in initiating tumours; several other epoxides that bind to DNA are of lesser importance. This is probably because the configuration of the 9,10-epoxides is relatively stable to attack by the enzyme, epoxide hydratase, and is presumably the most

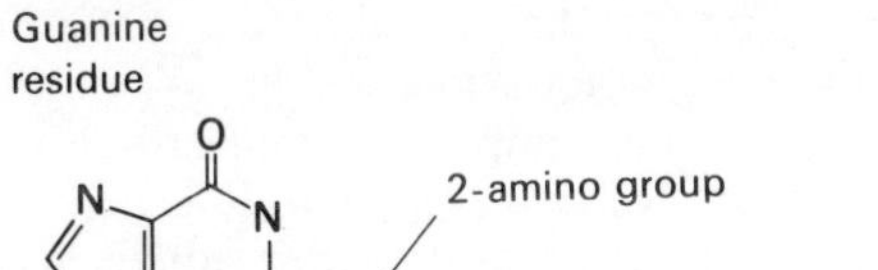

Fig. 38.19. **Formation of adduct between benzo[a]pyrene metabolite and DNA.**

favourable structure for fitting into the DNA helix.

Binding of this electrophile with bases on the DNA subsequently causes anomalous base-pairings during further replication, e.g. thymidine will pair with guanine alkylated at O–6 or N–7 and guanine will pair with thymidine alkylated at O–4. Such base-pairing will clearly give rise to synthesis of incorrect base sequences and therefore to mutations.

38.11 Induction of enzyme systems involved in oxidative metabolism of carcinogens

The discussion of the oxidative metabolism of potential carcinogens into the active form is clearly an important aspect of carcinogenesis, although carcinogens do possess the sinister property of being able to induce the enzyme system involved in their own metabolism. In Chapter 36, the inducing effects of drugs such as phenobarbitone were described, and carcinogens, such as methylcholanthrene, possess similar inductive effects. Despite the similarity of the end result, i.e. an increased rate of metabolism, the mechanisms for achieving this result are not identical since, as mentioned previously, the cytochrome P450 induced is different and also membrane proliferation, a characteristic of phenobarbitone induction, does not occur following carcinogen induction(*Table* 38.4).

Studies on mice have demonstrated that the inducibility of the system is under genetic control, e.g. induction of oxidative metabolism of benzo[a]pyrene by methylcholanthrene is very effective in C57 mice, but ineffective in DBA mice (*Fig.* 38.20). Experiments on 14 different strains of mice showed that the degree of induction in a strain correlated well with the carcinogenic index (*Fig.* 38.21). From these experiments, Niebert proposed that the genome contained a specific A_h locus (A_h = aryl hydroxylase) that could be readily transcribed in some strains but not in others. The mechanism proposed (*see Fig.* 38.22) is similar to the one proposed for

Table 38.4 **Comparison of induction in the endoplasmic reticulum by a drug (phenobarbitone) and a carcinogen (methylcholanthrene)**

Site of action	Phenobarbitone	Methylcholanthrene
Protein of endoplasmic reticulum	+++	0
Membranes of endoplasmic reticulum	+++	0
Cytochrome	+++ (P450) $P450_b$ or $P450_{B1}$	++ (P448) ($P450_c$ or $P450_{B2}$)
General oxidative drug metabolism	+++	0
Oxidative metabolism of polycyclic hydrocarbons and some related compounds (e.g. biphenyl)	+	+++

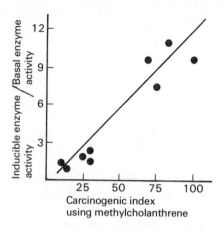

Fig. 38.21. **Relationship between the carcinogenic index for methylcholanthrene and the inducibility of the oxidative enzyme system in several strains of mice.**

control of the bacterial operon by Jacob and Monod.

The fact that the enzymes responsible for oxidative metabolism are likely to be under genetic control has important implications for the occurrence of cancer in man. It is, therefore, possible that some individuals may be more susceptible to carcinogens because their genetic makeup has resulted in enzyme systems that are more easily induced. Support for this concept of enzyme inducibility has come from study of leucocytes from both

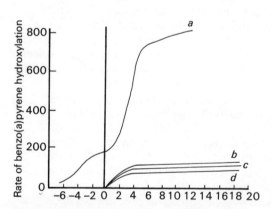

Fig. 38.20. **Induction of oxidative carcinogen metabolism by methylcholanthrene in different strains of mice.**

Methylcholanthrene (80 mg/kg) injected 24 h before measurement of enzyme activity.
(*a*) C57 + methylcholanthrene; (*b*) C57; (*c*) DBA; (*d*) DBA + methylcholanthrene.

Table 38.5 **Inducibility of polycyclic hydrocarbon hydroxylase and incidence of lung tumours in human subjects**

Number	Subject	Hydroxylase induction in leucocytes		
		Low	Inter	High
83	Healthy controls	44·7	45·9	9·4
46	Tumour controls	43·5	45·6	10·9
50	Lung cancer	4·0	66·0	30·0

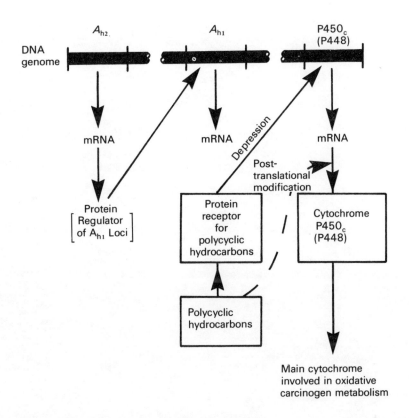

Fig. 38.22. **Genetic control of the A_h loci, where A_h is the aryl hydroxylase.**

groups of cancer patients and control subjects. The inducibility of the enzyme system metabolizing benzo[a]pyrene was studied by incubating the cells with methylcholanthrene; the inducibility was then rated as low, intermediate or high. The degree of induction in lung cancer patients was found to be much greater than that in control patients (*Table* 38.5).

These experiments could possibly provide an important scientific basis for the observation that some smokers succumb to lung cancer in middle age, whereas others survive into old age with little or no sign of the disease. It is possible that those who contract lung cancer possess the genetic pattern for effective induction of the enzyme systems that metabolize the polycyclic hydrocarbons produced from tobacco, whereas the induction capacity is very limited in those who survive.

Part 5

Biochemical basis of diagnosis—disease and its treatment

Chapter 39 Biochemical diagnosis

The possibility that enzyme determinations might be a useful aid to diagnosis was realized in the early part of this century when determinations of amylase activity were made on samples of urine. Later, in the 1930s, measurements were made of alkaline and acid phosphatases in blood and plasma, but the subject began to develop and extend to a wide range of enzymes about 25 years ago when more sophisticated methods of analysis were devised. Many of the methods used to determine enzyme activities in plasma have proved valuable for investigations of liver and heart disease, and are now in current use in almost every hospital.

Most investigations of enzyme activity in the plasma are designed to determine increases in the activity of particular enzymes resulting from leakage of enzymes into the plasma from organs that have suffered tissue damage, for example following coronary thrombosis.

39.1 Principles of methods used

Many enzymes arising from a variety of sources can be demonstrated in human plasma or serum. Some are secreted into the plasma and carry out their normal functions there, for example the coagulation enzymes and cholinesterase. For this group, injury to the organ producing the enzyme under investigation would cause a fall in activity.

A second and more important group, including many of the enzymes that play major roles in metabolism, are intracellular, and are only found in the plasma in significant quantities when the cells are damaged, so that consequently the cell membranes become leaky and even completely haemolysed. The measurement of enzyme activity in this group will normally be related to the number of cells damaged and thus to the extent of the lesion. These enzyme measurements are valuable in clinical diagnosis, both in the initial stages of the disease and during the period of recovery and repair.

39.2 Typical enzymes determined in the plasma

Enzyme activities should normally be quoted in units suggested by the Enzyme Commission

Table 39.1 **Enzymes frequently measured in plasma**

Name of enzyme	EC number	Abbreviation
Acid phosphatase	3.1.3.2	ACP
Aldolase	4.1.2.13	ALD
Alkaline phosphatase	3.1.3.1	AP
α-Amylase	3.2.1.1	
Choline esterase	3.1.1.7	CHE
Creatine phosphokinase	2.7.3.2	CPK(CK)
Glutamate dehydrogenase	1.4.1.2	GLDH
Glutamate–oxaloacetate transaminase	2.6.1.1	GOT
Glutamate–pyruvate transaminase	2.6.1.2	GPT
γ-Glutamyl transferase	2.3.2.2	γ-GT(GGTP)
3-Hydroxybutyrate dehydrogenase	1.1.1.30	HBDH
Leucine aminopeptidase	3.4.11.1	LAP
Lactate dehydrogenase	1.1.1.27	LDH
Lipase	3.1.1.3	
Sorbitol dehydrogenase	1.1.1.14	SDH

Table 39.2 **Normal enzyme activities in human serum**

Enzyme	Activity, mU per ml serum			
	Upper limit	Mean	Standard deviation	Range
Glutamate–oxaloacetate transaminase	12	10·6	± 2·5	4·7–18·3
Glutamate–pyruvate transaminase	12	7·3	± 3·1	2·3–13·9
Lactate dehydrogenase	195	111·0	± 25	66·0–164
Alkaline phosphatase	30	—	—	—
Acid phosphatase	11 (37 °C)	—	—	—
Amylase	4000 (37 °C)	—	—	—
Creatine phosphokinase	1	—	—	—
Glutamate dehydrogenase	0·9	—	—	—
Sorbitol dehydrogenase	0·4	—	—	—
Aldolase	3·1	—	—	—
3-Hydroxybutyrate	140	—	—	—
γ-Glutamyltransferase	28 (δ) 18 ($\female$)	—	—	—
Lipase	140	—	—	—
Choline esterase	1900–3800	—	—	—

Note: Determinations were made at 25 °C unless otherwise stated.

of the International Union of Biochemistry. The Commission defined an enzyme unit (U) as that amount of enzyme which, at 25 °C, catalyses the reaction of one micromole (μmol) substrate per minute. A temperature of 30 °C was originally recommended, but 25 °C is convenient and is more frequently used. For serum or plasma, the enzyme activity is normally recorded in mU/ml. Those enzymes whose activities in plasma are commonly increased are listed in *Table* 39.1. The 'upper limits' of the activities of these enzymes in serum are listed in *Table* 39.2, plus the normal ranges of activity for some of these. The 'upper limit' figure is a useful guide to the investigator, since activities above these figures must be regarded as indicating that tissue damage is likely to have occurred.

39.3 Distribution of enzymes in tissues and serum patterns

If the concentration of a particular enzyme in a tissue is normally high, then it is clear that damage to the tissue is likely to cause release into the plasma of a high concentration of this enzyme. Conversely, severe damage to a tissue containing a low concentration of a particular enzyme will cause only a modest increase of enzyme activity in the plasma. It is, therefore, apparent that determination of the type of enzyme released into the plasma coupled with activity measurements will provide important information about the extent and nature of pathological damage to any tissue. Patterns of activities in normal tissues are an important guide to the nature of the enzymes likely to be released, the more important of which are shown in *Figs*. 39.1–39.6. It will be noted, for example, that both transaminases are very active in the liver. Heart and muscle glutamate–oxaloacetate transaminase activities are similar, but glutamate–pyruvate transaminase is much more active in liver than in heart. Acid phosphatase is much more active in the prostate than in any other organ, whereas creatine phosphokinase is much more active in skeletal muscle than in liver. Determinations of several enzymes will, therefore, be necessary to provide information about the site and extent of pathological damage.

The nature of the enzyme released as a consequence of damage will depend on its subcellular localization. If the damage is minimal, only cytoplasmic enzymes will leak

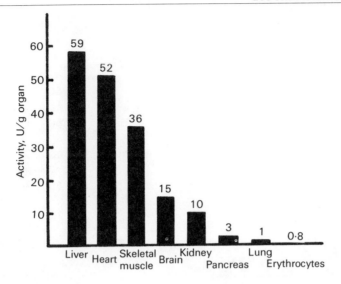

Fig. 39.1. **Glutamate–oxaloacetate transaminase activities in human organs.**

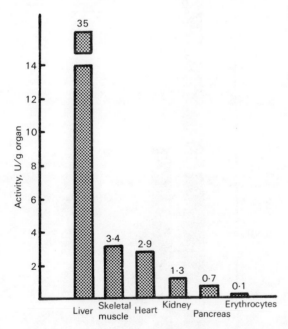

Fig. 39.2. **Glutamate–pyruvate transaminase activities in human organs.**

out, but if extensive damage occurs, mitochondrial enzymes will be released (*see Fig.* 39.7).

Provided several criteria are met, the pattern of released serum enzymes should be identical to the enzyme pattern in the organ concerned. These criteria are listed below.

a. The damage must be sufficient to destroy the intracellular membranes, so releasing all soluble enzymes from the cells into the plasma.

b. Onset of the damage must be acute, so that the normal enzyme pattern in the organ is not changed as a result of the injury and elimination of enzymes from the plasma has not commenced.

c. The damage must have occurred in a single organ, since enzyme output from other organs would produce a different pattern of enzyme release.

Following release into the plasma, enzymes are gradually eliminated. Some with a small molecular weight, like amylase, may be excreted in the urine, but the majority undergo normal catabolic degradation although at very different rates (*see Table* 39.3).

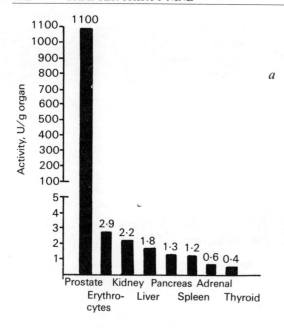

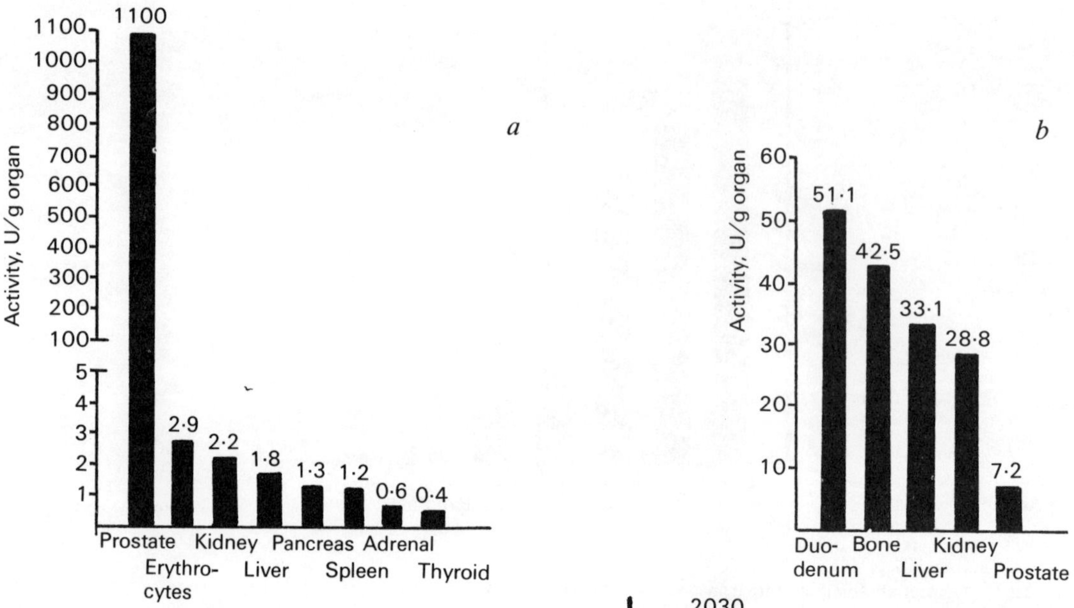

Fig. 39.3.   **Acid and alkaline phosphatase activities in human organs.**
(*a*) Acid phosphatase; (*b*) alkaline phosphatase.

Table 39.3 **Half-lives of enzymes in plasma**

Enzyme	Half-life ($t_{1/2}$)
Glutamate–oxaloacetate transaminase	17 ± 5 h
Glutamate–pyruvate transaminase	47 ± 10 h
Glutamate dehydrogenase	18 ± 1 h
Lactate dehydrogenase (hydroxybutyrate dehydrogenase)	113 ± 60 h
Creatine phosphokinase	≈ 15 h
Alkaline phosphatase	3–7 days
γ-Glutamyltransferase	3–4 days
Choline esterase	≈ 10 days
Amylase	3–6 h
Lipase	3–6 h

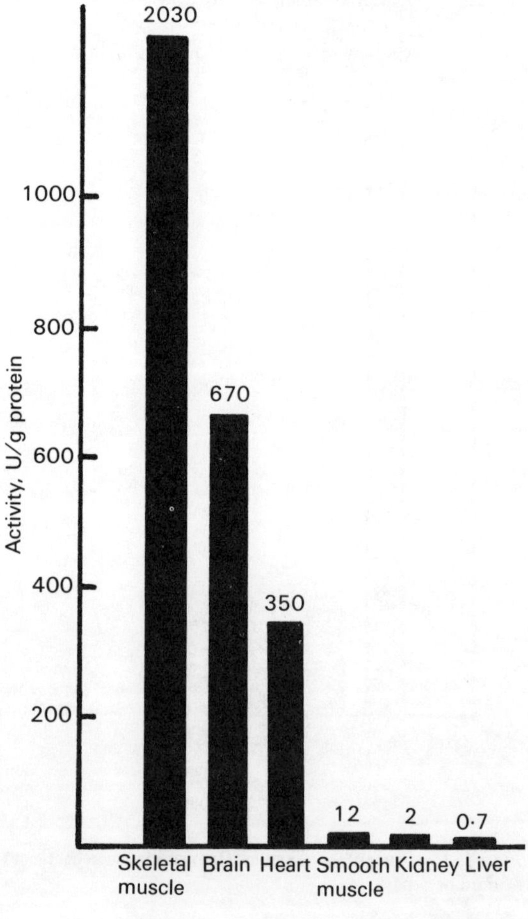

Fig. 39.4.   **Creatine phosphokinase activities in human organs.**

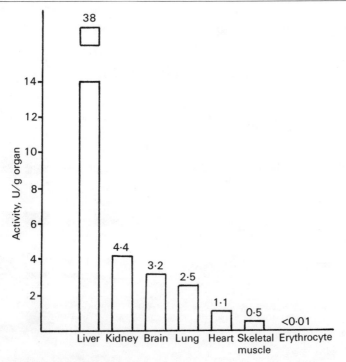

Fig. 39.5.   **Glutamate dehydrogenase activities in human organs.**

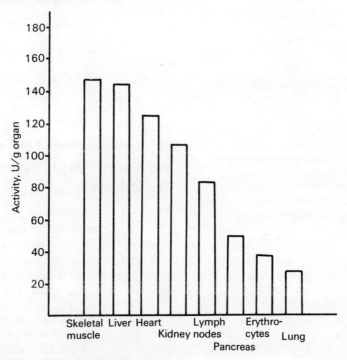

Fig. 39.6.   **Lactate dehydrogenase activities in human organs.**

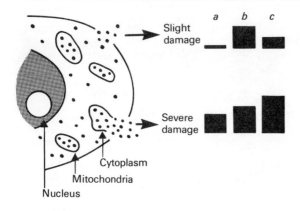

Fig. 39.7.   **Comparison of effects of slight and severe damage on enzyme release.**
(*a*) Glutamate dehydrogenase; (*b*) glutamate–pyruvate transaminase; (*c*) glutamate–oxaloacetate transaminase.

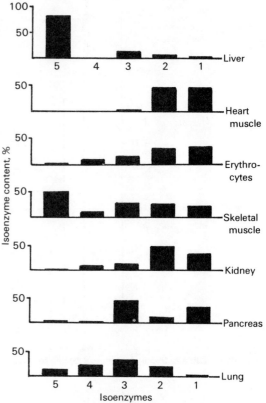

Fig. 39.8.   **Isoenzymes of lactate dehydrogenase in various tissues.**

39.4 Isoenzymes

Isoenzymes are different forms of an enzyme which catalyse the same reaction, but which exhibit different physical properties. The best studied example is that of lactate dehydrogenase which exists in five forms that may be separated by electrophoresis on cellulose acetate or by chromatography on DEAE-cellulose or DEAE-Sephadex.

The enzyme is a tetramer and composed of α and β subunits: four α units, α_4, mixtures of α and β, $\alpha_3\beta$, $\alpha_2\beta_2$, $\alpha\beta_3$ or four β units, β_4. The two subunits are coded by different genes and the two genes are expressed to different extents in different tissues. Heart muscle produces mainly α subunits, while skeletal muscle produces mainly β subunits.

Most organs possess entirely different lactate dehydrogenase isoenzyme patterns; consequently even when total lactate dehydrogenase activities are very similar in organs under investigation, e.g. liver and heart (*see Fig.* 39.6), the different isoenzyme patterns enable the source of the enzyme released in the plasma to be identified (*Fig.* 39.8).

39.5 Examples of the use of measurements of serum enzymes in diagnosis

a. Study of liver disease

Studies of plasma enzymes have been valuable in the diagnosis of liver disease. For example, large increases in transaminases and other enzymes occur during acute viral hepatitis (*Fig.* 39.9). Enzymes are sensitive indicators since elevation of serum enzymes frequently precedes the onset of jaundice. The activity of glutamate–pyruvate transaminase is always greater than that of glutamate–oxaloacetate

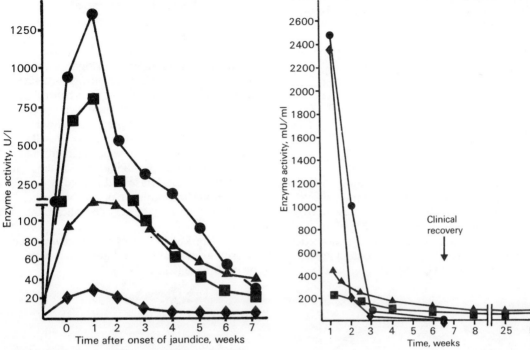

Fig. 39.9. **Serum enzyme activities during acute viral hepatitis.**
(—◆—) Glutamate dehydrogenase; (—▲—) γ-glutamyltransferase; (—■—) glutamate–oxaloacetate transaminase; (—●—) glutamate–pyruvate transaminase.

Fig. 39.10. **Recovery of serum enzymes to normal levels after acute hepatitis.**
Glutamate–oxaloacetate transaminase: (●) rapidly resolving; (■) slowly resolving. Glutamate–pyruvate transaminase: (◆) rapidly resolving; (▲) slowly resolving.

transaminase and increases of γ-glutamyl transferase and glutamate dehydrogenase are much less than those of the transaminases. Enzyme activities can also be used to study the progress of recovery (*Fig.* 39.10).

In chronic hepatitis, elevated enzyme levels can occur regularly over periods of several months (*Fig.* 39.11). Increases in liver enzymes also occur following liver damage caused by chemicals such as halothane, thiamazole or salicylates, but the patterns are variable and cannot be directly related to the chemical causing the toxic effect.

b. Study of heart disease
The effect of myocardial infarction on the release of enzymes from heart has been extensively studied. The plasma enzyme

activities of creatine phosphokinase, glutamate–oxaloacetate transaminase, lactate dehydrogenase and aldolase begin to rise within 2–8 hours of the infarction, reach a maximum 24–72 hours later and generally return to normal after 7 days, although some enzyme activities are close to normal values after 2–3 days (*Fig.* 39.12).

Average increases in enzyme levels after an infarction are shown in *Table* 39.4. In man, a correlation can be demonstrated between the height of the elevation of serum enzymes and the extent of the infarction, in so far as it can be estimated by electrocardiography. Reservation must be applied, however, because it is not always possible to be sure that the transient maximum of enzyme elevation has been recorded. In fact, mortality also correlates with the height

Table 39.4 **Average increase of enzyme activities in myocardial infarction**

Enzyme	Upper limit of normal, mU/ml	Average elevation, mU/ml
Creatinine phosphokinase	1	3–30
Glutamate–oxaloacetate transaminase	12	50–150
Lactate dehydrogenase	175	250–800
Aldolase	6	15–30

and duration of the enzyme elevation. A rise of glutamate–oxaloacetate transaminase above 150 mU/ml and lactate dehydrogenase above 800 mU/ml is considered prognostically unfavourable. Reinfarction usually causes the enzyme level to remain at the existing level and a second peak is not observed (*Fig.* 39.13).

c. Study of muscle disease

Strenuous physical work causes increase in enzyme activities in normal man. After 5 hours rigorous exercise, the aldolase levels are doubled and the glutamate–oxaloacetate transaminase and lactate dehydrogenase levels increase by about 50 per cent over control values. Creatine phosphokinase is, however, more sensitive and is elevated even under relatively small stress. After severe muscle trauma, for example following accidents and surgical operations, small increases in creatine phosphokinase, aldolase and glutamate–oxaloacetate transaminase occur, but secondary liver damage may complicate the observations.

Marked increases do, however, occur in muscular dystrophys and these vary with the type of disease (*Table* 39.5).

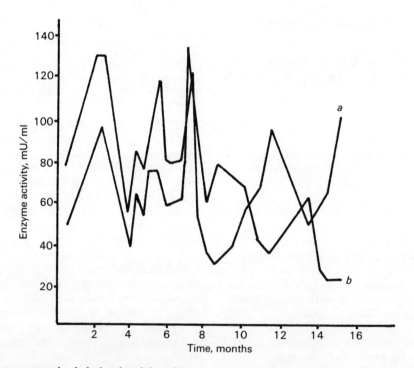

Fig. 39.11. **Serum enzyme levels during chronic hepatitis.**

(*a*) Glutamate–oxaloacetate transaminase; (*b*) glutamate–pyruvate transaminase.

Table 39.5 **Plasma enzyme activity in different forms of muscular dystrophy**

| Disease type | Enzyme activity, mU/ml, of | | | |
	Aldolase	Glutamate–oxaloacetate transaminase	Creatine phosphokinase	Glutamate–pyruvate transaminase
— (Normal)	3·1	12	1	12
Duchenne	14·1	48·6	8·3	40·9
Limb–girdle	6·5	21·8	1·5	13·8
Facio-scapulo humeri	6·3	34·8	0·75	19·9

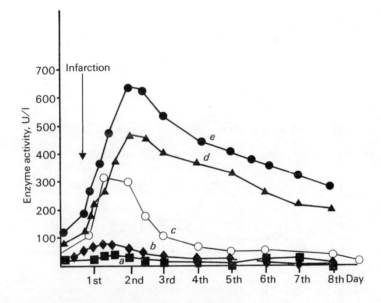

Fig. 39.12. **Serum enzyme levels following a myocardial infarction.**
(*a*) Glutamate–pyruvate transaminase; (*b*) glutamate–oxaloacetate transaminase;
(*c*) creatinine phosphokinase; (*d*) hydroxybutyrate dehydrogenase; (*e*) lactate
dehydrogenase.

d. Study of kidney diseases

Serum enzyme measurements have, as yet, not
made substantial contributions to the study
of renal disease. Non-specific small increases
in enzyme activities are found in about half
the diagnosed cases. Lactate dehydrogenase
is most frequently raised, glutamate-
oxaloacetate transaminase less often, although
no direct correlation exists between the extent
of the enzyme elevation and other symptoms.

e. Study of pancreatic diseases

In acute pancreatitis, serum amylase activities
begin to rise within 3–6 hours and normally
reach a maximum after 20–30 hours. Rises up
to 60 U/ml can occur and, in acute necrosis
of the gland, the serum amylase may rise to
120 U/ml and this may be demonstrated for up
to 48–72 hours. Released enzymes are
excreted in the urine where they may be
determined. Long periods of elevated serum

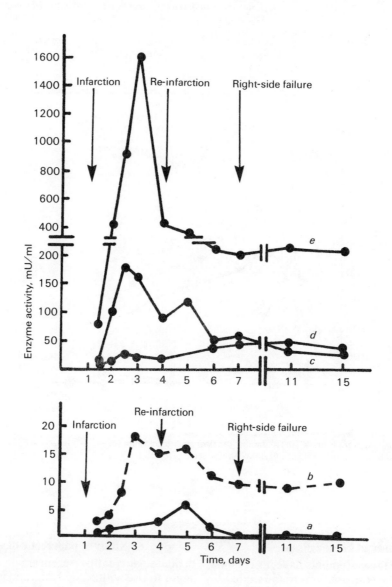

Fig. 39.13. **Serum enzyme levels following a myocardial reinfarction.**
(*a*) Creatinine phosphokinase; (*b*) aldolase; (*c*) glutamate–oxaloacetate transaminase; (*d*) glutamate–pyruvate transaminase; (*e*) lactate dehydrogenase.

amylase levels indicate necrosis of the pancreas.

Raised amylase levels are not, however, always reliable in diagnosing primary acute pancreatitis, since secondary damage may occur following perforated ulcers and diseases of the bile duct. Acute pancreatitis causes raised levels of glutamate–pyruvate transaminase, glutamate–oxaloacetate transaminase and lactate dehydrogenase, but these increases are much less marked than those observed in liver disease.

f. Study of bone disease

Alkaline phosphatase originates in the osteoblasts and osteoclasts of bone. It is transferred to the plasma and excreted in the bile and, therefore, in the absence of hepatobiliary disease, the alkaline phosphatase level is directly related to the functional activity of the osteoblasts, the main producers of the enzyme.

The activity of the enzyme (normally 30 mU/ml) rises to very high levels (130–1400 U/ml) in Paget's disease and in rickets where it can reach similar values. Measurements in plasma are a sensitive indicator, since increase in alkaline phosphatase levels are apparent 1–2 months before the disease becomes clinically manifest. Very high levels of alkaline phosphatase are also caused by bone tumours and by primary hyperparathyroidism.

g. Studies of tumours

Hopes that measurements of serum enzymes would be of value in diagnosis of malignant

tumours have not been realized. Serum acid phosphatase is, however, of value in the diagnosis of prostate carcinoma and, as noted above, primary and secondary tumours of the skeletal system cause raised levels of alkaline phosphatase. Characteristic patterns showing increased levels of several enzymes are connected with liver metastases, but these measurements are difficult to use in accurate diagnosis (*Fig.* 39.14).

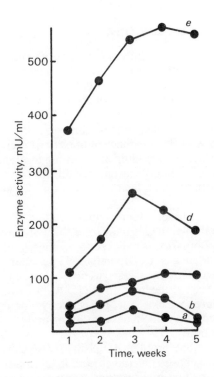

Fig. 39.14. **Serum enzyme levels in liver metastases.**
(*a*) Glutamate dehydrogenase; (*b*) glutamate–pyruvate transaminase; (*c*) alkaline phosphatase; (*d*) glutamate–oxaloacetate transaminase; (*e*) lactate dehydrogenase.

Chapter **40** An example of metabolic disturbance: obesity

40.1 Introduction

It is sad to reflect that, in the modern world where millions of people are suffering from serious malnutrition, many in the affluent societies are suffering from obesity. Affluence is an important cause of the condition since it is generally established that this correlates closely with the incidence of obesity. It is clear that increased spending on fattening foods and the use of so many mechanical aids for travel and work together lay the foundations for an obese society.

The problem is widespread. For example, in Germany, one-third of the adult population is believed to be obese and the incidence is sex dependent: 42 per cent of adult women are assessed as obese, but only 19 per cent of the men. In the USA, however, the figures are closer: 42 per cent of women as compared with 36 per cent of the male population.

The problem of obesity is also class related as based on studies both in Germany and in the United Kingdom. In the UK, it has been established in a recent survey that 21 per cent of the upper class could be classed as obese, 19 per cent of the middle class, but 52 per cent of the lowest social class. It would thus appear that 'affluence' of individuals is not the only or even the main factor when smaller groups of the population are considered.

Some comparisons between similar countries are difficult to explain. For example, wealthy German businessmen are generally much more obese than similar groups in other countries.

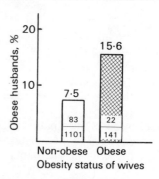

Fig. 40.1. **Correlation of obese husbands and wives.**
Data taken from 1242 spouse pairs of single continuous marriages. ☐ Non-obese, obesity index < 120; ▩ obese, obesity index < 120. $p < 0.01$.

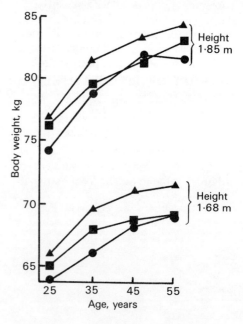

a

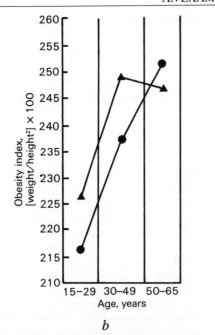

b

Fig. 40.2.   Increase of incidence of obesity with age.

(*a*) Average body weight from life insurance data.
(▲) America, 1935–54; (●) Scotland, 1926–47;
(■) Scotland, 1950–71. (*b*) Obesity index in a study in the
UK of 794 females (●) and 540 males (▲).

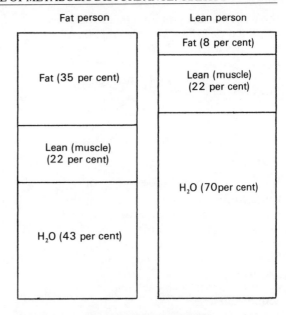

Fig. 40.3.   **Composition of the human body.**

obese adult of similar age and sex. However, it
is not easy to measure the total body fat and
the range between lean and typical obese
individuals is shown in *Fig.* 40.3.

    To overcome this problem, various
formulae have been devised which relate
height and weight, the simplest being to divide

The detailed 'Framingham study'
carried out in the USA to attempt to establish
factors responsible for coronary heart disease,
noted that a strong correlation existed between
obese husbands and wives (*Fig.* 40.1). It is
possible that obese persons may find obesity
attractive in the opposite sex, but,
alternatively, unwise eating habits after
marriage may encourage both partners to
become obese concurrently!

    It is generally observed and proved
scientifically that obesity increases with age
and this fact has been demonstrated in many
investigations (*Fig.* 40.2).

40.2   Measurement of obesity

An accurate measurement of obesity really
requires that the total body fat be measured
and compared with that for a normal non-

Table 40.1 **Obesity index**

Index	Correlation coefficient	
	Male	*Female*
1. Weight/Height	0·54	0·71
2. Quetelet index: Weight/Height^2	0·55	0·70
3. Rohrer index: Weight/Height^3	0·50	0·66
4. Ponderal index: Weight$^{1/3}$/Height	0·52	0·68
5. Sheldon index: Height/Weight$^{1/3}$	0·52	0·69
6. Nicholson–Zilvas index: Height^3/Weight	0·51	0·69

$$\text{Body fat} = \frac{\text{Weight} - \text{Lean body mass}}{\text{Weight}}$$

where

$$\text{Lean body mass} = \frac{\text{Total body weight}}{0.73}$$

the weight in kilogrammes by the height in centimetres. Other formulae that have been used are shown in *Table* 40.1, but it will be noted that none give a correlation with total body fat which is superior to a simple division of weight by height.

40.3 Relation of water loss to obesity

It is important to appreciate that the total body water forms a major part of the total body weight and some weight losses, especially in the short term, are simply a result of water loss and of no real lasting benefit to the obese. Many unscientific, or quack, slimming diets merely cause a relatively large quantity of water to be lost, giving the impression of genuine fat loss.

Table 40.2 **Relationship of starvation, low calorie diet and obesity**

Diet	Weight loss, g/day	N loss, g/day	Ketone, g/day	Fat loss, g/day
1. Starved	751	8·1	9	243
2. 800 cal fat-carb. (mixed diet)	278	1·6	0	165
3. 800 cal very low carb. (ketogenic diet)	467	2·9	3	165

Subjects are 135–239 per cent above desirable weight.
Note: Difference ≡ water loss.

     An example is shown in *Table* 40.2 in which the effects of three diets have been tested on obese volunteers. Starvation causes a large total weight loss and large fat loss, but also causes a substantial loss of body protein and the production of ketone bodies. Diet 3 appears to cause a much greater weight loss than diet 2 but this is almost entirely due to loss of water since the quantity of fat loss (165 g) is the same after feeding on either diet. Although it appears to be least effective, diet 2 would be

preferable to diet 3 because it causes the same fat loss but provides no ketone body formation and the minimum loss of nitrogen (1·6 g/day).

40.4 The fundamental causes of obesity

When considered in basic terms, the obese condition is simply the result of a disturbed relationship between three components of energy: the input as food, the energy expenditure, such as muscle activity and body maintenance and energy storage, primarily as fat. The relationships are shown diagrammatically in *Fig.* 40.4. The problem of obesity therefore arises from an excessive food intake associated with reduced energy expenditure. It is not always realized that the regulation must be very precise as can be judged from the following calculations: the average human consumes about 20 tons of food in a lifetime and a 1 per cent error in energy balance could cause an increase in weight of 50 kg, or convert a 70-kg adult into one weighing 120 kg. The weight of very few people, even if they are obese, rises to such an extent and it demonstrates how precise the regulation must be to maintain a stable weight.

     Although the problem of obesity illustrated in *Fig.* 40.4 appears simple, there are in fact many complicated and contributing factors especially in a normal human environment. Theories of obesity are generally based on one of two hypotheses; obesity is

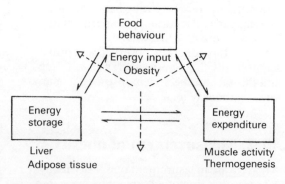

Fig. 40.4. **Inter-relationships of energy input and expenditure in obesity.**

believed to be caused either by a metabolic defect or by environmental, psychological or sociological pressures. These hypotheses are sometimes described as the pull and the push theories and are illustrated in *Fig.* 40.5. The 'pull theory' supposes that certain metabolic defects cause the subject to consume much more food than he or she really requires, whereas the 'push theory' presumes that pressures on the individual from outside the body cause the excessive consumption of food.

These two fundamental concepts can be extended further as shown in *Fig.* 40.6. Metabolic defects are complex and are discussed in subsequent sections. Obesity may also arise from both 'psychological' and 'sociological' origins acting together. For example a stressful sociological disturbance may lead to psychological

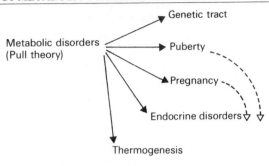

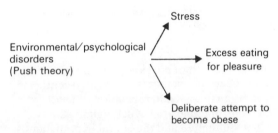

Fig. 40.6. **Facts of metabolic causes and environmental causes of obesity.**

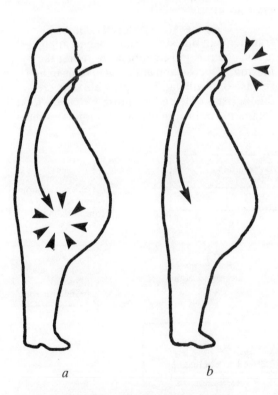

a   *b*

Fig. 40.5. **The pull and push theories of obesity.** (*a*) The 'pull theory' postulates that a metabolic disturbance is mainly responsible; (*b*) the 'push theory' postulates that external or environmental factors are mainly responsible.

disturbances that could be demonstrated as a pattern of excess eating. The obese state is further complicated by the fact that metabolic defects and environmental pressures often occur simultaneously so that they complement each other and exacerbate the condition. Alternatively, many believe that a psychological disturbance that leads to a pattern of eating giving rise to obesity may eventually cause metabolic disturbances.

40.5 Regulation of food intake

It is quite clear that an overriding factor in the development of obesity is the excessive food intake, and an understanding of the obese condition clearly necessitates a discussion of the factors regulating the quantity of food consumed.

The regulation is complex and only partially understood. The intake of food is known to be controlled by the hypothalamus

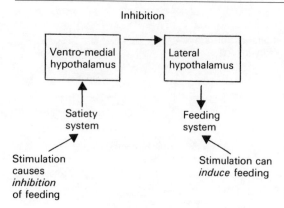

Fig. 40.7. **Role of the hypothalamus in the short-term regulation of food intake.**

and, as a result of physiological experiments, it has been shown that there are two centres in this organ involved in regulation of feeding: the ventromedial area which signals satiety when stimulated and the lateral hypothalamus which stimulates feeding. The satiety region can also cause inhibition of the feeding system (*Fig.* 40.7). This concept of a regulation mechanism for food intake is useful, but it takes no account of the important effects that are caused by the components and the concentration of ingested foods in the regulation of food intake.

Two main types of regulation of these centres have been established: the 'short-term' control that regulates food intake at one meal or over a short period of a few hours and the 'long-term' control that regulates food intake over periods of weeks or months. Short-term regulation is much better understood than long term and occurs in two phases, phase I and phase II (*Fig.* 40.8).

Phase I regulation is dependent on the presence of food in the gastrointestinal tract and the mechanism of control is believed to involve the release of the gastrointestinal hormones cholecystokinin and enteroglucagon. These hormones pass into the blood and stimulate the satiety centre which leads to inhibition of the feeding centre (*Fig.* 40.8).

As digestion proceeds small fragments of digested foods, such as amino acids, glucose and chylomicrons, enter the plasma. Glucose and amino acids cause the release of insulin and it is possible that the increase of the concentration of these food fragments or insulin in plasma stimulates the

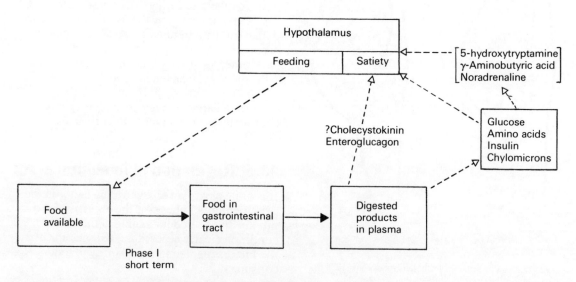

Fig. 40.8. **Mechanisms of regulation of food intake in the short term.**
There are both long-term and short-term regulations and the short-term regulation can be divided into two phases: phase I and phase II.

satiety centre in the hypothalamus to inhibit the feeding centre. Alternatively, the food fragments stimulate the release of intermediate transmitters, such as 5-hydroxytryptamine, γ-aminobutyric acid or noradrenaline, which stimulate the satiety centre. At present it is unclear whether stimulation of the satiety centre is caused directly by the nutrients in the plasma or by intermediate intervention of the transmitters.

These mechanisms have been established in animal experiments but, although similar mechanisms are believed to regulate human eating, it is also controlled by strong habits and customs. It has been shown, for example, that control groups of young cadets cannot regulate their energy intake in relation to energy demands during short periods of 24 hours or so (*see Fig.* 40.9).

Animals such as rats normally regulate their food intake for energy purposes much more accurately.

At the beginning of the chapter, it was noted that animals and most people are able to regulate their energy intake over the long term, i.e. over several months or years. Although several mechanisms by which this regulation is achieved have been proposed, no theory is as yet acceptable that can explain all the facts. The main problem is that all the normal nutrients found in plasma, such as glucose and amino acids, fluctuate in concentration from hour to hour so that it would be quite impossible for regulation to be achieved over long periods by changes in the concentration of any of these nutrients.

It has been postulated that certain hormones may be involved. Attention has

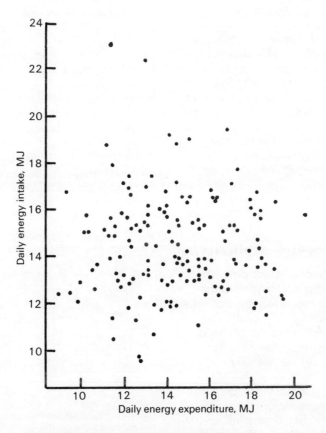

Fig. 40.9.   **Relation between daily energy intake and daily energy expenditure in army cadets.**

been focused on the possible role of progesterone since, during pregnancy in humans and animals, it has been observed that the body weight increases considerably more than that of the fetus. Furthermore, female rats put on weight after an injection of progesterone, but the ovaries are essential for the effect which is not observed in males or ovarectomized females. Prostaglandins have also been suggested as regulators and it has been postulated that they could be liberated steadily from the adipocytes and indicate, to the hypothalamus, the fat loading of the cells. They are also known to be one of the regulators of the release of fatty acids from adipose tissue (cf. Chapter 19). Overproduction of prostaglandins could, therefore, cause the retention and build up of body fat by inhibiting release of fatty acids.

Finally, an ingenious theory has been proposed which does not link fat stores with hypothalamus function by hormones or regulators. Instead, it is proposed that the hypothalamus itself can act as a microcosm of the body fat stores, building up small stores of fats as the adipose tissue becomes loaded with fat and becoming depleted as the quantity of fat in the stores is reduced. The fat stores of the hypothalamus would, according to this theory, regulate feeding patterns. This attractive hypothesis so far lacks experimental verification and, at present, details of the processes involved in the long-term regulation of the body energy stores remain a mystery.

40.6 The adipocytes in obesity

In view of the fact that the major proportion of the body fat is stored in special cells of the adipose tissues, the adipocytes, these cells have been studied very carefully. The important question can be posed: Is obesity due to an increased number of adipocytes or is it mainly due to excess loading of the normal number of adipocytes? It has been suggested that increased numbers of adipocytes, possibly arising as a result of genetic factors or due to an early pattern of feeding during infancy, could be a major component in the development of obesity.

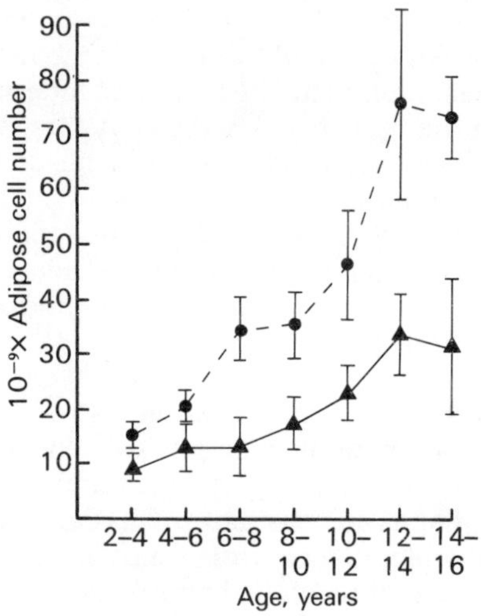

Fig. 40.10. **Increase of numbers of adipocytes with age in normal and obese children.**
(– – –) Obese; (——) normal.

Many experimental observations have been made of adipocytes in obese individuals and it is generally concluded that both the numbers and fat loading of the adipocytes increase significantly in the obese condition (*Fig.* 40.10). Furthermore, it appears that several forms of obesity are often characterized by the appearance of large numbers of adipocytes in early life and they can be progressively loaded with fat during growth and development (*Fig.* 40.11).

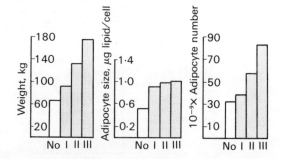

Fig. 40.11. **Relation between adipocyte size and number and the total body weight.**
No = normal weight; I, II and III represent progressive states of obesity.

Numbers of adipocytes can, therefore, exert an important effect on obesity, but the reason that some individuals possess more of this type of cells than others is, as yet, unresolved.

40.7 The biochemical changes observed in obesity

Several significant metabolic changes take place in obese individuals that are reflected in plasma concentrations and these are summarized in *Table* 40.3. Most characteristic

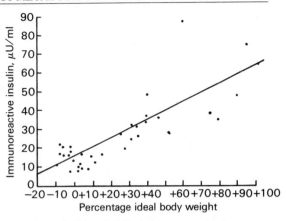

Fig. 40.13. **Relation between the concentration of circulating insulin and obesity.**
$r = +0.72; p < 0.001.$

Table 40.3 **Biochemical changes in obesity**

1. Numbers of adipocytes	↑
2. Synthesis of cholesterol	↑
3. Synthesis of fatty acids	↑
4. Plasma concn of fatty acids	↑
5. Plasma concn of triacylglycerols	↑
6. Plasma concn of *insulin* (pancreatic hypertrophy)	↑

is the altered pattern of lipid metabolism: the plasma concentration of β-lipoproteins transporting lipids is increased (cf. Chapter 19), as is the concentration of free fatty acids and is shown in *Fig.* 40.12. It can also be demonstrated that the inhibition of fatty acid release by an infusion of glucose is much less marked in obese persons and that synthesis of fat from carbohydrate in the liver is also

increased. The reasons for all these changes are not clear but insulin may play a central role.

The concentration of circulating insulin is markedly increased in obese individuals and a close correlation between its concentration in plasma and the degree of obesity may be demonstrated (*Fig.* 40.13). The following sequence of events is believed to occur: excess food intake, particularly carbohydrate loads, leads to overproduction of insulin which in turn causes hypertrophy and hyperplasia of the pancreas. This development of the pancreas causes further production of insulin which will stimulate production of fatty acids in the liver from carbohydrate and increase fat synthesis in adipose tissue. This clearly leads to several of the symptoms of obesity.

The effect of insulin on the metabolism of many tissues is, however, modified by the fact that the tissues become unresponsive. This is presumably caused by changes in the insulin receptors on these tissues, which eventually respond feebly to insulin, a fact that may be demonstrated in genetically obese mice (*Fig.* 40.14). This failure of the tissues to respond to insulin is a serious situation since the tissues will then tend to behave as though the supply of energy-providing foods, such as glucose, is quite

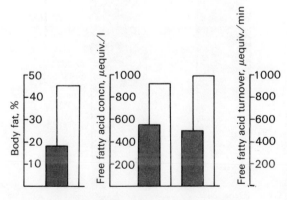

Fig. 40.12. **Plasma concentration of fatty acids and turnover in normal and obese persons.**
□ Obese; ▨ control.

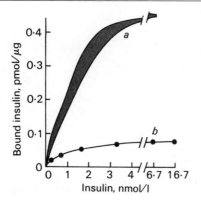

Fig. 40.14. Binding of insulin to fat cell plasma membranes of lean and obese mice.
(*a*) Lean mice; (*b*) obese mice.

inadequate, whereas in the obese the reverse is clearly the case. The metabolic regulation, therefore, partially breaks down and the tendency to become obese is increased.

40.8 The role of thermogenesis in obesity

For a considerable time many lay people have been well aware that some individuals can consume a great quantity of food without becoming obese, whereas others of similar age and sex rapidly become obese whilst consuming this same quantity of food. It is therefore remarkable, in view of these long-standing observations, that only relatively recently has the subject been studied scientifically.

One of the first careful studies to be recently described was made by Sims in the USA. His first experiments on feeding special diets to medical students proved unsuccessful since the students were unreliable in adhering to rigid dietary regimes over long periods! He therefore switched to study prisoners in the State Penitentiary. Groups of volunteers were fed diets containing very large amounts of energy so that they received about 10 000 cal/day for 4–6 months. All the subjects showed changes typical of the obese condition, the adipocyte loading with fat increased, and their

plasma concentrations of cholesterol, free fatty acids and insulin were all increased. It was very significant, however, that the weight gain of the subjects varied widely, from 12 to 25 per cent of their original weight despite the fact that they were all on similar diets. These results, scientifically obtained, confirmed the experience of many lay people that obesity is not simply related to the quantity of food eaten. Several possible reasons for this finding were considered, including variations in water retention and varied faecal loss of energy, but these reasons could not be substantiated. Sims considered the reasons that different individuals used different quantities of energy could be explained on the basis of variations in the conversion of energy to heat or 'thermogenesis'.

On theoretical grounds, two British nutritionists came to a similar conclusion, namely that obesity arises primarily because the group tending to obesity uses a smaller proportion of their energy for heat production. The proposal is illustrated diagramatically in *Fig.* 40.15. It is clear from this diagram that, in obese individuals, more energy is available from ingested food for storage as fat.

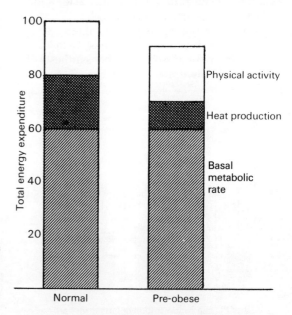

Fig. 40.15. Energy expenditure in normal and obese persons.

It was decided to test these theories by animal experiments and to attempt to establish the mechanism of heat production by thermogenesis. Although apparently simple, this is not easy since animals such as rats do not normally eat excessively to become obese. Recently, however, it was observed that if rats are fed on special diets containing many of the special foods, such as chocolate biscuits or Mars bars, that humans find attractive, the rats will eat excessively and become obese.

Table 40.4 **Role of thermogenesis: brown fat in obesity**

	Control	*'Cafeteria'*
Body wt gain, g	103	131
Body fat gain, g	40	66
Energy intake, kJ	6480	11 670
Energy gain, kJ	1790	2230
Efficiency of gain (kJ gain/kJ eaten)	0·36	0·22
Brown fat (Wt), mg	266	692

Rats are fed on diet for 21 days.
From Rothwell N. J. and Stock M. J., 1979, *Nature,* Vol. 281, p. 31.

This type of diet was called, by the experimentalists in London the 'cafeteria' diet because all the diet was obtained from the students' cafeteria. An experiment comparing the response of rats fed on normal stock diet with those fed 'cafeteria' diet is shown in *Table* 40.4. It will be noted that, over the duration of the experiment, the total energy intake in the 'cafeteria' group was nearly twice that taken in the control group. The weight gain, energy gain and body fat gain were all greater in the 'cafeteria' group than in the control group, but none of these measurements appeared to increase in proportion to the energy intake. This fact may be clearly demonstrated by calculating the efficiency of energy gain which was much greater in the control group (0·36) than in the 'cafeteria' group (0·22). The reason for this was traced to the development of large quantities of 'brown fat' in the 'cafeteria'-fed group.

'Brown fat', which is 'brown' due to its high mitochondria content, has been known for many years to be an important source of heat or 'thermogenesis' in the newborn and hibernating animals, but it was not thought to be of much significance in normal animals or non-hibernating animals. The brown fat produces heat by oxidation in the mitochondria which, because oxidative phosphorylative is effectively uncoupled, produce no or little ATP; most of the energy is, therefore, converted to heat. These experiments demonstrate, however, that animals such as rats continue to use brown fat in the adult and, furthermore, can regulate formation of brown fat very effectively.

It is not known whether humans can regulate their quantities of brown fat, but it is clear that theories of obesity could effectively be based on an important role of the brown fat (*Fig.* 40.16). If the adult human cannot regulate the quantity of brown fat or respond to his diet as can the rat, then lean individuals may possess large quantities of brown fat and obese individuals only small quantities. On the other hand, if the quantities of brown fat can be regulated in the human as in the rat, the lean individual could be assumed to respond in a similar way to the rats by the production of more brown fat, whereas obese individuals may respond much less efficiently.

Naturally the theory of the relation of brown fat to obesity would lead to the conclusion that obese animals or humans

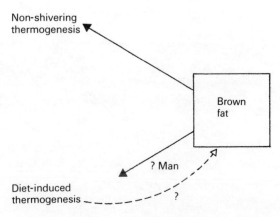

Fig. 40.16. **Role of brown fat in the production of body heat or thermogenesis.**

would suffer from cold to a much greater extent than lean persons, and this has been clearly demonstrated for genetically obese mice who die at temperatures well tolerated by normal mice.

We can, therefore, speculate that obesity may be a result of a desirable genetic trait: during man's early development, he would probably have had to develop resistance to both extreme cold in winter and food deprivation at any time. The lean individual has developed as the type best able to be efficient in thermogenesis and to withstand the cold, whereas the obese person has developed, being poorly adapted to withstand cold through inefficient thermogenesis, but well prepared, with good energy stores, to withstand food scarcity.

Chapter 41 Biochemical genetics: inborn errors of metabolism

41.1 Introduction

The basic concepts of genetics were correctly described by Mendel, in 1866, as a result of experiments on the breeding of peas, studying aspects such as their colour and their shape (i.e. wrinkled or round). He proposed that inheritance was based on specific units, the genes, that each unit existed in two forms called 'alleles' and that one member of each pair of alleles was usually dominant to the other, although some characteristics of the offspring were dependent on the two alleles.

Mendel's experiments, which were not widely known until 1900, were extended into studies of many other species of plants and animals but, normally, investigators evaluated genetic inheritance of physical characteristics such as height, flower colours and eye colours.

Not long after this, however, Garrod, working at Oxford and St Bartholomew's Hospital, London, described, in 1908, a condition in which the urine of certain individuals rapidly darkened when exposed to air. It was called 'alkaptonuria' and was caused by an abnormality of tryosine metabolism. Furthermore, it appeared to occur specifically in certain families and was not as common in males as in females.

Within a relatively short time, other conditions such as cystinuria, albinism and haematoporphyria were described by Garrod in a classic book *Inborn Errors of Metabolism*. He correctly deduced that certain metabolic abnormalities were, just as eye colour and anatomical form, genetically linked. Unfortunately, at that time the understanding of cellular metabolism was very limited, so that it was difficult to understand the significance of these important observations and the research was virtually ignored for nearly 30 years. Gradually, however, through wide-ranging studies, especially on the genetics of micro-organisms by Beadle, it became clear that genes controlled the synthesis of enzymes and that any genetic abnormality could have a significant effect on the activities of important enzymes. The general concept may be stated thus:

Altered genetic information → Abnormal structure of host protein → Abnormal chemical activity of the protein

The abnormalities can vary widely in severity, from causing only a minor embarrassment to the individual to becoming incompatible with life and leading to death of the fetus or even the ovum.

41.2 Genetic defects in metabolic processes

Although early research established clearly that defective enzyme synthesis was a potential cause of a defect in metabolism it should be appreciated that this is not the only possible mechanism by which metabolic pathways can be impaired. Many such mechanisms probably exist and are illustrated below.

a. Enzyme synthesis
The simplest impairment is a complete inability to synthesize a particular enzyme, although many other possibilities can occur. For example, an enzyme can be synthesized in limited amounts or the enzyme could be

synthesized with a much reduced affinity for the substrate, so the kinetic characteristics are drastically altered.

b. Transport mechanisms

'Permease' enzyme systems or cell surface receptors may be changed as a result of genetic effects. These could lead to inability to transport molecules through extracellular or intracellular membranes or to impairment of binding of important, for example hormonal, regulators.

c. Changes in rate-limiting step

As a consequence of alterations in enzyme activity which occur as described above, the rate-limiting step of essential metabolic pathways may be changed. For example, if the pathway $A \rightarrow F$ is normally regulated by the rate-limiting enzyme X_2, genetic changes in the activity of X_4 may cause this to become the rate-limiting stage:

$$A \xrightarrow{X_1} B \xrightarrow{X_2} C \xrightarrow{X_3} D \xrightarrow{X_4} E \xrightarrow{X_5} F$$

Original rate-limiting step   New rate-limiting step

This could cause significant changes in the regulation of the whole pathway, for example in hormonal or feedback control.

d. Changes in feedback control

As an alternative to the mechanism proposed in (*c*) the structure of enzyme X_2, which catalyses the original rate-limiting step, could be subtly changed so that, although the rate of the reaction catalysed is unchanged, it could respond quite differently to feedback control.

$$A \xrightarrow{X_1} B \xrightarrow{X_2} C \xrightarrow{X_3} D \xrightarrow{X_4} E \xrightarrow{X_5} F$$

Normal feedback control on activity of X_2

As a consequence regulation could be seriously impaired.

e. Formation of new products

Genetic changes in an enzyme could, as an alternative to causing changes in substrate affinity, cause the enzyme to produce an entirely different product, with the accompanying possibility of serious derangements of normal metabolic processes:

$$A \xrightarrow{X_1} B \xrightarrow{X_2} C \xrightarrow{X_3} D \xrightarrow{X_4} E$$
$$\searrow \textcircled{X_3}$$
$$\searrow F \rightarrow G$$

where

X_3 = normal enzyme

$\textcircled{X_3}$ = genetically altered enzyme catalysing the formation of F from C instead of D from C

41.3 Clinical manifestations of metabolic errors

Degrees of severity

Asymptomatic

Disorders of this type are of no consequence because the metabolic change is minor or easily taken over by several alternative pathways. Conditions of this type may be very numerous in any one individual and are, clearly, very difficult to detect.

Asymptomatic except for accidental circumstances

The patient may not notice any abnormality until a traumatic event, such as severe injury, occurs. Deficiencies of certain blood-clotting factors fall into this category.

Alternatively, the condition may not be demonstrated until the patient is treated with a particular drug. Certain drug-induced haemolytic anaemias fall into this category.

Mild to moderate

The patient is clearly aware of conditions of this category, but is inconvenienced only to a mild extent.

Severe to lethal

Conditions of this type usually turn up very early in life. They are very serious and often lead to early death or severe, e.g. mental, abnormalities.

Age of onset

Some conditions, such as phenylketonuria, are manifested at birth whereas others, such as gout, are not usually demonstrated until the patient is past 45 years of age.

41.4 Therapeutics

For many genetic abnormalities, very little early treatment is available. It would clearly be desirable for the abnormal gene to be replaced, possibly by some form of genetic engineering, but this form of treatment must lie well in the future.

In certain conditions, however, treatment is available and examples are shown below.

a. Limiting the precursor that undergoes toxic accumulation

A good example of this treatment is provided for phenylketonuria, which is caused by an abnormality of tryosine metabolism. The dangerous metabolite phenylpyruvic acid is formed from phenylalanine and, by limiting the amount of this amino acid in the diet, the formation of the toxic metabolite can be drastically reduced.

b. Supply of missing metabolite

If the end-product of a metabolite sequence is lacking, as a result of the lack of an enzyme involved in its production, then this can sometimes be supplied.

c. Use of drugs in treatment

Certain individuals possess a genetic trait that makes them very sensitive to certain drugs. Examples are drugs that can cause haemolytic anaemia and those that strongly inhibit cholinesterase. Avoidance by these individuals of drugs of this type is obviously very desirable.

d. Excess storage

Some genetic conditions, such as Wilson's disease, lead to an excess store of a metal, in this case copper, in the tissues. Removal of the excess deposit of metal can often be achieved by means of metal chelating agents (cf. Chapter 35).

41.5 Typical metabolic disorders

A summary of typical metabolic disorders is shown in *Table* 41.1. It will be noted that they can involve specific proteins, for example in blood and those in almost every aspect of metabolism, transport processes and metal storage.

In the following sections we shall discuss typical examples of these diseases in detail or, alternatively, make reference to other chapters in which they have been described.

Table 41.1 **Summary of typical metabolic disorders**

Associated with	*Disorder*
1. Blood-related tissues	Haemoglobinopathies and thalassaemias Drug-induced haemolytic anaemias Methaemoglobinaemias Blood-clotting factor abnormalities Porphyrias
2. Carbohydrate metabolism	Diabetes Pentosuria Fructosuria Glycogen deposition (Von Gierke's disease) Galactosaemia
3. Amino acid metabolism	Phenylketonuria Tyrosinosis Alkaptonuria Albinism Hyperoxaluria Maple syrup urine disease (abnormal excretions of branched-chain amino acids, e.g. leucine, isoleucine, valine)
4. Lipid metabolism	Hyperlipidaemias Gaucher's disease (excess cerebrosides in cells)
5. Purine/pyrimidine metabolism	Gout Xanthinuria β-Aminobutyric aciduria
6. Metal metabolism	Wilson's disease Haemochromatosis Periodic paralysis
7. Renal tubule transport	Renal glycosuria Renal tubular acidosis Glycinuria Cystinuria Hartnup disease

41.6 Genetic defects of blood proteins

The fact that haemoglobin could exist in at least two forms was described over 100 years ago by Von Korber. He noted that the fetus and newborn infant possessed a form of haemoglobin (F) which was different from that of the adult (A). Normally, the F form of haemoglobin, constituting 60–90 per cent of the total, rapidly disappears within 3–4 months, so that only the A form remains.

However, in certain individuals the F form of haemoglobin may persist into adult life and this causes a serious type of anaemia, called 'thalassaemia', which is widespread in the Mediterranean, particularly in Cyprus. Normal adult (A) haemoglobin is composed of four peptide chains , two α chains and two β chains, usually denoted as $\alpha_2\beta_2$. In fetal (F) haemoglobin, the β chains are replaced by γ chains, so that the haemoglobin is $\alpha_2\gamma_2$. In β-thalassaemias, this fetal form may be synthesized into adult life, although other abnormalities such as defective synthesis of β chains can also occur. In β-thalassaemias, the synthesis of one type of chain, such as the γ or β, is greatly in excess of that of the α chain, so that haemoglobins are composed entirely of γ chains, γ_4, or entirely of β chains β_4.

An important observation on abnormal genetically related haemoglobin was made in 1910 by Herrich, who observed that the red blood cells of Negroes in New York and in the West Indies assumed a curved shape, similar to that of a sickle. It gave rise to a form of anaemia described as 'sickle-cell anaemia'. Careful study of the distribution of this condition showed that it was endemic in all countries of the world where malaria was widespread and it was demonstrated to help in resisting malarial infections.

Studies of the haemoglobin isolated from patients with sickle-cell anaemia showed that it differed from that of normal haemoglobin, but the subtle nature of the difference was not established until sophisticated methods for the analysis of amino acid sequence of peptide chains became available. Classical analytical studies on haemoglobins were carried out by Ingram, in

Amino acid sequence of β chain

Haemoglobin		Position 6 7
A	Val – His – Leu – Thr – Pro –	**Glu** – **Glu** – Lys
S (Sickle cell)	Val – His – Leu – Thr – Pro –	**Val** – **Glu** – Lys
C	Val – His – Leu – Thr – Pro –	**Lys** – **Glu** – Lys
G	Val – His – Leu – Thr – Pro –	**Glu** – **Gly** – Lys

Fig. 41.1. **Comparison of the structure of various haemoglobins.**

1958, who showed that a peptide of normal haemoglobin and a peptide chain of sickle-cell haemoglobin differed in just one amino acid, valine replacing the glutamic acid residue. The location of the changed amino acid was found to be in position 6 of the β chains of haemoglobin. The R group of valine is clearly not charged, whereas that of glutamic acid has a negative charge; consequently, sickle-cell haemoglobin has a lower negative charge than normal haemoglobin. Other haemoglobins, C and G, also differed in just one or two amino acids (*see Fig.* 41.1). It is remarkable that such a small change in the amino acid composition should cause such a dramatic change in the properties of the haemoglobin molecule. The introduction of a valine in place of a glutamic acid residue only requires the change of an adenine for an uracil as the second base of its triplet codon, as will be seen from studying the genetic code (*see* Appendix 29). The sickle-cell haemoglobin model, therefore, provides a very clear example of the basic mechanism of human genetics which could no doubt be of general application (*Fig.* 41.2).

Mutation change in base sequence in DNA

$\downarrow$

Changed base sequence in mRNA

$\downarrow$

Synthesis of protein containing abnormal amino acids

$\downarrow$

Altered histological function, e.g. sickle-cell anaemia.

Fig. 41.2. **Sequence of events in the generation of a genetically dependent biochemical abnormality.**

Defective synthesis of plasma proteins, such as those involved in blood clotting, can also occur as a result of genetic defects. In these patients blood clotting is inadequate. One of the best known of these conditions is haemophilia which is described in detail in Chapter 26.

41.7 Carbohydrate metabolism

The best documented metabolic errors involving carbohydrate metabolism are those associated with glycogen metabolism, the essential stages of which are shown in *Fig*. 41.3 (also cf. Appendices 14, 15).

A disease that causes a severe disturbance in glycogen metabolism was described by Von Gierke in 1929; this disease is shown in early infancy and is characterized by convulsions and massive liver enlargement. The disease, sometimes called 'type I deficiency', is caused by a lack of, or serious deficiency in, the enzyme glucose 6-phosphatase, as a result of which glycogen cannot be broken down to form blood glucose. Synthesis of glycogen is not, however, impaired because the enzyme hexokinase functions normally. The deficiency of the enzyme therefore causes massive accumulation of glycogen in the liver, and this leads to liver enlargement and a serious drop in the level of blood glucose (a hypoglycaemia) which causes the convulsions.

Another defect, called 'type II' or 'Cori glycogen storage disease', is caused by a deficiency of the debranching enzyme. This will also lead to excessive glycogen storage and inhibition of breakdown; consequently, the clinical symptoms will be almost identical in the two conditions.

Another important example of an impairment of carbohydrate metabolism resulting from a genetic defect, is that of a deficiency of galactose 1-phosphate uridyl transferase. This defect is relatively common and as much as 1 per cent of the population may be carriers. The disease appears to be incompletely recessive, because some individuals can show modest increases in circulating galactose and some predisposition to the development of cataracts. The deficiency of the enzyme normally occurs in several tissues, such as liver, kidney and small intestine. This enzyme is essential for subsequent metabolism of galactose, either for storage as glycogen or for metabolism to liberate energy (*Fig*. 41.4). As a consequence

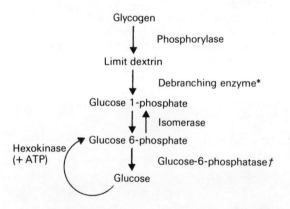

Fig. 41.3. **Glycogen degradation and metabolic defects.**

* Deficiency leads to Cori type II disease; † deficiency leads to Von Gierke's disease.

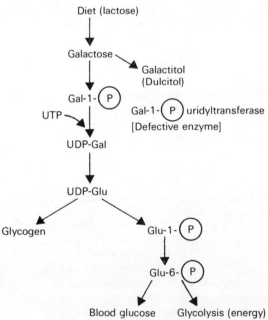

Fig. 41.4. **Galactose metabolism and metabolic defects.**

of this block, galactose accumulates in the blood and is excreted in the urine; galactose 1-phosphate and galactose accumulate in the tissues.

The most serious consequence of this disease is the development, during the first few months of life, of cataracts in the eye lens. The reason for the development of cataracts is uncertain, but the reduction of the galactose present in high concentrations in the lens to galactitol (dulcitol) could be a major cause. This product, formed from galactose, can be neither metabolized nor diffuse from the lens. It thus greatly increases the osmotic pressure in the lens causing an influx of fluid and eventually leading to opacification.

The enzyme lactase can also be deficient in the intestinal mucosa and lactase production in the adult is inherited as an autosomal dominant factor. The condition of lactose deficiency is described in Chapter 16. Caucasian communities in Europe have approximately 10 per cent incidence of hypolactasia, whereas Orientals, Australian Aborigines and North American Indians may have 90 per cent of their population with the condition.

41.8 Defects of amino acid metabolism

We noted in Section 41.1 that one of the first discoveries of a genetic error of metabolism was that described by Garrod, who observed that the urine produced by some individuals turned to a very dark colour on standing. The defect was ultimately shown to be caused by the lack of an enzyme involved in normal tryosine metabolism and the defect was described as 'alkaptonuria'. The step involved is catalysed by an enzyme which cleaves the opening of the benzene ring to produce metabolites entering the citric acid cycle. If this enzyme is absent, homogentisic acid is the end-product and this polymerizes to give the dark-coloured substances observed (*Fig.* 41.5).

Later, in 1934, a much more serious disease associated with defective phenylalanine metabolism was described by a Norwegian

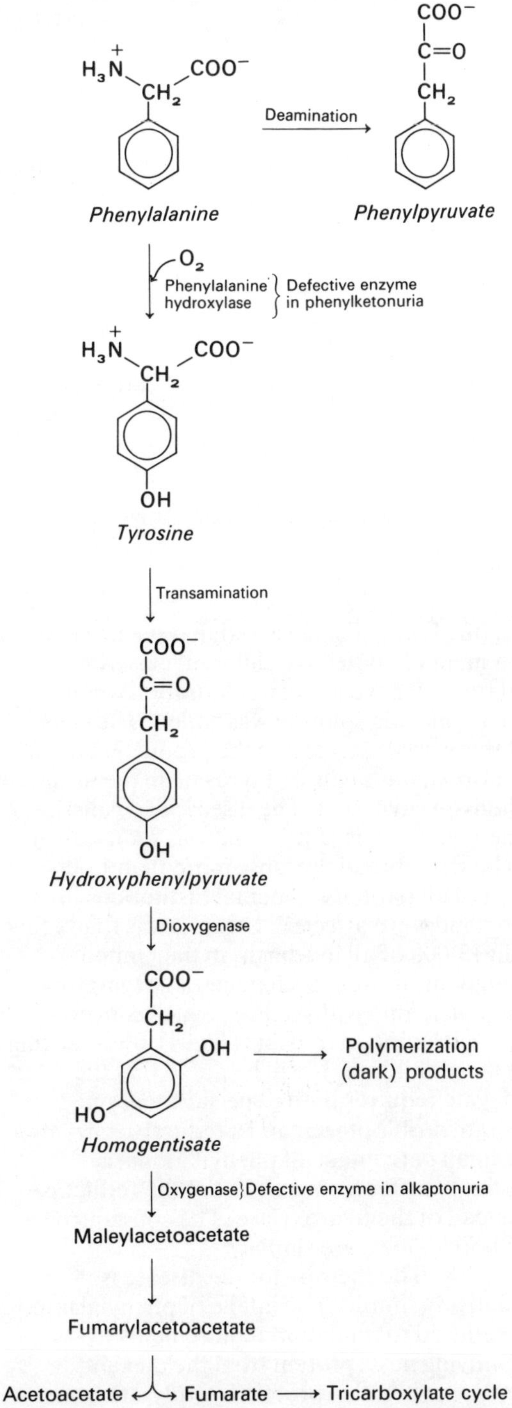

Fig. 41.5. **Essential stages in phenylalanine and tyrosine metabolism.**

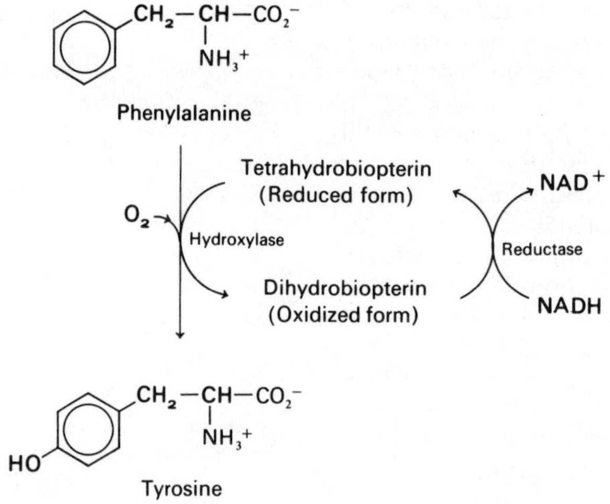

Fig. 41.6.   **Conversion of phenylalanine to tyrosine.**

doctor, Follig. He observed that the urine of ten mentally defective children, closely related, all gave a green colouration when ferric chloride solution was added. The cause of this was also traced to the production of an abnormal metabolite of tyrosine in the urine, phenylpyruvic acid. The disease was called phenylketonuria (PKU) and was soon shown to be widespread, because more than 1 per cent of all patients in mental institutions in England were affected. Later, it was found that 1 in 15 000 of all live births in the United Kingdom showed its characteristic symptoms. Phenylalanine hydroxylase, which converts phenylalanine to tryosine (*Fig.* 41.6), is lacking in these children. It will be noted that the enzyme requires both a special coenzyme, tetrahydrobiopterin, and a reductase enzyme. A small percentage of phenylketonuria patients suffer from a defect of the reductase instead of the hydroxylase. The consequences of both defects are similar.

The therapy for the disease is relatively simple. The intake of phenylalanine is reduced to minimum requirements by removing most protein from the diet and replacing it with a protein hydrolysate from which the phenylalanine has been removed. Phenylalanine cannot be completely removed from the diet, because it is an essential amino acid and is required for incorporation into proteins. This treatment must be started very early since damage, and particularly brain damage, cannot be reversed. Older untreated patients usually have an IQ of less than 50 and half are below 20, whereas that of patients who receive early treatment is about 90. Early treatment clearly depends on accurate diagnosis soon after birth and several methods have now been described which utilize a small sample of urine or blood.

The reason that phenylalanine is toxic for the brain is still unclear and many theories have made little advance in the explanation of the mechanism. One which has received much support is that the deaminated product of phenylalanine (phenylpyruvate) formed in the brain causes the damage (*Fig.* 41.6). It is postulated that phenylpyruvate will antagonize, possibly by competitively inhibiting the utilization of the pyruvate. Pyruvate is essential for many important functions in the brain, including the energy supply (cf. Chapter 33) and it is thus very likely that interference with pyruvate metabolism will cause severe brain damage.

Homocystinuria

In 1962, some patients in mental institutions in Ireland and the USA were shown to excrete a

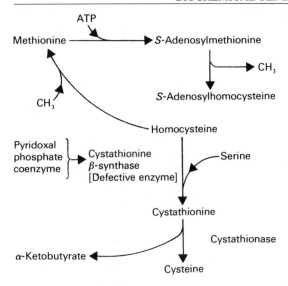

Fig. 41.7.   **The metabolism of homocysteine.**

large concentration of homocysteine in their urines. They were also shown to suffer from dislocation of the lens and osteoporosis. Homocysteine is formed from methonine and is normally converted to cysteine (*Fig.* 41.7).

Cystathionine β-synthase is usually the defective enzyme in the condition, although a few patients possess other defective enzymes. It will be noted that pyridoxal phosphate is a coenzyme and some patients respond to large doses of this vitamin. These and other studies have shown that several defective forms of the enzyme are likely to occur in which reduced affinity for the coenzyme could clearly be important.

Cystinuria

This condition, described by Garrod in 1908, which is characterized by the excretion of large quantities of cysteine, ornithine and lysine in the urine, is caused by a failure of the kidney transfer mechanism for these amino acids and is also manifested in the intestine. It has been described in Chapter 30.

41.9 Lipid metabolism

The disorders of lipid metabolism, so far described, mainly cause abnormal lipid storage, especially in the brain and central nervous system.

Deficiencies of specific lysosomal enzymes cause accumulation of a variety of glycolipids, as described in Chapter 33. Although these diseases are characterized by abnormal lipid storage, many of the enzymes which are deficient catalyse the splitting of carbohydrate residues from the glycolipids.

41.10 Purine/pyrimidine metabolism

Several genetic abnormalities involve the metabolism of purines and pyrimidines; the best known of these is 'gout' which is so well known that it has passed into popular literature. Gout is characterized by an attack of arthritis in a joint, often that of the big toe, and is caused by precipitation of sodium urate crystals. The plasma becomes supersaturated and the crystals are taken up by the phagocytic leucocytes. Urate is the end-product of purine metabolism in man and its formation is catalysed by the enzyme xanthine oxidase (*Fig.* 41.8).

There does not, in most cases, appear to be a single abnormality and several stages of purine metabolism may be involved. It is not caused simply by an increased rate of degradation of purines, since increased synthesis from labelled glycine can be demonstrated in many cases of gout. However, diets rich in purine always exacerbate the condition and this fact is probably responsible for the association of gout with high living.

Treatment with the drug allopurinol, which is an analogue of hypoxanthine (*Fig.* 41.9), is usually effective. It inhibits the action of xanthine oxidase and also *de novo* synthesis of purines by inhibitors of amidophosphoribosyltransferase.

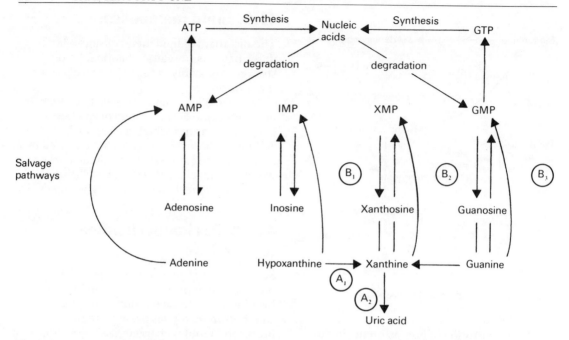

Fig. 41.8. **Pathways of synthesis and degradation of nucleotides.**

A_1 and A_2 are reactions catalysed by xanthine oxidase; B_1, B_2 and B_3 are reactions catalysed by hypoxanthine–guanine phosphoribosyltransferase (HGPRT).

The biochemistry of one rare form of gout is understood in more detail. It is called the 'Lesch–Nyhan syndrome' after the discoverers of the disease. Unlike the common form of gout which appears late in life, this form, characterized by self-mutilation including chewed lips and fingers, is shown in the first years of life. The condition is caused by a deficiency of the enzyme hypoxanthine–guanine phosphoribosyltransferase (HGPRT). This enzyme catalyses the synthesis of phosphorylated metabolites inosine mono-phosphate (IMP), xanthosine monophosphate (XMP) and guanosine monophosphate (GMP) from the respective nucleotides in the salvage pathway of nucleotide synthesis (*Fig.* 41.8).

The fact that deficiency of this enzyme plays a vital metabolic role, has helped to support the importance of salvage pathways in nucleotide metabolism.

Allopurinol   Hypoxanthine

Fig. 41.9. **The comparison of the structures of allopurinol and hypoxanthine.**

Chapter 42 Immunology

42.1 Introduction

Some understanding of immunology probably existed long ago amongst the populations of India and China, because they were aware that inoculation by live organisms causing smallpox could protect against subsequent attacks. Modern immunology, however, really started with the demonstration by Jenner in 1798 that non-virulent cowpox vaccination would protect against human smallpox.

During the ensuing 150 years or so, greater understanding of immunology, the main process by which the body fights disease, was developed, but in the 1960s a major breakthrough was achieved, mainly as a result of the work of Porter and Edelman who elucidated the chemical structures of the proteins involved, the immunoglobulins or antibodies. Since this period the subject has greatly expanded and it is now difficult to overestimate its importance in medicine.

Although the understanding of the immunological process involved in the resistance to infection by invading micro-organisms and viruses is still clearly of paramount importance, more recent research has demonstrated many other important aspects of the subject. For example, the body can sometimes manufacture antibodies against its own tissues, 'autoimmunity', which may cause severe tissue damage; also, an understanding of immunology is vital for the development of correct procedures during the exciting modern techniques of transplant surgery.

Immunological techniques have also proved invaluable for locating certain large molecules or small molecules bound to carriers in tissues: antibodies are prepared with a fluorescent label and added to the tissue sections. The very great specificity of the antibodies will interact with only one type of molecule which is then readily located by fluorescence. This technique is therefore known as 'immunofluorescence', and forms part of the wider field of study of 'immunocytochemistry'.

42.2 Antigens and antibodies

Whenever a foreign protein gains access to the blood stream, it will cause the formation of special proteins in the plasma of the host. The invading protein molecule is termed an 'antigen' and the special protein produced, which on electrophoresis migrates with the γ-globulin fraction, is termed an 'antibody'. The antigenic proteins can be pure proteins, composed of amino acids only, or more commonly they can also contain carbohydrates (glycoproteins), lipids (lipoproteins) or nucleic acid (nucleoproteins). Some, but not all, pure carbohydrates (polysaccharides) are effective antigens, but lipids are ineffective. In the normal course of infection, the foreign protein is often a bacterial cell, a cell fragment or a viral particle, although other proteins, such as egg albumin, will also cause the production of antibodies. Proteins of this latter type often do not enter the blood, but many food proteins are believed to cross the intestinal mucosa and give rise to 'allergic' conditions because they initiate the production of antibodies.

The several important properties of antigens and antibodies, and their interactions are discussed in the following sections.

a. Primary and secondary responses

If an animal is injected with an antigen, for example a toxin of *Staphylococcus*, the antibody concentration in the plasma increases, after about 5 days, from zero up to a maximum which then steadily decreases. If, however, the animal is again injected with a similar dose of antigen, the antibody response is much faster and much greater (*Fig.* 42.1).

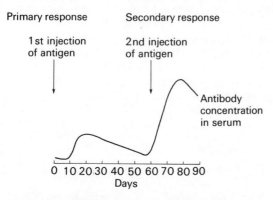

Fig. 42.1.   **Primary and secondary responses during antibody production.**

b. Antigen–antibody precipitation: the precipitin reaction

This can be clearly demonstrated by an example. A rabbit is injected several times with a solution of egg albumin (the antigen) to build up a high concentration of the antibody to egg albumin in the plasma. A sample of the plasma is then collected and the remainder of the experiment carried out, *in vitro*, in a test tube. A constant volume of the diluted serum is placed in each test tube and graded quantities of the antigen, egg albumin, are added. This causes precipitation of the antigen–antibody complex, but on continued addition of egg albumin, the precipitate eventually dissolves (*Fig.* 42.2).

This experimental procedure can be used to determine both the antibody content of the serum and to indicate the valency of the

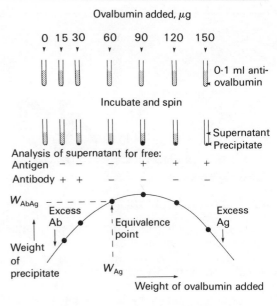

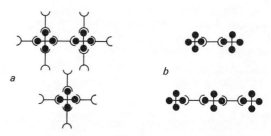

Fig. 42.2.   **Precipitation of antigen–antibody complexes.** Ab, antibody; Ag, antigen.

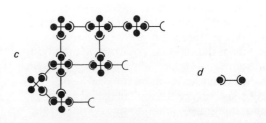

Fig. 42.3.   **Multivalency of antibodies.**
(*a*) Excess antibody; (*b*) excess antigen; (*c*) precipitate; (*d*) monovalent antigen—cannot cross-link.

Tetravalent antigen;   bivalent antibody.

antigen. Valencies of antigens can vary over a wide range, for example egg albumin has a valency of 10 and thyroglobulin a valency of 40; antibodies, however, must possess at least two valencies. Typical complexes are shown in *Fig.* 42.3. The precipitate probably consists of a large three-dimensional lattice as shown in the figure; but when conditions of excess antibody or antigen exist, smaller soluble complexes are formed. More recently, antibody–antigen reactions are usually studied in gel plates, a method first developed by Ouchterlony, after whom the plates used ('Ouchterlony' plates) are named. Small cavities are cut in an agar-gel plate and samples of the antigens and antibodies under test placed in them. The proteins slowly diffuse into the gel and, on meeting, precipitate to produce a clear line (*Fig.* 42.4).

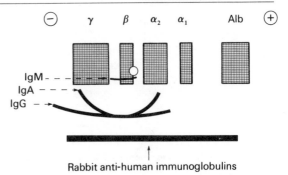

Fig. 42.5. **Immunoelectrophoresis for the separation of antibodies.**

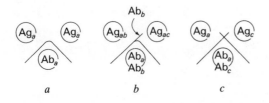

Fig. 42.4. **Use of Ouchterlony plates for the demonstration of antibody cross-reactions.**
Ag_a, Ag_{ab}, Ag_{ac}, Ag_c are all antigens; Ab_a, Ab_b, Ab_c are all antibodies. (*a*) Identity; (*b*) partial identity; (*c*) non-identity.

The method is valuable in testing the relationship between antigens and antibodies. Typical demonstrations of identical, partially identical and non-identical antigens are shown in *Fig.* 42.4. As a variation to this procedure, stimulation of the antigen to make it move towards the antibody can be brought about by electrophoresis. This technique is termed 'immunoelectrophoresis' and is valuable for separation of both antigens and antibodies (*Fig.* 42.5).

The study of antigen–antibody reactions has been greatly aided by the use of radioactively labelled antigens or antibodies and the adaptation of the principle has found wide and important applications in the field of radioimmunoassay. The method can be

illustrated by the following example. In order to develop a method for determination of a peptide hormone, for example insulin, antibodies are first raised to the hormone (insulin) in a rabbit. These antibodies are then separated from the serum and reacted *in vitro* with radioactive insulin (the antigen) to form an antigen–antibody complex. To determine the insulin concentration in a sample of serum, it is added to the anti-insulin–insulin (labelled) complex when some of the radioactive insulin will be displaced. The quantity of radioactivity remaining associated with the antibody will be proportional to the 'cold' insulin added (*Fig.* 42.6).

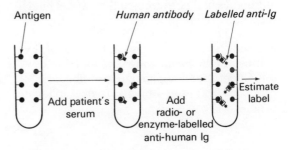

Fig. 42.6. **Use of radioimmunoassay for the determination of small quantities of peptide hormones, e.g. insulin.**

c. Specificity of the antigen–antibody reaction

Early in investigations of antibody–antigen reactions, it was observed that antibodies

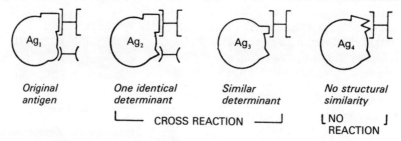

Fig. 42.7. **Specificity and cross-reactions of antibodies.**
Ⴙ, Ⴑ antibodies.

exhibited a high degree of specificity, but certain cross-reactions occurred. For example, antibodies raised in rabbit serum, by injection of chicken egg albumin as antigen, gave a limited precipitation reaction with duck egg albumin, whilst antibodies raised in rabbit serum to bovine serum albumin had very little reaction with human serum albumin. Diagrammatic representation of cross-reactions is shown in *Fig.* 42.7.

The great specificity of antibodies was clearly demonstrated by the elegant early work of Landsteiner. He used a wide variety of small organic molecules, such as *m*-aminobenzene sulphonate, which he coupled to proteins by a diazotization reaction (*Fig.* 42.8). These coupled proteins were then used as antigens in the production of antibodies. Once these antibodies had been formed they would bind to the small organic molecule, the hapten, very readily. This enabled the specificity of the antibody binding to be tested, for example by comparing the binding of *o*-, *m*- and *p*-aminobenzene sulphonate to antibodies produced against *m*-aminobenzene sulphonate (*Fig.* 42.9).

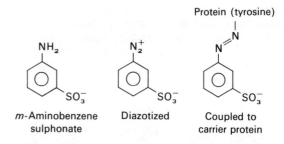

Fig. 42.8. **Coupling of 'haptens' to protein.**

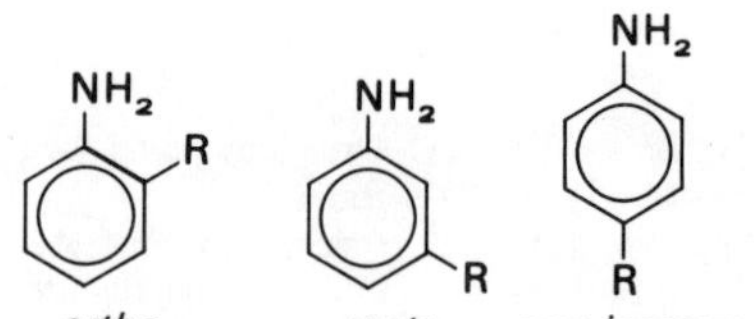

	ortho	*meta*	*para* isomers
R = sulphonate	+ +	+ + +	±
R = arsonate	−	+	−
R = carboxylate	−	±	−

Fig. 42.9. **Use of 'haptens' to evaluate antibody specificity.** Effect of variations in hapten structure on strength of binding to *m*-aminobenzene sulphonate antibodies.

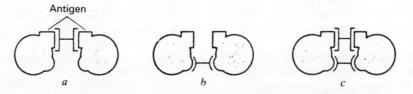

Fig. 42.10. **Avidity of antibody–antigen reactions.**
(*a*) Moderate; (*b*) strong; (*c*) very strong.

d. Antibody–antigen affinity

The combination of antibody with an antigen or hapten is reversible and the complex may dissociate. This will depend on the strength of the binding which is given by the expression:

$$Ab + Ag \rightleftharpoons Ab \cdot Ag$$

$$K = \frac{[Ab \cdot Ag]}{[Ab][Ag]}$$

where Ab is the antibody and Ag the antigen.

The term 'avidity' is often used to describe the strength of antibody–antigen binding and multivalency is important in giving strength to the binding (*Fig.* 42.10).

The forces holding antigens and antibodies together are those commonly occurring between proteins, and include electrostatic interactions, hydrogen bonding, hydrophilic interactions and van der Waals forces.

42.3 Antibody structure: the immunoglobulins

The basic structure of the antibody molecule was first elucidated by Porter in the early 1960s and was shown to be composed of two heavy chains of molecular weight 50 000 and two lighter chains of molecular weight 25 000 linked by disulphide bridges (*see Fig.* 42.11). The heavy chains are 'hinged' and can rotate through a wide angle. Antibodies migrate with the γ-globulins during electrophoresis of plasma proteins and are, therefore, frequently referred to as immunoglobulins.

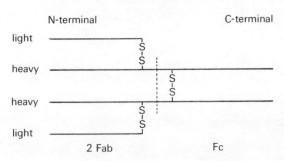

Fig. 42.11. **Schematic structure of antibody molecules.**

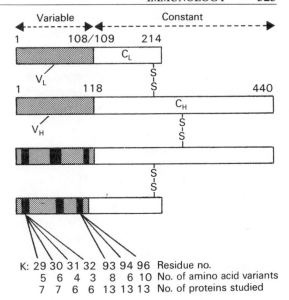

K:	29	30	31	32	93	94	96	Residue no.
	5	6	4	3	8	6	10	No. of amino acid variants
	7	7	6	6	13	13	13	No. of proteins studied

Fig. 42.12. **'Constant' and 'variable' regions of antibody molecules.**

▨ Variable regions, V_L, V_H; ☐ constant regions, C_L, C_H; ■ hypervariable region with numbers of amino acid variants indicated.

If all antibodies have a standard structure, the question immediately arising is how they exert their great range of specificities. The answer lies in the ability of the cells to synthesize a wide range of antibodies with different amino acid sequences. The variability of sequence is, however, confined only to certain regions of the molecules, described as the 'variable' regions (V regions), the remainder of the polypeptide chains retaining a relatively constant amino acid composition ('constant' or C regions). Some parts of the 'variable' regions of both heavy (H) and light (L) chains of different antibodies show very extensive changes in amino acid composition and are described as 'hypervariable' (*see Fig.* 42.12). Each L chain consists of two 'domains' of approximately equal size. The N-terminal domain has a variable amino acid sequence which is different in every L chain studied. The C-terminal domain is constant but can be one of two types, κ and λ, in any one molecule.

Folding of the chains is believed to align the hypervariable regions, sometimes known as 'hot-spots', with the antigen to form the antigen–antibody complex (*Fig.* 42.13).

The relatively constant regions of the immunoglobulins that are not concerned with antigenicity, also show extensive heterogeneity throughout different tissues of the body and the study of immunoglobulins has demonstrated the existence of several different types of light and heavy chains. The concept of 'constant' and 'variable' regions of the chains is, therefore, relative.

Immunological methods, using Bence–Jones proteins that are secreted in high concentration in the pathological condition of 'myeloma', have been used to demonstrate the existence, as mentioned above, of two types of light chains, termed 'κ' and 'λ'.

In man, approximately 65 per cent of the immunoglobulins are of the κ type and 35 per cent of the λ type. Five different types of heavy chain have been identified: γ, α, μ, δ or ϵ. Each heavy-chain type can be associated with either κ or λ light chains and the main classes of immunoglobulins are consequently described as IgG, IgA, IgM, IgD, and IgE, according to the type of heavy chain present. Their main properties are summarized in *Table* 42.1.

IgG is the immunoglobulin that occurs in the highest concentration, particularly in the extravascular fluid where it combats invading micro-organisms and the toxins they produce. It is also important in causing complement activation which will be discussed, in more detail in Section 42.6. IgG is composed of three fragments, produced on

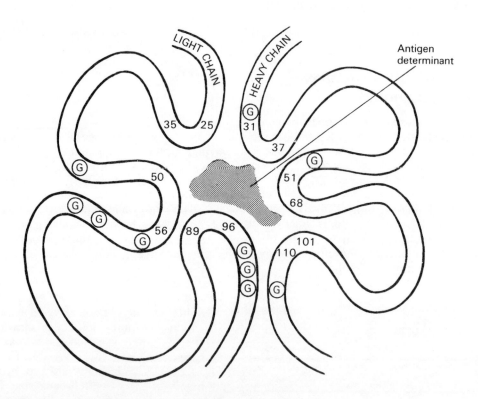

Fig. 42.13. **Special arrangements of antibody chains to form antibody–antigen complex.**
Alignment of antibody chains to allow close association between hypervariable domains and the antigen. G is glycine which enables correct folding of the chains of amino acids into β-pleated sheet.

Table 42.1 **Properties of major human immunoglobulins**

Property	IgG	IgA	IgM	IgD	IgE
Mol wt	150 000	160 000 (polymers)	900 000	185 000	200 000
Number of 4-peptide units	1	1 or 2	5	1	1
Heavy chains	γ	α	μ	δ	ϵ
Light chains	κ or λ	κ or λ	κ or λ	κ or λ	κ or λ
Molecule formed	$\gamma_2\kappa_2$ or $\gamma_2\lambda_2$	$(\alpha_2\kappa_2)_{1-3}$ or $(\alpha_2\lambda_2)_{1-3}$	$(\mu_2\kappa_2)_5$ or $(\mu_2\lambda_2)_5$	$\lambda_2\kappa_2$ or $\delta_2\lambda_2$	$\epsilon_2\kappa_2$ or $\epsilon_2\lambda_2$
Valency for antigen binding	2	2	5(10)	?	2
Total immunoglobulin in plasma, %	80	13	6	1	0·002
Carbohydrate content, %	3	8	12	13	12

papain digestion, of which the two Fab fragments are responsible for binding of antigen or hapten and the one Fc fragment for complement binding. The IgG group of immunoglobulins can be classified into four subclasses, IgG1, IgG2, IgG3 and IgG4, depending on differences in their heavy chains which can be described as γ_1, γ_2, γ_3, γ_4. All of these heavy chains show homology and have certain structural relationships in common.

IgA appears selectively in mucous secretions, such as saliva, tears, nasal fluid, in secretions of the gastrointestinal tract, and in sweat. Its main function is clearly to defend the body surfaces against bacterial and viral attack.

IgM is characterized by a much larger molecular weight than that of the other immunoglobulins. Furthermore, it possesses a very high valency and is, therefore, a very effective agglutinating agent. It appears very early in the immune response, mainly in the blood stream, and is generally regarded as the body's first line of defence against bacteraemia.

IgD is an immunoglobulin with a very short half-life of about three days. Its function is less clear than that of the other immunoglobulins but since it is found on the surface of lymphocytes, it is likely that it plays a role in the control of lymphocyte activation and suppression.

IgE is present in very low concentrations in the plasma, synthesis being carried out by very few plasma cells. The serum level rises considerably, especially after infection with certain parasites, particularly the helminths, and this may be connected with the release of histamine from mast cells having fixed IgE antibody. This immunoglobulin may also be associated with certain forms of allergy.

42.4 Antibody synthesis: roles of macrophages and lymphocytes

a. Role of macrophages

The thymus contains many macrophages and it can be established, by experiments in which cells are cultured *in vitro*, that this cell type must play an important role in the immune response. If the macrophages are removed from mixed cultures of lymphocytes and macrophages, the immune response is considerably reduced.

The precise role played by the macrophage is not fully understood, but it is very likely that the macrophages actively phagocytose the antigen. Antigens may be processed, possibly by partial degradation in the lysosomes by lysosomal hydrolases (cf. Chapter 3) and then returned, in a partially degraded state, to the surface of the macrophage cells. In this form the degraded antigen is a powerful immunogen and it is sometimes termed 'super-antigen'. A macrophage product, possibly a type of RNA, is also likely to be associated with the processed antigen on the cell surface.

b. Role of lymphocytes:
T and B-lymphocytes

The primitive lymphoid cells from the bone marrow play a very important role in the synthesis of the antibodies. The lymphocytes fall into two groups: T-lymphocytes, so called because they undergo processing in the thymus gland, and B-lymphocytes which are named after the Bursa of Fabricus gland in the intestinal epithelium, where the cells are processed. The Bursa gland was originally discovered in the chicken and the equivalent in man or other mammals has not been established. Nevertheless, the existence of B-type lymphocytes is well established and, in mammals, B-lymphocytes are likely to be processed in the bone marrow. Both populations of lymphocytes are stimulated by antigens which cause morphological changes and production of antibodies. The stimulated lymphocytes then move through the lymphatic system and are taken up by lymph nodes where they induce the formation of additional antibodies by initiating cell proliferation.

Each lymph node is composed of two distinct regions, the cortical region and the follicular region, which respond to the antigenic stimulation by the production of different cell types. The cortical region responds to produce T-cells, whilst the B-cells proliferate from the central or follicular region of the glands. The T-cells diffuse rapidly from the cortical region into the plasma, whilst the B-cells migrate to the medullary cords of the

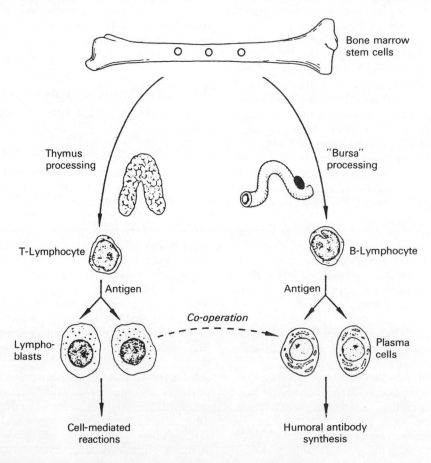

Fig. 42.14. **Formation of B and T-lymphocytes and their conversion to lymphoblasts or plasma cells.**

gland where they synthesize most of the antibody molecules. The role of the B-cells is more thoroughly understood than that of the T-cells and it is clear that their principal function is the synthesis of the main bulk of the antibody molecules. On the surface of the B-cells antibodies of the IgM type are present as receptor molecules (*Fig.* 42.14). It is generally believed that each lymphocyte is capable of making only one type of antibody, and that its receptor molecules are related to the type of antibody synthesized.

After stimulation with an antigen, the B-type lymphocytes are converted into large 'blast' cells which then undergo rapid division to produce a 'clone' of plasma cells possessing a well-developed endoplasmic reticulum and containing many ribosomes. The system is very active in producing antibody molecules which are then exported from the cells. The division of cells into a 'clone', i.e. a cluster of identical cells, ensures that they produce only the one type of antibody associated with the original stimulated cell. The sequence of development of the different cell types is shown in *Fig.* 42.15.

The role of the T-cells, although also produced from primitive stem cells in the bone marrow, is more complex and less well understood than that of the B-cells. They

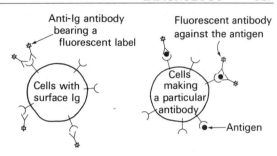

Fig. 42.16. **Receptors on lymphocytes, using a fluorescent antibody technique for detecting Ig or specific antibody-forming cells.**

possess a variety of functions, some of which are closely related to those of B-cells, but others which are unrelated.

Although it is firmly established that the surface of the B-cells contains many active receptors (*Fig.* 42.16), it is much more difficult to demonstrate the existence of receptors on T-cells. This may be on account of their small

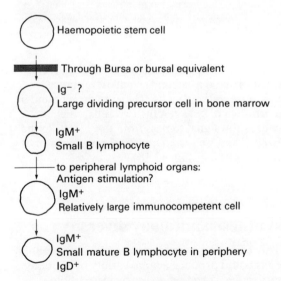

Fig. 42.15. **Development of B-lymphocytes in mammals.**

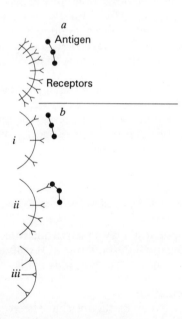

Fig. 42.17. **Comparison of receptors on B and T-lymphocytes.**
(*a*) B-cell: large number of closely packed receptors; potentially high avidity; sufficiently exposed to be susceptible to anti-Ig. (*b*) T-cell: possible reasons for poor binding of antigens. (i) Few receptor molecules; (ii) receptors rapidly shed; (iii) receptors 'buried' in plasma membrane.

numbers and their lability or due to the fact that they are buried in the plasma membrane (*Fig.* 42.17).

It is also possible that T-cell antigen receptors are of a specialized type, i.e. they are composed of the same V regions as other antibodies in B-cells, but they possess a distinct C region and have no light chains.

T-cells, like B-cells, are arranged in clones of different specificities and, furthermore, possess 'memory' of challenge by earlier antigens.

It is currently believed that T-cells carry out the following functions:

i.   They regulate immune reaction by other cells

ii. They can interact with an antigen to produce cellular immunity, i.e. transfer of the cells from one animal to another can transfer immunity. This concept is important in inducing immunity against a disease, such as tuberculosis, immunity being achieved by transfer of cells, but not serum, from an immune donor

iii. Some T-cells can become 'cytotoxic' and attack antigens on the surface of foreign cells

iv. Unlike B-cells which liberate the protein antibodies into the circulation, T-cells exert their action by the production of specific proteins on their surface.

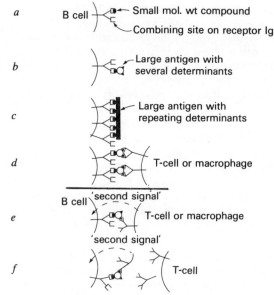

Fig. 42.18. **Possible mechanisms of B and T-cell cooperation.**
The antigen matrix hypothesis: (*a*) a low-molecular-weight compound cannot cause cell induction; (*b*) a larger antigen, with several determinants; (*c*) a very large polymeric antigen can bind firmly to a cell; (*d*) receptors from T-cells can, by reacting with 'carrier' determinants on an antigen, 'line up' and array small molecules into an immunogenic matrix. *The second signal hypothesis:* (*e*) two cells come together around a 'sandwich' of antigen, the B-cell then being induced by a non-specific signal from the other cell; (*f*) the specific receptor from a T-cell binds first to the carrier determinants of the antigen and then 'signals' the B-cell.

c. Cooperation between B- and T-cells

T-cells appear to play an important role in regulating the differentiation of B-cells for antibody synthesis; exactly how this is achieved is, however, still a matter of current research and speculation.

Two main theories have been proposed. In one, it is suggested that small antigen molecules cannot induce antibody formation without the intervention of T-cells. As shown in *Fig.* 42.18, the T-cells can align a series of small antigen molecules for presentation to B-cells and so trigger their response.

A second postulate is that the T-cells must provide a 'second signal' to initiate a response in the B-cells. The antigen is bound in a 'sandwich' between the two cells which causes the signal to pass from the T-cells to the B-cells and, in turn, stimulates proliferation of the B-cells (*see Fig.* 42.18).

42.5 Control of antibody synthesis: antibody diversity

There are two possible ways in which lymphoid cells could produce a large variety of antibody molecules. The first way is by instruction, i.e. the antigen would act as a template for the formation of the correct antibody; the second way is by selection from a vast range of pre-existing antibodies.

Although a long debate, ranging over many years, was conducted on the merits of these two theories, it is now generally agreed that the selection theory is much more likely to be true than the instruction theory. The theory of selection was initiated and strongly supported by Burnett, who postulated that immunocompetent cells were committed to making only one type of antibody. The evidence for this theory is based on several observations.

a. The tertiary structure of proteins depends on their primary structure and different types of folding of a particular amino acid sequence are extremely unlikely. It follows, therefore, that a large number of different amino acid sequences must be produced during the formation of different antibodies and these must be generated from different regions, or 'genes', of the DNA.

b. Antibody-forming cells have little or no antigen in them, as would be required by a template theory.

c. Immunocompetent cells, the B-lymphocytes, would, under the instruction theory, have to be capable of making a vast range of antibodies depending on the instruction. This has not been found to be the case and cells are committed to making only one type of antibody.

Antibody production follows the normal pattern of protein synthesis and, therefore, each different antibody must originate from a gene or DNA. The diversity of antibodies could arise on a genetic basis, i.e. coding for antibodies is already present in the DNA (the germ-line theory) or, alternatively, mutations could arise to give new genes during the lifetime of the individual (the somatic mutation theory).

The problem of which mechanism is most likely to represent the true situation is still unresolved. The two theories are not mutually exclusive, since the germ line could provide a limited series of alternative

antibodies with somatic mutations initiating more extensive modifications.

Direct operation of the germ-line theory appears unlikely, however, from simple calculations. The human haploid DNA contains about 10^7 genes. Humans can manufacture about 10^6 different antibodies and taking into account the large number of other proteins synthesized in the cells, this would clearly be impossible using 10^7 genes. This problem may, however, be circumvented by $V_L + V_H$ combinations, i.e. by combinations of, for example, 10^3 genes which code for the hypervariable regions of the light chains with 10^3 genes which code for those regions of the heavy chains. The two theories are illustrated in *Fig.* 42.19.

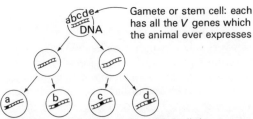

Fig. 42.19. **Antibody diversity: genetic and somatic mutation theories.**
(*a*) Germ-line theory; (*b*) somatic mutation theory.

The following evidence is usually put forward in support of the somatic mutation theory.

a. The large number of genes required for the germ-like theory.

b. The maintenance through many generations of specific coding sequences, some of which (e.g. those producing antibodies to unusual haptens) could have had no survival value in the past. Genetic drift would certainly have spoiled several of these sequences in the course of time.

c. The same characteristic marked sequence of amino acids has been found in the V regions of all the many antibodies (possibly thousands) produced by an animal such as the rabbit. It seems very unlikely that code for these chemical sequences would be repeated in the DNA of many genes, as would be required by the germ-line theory.

More recently, however, support for the germ-line theory has been provided by the discovery of *J* or joining genes in DNA. The coding gene for the V region of the immunoglobulin has been found to terminate at amino acid 95 and not at 108, the end of the variable region. The short region between 95 and 108 is coded for by short sections of DNA described as *J* genes, 5 of these having been described so far. Combinations of 300 short *V* genes with 5 different *J* genes allows coding for 1500 continuous *V* genes and demonstrates how the number of antibodies could be significantly increased by different combinations. Recently, the existence of *D* or diversity genes has been described. These are short sequences of bases (up to 40) which code for short peptide chains of 0–14 amino acids. The combination of peptides coded for by the *D* genes, with peptides coded for by *J* genes would greatly increase the number of possible amino acid sequences in the antibodies. Recombination also occurs so that stretches of DNA are deleted before transcription occurs (*Fig.* 42.20).

The possibility that mutations sometimes occur cannot, of course, be ruled out and may be responsible for a number of 'hot-spots', the points on the immunoglobulin at which mutations occur frequently.

The occurrence of somatic mutations could greatly aid the development of new

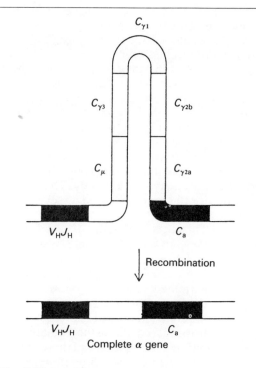

Fig. 42.20. **Role of V and J genes in regulation of antibody synthesis and diversity.**
Recombination allows close association of $V_H J_H$ and C_α genes.

antibodies, but clearly combinations of peptides produced by different genes demonstrates that the mutations need not be as numerous as would be required by a simple somatic mutation theory.

42.6 Complement

Complement is a system of at least eleven proteins present in the plasma that are normally inert, but are activated by a complex series of cascade reactions, analogous to those used in blood clotting (*see* Chapter 26). Complement is often studied using red blood cells coated with antibody, since the final stage in the activation is the formation of an active complex that causes loss of structure of the membrane and leakage of haemoglobin. In the normal defence of the body, bacteria take the place of the red blood cells and the

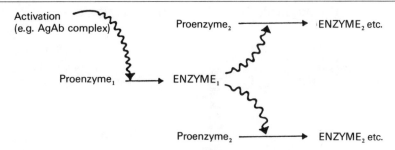

Fig. 42.21. **Basic concept of complement system.**

complement system plays a very important role in killing invading bacteria.

The initial activation is of an antigen–antibody complex, particularly that bound to a membrane, which then transfers an enzyme precursor to an active enzyme; the process is then continued (*Fig.* 42.21).

Only the IgG and IgM classes of immunoglobulins bind complement and subclasses of Ig bind with different affinities, IgG1, and IgG3 fixing complement effectively but IgG4 binding poorly.

The complement can be divided into a number of components and the first component of complement, C1, is composed of the three subcomponents C1q, C1r and C1s.

After attachment of the IgM or IgG antibodies to the surface membrane of the cells, C1q binds to the Fc regions of the immune complex. C1g is bound to C1r and C1s by means of calcium. C1s is converted to an active peptidase that possesses esterase activity and activates the fourth and second components of the complement, C4 and C2, causing them to bind to a hydrophobic site on the membrane (*see Fig.* 42.22). This complex of C1, C4 and C2 has proteolytic activity and attacks the third component C3, so producing a peptide fragment which is the C3a and C3b component. This component possesses a hydrophobic region that enables it to attach to the membrane.

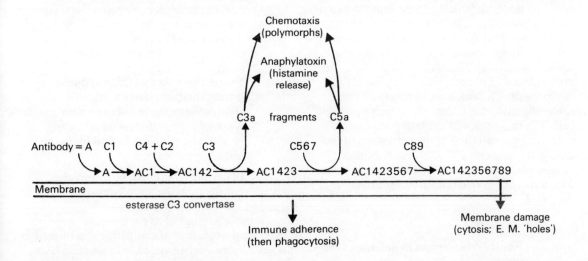

Fig. 42.22. **Complement activation.**

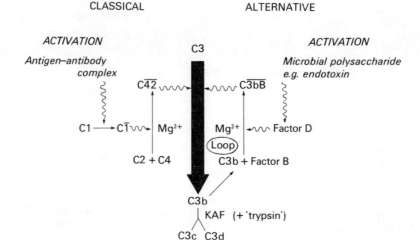

Fig. 42.23. **Alternative pathway of complement activation.**
KAF, conglutinogen activating factor.

Activation of C3 can take place by an entirely separate pathway, described as the alternative pathway (*see Fig.* 42.23). This involves initial activation by a microbial polysaccharide or endotoxin and the transformation of C3b into an active form by the addition of factor B, factor B then producing a new C3 convertase, C3bB.

Formation of the factor C3a then brings about splitting of the fifth component into C5a and C5b which, in turn, bind the sixth and seventh components, C6 and C7, on the membrane. Finally the terminal components, C8 and C9, are bound. The binding of these components then initiates lysis of the cell, the potent agent in the cytolysis being the eighth component, C8, which is extremely effective since one molecule is capable of lysing one red cell. The main role of the ninth component, C9, is the potentiation of the activation of the eighth component, C8, although the other components may be involved in binding of C8. The sequence of reactions is summarized in *Fig.* 42.22.

Role of complement in defence of the body

The complement system plays many important roles in the body's defence against disease which are summarized below.

a. The complement system causes bacteriolysis in Gram-negative organisms by allowing lysozyme to react with the plasma membrane.

b. The complement system facilitates phagocytosis of a micro-organism by coating with antibody and complement. C3 may play a major role in the process by providing a site for macrophage binding and C3b is also likely to be involved by triggering the release of lysosomal enzymes from macrophages.

c. Fragments C3a and C5a produced during the complement cascade stimulate the acute inflammatory response. They do this by attracting phagocytic neutrophils and initiating histamine release.

In certain situations, however, the activation of the complement system can be turned against the individual and cause considerable damage. For example, in paroxysmal nocturnal haemoglobinuria the red blood cells are very susceptible to reactive lysis as a result of an unusual ability to fix the complex C567.

42.7 Immunity to infection

The application of the science of immunology to the understanding of the defence mechanism against infectious diseases is clearly of major importance in medicine.

The fact that individuals who have recovered from a disease are usually immune to reinfection was an important, and early, observation in the history of mankind. Mechanisms of resistance are, in fact, complex and are sometimes classified as 'innate' and 'acquired'. In the former group are included general mechanisms not dependent on specific antibody effects whereas specific antibodies are classified in the 'acquired' group. Contributions made by the various processes to these two groups are considered below.

Innate group

'Humoral agents'

In this category are included secretions from the skin, such as lactic acid and fatty acids, which inhibit the growth of bacteria. In addition there are several 'natural antibodies' in the plasma which help in the reaction with pathogens. Their origin is uncertain, but they may have arisen from interactions of lymphocytes with natural antigens, such as dust or food components. Antibodies of this group are normally of the IgM type. Much interest during recent years has arisen from the discovery of 'interferon': this is a protein which is released after a viral infection and inhibits virus replication.

Phagocytic cells

Substances are released from wounds and inflamed sites that bring about the congregation of phagocytic cells. These cells then attach to particles, such as bacterial fragments or virus particles, and engulf them forming a heterophagic vacuole, the contents of which are digested by lysosomal action as described in Chapter 3. The problem of recognition of a foreign invading particle is helped by combination of the antibody with the

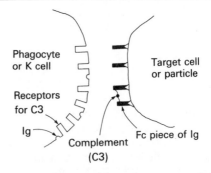

Fig. 42.24. **Attachment of phagocytic cells to larger cells.**

particle, the complex formed then being effectively attacked (*see Fig*. 42.24). Some lower forms of life, for example invertebrates, do not require formation of antibody complexes in the regulation of phagocytosis, so some other methods of recognition must be involved; these may also persist in higher animals.

As discussed in Section 42.6, complement also plays an important role in phagocytosis. Activation of the complement system causes bacteriolysis and fragments released during the complement cascade attract phagocytic cells. Furthermore, particles bound to antibody and complement are phagocytosized with great avidity.

Acquired immunity

Under this classification, the various responses to invading antigens that produce antibodies by mechanisms described earlier are grouped.

After formation, antibodies can help resistance to invading organisms by several mechanisms.

a. Many bacteria produce dangerous protein toxins. They are frequently rendered inactive by antibodies which combine with them forming inactive complexes that are rapidly phagocytosed

b. Bacteria and virus-infected cells are lysed in the presence of complement following reaction with antibodies, in particular IgM and IgG

c. Antibodies play an important role in stimulating phagocytosis and antibody–bacteria complexes are actively phagocytosed.

The role of cells in the immune process

The importance of cells in the immune process became clear when it was observed, about 40 years ago, that immunity, under certain conditions, could be transferred to normal animals via lymphoid cells, but not via antiserum. This type of immunity came to be referred to as 'cell-mediated immunity'. Two types of cell-mediated immunity can be distinguished. In type I, the immune cells possess direct complement-independent cytotoxic effects on target cells. In type II, the immune cells liberate lymphokines on contact with antigens. These are a heterogeneous group of non-specific proteins, of molecular weight 35 000–150 000, which attract polymorphs and monocytes to their site of release, bring about aggregation of macrophages and induce them to phagocytose.

Combination of cells and antibodies involved in the protection process depends on the fact that effective cells have receptors for the Fc part of IgG and sometimes also for the C3 component of complement. For example, small particulate antigens may be coated with IgG and then cross-linked with complement (C3). The structural pattern obtained on the surface of the largest cell is assumed to possess the complementary structure necessary for fixation to the phagocytic cell (*see Fig*. 42.24).

Protection against viruses

Macrophages of the alveoli and those lining the small blood vessels of liver, bone marrow and spleen are the first line of defence against viruses. Cell-mediated immunity is important in the protection process especially in viruses, such as herpes and measles, which pass directly from cell to cell. This acts in two ways: by destruction of infected cells with cytotoxic T-lymphocytes and by the liberation, from sensitized T-cells, of lymphokines which attract macrophages to the site of infection and initiate their activation.

Some viruses, such as measles, mumps, chickenpox and smallpox, do not undergo antigen variation and must, therefore, rely on new non-immunized hosts for their propagation. Others, such as influenza and polio, produce radically new antigenic viruses every few years with consequent epidemics. Immunity to an older strain may be harmful. The reasons for this are not fully understood, but the new antigen may cause production of the original influenza antibody and suppression of formation of antibodies to the new virus.

Protection against bacteria

Formation of antibodies is important in the protection against many bacteria such as streptococci, pneumococci, and *Meningococcus*. These antibodies bind with toxins and, with complement, they attract macrophages and stimulate phagocytosis. Cell-mediated immunity is also important in protection against organisms involved in, for example, tuberculosis, typhoid, syphilis and leprosy.

Summary

The reactions involved in protection against infection are summarized in *Fig*. 42.25.

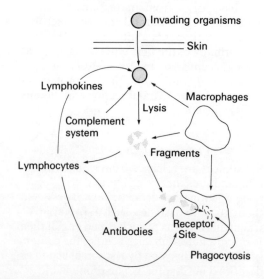

Fig. 42.25. **Reactions involved in the defence against infecting organisms.**

42.8 Allergy, autoimmune diseases and transplantation

a. Allergy

A large proportion of the population, possibly 10 per cent, suffer from localized anaphylactic reactions following exposure to allergens, such as grass pollen or house mites, animal dandruff, or foods like strawberries.

The term 'anaphylaxis', devised as long ago as 1902, means the opposite of prophylaxis, emphasizing the fact that consequences of immune reactions could be harmful. The phenomenon of anaphylaxis was originally observed as a result of experiments on animals, such as the guinea-pig. It was discovered that, if a guinea-pig was injected with 1 mg of a protein such as egg albumin, no effect was observed, but if the dose was repeated two to three weeks later, the animal began to wheeze and could die from asphyxia within a few minutes. The death is said to be caused by 'anaphylactic shock'. This condition is primarily due to the release of histamine, of other vasoactive amines such as serotonin and of 'slow-reacting substances' that are now known to be identical with the leukotrienes present in mast cells.

The main reason for the condition is believed to result from initiation of the synthesis of increased quantities of IgE bound to plasma cells. The antibody attaches to specific sites on the mast cells and further doses of the antigen cause liberation of histamine and other cell components (*see* Fig. 42.26). These substances cause contraction of smooth muscle, and dilatation and increased permeability of small blood vessels.

It is generally believed that the allergic response is a mild form of anaphylatic shock. Certain individuals have a tendency to produce relatively high concentrations of IgE and, consequently, have anaphylactic tendencies to allergens in the environment. These tendencies are frequently inherited.

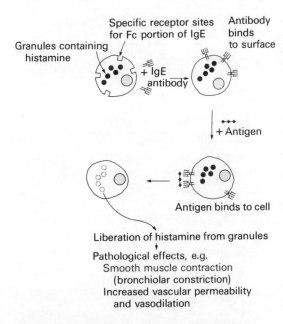

Fig. 42.26. **Possible basis of histamine release in anaphylaxis.**

b. Autoimmune diseases

A large number of diseases are believed to be caused by autoimmune reactions, i.e. the body produces antibodies against part of its own tissues. These may be localized in the joints, lymph nodes, spleen, liver, lungs, gastrointestinal tract or kidneys. Examples are systemic lupus erythematosis in which antibodies are produced to nucleoproteins of the cell nuclei, and pernicious anaemia in which antibodies are believed to be produced against the intrinsic factor required for the absorption of vitamin B_{12}. Many normal individuals have low titres of antibody against some of their own tissues and the condition is dangerous only if excess antibodies are formed. The condition of autoimmunity is more common in older than in young people and in women than in men. There is often a strong genetic tendency to autoimmune diseases.

The causes of autoimmune diseases are not clear and several theories have been advanced to explain their existence. These fall into two main groups. In the first theory, the environment is believed to be the main cause. It is suggested that mechanical or chemical damage, or infectious diseases in some way cause an altered presentation of normal cell

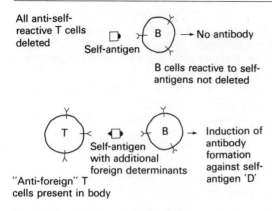

components or fragments to the immune system. Particularly receptive T-cells may also be necessary to trigger the response as illustrated in *Fig.* 42.27. The second theory focuses attention on a defect in the lymphoid system. There are several possible ways in which this could occur but the possibility considered most likely is that a failure to suppress DNA transcription occurs. If this transcription involves the production of antibody to normal tissues, then serious autoimmune diseases may clearly result and give rise to the formation of 'forbidden clones'.

Fig. 42.27. **Possible mechanism of autoimmune reactions.**

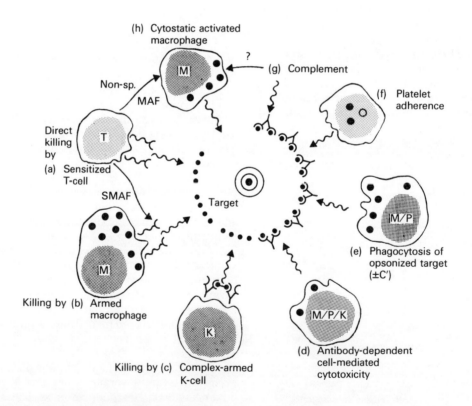

Fig. 42.28. **Target cell destruction in graft rejections.**

●—● Antigen; Y antibody; ⌒ Fc binding site; M macrophage; P polymorphs; K K-cell; SMAF, specific arming factors; MAF, macrophage activating factor.

c. Transplantation

The possibility of transplanting whole organs, such as a kidney or heart, from a deceased donor to a recipient is one of the major steps forward in modern surgery. It had been known for many years, however, that unless the organ to be grafted (e.g. a piece of skin) was transplanted from an individual genetically identical to the recipient, rejection of the transplant was likely to occur. The problem does not arise, of course, when a transplant is made from one part of an individual to another and, in the treatment of serious burns, it is common to prepare skin grafts in this fashion.

The sequence of graft rejection by unlike individuals usually follows a typical pattern. In mice, a skin graft may become vascularized after a few days, but after a week or 10 days the new skin graft becomes infiltrated with lymphocytes and soon afterwards necrosis occurs, with the result that the graft is completely sloughed off. There is extensive evidence that graft rejection has an immunological basis. Thus a second graft is rejected more strongly, the reaction being specific and similar to that of anaphylatic shock; antibodies are also produced in the plasma of the recipient.

It is quite clear that the specificity of the rejection system is under genetic control. In mice, at least 20 regions on the genome have been recognized as coding for antigens involved in graft rejection. A region, known as 'H-2', is particularly important since it controls the strongest transplantation antigen, and constitutes the major 'histocompatability complex'.

Several mechanisms may be involved in graft rejection and it is clear that lymphocytes are likely to play an important role, although antibodies may also play a part. In acute early rejection, 10 days after transplantation, dense infiltration of the graft occurs and this is mediated by T-lymphocytes. In late rejection, for example in the rejection of a kidney transplant, the immunoglobulins bind to the arterioles and glomerular capillaries. The macrophage deposit causes platelet aggregations which lead to renal failure. Typical mechanisms are shown in *Fig*. 42.28.

Graft rejection can be prevented by tissue matching, i.e. typing individuals with close genetic similarities, or by the use of 'immunosuppressive drugs'. These drugs are very similar to those used in cancer therapy, described in Chapter 38, and are usually nucleic acid inhibitors. They strongly inhibit cell division and, thus, the formation of lymphocytic clones which prevents the initiation of lymphocytic action against the transplant and the formation of antibodies to the graft.

Chapter 43 Principles of chemotherapy

The word 'chemotherapy' literally means therapy by chemicals and could, therefore, be considered to include therapy by any chemicals, including drugs, such as aspirin, or hormones, such as insulin, or steroids.

Traditionally, however, in the 1920s and 1930s the term 'chemotherapy' began to be used solely in describing the attack on invading micro-organisms, mainly bacteria or protozoa, by specific chemicals. More recently the term is also widely used in describing drugs that are used in the treatment of cancer. Drugs that have been developed during the past 50 years have proved very effective against most species of bacteria and protozoa, but they have limited success against tumour growth and virtually no success against viruses.

43.1 Historical background

Chemotherapy has a very recent origin and, although some drugs were available in earlier years, the first really useful drugs, the sulphonamides, were produced less than 50 years ago in 1936. Although Western European societies were very successful in many fields of scientific discovery they made very little progress during the fifteenth to nineteenth centuries in the field of chemotherapy. For most of this period, physicians persevered with the use of various concoctions of mercury, antimony and arsenic that were first used in the Middle Ages. These had some value for external application to skin infections but were far too toxic for use internally.

During this period, however, many native communities, particularly those in South

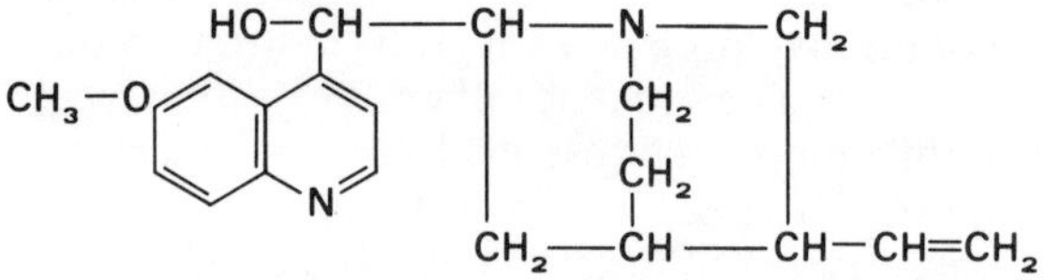

Fig. 43.1. **The structure of quinine.**
Discovered 1513; structure elucidated 1942; synthesized 1945.

America, had discovered that extracts of certain plants possessed really effective chemotherapeutic properties. The Spaniards, following their invasion of Peru in 1513, observed that the natives were using extracts of the bark of certain trees for the treatment of fevers. This bark was taken to Spain in 1640 for the treatment of the Countess of Cinchona and was thence known as 'Cinchona bark'. The extract contains at least 30 different alkaloids, the most important of which is quinine whose structure is shown in *Fig.* 43.1. It possesses antipyretic properties and is, thus, a valuable treatment for many fevers, although more importantly it possesses a strong chemotherapeutic action against the malaria parasite. Since that time the drug has been used extensively throughout the world for the treatment of malaria and, although now generally superceded by other drugs, is still occasionally used today. It is therefore of great historic interest as the first really effective chemotherapeutic drug.

Soon after the discovery of Cinchona bark, another very effective extract from the root of a small shrub, ipecuanha, came to light. This was brought from Brazil in 1625 and purchased in 1670 by the French government

540

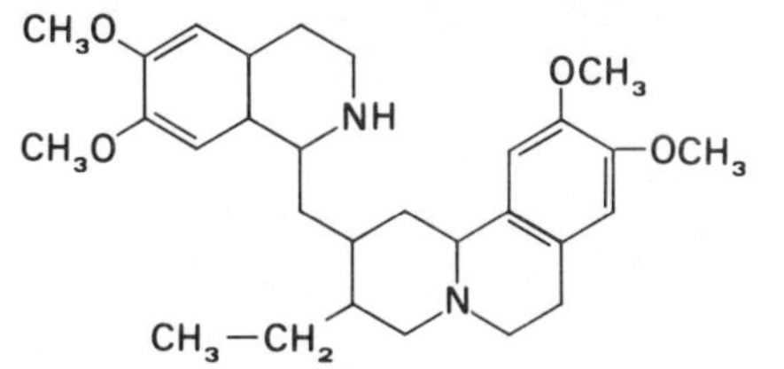

Fig. 43.2. **The structure of emetine.**
Discovered 1625; structure elucidated 1948; synthesized 1950.

for 1000 Louis d'or. The extract contains the alkaloid emetine (*Fig.* 43.2) which is very effective in the treatment of amoebic dynsentry. Like quinine it proved valuable for many hundreds of years.

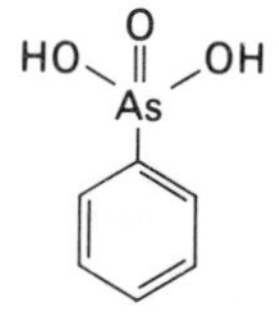

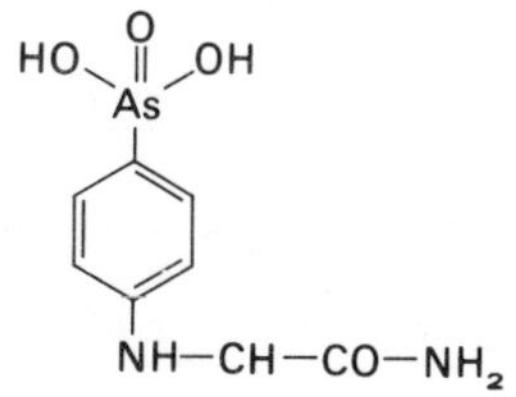

Very little progress in chemotherapy was made for 300 years until the famous German, Paul Ehrlich, became involved in the early years of the twentieth century. His efforts were stimulated by the German colonization of Africa. Great suffering was caused by trypanosomiasis (sleeping sickness) which resulted from infection with the protozoan trypanosome carried by infected tsetse flies. German chemists were encouraged to attempt the development of new drugs for controlling the scourge in Africa. Ehrlich's great strength lay in his conviction that it was possible to successfully treat infection by 'chemotherapy', at a time when few believed this to be possible.

The philosophy behind Ehrlich's approach was to use the original remedies of alchemists and Ehrlich made arsenic his first choice. By organic chemical combination he attempted to convert inorganic arsenic to an organic form that was much less toxic. In 1905, he produced the first arsenical drug, atoxyl, effective against trypanosomiasis (*Fig.* 43.3), but his greatest success came a few years later, in 1910, with the synthesis of 'salvarsan' which was effective against the spirochaete causing syphilis (*Fig.* 43.3). He called this '606' and it is believed to be the six hundred and sixth compound synthesized by Ehrlich. The discovery of an effective drug for the treatment of syphilis really established the great value of 'chemotherapy' in the treatment of disease. Later, in 1919, a much more effective drug against trypanosomiasis was synthesized—

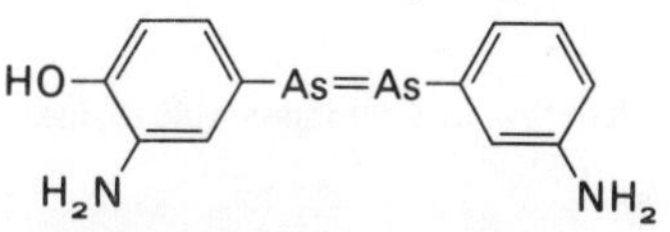

Fig. 43.3. **Arsenicals used in the treatment of trypanosomiasis and syphilis.**

tryparsamide (*Fig.* 43.3). Despite the great success of these drugs against invading protozoa and the syphilis-causing spirochaete, no drug had then been discovered suitable for treating systemic bacterial infections and indeed progress came about by a strange accident.

The success achieved using chemical treatment encouraged the almost haphazard testing of numerous chemicals synthesized by the chemical industry. Many of the compounds tested were dyestuffs and one of these, the red dye prontosil (*Fig.* 43.4), proved effective in the treatment of septicaemia caused by staphyl-ococci in mice. This discovery could have led to the synthesis of many related or larger molecules without the fortunate discovery that prontosil's effectiveness was, in fact, due to its reduction, in the liver, to sulphanilamide (*Fig.* 43.4). This colourless compound proved extremely effective in treating a wide range of bacterial infections in animals and man. Numerous derivatives of sulphanilamide, such as sulphathiazole and sulphaguanidine, were soon synthesized, many of these proving rather more effective than sulphanilamide and were soon widely used in the treatment of human disease.

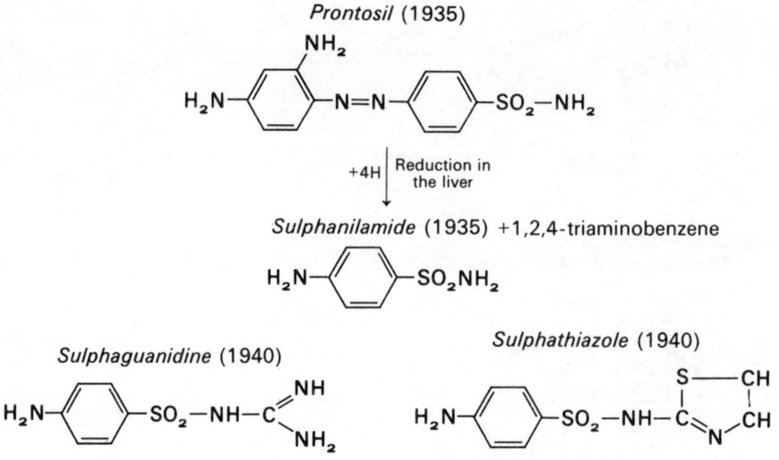

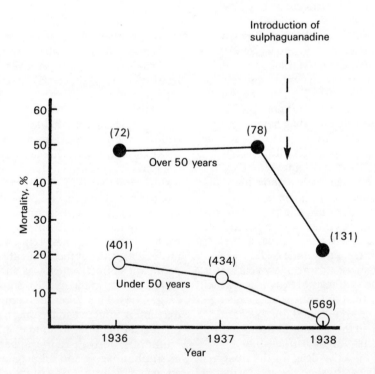

Fig. 43.4. **Prontosil and sulphonamide drugs.**

Fig. 43.5. **Death rate from pneumonia before and after the introduction of sulphonamides.**
Figures in brackets refer to numbers of cases.

The discovery of these drugs had an enormous effect on prognosis for human sufferers and this is demonstrated by the dramatic drop in the deaths from pneumonia (*Fig.* 43.5). With the discovery of the sulphonamides, chemotherapy had taken a tremendous step forward, although it is ironic to reflect that pure sulphanilamide had been known since 1852 but no one had thought of testing such a simple compound. Countless millions could have been saved from an early death by its use during the ensuing 100 years.

However dramatic the discovery of the sulphonamides was, this group of drugs was soon superceded by an even more important discovery. In 1930, Alexander Fleming, working at St Mary's Hospital, London, observed that penicillin moulds of the type found on mouldy bread could antagonize the growth of pathogenic bacteria in petri dishes. He correctly surmised that the mould was producing a substance, later called 'penicillin', that was secreted and prevented the bacteria from growing. He was not able to purify the substance, but during the late 1930s Florey and Chain in Oxford eventually achieved the purification and characterization of the famous chemotherapeutic drug penicillin (*Fig.* 43.6). This drug proved valuable against a wide range of bacteria causing disease in man and possessed the added advantage that it was virtually completely non-toxic. Drugs, such as penicillin, produced by micro-organisms are known as 'antibiotics'.

The discovery of penicillin triggered research into compounds produced by an enormous number of naturally occurring moulds and numerous other 'antibiotics' of great value were isolated. Of particular importance are chloramphemicol and streptomycin, the latter being extremely effective against tuberculosis which has led to the eradication of this very serious disease from the UK and many other countries. Research for new antibiotics is, nevertheless, still continuing today.

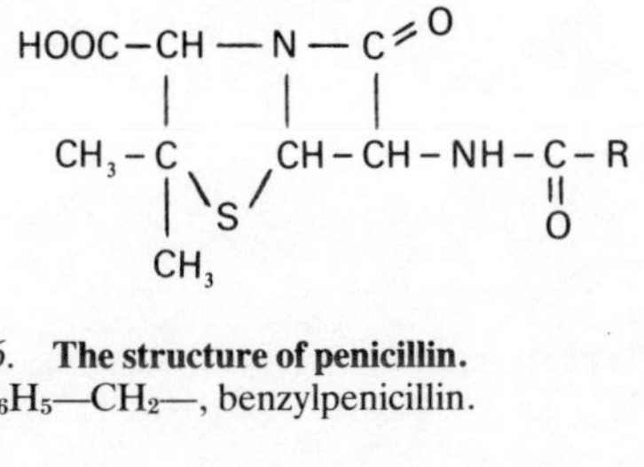

43.2 Origins of new drugs

Although the drugs so far discovered have proved very effective against a wide range of protozoa and bacteria, investigations and searches for new drugs must continue. There are two main reasons for this:

a. Bacteria and protozoa frequently develop resistance to existing drugs, especially if they are exposed to low doses over a prolonged period

b. So far only limited success has been achieved in cancer chemotherapy and very little success in virus chemotherapy.

Below are listed the approaches that may be adopted in the search for new drugs.

i. Study of plant extracts

The original effective drugs, quinine and emetine, were extracted from plants but since their discovery few useful plant compounds possessing chemotherapeutic activity have been found. The investigational problem is a daunting one, since there are 75 000 species of plants in South America alone, each plant producing many different chemicals. Their study and isolation, which often involves laborious and difficult techniques, has so far made little progress but it is possible that some new effective drugs, synthesized by plants, will eventually be discovered.

ii. Synthesis of metal or non-metal derivatives

Ehrlich's plan was to synthesize derivatives of metalloids, such as arsenic, that were effective but much less toxic than the original metalloid. Despite Ehrlich's success it is generally agreed that this line of investigation is unlikely to be profitable.

Fig. 43.6. **The structure of penicillin.**
If R = C_6H_5—CH_2—, benzylpenicillin.

iii. Haphazard testing

Some programmes have been initiated, particularly in the USA, for the testing of all known chemicals for antitumour activity. Although this procedure is unlikely to lead to the discovery of a new revolutionary drug, since it only involves testing compounds of known chemical structure, it could possibly indicate types of molecule likely to be effective.

iv. Natural antibiotics

As indicated above, work continues on the study of the properties of compounds produced by moulds and fungi.

v. Design of antimetabolites

The discovery that the sulphonamide series of drugs was effective because the structural similarity to the normal metabolite blocked its utilization, encouraged research into the possibilities of synthesizing compounds that were close structural analogues of the normally occurring metabolite. The most fruitful line of research in this field has been the development of coenzyme analogues and nucleic acid precursors.

The principle may be illustrated as follows: if the structure of the normal coenzyme is:

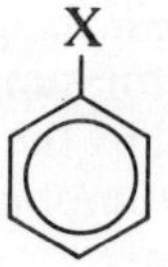

then analogues may be synthesized with many related structures, e.g.

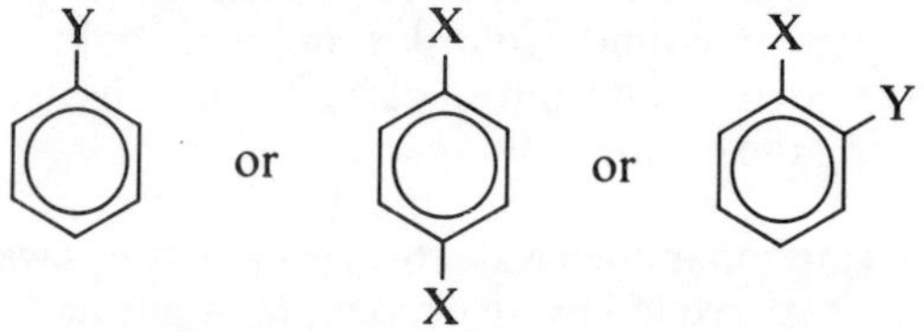

Some analogues, for example folic acid analogues and purine and pyrimidine analogues, have proved valuable in cancer therapy.

43.3 Why are chemotherapeutic drugs effective?

For many drugs the mode of action is quite unknown and this is still true for several of the old established drugs, such as quinine and emetine, which have been known for hundreds of years and for the trypanocidal drug, suramin (*Fig.* 43.7), which was discovered in 1920.

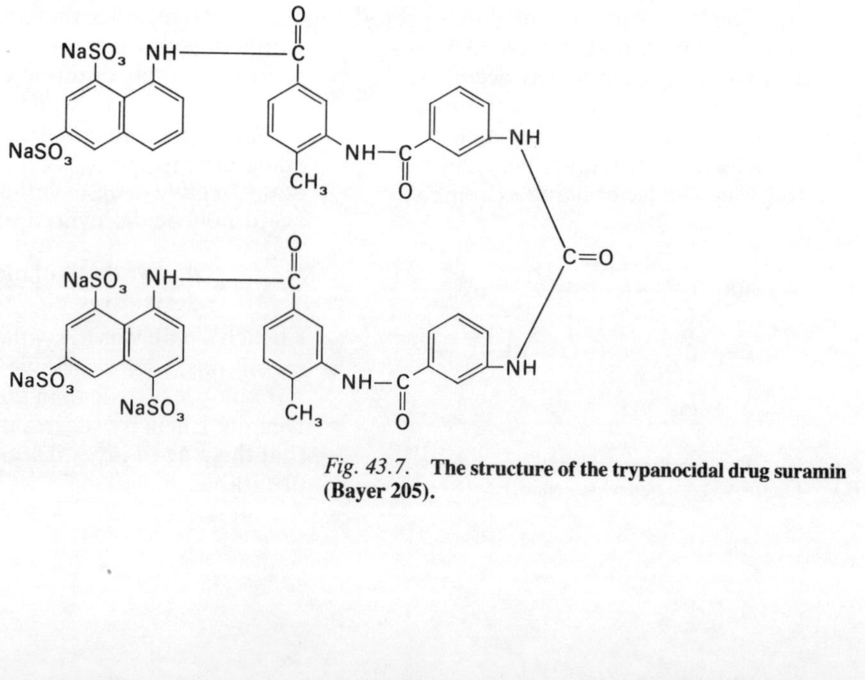

Fig. 43.7. **The structure of the trypanocidal drug suramin (Bayer 205).**

Other drugs are effective because metabolic pathways in bacteria and mammalian cells often differ. For example, penicillin is effective because it blocks the formation of special molecules used in bacterial cell wall synthesis for which there is no counterpart in humans.

Many successful chemotherapeutic drugs act most effectively on rapidly dividing cells: invading bacteria usually divide very rapidly and cell division in tumours is often rapid. Cell division involves increased rates of DNA synthesis, RNA synthesis and protein synthesis, and drugs that interfere with any of these processes are often very effective. Drugs of this type are, however, not without dangers to the host since they will block important cell division processes normally occurring in healthy tissues, e.g. in the intestinal mucosa and bone marrow.

43.4 Relation of chemical structure to chemotherapeutic activity

For some drugs, particularly those designed as antimetabolites, the relation of structure to activity is understood, at least

to a limited extent, but for many others it is completely unknown. Frequently, small modifications in the chemical structure of the drug will reduce or considerably increase its efficiency as a chemotherapeutic drug, and there are several possible reasons for this. The modified drug may be less accessible to the site of action, the rate of metabolism by the host or the rate of excretion may be increased, or the mechanism of interference in the specific metabolic pathway may be changed.

It must, therefore, be appreciated that complex inter-relationships almost always exist between the hosts own drug metabolism and the effect on the invading micro-organism or tumour. The bacteriostatic effects on activity of a slight structural change in the case of the acridines are shown in *Fig.* 43.8.

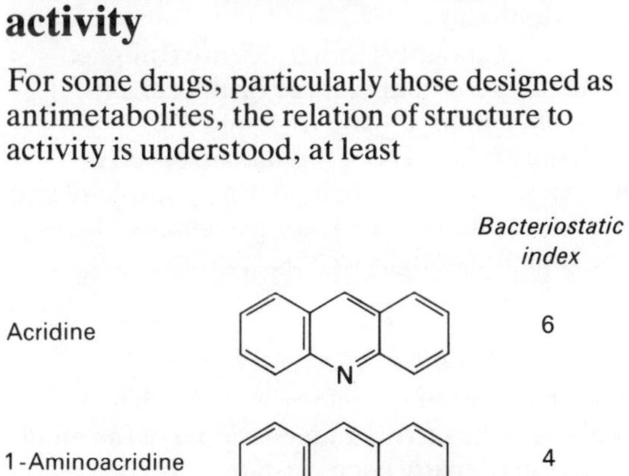

43.5 Summary of mode of action of chemotherapeutic drugs

As explained above (*see* Section 43.3), the mode of action of all chemotherapeutic drugs is not known, but when the action is understood it normally falls into one of the following categories:

- *a.* Compounds that bind to protein, particularly sulphydryl proteins
- *b.* Compounds that attack cell walls and membranes or block membrane synthesis
- *c.* Compounds that block protein synthesis
- *d.* Compounds that bind to nucleic acids
- *e.* Compounds that block nucleic acid synthesis.

We shall consider below examples of drugs that fall into each of these groups in more detail.

a. Drugs that bind to proteins

The arsenical drugs developed by Ehrlich were either pentavalent arsenical compounds or compounds formed by the binding of two trivalent arsenical molecules as in salvarsan. It has been established that all these compounds are inactive as such and that all must first be metabolized by the host, usually

Fig. 43.8. **The effect of structural changes on activity: bacteriostatic effects of substituted acridines.**

in the liver, to form trivalent arsenicals of the general structure R—As=0. The pentavalent arsenicals are reduced and salvarsan is oxidized to the relevant compound.

Arsenical compounds of this type are known to bind strongly to —SH groups of proteins and to inhibit —SH enzymes *in vitro*:

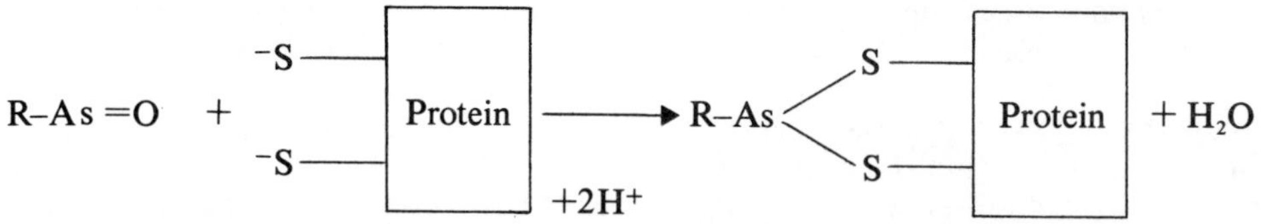

$$R-As = O \quad + \quad \begin{array}{c} {}^{-}S \longrightarrow \\ \\ {}^{-}S \longrightarrow \end{array} \boxed{\text{Protein}} \longrightarrow R-As \begin{array}{c} S \longrightarrow \\ \\ S \longrightarrow \end{array} \boxed{\text{Protein}} + H_2O$$
$$+2H^+$$

It is, therefore, very likely that some vital —SH groups in the trypanosome proteins are attacked by these drugs. One important —SH enzyme is glyceraldehyde phosphate dehydrogenase and its inhibition would cause the cessation of glycolysis upon which the typanosome almost exclusively relies for its energy supply. Consequently an attack on this enzyme could be vital, although it is possible that other —SH proteins, such as those in the membrane, could also be vulnerable.

b. Drugs that attack bacterial cell walls and membranes

To understand the action of these drugs, it is first necessary to describe the structure of the bacterial cell wall which is very complex and composed of many molecular units not found in animal cells. The cell wall forms an envelope surrounding the much more delicate cytoplasmic membrane.

Gram-positive and Gram-negative staining bacteria differ in the composition of their cell walls:

	Gram positive	Gram negative
Peptidoglycan	+	+
Teichoic acid	+	−
Lipopolysaccharide	−	+
Lipoprotein	−	+
Phospholipid	−	+
Protein	+/−	+
Polysaccharide	+/−	−

Peptidoglycans consist of an oligosaccharide backbone chain linked by short branched peptide chains of 7–12 amino acid residues. The saccharide chains consist of alternating residues of *N*-acetylglucosamine and *N*-acetylmuramic acid. The units are 1–4 β-linked and form very long chains, often being composed of 500 units (*Fig.* 43.9).

Teichoic acids are a heterogenous group of phosphorous-containing compounds which can associate with the cytoplasmic membrane. They are composed of chains of 30 residues of glycerol or phosphate and often linked to the muramyl residues of the peptidoglycans.

Lipopolysaccharides are complex molecules containing three distinct chemical regions; a lipid, mostly with solubilized diglucosamine, a core region composed of hexoses, heptose, ethanolamine phosphate and pyrophosphate and 'O' antigen chains which are repeating oligosaccharide units.

Inhibitors of cell wall synthesis

The structure of the bacterial cell wall is an important key to an understanding of the mode of action of many bacteriostatic and biochemical drugs because it is known that several effective drugs inhibit cell wall synthesis. As a result of the inhibition of bacterial wall synthesis, caused by inhibition of peptidoglycan synthesis, bacterial growth ceases, the bacterial culture becomes less viable and the cytoplasmic membrane is very sensitive to osmotic changes and thus readily undergoes lysis.

Examples of these drugs are shown below

i. Cycloserine

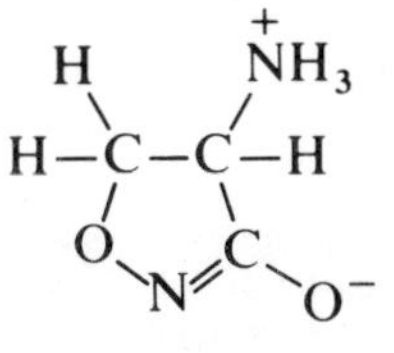

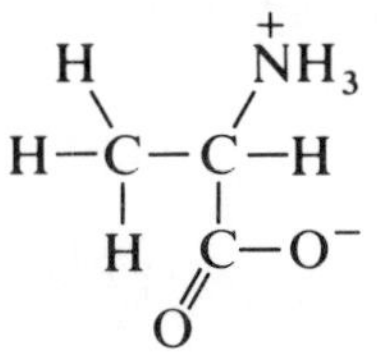

H—C—C—H with NH_3^+, O, N, C, O⁻ (ring structure)

Cycloserine

H—C—C—H with NH_3^+, H, C—O⁻, O

D-Alanine

Cycloserine inhibits bacterial wall synthesis and the action can be prevented by the addition of D-alanine into the medium. It is therefore believed that cycloserine inhibits the incorporation of D-alanine into the peptidoglycan of the cell wall. D-Alanine must first be formed from L-alanine by alanine racemase and cycloserine inhibits this step as well as the synthetase which incorporates the D-alanine into peptides (*Fig.* 43.10). In view of the structural similarity between cycloserine and D-alanine, it is clear that this drug is an excellent example of an antimetabolite.

The mode of action of cycloserine is indicated in *Fig.* 43.10.

ii. β-Lactam antibiotics—penicillins and cephalosporins

Penicillins

In early studies of the action of penicillin it was observed that damage appeared to be located in the outer cell walls of the bacteria and that the cells were much more sensitive to osmotic disturbances.

| D-Cycloserine |

o-Carbamyl-D-serine

Alafosfalin

1   2 L-Alanine ⇌ 2 D-Alanine

Alanine racemase

| D-Cycloserine |

2   2 D-Alanine + ATP → D-Ala-D-Ala + ADP + P_i

D-Alanine : D-Alanine ligase (synthetase)

3   D-Ala-D-Ala + UDP-MurNAc-tripeptide + ATP →
UDP-MurNAc-pentapeptide + ADP + P_i

D-alanyl-D-alanine adding enzyme

Fig. 43.10. Mode of action of cycloserine.
(1) Inhibition of alanine racemase; (2) inhibition of the peptide synthetase.

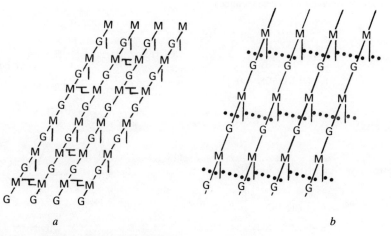

a   *b*

Fig. 43.9. **Peptidoglycan structure of bacterial cell walls.**

Schematic representation of a peptidoglycan monolayer of (a) *E. coli* and (b) *Staphylococcus aureus.* Glycan chains are composed of *N*-acetylglucosamine (G) and *N*-acetylmuramic acid (M). The vertical lines attached to *N*-acetylmuramic acid represent the tetrapeptide —L-Ala—D-Glu—L-Lys—D-Ala. In (a) the horizontal lines which join two muramic acid residues represent two tetrapeptides joined by a direct peptide bond. In (b) the cross-link between the tetrapeptides is composed of a pentaglycine bridge, represented by dots.

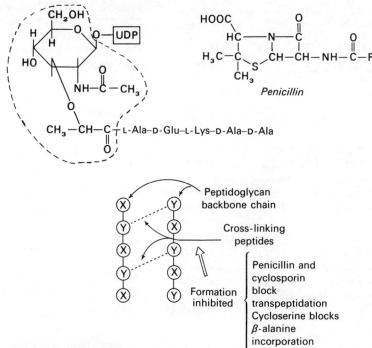

Fig. 43.11. **Role of cycloserine and β-lactams in inhibiting bacterial cell wall synthesis.**

Ⓧ *N*-acetylglucosamine residue; Ⓨ *N*-acetylmuramic acid residue (also indicated by dotted line on UDP-*N*-acetylmuramyl pentapeptide).

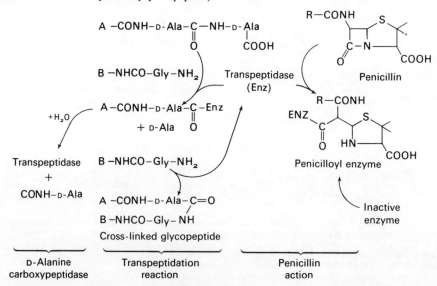

Fig. 43.12. **Mode of action of penicillin – inhibition of transpeptidation.**

A—the end of the main peptide chain of the glycan strand; B—the end of the pentaglycine substituent from an adjacent strand.

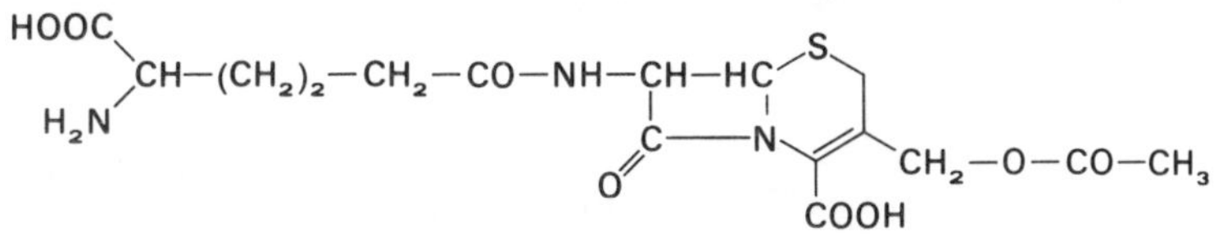

Fig. 43.13. **Cephalosporin.**

Further research indicated that penicillins inhibited the biosynthesis of the peptidoglycan·component of the cell walls of bacteria. Penicillin also acts as a analogue of the normal substrate and is believed to act as an inhibitor of the transpeptidation enzyme which is necessary to link the dipeptide D-Ala-D-Ala to form the peptide chain of the peptidoglycan (*Figs*. 43.11 and 43.12).

Cephalosporins
The cephalosporins are a group of antibiotics related to the penicillin but discovered more recently. They contain a fused β-lactam dehydrothiazine ring system (*Fig*. 43.13). They are also believed to act like penicillin by inhibiting peptidoglycan synthesis but this mechanism of action is likely to be enhanced by a lytic process. This lysis kills the bacteria and is stimulated by the β-lactam group of drugs.

The mechanism of the control of triggering of lysis is unclear. It could be dependent on the accumulation of cell wall precursors in the membrane or cytoplasm or the production of inadequately linked peptidoglycan. The final result is that autolysis is actively switched on and lysis ensues.

A whole family of β-lactam antibiotics have been developed during the last 50 years, many of which are of great value in clinical treatments and they are illustrated in *Fig*. 43.14.

Cytoplasmic membranes

The cytoplasmic membranes of bacteria are similar in composition and structure to those of eukaryotic cells (cf. Chapter 2). They are therefore attacked by the same agents which attack most membranes. These agents can:

i.   Produce a major disorganization of the membrane

ii. Produce specific changes in permeability towards specific ions

iii. Produce a channel or port in the membrane allowing diffusion of substances through the membrane

iv. Inhibit a membrane-bound enzyme involved in transport process

v.   Inhibit enzymes involved in the synthesis of essential components of membranes.

Many detergents will cause disintegration or increase the permeability of membranes. Detergents can be one of three types: anionic, cationic or non-ionic as illustrated in *Fig*. 43.15. Although, in sufficient concentration, many detergents are bactericidal, cationic detergents, as typified by the cetyl pyridinium group of compounds, have proved most valuable, causing maximum damage to bacterial membranes with minimum damage to the host cells. They are very effective for external application to wounds but cannot, however, be used internally, on account of their haemolytic effects on red blood cells. Non-ionic surface active agents which contain a non-ionizable hydrophilic group have little or no antibacterial activity.

Several cyclic peptides, such as valinomycin and gramicidin, also cause greatly increased membrane permeability. They are believed to become incorporated into the membranes and generate new pores (*Fig*. 43.16). Although of limited chemotherapeutic value, they have proved most useful for the study of membranes and their permeability properties.

c. Drugs that inhibit protein synthesis

The chemical structures of typical drugs that inhibit protein synthesis are shown in *Fig*. 43.17. Many of the antibiotics that inhibit

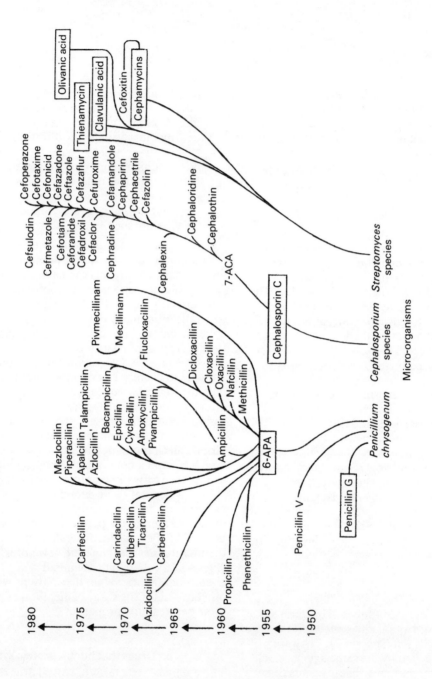

Fig. 43.14. β-**Lactam antibiotics and** β-**lactamase inhibitors.**
Naturally occurring compounds are in rectangles.

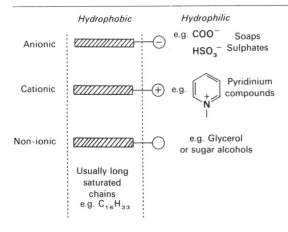

Fig. 43.15. **Types of detergents or surface active agents.**

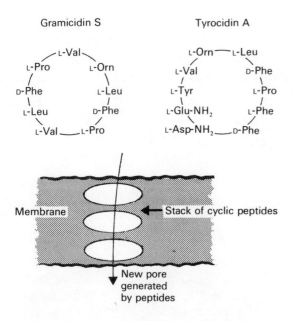

Fig. 43.16. **Cyclic peptides and their effect on membrane permeability.**

the growth of bacteria are known to inhibit protein synthesis. They accomplish this inhibition by binding to either the 30-S or 50-S subunit of the ribosome as indicated in *Fig.* 43.18. In view of the complex nature of the ribosome's structure, the mode of action is only understood in detail for a few drugs, but research in this field has demonstrated that drugs may act at certain steps in protein biosynthesis as shown below.

i. Drugs that bind to the 30-S subunit

α. Binding at a site where the aminoacyl-tRNAs attach. Streptomycin and tetracycline act in this way.

β. Some drugs cause misreading of the mRNA message. Under certain conditions streptomycin will have this effect. Note that drugs do not necessarily possess a single mode of action.

ii. Drugs that bind to the 50-S subunit

α. Drugs can block binding of initiator tRNA so that movement of the ribosomes along the mRNA is blocked. Cycloheximide and streptomycin have this effect.

β. Drugs can inhibit the synthesis of the peptide bond, and chloramphenicol inhibits this stage.

γ. Drugs can cause release of short chain peptides and the antibiotic puromycin has this effect. In fact this drug is far too toxic for use as a chemotherapeutic drug, but interest in it rests in the fact that the detail of its mechanism of action has been elucidated. The drug apparently has sufficient structural similarity to the terminal residue of a tRNA to deceive the ribosome into introducing a few amino acid residues into the puromycin structure. These short-chain puromycin peptides are rapidly released and no complete peptides are formed. The action is shown in *Fig.* 43.19.

δ. Prevention of polysome formation. Emetine is believed to block the attachment of ribosomes to mRNA to form polysome units.

d. Drugs that bind to nucleic acids

The extremely toxic gas, sulphur mustard (a mustard gas, *see Fig.* 43.20) was first used during the 1914–18 World War causing very serious blistering of the skin of those coming

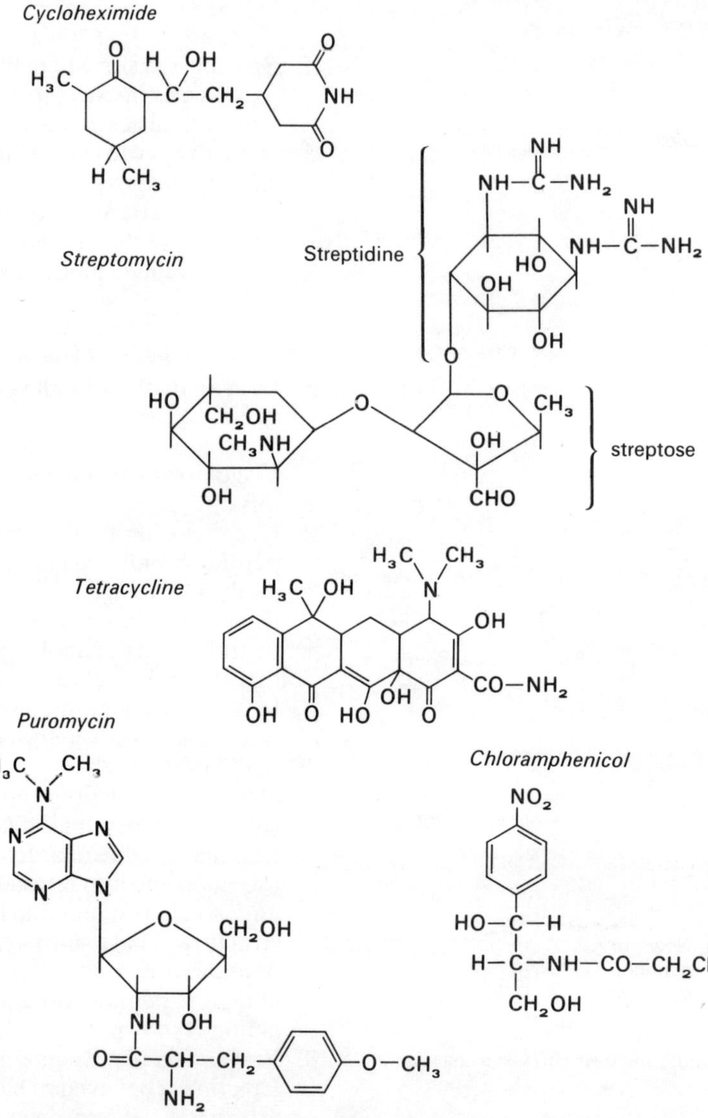

Fig. 43.17. **Protein synthesis inhibitors.**

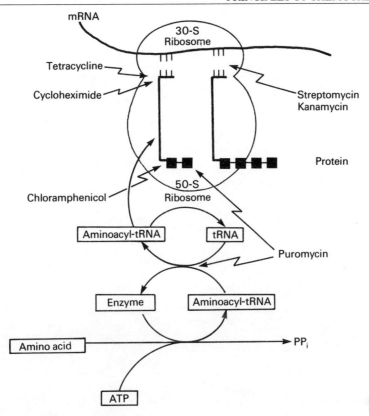

Fig. 43.18. **Mechanisms of inhibition of protein synthesis.**

into contact with it. The possibility that the gas could be used again during the 1939–45 World War was considered a serious possibility and much research was directed towards understanding its properties and action. Fortunately the gas was never used, but the research, almost as a by-product, produced a related group of compounds, the nitrogen mustards, in which the sulphur atom was replaced by nitrogen. This group of compounds was extremely interesting because they were much less toxic than sulphur mustards, but they strongly inhibited cell division, suggesting that the nitrogen mustards could be useful in the treatment of some tumours and, indeed, their use in cancer treatment started during the late 1940s and early 1950s with some degree of success. Nitrogen mustard was, in fact, the first chemotherapeutic drug with a proven effect on

tumour growth to be used. The discovery of this drug demonstrates a remarkable event: the study of a compound used in the destruction of man was turned into the discovery of a drug that could help to cure one of the most serious of all diseases.

Nitrogen mustard itself proved relatively toxic and during the 1950s numerous derivatives were synthesized, many of which were less toxic and more effective against tumour growth. The structures of some of these are shown in *Fig.* 43.20. The most successful of this group has been cyclophosphamide which, in conjunction with other drugs, is successfully used in cancer treatment today. The nitrogen mustards are powerful alkylating agents with strong affinity for the nitrogen and hydroxyl groups of the purine bases of nucleic acids. Their two side chains will powerfully cross-link two separate

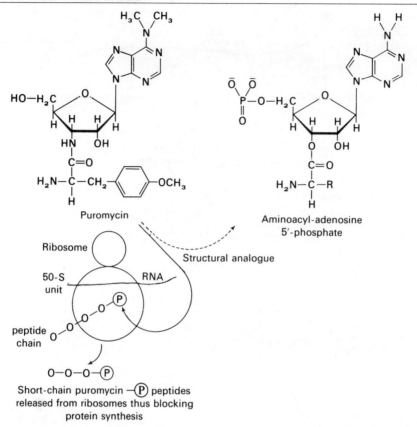

Fig. 43.19. **Mode of action of puromycin.**

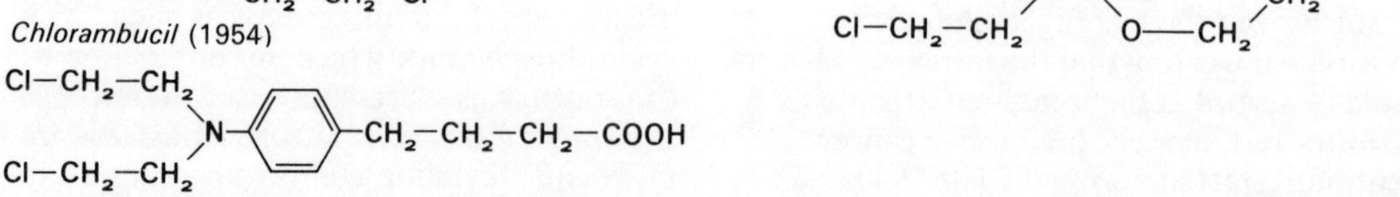

Sulphur mustard

Nitrogen mustard

Chlorambucil (1954)

Nitromin (1951—Japan)

Endoxan (Cyclophosphamide) (1952—Germany)

Fig. 43.20. **Sulphur mustard and nitrogen mustards.**

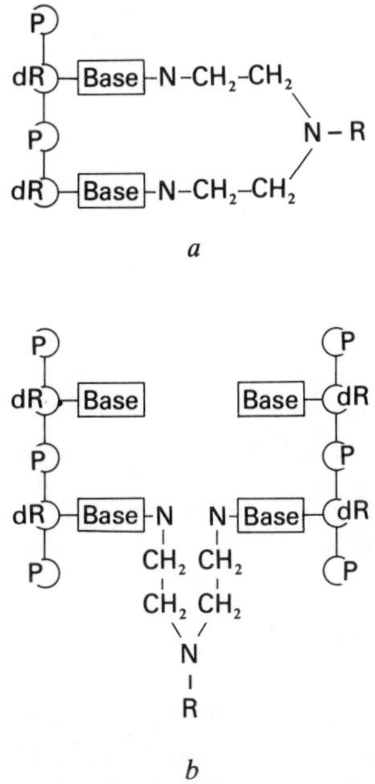

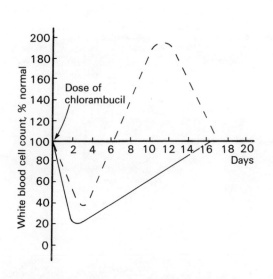

Fig. 43.21. **Cross-linking of DNA chains by nitrogen mustards.**
(*a*) Intra-chain linking; (*b*) inter-chain linking. Link to N-7 of guanine commonly occurs.

nucleic DNA strands known as 'cross-chain linking', or, alternatively, they will link two parts of the same DNA strand known as 'intra-chain linking'. Such distortions, shown in *Fig.* 43.21, effectively prevent DNA replication and thus inhibit cell division.

The disadvantage of drugs of this type is that they will also inhibit the division of non-tumour cells and therefore seriously inhibit division in the bone marrow so reducing dramatically the number of circulating lymphocytes (*Fig.* 43.22). They must, therefore, be used with great care in clinical medicine.

e. Drugs that inhibit nucleic acid synthesis

An outline of the sequence of events involved in the biosynthesis of nucleic acids is given in *Fig.* 43.23. The possibility of designing drugs that could inhibit DNA synthesis is particularly attractive in cancer therapy and many successful drugs have been developed which act at one of the points shown.

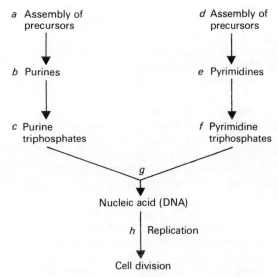

Fig. 43.22. **Effect of chlorambucil on white blood cell count.**
(– – –) Neutrophils; (——) lymphocytes.

Fig. 43.23. **Outline of nucleic acid synthesis and possible sites of attack by chemotherapeutic drugs.**
These sites are indicated as stages *a, b, c, d, e, f, g, h.*

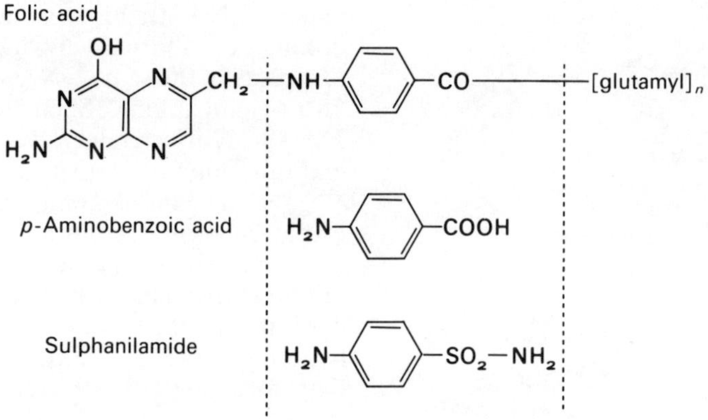

Folic acid

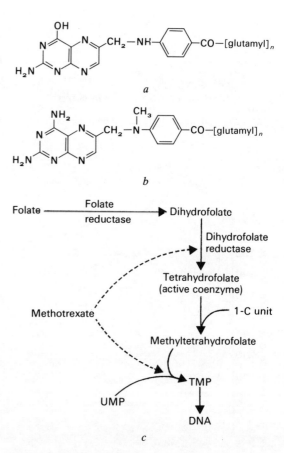

Fig. 43.24. **Relation between folic acid, p-aminobenzoic acid and sulphanilamide.**

i. Assembly of precursors (stages a and d)

Nearly all drugs so far developed inhibit purine rather than pyrimidine synthesis, i.e. act at stage a.

The first successful group of drugs described above was the sulphonamides which were discovered by accident. It was soon observed that they were antagonized by p-aminobenzoic acid (*Fig.* 43.24) and that they were, in fact, antimetabolites for p-aminobenzoic acid which is the normal metabolite. Later it was discovered that p-aminobenzoic acid is a precursor of folic acid, which, in the tetrahydro form, is an essential coenzyme for purine synthesis. The sulphonamides therefore interfere with the assembly of precursors into purines (*Fig.* 43.23). Mammals cannot synthesize folic acid, their source being folic acid synthesized by plants, so sulphonamide drugs have no effect on folic acid metabolism in human cells. It therefore follows that sulphonamide drugs will not antagonize the growth of tumour cells.

However, many analogues of folic acid itself have been tested and one of the most successful of these is methotrexate. This drug blocks the utilization of folic acid since it competitively inhibits the conversion of dihydrofolate to tetrahydrofolate (*Fig.* 43.25). Methotrexate has, in fact, been used successfully for many years in cancer chemotherapy.

Fig. 43.25. **Folic acid antagonists and action of methotrexate.**
(*a*) Folic acid; (*b*) methotrexate; (*c*) mode of action of methotrexate.

Glutamine $H_2N-CO-CH_2-CH_2-CH-COO^-$
                                          $\overset{|}{NH_3{}^+}$

Azaserine $N=\overset{+}{N}-CH_2-CO-CH_2-CH-COO^-$
                                          $\overset{|}{NH_3{}^+}$

'DON' $N\equiv\overset{+}{N}-CH_2-CO-CH_2-CH_2-CH-COO^-$
                                          $\overset{|}{NH_3{}^+}$

Fig. 43.26. **Analogues of glutamine.**

Purine synthesis also requires a supply of glutamine and inhibition of glutamine by analogues, such as azaserine or 'DON' (*Fig.* 43.26), has been tested. These are, however, generally less effective than folic acid analogues.

ii. Analogues of purines and pyrimidines (stages b, c, e, f)

Purine and pyrimidine ring structures allow great scope in the design of antimetabolites. Many drugs have been synthesized in which —SH groups, azo nitrogen and halogens have been substituted (*Fig.* 43.27). As an alternative to changes in the purine or pyrimidine ring system, it is also possible to synthesize drugs containing other sugar residues, e.g. cytosine arabinoside (*Fig.* 43.27), and this is a successful clinical drug. Although many of these drugs can be demonstrated to inhibit division of cells in culture and to be effective clinically, it is often difficult to precisely assess their mode of action.

Although these drugs may be incorporated into the nucleic acids and inhibit cell division (stage *h*), they may also inhibit earlier phosphorylation stages. For example, formation of the diphosphate of the antimetabolite may inhibit formation of the naturally occurring triphosphate of the normal purine or pyrimidine, i.e. the action is at stages *c* or *f*. Alternatively, triphosphates may be formed which cannot be incorporated into nucleic acids, i.e. inhibition occurs at stage *g*.

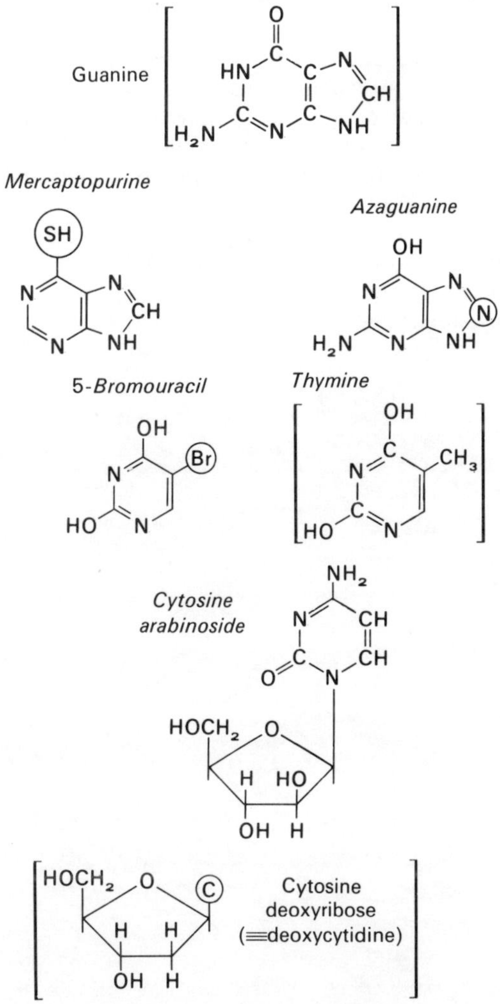

Fig. 43.27. **Purine and pyrimidine analogues.** Normal metabolites are shown in brackets.

43.6 Problems in cancer therapy – multiple drug therapy and effect of drugs on cell cycle

Although single drugs are usually used successfully to treat bacterial or protozoal infections, it is rare that any single drug can be used successfully in cancer therapy; resistance is usually rapidly acquired and then a remission occurs. Modern treatments nearly always use multiple therapy with a combination of different drugs given at specific time intervals.

       The greater success of multiple therapy is probably a consequence of different drugs acting at different stages of the cell division cycle: if the drugs selected act at different stages (*see* Section 43.5), better results are achieved (*see Fig.* 43.28).

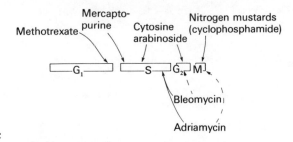

Fig. 43.28. **Effect of chemotherapeutic drugs used in cancer therapy on different stages of the cell cycle.**
The solid lines represent established sites of action; the dotted lines are probable sites of action, not yet firmly established. For description of the stages of the cell cycle *see* Chapter 8.

Appendices

Appendix 1

Commonly occurring mono- and disaccharides

1. Monosaccharides

(a) Pentoses

α-D-Ribose

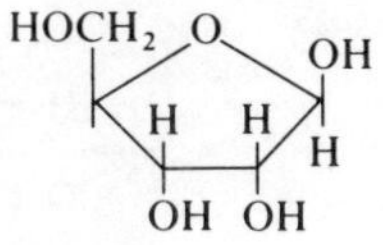

$HOCH_2$ O OH

H   H
H
OH OH

D-2-Deoxyribose

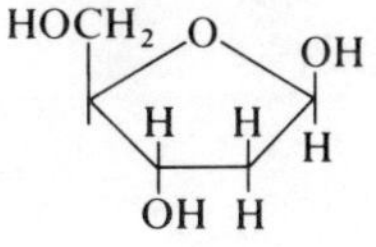

$HOCH_2$ O OH

H   H
H
OH H

(b) Hexoses

α-D-Glucose   α-D-Galactose

'Pyranose rings'

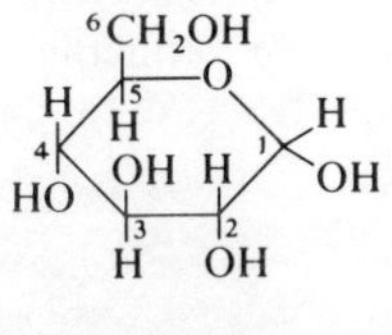

$^6CH_2OH$
O
H   5   H
4   OH H   1   OH
HO   3   2
H   OH

CH_2OH
O
HO   H   H
H   OH H   OH
H   OH

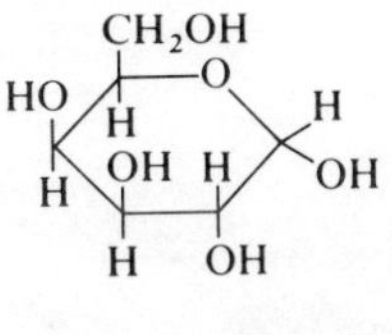

α-D-Fructose

'Furanose ring'

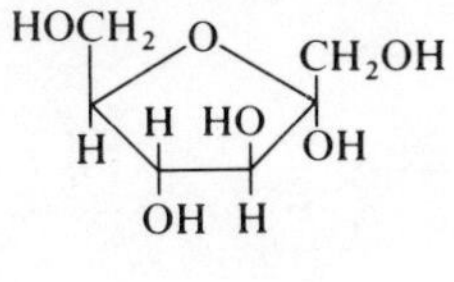

$HOCH_2$ O CH_2OH

H   HO   OH
H
OH H

Appendix 1 (*continued*)

2. Disaccharides

Maltose

Glucose $\xrightarrow{\alpha\text{-}1,4}$ Glucose

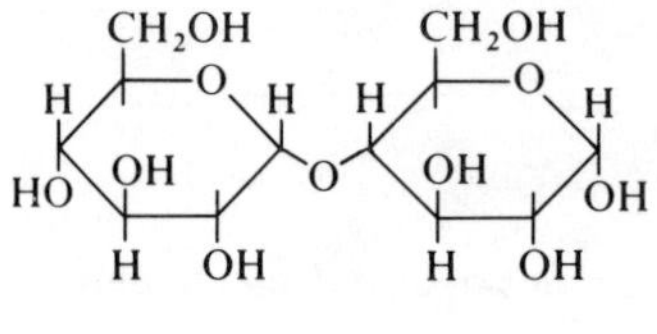

Cellobiose

Glucose $\xrightarrow{\beta\text{-}1,4}$ Glucose

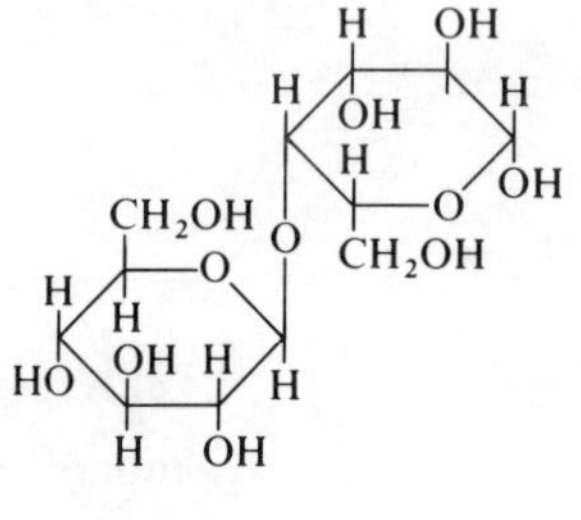

Lactose

Galactose $\xrightarrow{\beta\text{-}1,4}$ Glucose

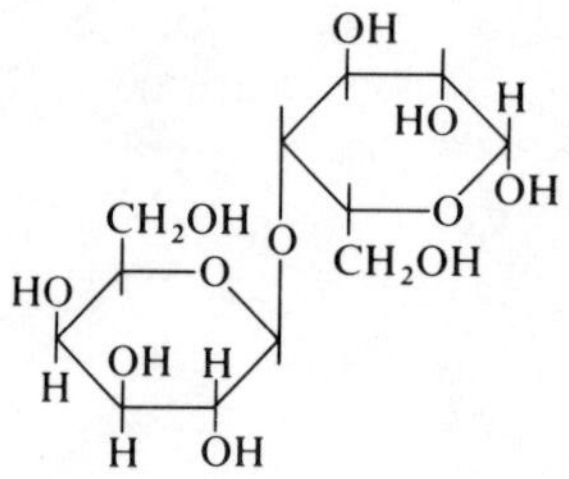

Sucrose

Glucose $\xrightarrow{\beta\text{-}1,2}$ Fructose

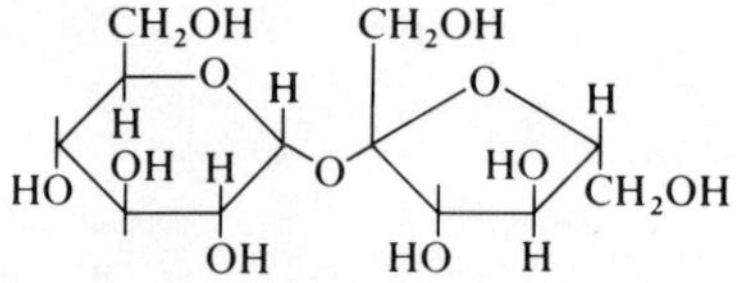

NOTE: Sucrose is not a reducing sugar because the reducing groups of glucose and fructose take part in the formation of the bond.

Appendix 2

Polysaccharides—starch and glycogen

Both starch and glycogen are composed of glucose units linked by 1,4 linkages:

In the amylopectin constituent of starch and in glycogen the chains of glucose units branch, using 1,6 links:

Starch: contains *'amylose'* a linear polysaccharide: $-$O$-$O$-$O$-$O$-$O$-$O$-$O$-$O

and *'amylopectin'* a branched polysaccharide:

Glycogen has a similar structure to amylopectin:

Branch points—1,6 links

Appendix 3

Mucopolysaccharides—proteoglycans

Basic structure

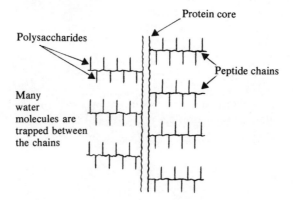

Polysaccharides are a repeating disaccharide unit composed of:

Glucuronic acid   *N*-Acetylglucosamine (structure shown)

or *N*-acetylgalactosamine

or *N*-acetylgalactosamine 4-sulphate

Hyaluronic acid contains repeating units of D-glucuronate + *N*-acetyl-D-glucosamine

Chondroitin contains repeating units of D-glucuronate + *N*-acetylgalactosamine 4-sulphate

Appendix 4

Naturally occurring α-amino acids

General formula:

$$R-CH\begin{array}{l}COO^-\\NH_3^+\end{array}$$

Aliphatic R		Abbreviation	Aromatic R		Abbreviation
H—	Glycine	Gly		Phenylalanine	Phe
CH_3—	Alanine	Ala			
$HSCH_2$—	Cysteine	Cys			
$HOCH_2$—	Serine	Ser			
$CH_3SCH_2(OH)$—	Threonine	Thr		Tyrosine	Tyr
$CH_3SCH_2CH_2$—	Methionine	Met			
$\begin{array}{l}H_3C\\ CH-\\H_3C\end{array}$	Valine	Val			
$\begin{array}{l}H_3C\\ CHCH_2-\\H_3C\end{array}$	Leucine	Leu	*Heterocyclic*	Tryptophan	Trp
$\begin{array}{l}C_2H_5\\ CH-\\H_3C\end{array}$	Isoleucine	Ile		Histidine	His
			Imino acids	Proline	Pro
				Hydroxyproline	Hyp

Acidic			Basic		
$^-OOC-CH_2$—	Aspartic acid	Asp	$H_3\overset{+}{N}-(CH_2)_4$—	Lysine	Lys
$^-OOC-CH_2-CH_2$—	Glutamic acid	Glu	$H_3\overset{+}{N}CH_2CH(OH)CH_2CH_2$—	Hydroxylysine	Hyl
			$HN=C\begin{array}{l}\overset{+}{N}H_3\\NH(CH_2)_3-\end{array}$	Arginine	Arg

Appendix 5

Lipid chemistry and classification

Fatty acids

Commonly occurring fatty acids can be classified into three groups:

Saturated, e.g.
  $C_{16:0}$ hexadecanoic = palmitic
  $C_{18:0}$ octadecanoic = stearic

Monounsaturated, e.g.
  $C_{18:1}$ octadecenoic = oleic ω-9

Polyunsaturated, e.g.
  $C_{18:2}$ octadecadienoic = linoleic ω-6, 9
  $C_{18:3}$ octadecatrienoic = linolenic ω-3, 6, 9
  $C_{20:4}$ eicosatetraenoic = arachidonic ω-6, 9, 12, 15

Position of double bonds counting from the CH_3 terminal carbon indicated by ω numbers

Note: Geometrical isomerism can occur in all unsaturated fatty acids with *cis* and *trans* isomers about each double bond

Most naturally occurring acids are '*cis*' acids

Esters of glycerol
Fats = triglycerides (mixture) or triacylglycerols

Glycerol: CH_2—OH   can be esterified to form:
  |
  CH—OH
  |
  CH_2—OH

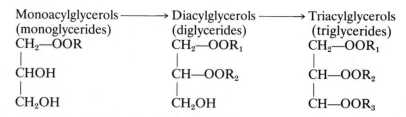

Monoacylglycerols ⟶ Diacylglycerols ⟶ Triacylglycerols
(monoglycerides)   (diglycerides)   (triglycerides)
CH_2—OOR CH_2—OOR$_1$ CH_2—OOR$_1$
  | | |
CHOH   CH—OOR$_2$ CH—OOR$_2$
  | | |
CH_2OH CH_2OH CH—OOR$_3$

The triacylglycerols are the most important of the group and form a major energy store of the body in fat depots. The most common fatty acids found in triacylglycerols are palmitic, stearic or oleic acids. Triacylglycerols can be composed of fatty acids of the same type, e.g. oleic acid, or mixtures of the three.

Phospholipids
(a) Glycerophospholipids

Appendix 5 (*continued*)

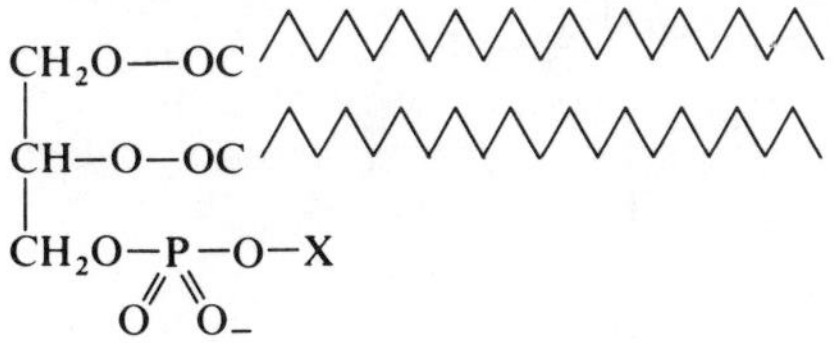

$$CH_2O-OC \diagdown\!\!\diagup\!\!\diagdown\!\!\diagup\!\!\diagdown\!\!\diagup\!\!\diagdown\!\!\diagup\!\!\diagdown$$
$$CH-O-OC \diagdown\!\!\diagup\!\!\diagdown\!\!\diagup\!\!\diagdown\!\!\diagup\!\!\diagdown\!\!\diagup\!\!\diagdown$$
$$CH_2O-P-O-X$$

Where X is choline, serine, ethanolamine or inositol

(*b*) *Sphingophospholipids*
see Appendix 6

Glycolipids (do not contain P)

Appendix 6

Phospholipids

Glycerophospholipids
of major importance in all cellular membranes

Structure

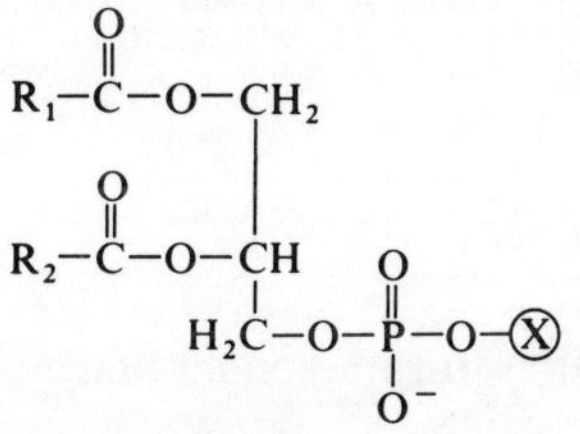

$$R_1-\overset{\overset{O}{\|}}{C}-O-CH_2$$
$$R_2-\overset{\overset{O}{\|}}{C}-O-CH$$
$$H_2C-O-\overset{\overset{O}{\|}}{\underset{\underset{O^-}{}}{P}}-O-\boxed{X}$$

Components
R_1 is usually a saturated or mono-unsaturated fatty acid, e.g. palmitic acid or oleic acid
R_2 is often a polyunsaturated fatty acid, e.g. linoleic acid.

when X = choline *phospholipid*

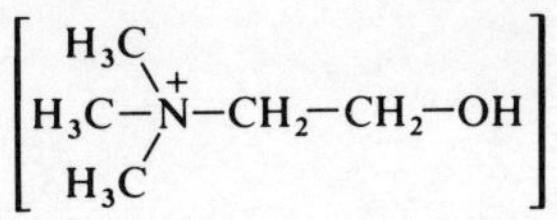

$$\left[\begin{array}{c} H_3C \\ H_3C-\overset{+}{N}-CH_2-CH_2-OH \\ H_3C \end{array} \right]$$ phosphatidylcholine

when X = ethanolamine   phosphatidylethanolamine

$$[H_3\overset{+}{N}-CH_2-CH_2-OH]$$

Appendix 6 (*continued*)

when X = inositol

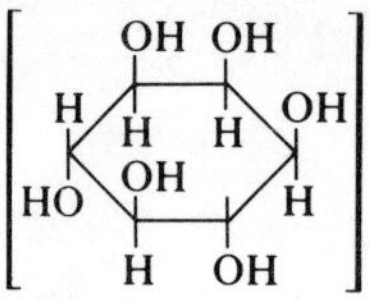

phosphatidylinositol

when X = serine

phosphatidylserine

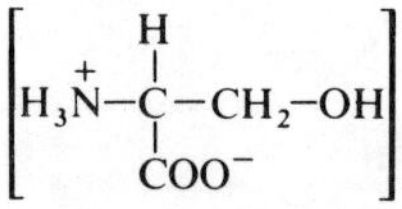

Sphingophospholipids
of minor importance in membranes of many tissues, but are
important in the nervous tissue

Structure

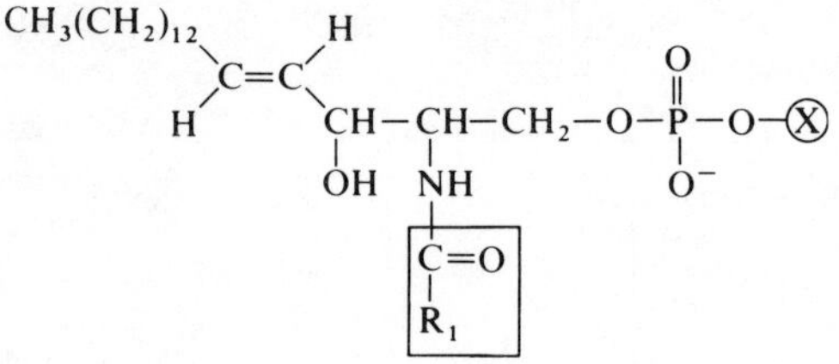

R_1 is usually a long chain saturated on mono-unsaturated fatty acid,
e.g. $C_{22:0}$ or $C_{24:1}$

Appendix 7

Unsaturated fatty acids and their metabolism

Numbering the atoms of fatty acids from the carboxyl end is the traditional method of nomenclature in chemistry texts. Chain lengthening of the fatty acids normally occurs from the —COOH group and thus biochemists prefer to number from the methyl end as this end is not lengthened, and use the omega (ω) or n-system adopted here.

Saturated and mono-unsaturated
$\Bigg\{$

Palmitic (16:0) $CH_3—(CH_2)_{14}—COOH$

Stearic (18:0) $CH_3(CH_2)_{16}—COOH$

Oleic acid (18:1) (ω-9)
$\overset{1}{C}H_3(CH_2)_7—\overset{9}{C}H=\overset{10}{C}H—(CH_2)_7—\overset{18}{C}OOH$

Poly-unsaturated
$\Bigg\{$

Linoleic acid (18:2) (ω-6, 9)
$\overset{1}{C}H_3(CH_2)_4\overset{6}{C}H=\overset{7}{C}HCH_2\overset{9}{C}H=\overset{10}{C}H (CH_2)_7—\overset{18}{C}OOH$

Linolenic acid (18:3) (ω-3, 6, 9)
$\overset{1}{C}H_3CH_2\overset{3}{C}H=\overset{4}{C}HCH_2—\overset{6}{C}H=\overset{7}{C}HCH_2—\overset{9}{C}H=\overset{10}{C}H—(CH_2)_7—\overset{18}{C}OOH$

Arachidonic acid (20:4) (ω-6, 9, 12, 15)
$\overset{1}{C}H_3(CH_2)_4—\overset{6}{C}H=\overset{7}{C}HCH_2—\overset{9}{C}H=\overset{10}{C}HCH_2\overset{12}{C}H=\overset{13}{C}HCH_2—\overset{15}{C}H=\overset{16}{C}H—(CH_2)_3\overset{20}{C}OOH$

Metabolism to longer chain fatty acids

chain elongation $\Big\downarrow$ $\quad\overset{\text{Desaturation}}{\longrightarrow}$

Linoleic acid (18:2) (ω-6, 9) → 18:3 (ω-6, 9, 12)
$\qquad\qquad\quad\downarrow\qquad\quad\downarrow$
$\qquad$ 20:2 → 20:3 (ω-6, 9, 12) → 20:4 (ω-6, 9, 12, 15)
$\qquad$ (ω-6, 9) $\qquad\qquad\qquad\quad\downarrow\qquad$ Arachidonic acid
$\qquad\qquad\qquad\qquad\qquad$ 22:4 → 22:5

Linolenic acid (18:3) (ω-3, 6, 9) → 18:4 (ω-3, 6, 9, 12)
$\qquad\qquad\qquad\qquad\qquad\downarrow$
$\qquad\qquad\qquad$ 20:4 (ω-3, 6, 9, 12)
$\qquad\qquad\qquad\qquad\downarrow$
$\qquad\qquad\qquad$ 22:5 (ω-3, 6, 9, 12, 15)

Appendix 8

Structure of the steroids and major inter-relationships

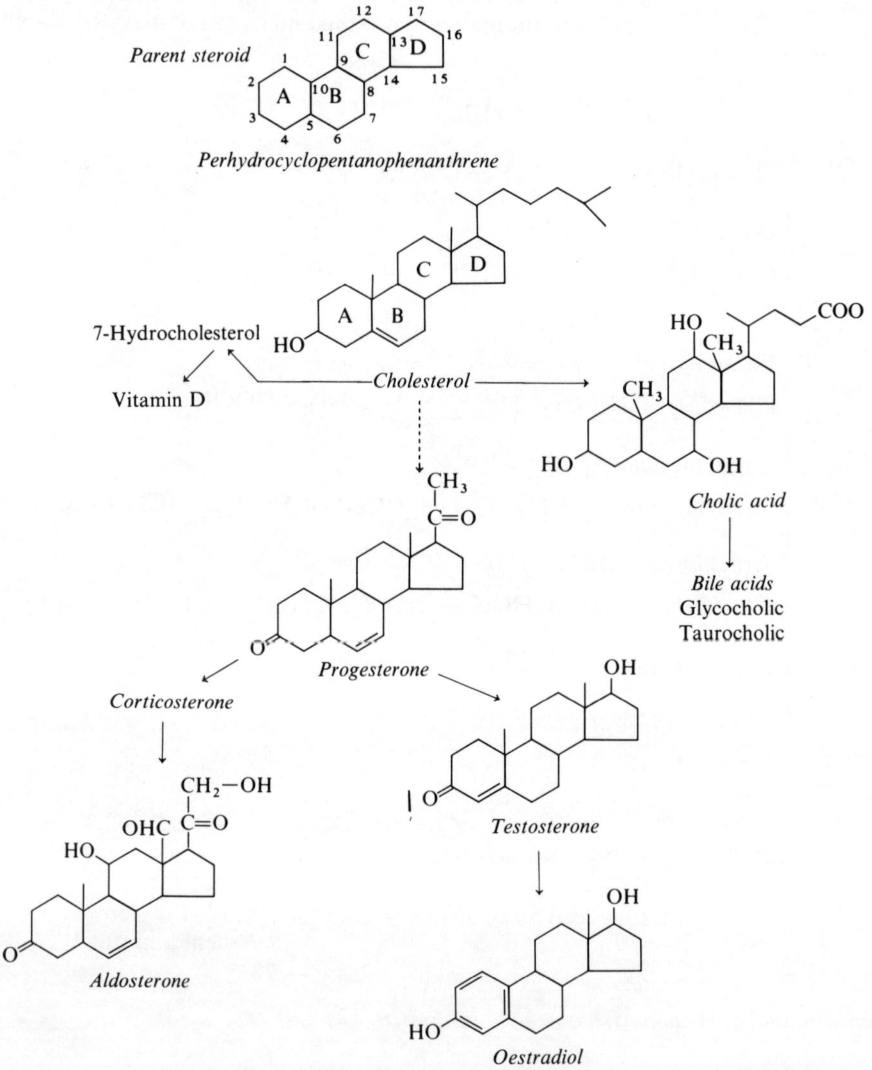

Parent steroid
Perhydrocyclopentanophenanthrene
7-Hydrocholesterol
Vitamin D
Cholesterol
HO
COO
CH_3
OH
Cholic acid
Bile acids
Glycocholic
Taurocholic
CH_3
C=O
Progesterone
Corticosterone
CH_2-OH
OHC
Aldosterone
Testosterone
Oestradiol

Appendix 9

Purine and pyrimidine bases

Pyrimidine

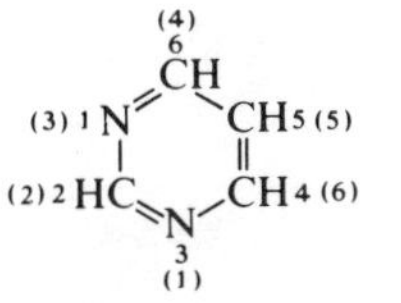

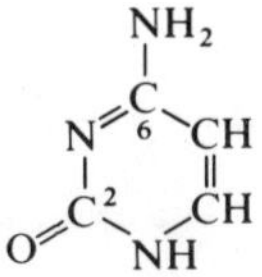

Cytosine
(2-oxy-6-amino-)

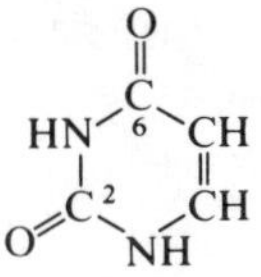

Uracil
(2,6-dioxy-)

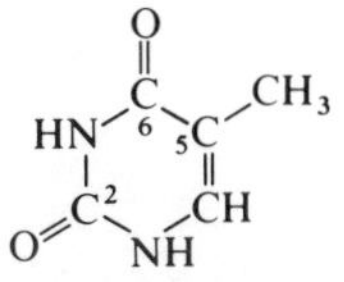

Thymine
(2,6-dioxy-5-methyl-)

Purine

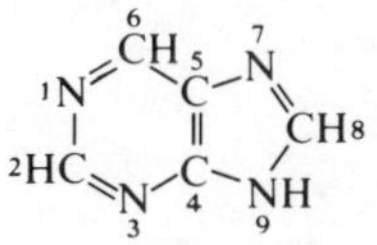

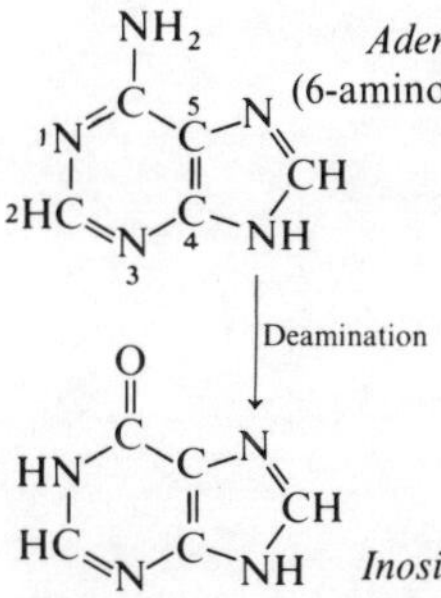

Adenine
(6-aminopurine)

Deamination

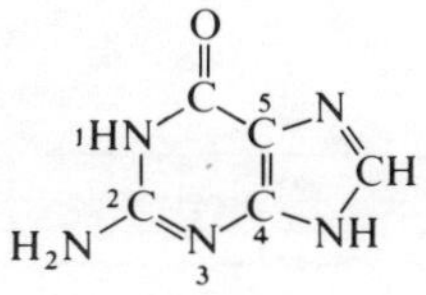

Guanine
(2-amino-6-oxypurine)

Inosine

Note: Ring numbering is according to the older Fisher system. The newer International system is shown in brackets, but the older convention is often used.

Appendix 10

Nucleosides—nucleotides

Cytidine monophosphate (CMP)

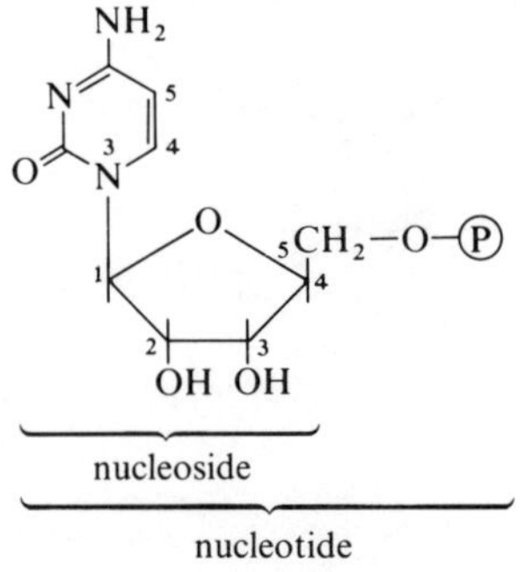

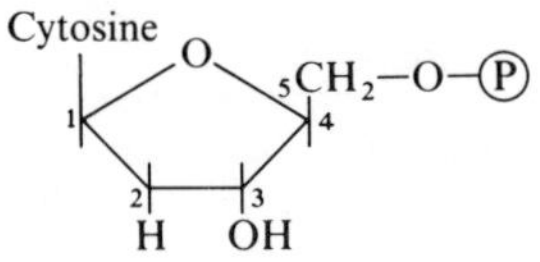

nucleoside

nucleotide

2-deoxyribose (dCMP)

Adenosine monophosphate (AMP)

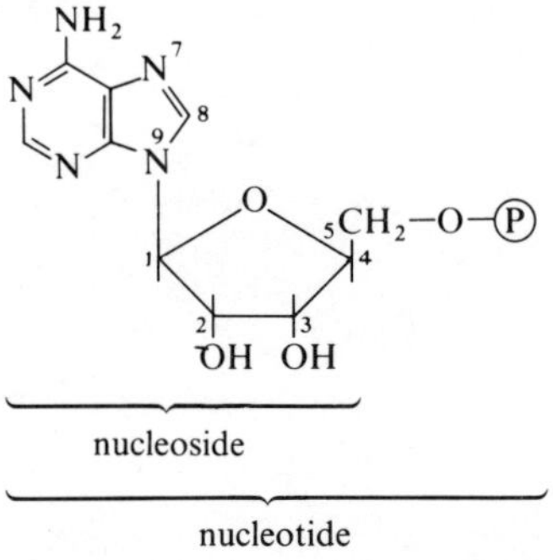

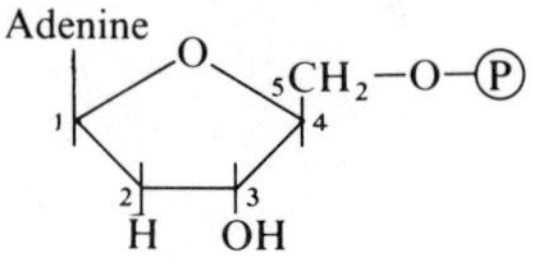

nucleoside

nucleotide

2-deoxyribose (dAMP)

Appendix 11

Nucleic acid structure

Appendix 12

Classification and numbering of enzymes

The first Enzyme Commission in its report in 1961, devised a system for classification of enzymes that also serves as a basis for assigning code numbers to them. These code numbers, prefixed by EC, which are now widely in use, contain four elements separated by points, with the following meanings.

    i.   The first number shows to which of the six main divisions (classes) the enzyme belongs

    ii. The second figure indicates the subclass

    iii. The third figure gives the sub-subclass

    iv. The fourth figure is the serial number of the enzyme in its sub-subclass.

The subclasses and sub-subclasses are formed according to principles indicated below.

The main divisions and subclasses are:

1. Oxidoreductase
To this class belong all enzymes catalysing oxidoreduction reactions. The substrate that is oxidized is regarded as the hydrogen donor. The

Appendix 12 (*continued*)

systematic name is based on donor:acceptor oxidoreductase. The recommended name will be dehydrogenase, wherever this is possible; as an alternative, reductase can be used. Oxidase is only used in cases where O_2 is the acceptor.

The second figure in the code number of the oxidoreductases indicates the group in the hydrogen donor which undergoes oxidation: 1 denotes a —CHOH group, 2 an aldehyde or keto-group and so on as listed in the summary.

The third figure, except in subclasses 1.11 and 1.15 indicates the type of acceptor involved: 1 denotes $NAD(P)^+$; 2, a cytochrome; 3, molecular oxygen; 4, a disulphide; 5 a quinone or related compound etc.

2. Transferases

Transferases are enzymes transferring a group, e.g. the methyl group or glycosyl group, from one compound (generally regarded as donor) to another compound (generally regarded as acceptor). The systematic names are formed according to the scheme donor:acceptor group transferase. The recommended names are normally formed according to acceptor group transferase or donor group transferase. In many cases, the donor is a cofactor (coenzyme) charged with the group to be transferred. A special case is that of amino transferase (*see below*).

Some transferase reactions can be viewed in different ways. For example, the enzyme-catalysed reaction

$$X - Y + Z = X + Z - Y$$

may be regarded either as a transferase of the group Y from Z or as a breaking of the X–Y bond by the introduction of Z. where Z represents phosphate or arsenate, the process is often spoken of as 'phosphorolysis' or 'arsenolysis', respectively, and a number of enzyme names based on the pattern of phosphorylase have come into use. These names are not suitable for a systematic nomenclature, because there is no reason to single out these particular enzymes from the other transferases, and it is better to regard them simply as Y-transferases.

The second figure in the code number of transferases indicates the group transferred: a one-carbon group in 2.1, an aldehydic or ketonic group in 2.2, a glycosyl group in 2.3 and so on.

The third figure gives further information on the group transferred, e.g. subclass 2.1 is subdivided into methyltransferases (2.1.1) hydroxymethyl- and formyltransferases (2.1.2) and so on; only in subclass 2.7, does the third figure indicate the nature of the acceptor group.

3. Hydrolases.

These enzymes catalyse the hydrolytic cleavage of C–O, C–N, C–C and some other bonds, including phosphoric anhydride bonds. Although the systematic name always includes hydrolase, the recommended name is, in many cases, formed by the name of the substrate with the suffix

Appendix 12 (*continued*)

. . . ase. It is understood that the name of the substrate with this suffix means a hydolytic enzyme.

A number of hydrolases acting on ester, glycosyl, peptide, amide or other bonds are known to catalyse not only hydrolytic removal of a particular group from their substrates, but likewise the transfer of this group to suitable acceptor molecules. In principle, all hydrolytic enzymes might be classified as transferases, since hydrolysis itself can be regarded as a transfer of a specific group to water as the acceptor. Yet, in most cases, the reaction with water as the acceptor was discovered earlier and is considered as the main physiological function of the enzyme. This is why such enzymes are classified as hydrolases rather than as transferases.

The second figure in the code number of the hydrolases indicates the nature of the bond hydrolysed: 3.1 are the esterases, 3.2 the glycosidases and so on (cf. the summary).

The third figure normally specifies the nature of the substrate, e.g. in the esterases the carboxylic ester hydrolases (3.1.1), thiol ester hydrolases (3.1.2), phosphoric monoesterases (3.1.3), in the glycosidases the *O*-glycosidases (3.2.1), *N*-glycosidases (3.2.2) etc. Exceptionally in the case of the peptidyl-peptide hydrolases, the third figure is based on the catalytic mechanism as shown by active centre studies or the effect of pH (cf. the summary for the full key).

4. Lyases

Lyases are enzymes cleaving C–C, C–O, C–N and other bonds by elimination, leaving double bonds or conversely adding groups to double bonds. The systematic name is formed according to the pattern substrate group-lyase. The hyphen is an important part of the name and to avoid confusion should not be omitted, e.g. hydro-lyase not 'hydrolyase'. In the recommended names, expressions like decarboxylase, aldolase, dehydratase (in case of elimination of water) are used. In cases where the reverse reaction is much more important, or the only one demonstrated, synthase (not synthetase) may be used in the name. Various subclasses of the lyases include pyridoxal-phosphate enzymes that catalyse the elimination of β- or γ-substituents from an α-amino acid followed by a replacement of this substituent by some other group. In the overall replacement reaction, no unsaturated end-product is formed; therefore, these enzymes might formally be classified as alkyl-transferases (EC 2.5.1). However, there is ample evidence that the replacement is a two-step reaction involving the transient formation of enzyme-bound α, β (or β, γ)-unsaturated amino acids. According to the rule that the first reaction is indicative for classification, these enzymes are correctly classified as lyases. Examples are tryptophan synthase (EC 4.2.1.20) and cystathionine β-synthase (EC 4.2.1.22).

The second figure in the code number indicates the bond broken: 4.1 are carbon-carbon-lyases, 4.2 carbon-oxygen-lyases and so on.

The third figure gives further information on the group eliminated (e.g. CO_2 in 4.1.1, H_2O in 4.2.1).

Appendix 12 (*continued*)

5. Isomerases

These enzymes catalyse geometric or structural changes within one molecule. According to the type of isomerization they may be called racemerases, *cis-trans*-isomerases, tautomerases, mutases or cyclo-isomerases.

In some cases, the interconversion in the substrate is brought about by an intramolecular oxidoreduction (5.3); since hydrogen donor and acceptor are the same molecule, and no oxidized product appears they are not classified as oxidoreductases, even if they may contain firmly bound $NAD(P)^+$.

The subclasses are formed according to the type of isomerism, the sub-subclasses to the type of substrates.

6. Ligases (synthetases)

Ligases are enzymes catalysing the joining together of two molecules coupled with the hydrolysis of a pyrophosphate bond in ATP or a similar triphosphate. The bonds formed are often high-energy bonds. The systematic names are formed on the system X:Y ligase (ADP-forming). In the recommended nomenclature, the term synthetase may be used if no other short term (e.g. carboxylase) is available. Names of the type 'X-activating enzyme' should not be used.

The second figure in the code number indicates the bond formed: 6.1 for C–O bonds (enzymes acylating tRNA), 6.2 for C–S bonds (acyl-SCoA derivatives) etc. Sub-subclasses are only in use in the C–N ligases (cf. the summary below).

Summary

1. Oxidoreductases

1.1 Acting on the CH—OH group of donors.
1.2 Acting on the aldehyde or oxo group of donors.
1.3 Acting on the CH—CH group of donors.
1.4 Acting on the CH—NH_2 group of donors.
1.5 Acting on the CH—NH group of donors.
1.6 Acting on NADH or NADPH.
1.7 Acting on other nitrogenous compounds as donors.
1.8 Acting on a sulphur group of donors.
1.9 Acting on a haem group of donors.
1.10 Acting on diphenols and related substances as donors.
1.11 Acting on hydrogen peroxide as acceptor.
1.12 Acting on hydrogen as donor.
1.13 Acting on single donors with incorporation of molecular oxygen (oxygenases).
1.14 Acting on paired donors with incorporation of molecular oxygen.
1.15 Acting on superoxide radicals as acceptor.
1.16 Oxidizing metal ions.
1.17 Acting on —CH_2 groups.
1.97 Other oxidoreductases

Appendix 12 (*continued*)

2. Transferases
2.1   Transferring one-carbon groups.
2.2   Transferring aldhyde or ketonic residues.
2.3   Acyltransferases.
2.4   Glycosyltransferases.
2.5   Transferring alkyl or aryl groups, other than methyl groups.
2.6   Transferring nitrogenous groups.
2.7   Transferring phosphorus-containing groups.
2.8   Transferring sulphur-containing groups.

3. Hydrolases
3.1   Acting on ester bonds.
3.2   Acting on glycosyl compounds.
3.3   Acting on ether bonds.
3.4   Acting on peptide bonds (peptide hydrolases).
3.5   Acting on carbon–nitrogen bonds, other than peptide bonds.
3.6   Acting on acid anhydrides.
3.7   Acting on carbon–carbon bonds.
3.8   Acting on halide bonds.
3.9   Acting on phosphorus–nitrogen bonds.
3.10   Acting on sulphur–nitrogen bonds.
3.11   Acting on carbon–phosphorus bonds.

4. Lyases
4.1   Carbon–carbon lyases.
4.2   Carbon–oxygen lyases.
4.3   Carbon–nitrogen lyases.
4.4   Carbon–sulphur lyases.
4.5   Carbon–halide lyases.
4.6   Phosphorus–oxygen lyases.
4.99   Other lyases.

5. Isomerases
5.1   Racemases and epimerases.
5.2   *Cis-trans* isomerases.
5.3   Intramolecular oxidoreductases.
5.4   Intramolecular transferases.
5.5   Intramolecular lyases.
5.99   Other lyases.

6. Ligases (synthetases)
6.1   Forming carbon–oxygen bonds.
6.2   Forming carbon–sulphur bonds.
6.3   Forming carbon–nitrogen bonds.
6.4   Forming carbon–carbon bonds.
6.5   Forming phosphate ester bonds.

(From *Enzyme Nomenclature* (1978) Recommendations of the International Union of Biochemistry, Academic Press, London.)

Appendix 13a

Enzyme kinetics and enzyme inhibition

Enzyme kinetics is a tool for the elucidation of reaction mechanisms. Many would regard it as the most important (and with impure preparations, frequently the only) means of investigating enzyme function. In addition to this primary role in biochemistry, kinetics of this type are used also in other ways, e.g. (i) analytically, to determine the concentration of a substrate, (ii) to study processes which follow a rate law indistinguishable from the Michaelis–Menten equation e.g. mediated transport, and (iii) to study selective enzyme inhibition as a rational approach to chemotherapy.

Michaelis–Menten (steady-state) kinetics

If we measure initial velocities (v_i), i.e. rates of reaction close to zero time when very little product has formed, and if we make the assumption that [EP] is negligible, Equation (1) may be written:

$$E + S \underset{k_{-1}}{\overset{k_{+1}}{\rightleftharpoons}} ES \overset{k_{+2}}{\rightarrow} E + P \tag{1}$$

Although true one-substrate reactions are rare (they are restricted to certain isomerizations), they provide a model for which rate equations can be obtained which are usable also for two-substrate reactions when the concentration of one of the substrates is kept constant.

One way of studying kinetics of enzyme-catalysed reactions is to use a $[S]_o$ (the initial value of $[S]$) much greater than $[E]_T$ (total enzyme concentration) (so-called 'Michaelis–Menten conditions') and to measure the initial velocities. This type of kinetic study (Michaelis–Menten kinetics) is used to provide a basis for mechanistic study of enzyme reactions and also analytically.

The rate (v_i) of an enzyme reaction proceeding according to Equation (1) is given by Equation (2):

$$v_i = k_{+2}[ES] \tag{2}$$

When $[E]_T$ is maintained constant, v_i increases with increasing $[S]_o$, but approaches a limiting value $= k_{+2}[E]_T$. This limiting value is called the maximum velocity (V_{max}). V_{max} is equal to $k_{+2}[E]_T$ for the simple model of Equation (1) and, in general, for all reactions obeying the Michaelis–Menten equation irrespective of the number of steps in the model, V_{max} is equal to $k_{cat}[E]_T$. The catalytic constant (k_{cat}) which is calculated from $V_{max}/[E]_T$ is, in general, an assembly of all of the rate constants of the forward steps of the kinetic model.

An enzyme catalysis is characterized under a given set of experimental conditions by k_{cat} and by a quantity K_m, the Michaelis constant. The latter is *defined* operationally as the value of $[S]_o$ when $v_i = V_{max}/2$. The equation may be written also in the form:

Appendix 13a (*continued*)

$$v_i = \frac{V_{max} \cdot [S]_o}{K_m + [S]_o} \qquad (3)$$

When

$$[S]_o = K_m, \ v_i = \frac{V_{max} \cdot [S]_o}{[S]_o + [S]_o} = \frac{V_{max} \cdot [S]_o}{2[S]_o} = \frac{V_{max}}{2}$$

The limiting value of v_i is reached when the enzyme is saturated with substrate but this strictly occurs only when $[S]_o = \infty$. Saturating concentrations of substrate are taken to be when $[S]_o \gg K_m$, but the shape of the hyperbolic substration curve means that even when $[S]_o = 9K_m$, v_i is still only 90 per cent of V_{max}. It is important to note that the ratio k_{cat}/K_m is now considered to be the most useful index of the catalytic effectiveness of an enzyme reaction and this ratio has superceded K_m as the best measure of 'specificity'.

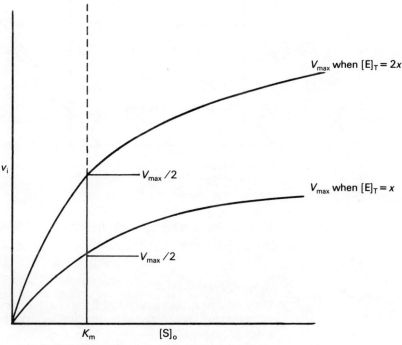

Note that $V_{max} \propto [E]_T$ and K_m is independent of $[E]_T$.

Experimental determination of K_m *and* V_{max}

These parameters are determined by measuring v_i at constant $[E]_T$, at a number of concentrations of the substrate. If the reaction involves more than one substrate, one of the substrates must be maintained at constant concentration while the concentration of the other is varied. V_{max} and

Appendix 13a (*continued*)

K_m can then be obtained by using linear transform of the equation as, for example, in:

The double-reciprocal (Lineweaver–Burk) plot

$$v_i = \frac{V_{max} \cdot [S]_o}{K_m + [S]_o}$$

$$\frac{1}{v_i} = \frac{K_m}{V_{max}} \cdot \frac{1}{[S]_o} + \frac{1}{V_{max}}$$

Compare $y = a\,x + b$

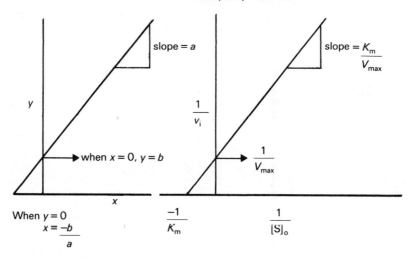

This is a simple traditional method of obtaining V_{max} and K_m, but many better methods now exist.

Reversible inhibitors

Only inhibitors that interact with the enzyme (or the enzyme substrate complex) are considered here, but inhibition by interaction with the substrate is possible also.

a. Inhibitors which increase K_m but leave V_{max} unchanged or 'competitive inhibition'

Appendix 13a (*continued*)

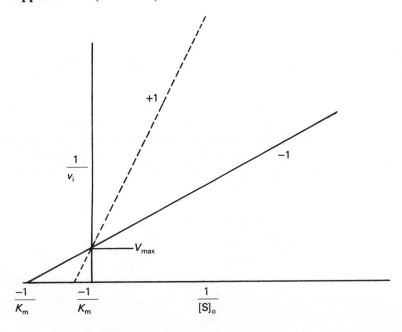

Competitive inhibitors include compounds that are structurally similar to substrates and therefore bind to the enzyme at the site at which substrate must bind for catalysis to occur. If an inhibitor molecule is bound to the substrate-binding site, the substrate cannot bind to this site and vice versa.

$$E + S \rightleftharpoons ES \rightarrow E + P$$
$$+$$
$$I$$
$$\updownarrow$$
$$EI \longrightarrow$$

This means that for a given substrate concentration, a smaller fraction of the enzyme will be in the form of ES when inhibitor is present than when it is absent. When $[S] = \infty$, however, $[ES] = [E]_T$ even when inhibitor is present and hence V_{max} is unchanged.

b. Inhibition which decreases V_{max} but leaves K_m unchanged or 'pure non-competitive inhibition'

A decrease in V_{max} without change in K_m may occur if the inhibitor binds to E and to ES **equally strongly**, and such that binding of the inhibitor prevents the enzyme substrate complex from breaking down to products.

Appendix 13a (*continued*)

$$E + S \rightleftharpoons ES \rightarrow E + P$$

$$\begin{array}{ccc} & + & & + \\ & I & & I \\ K_i \updownarrow & & \updownarrow K_{si} \\ EI + S & \rightleftharpoons & ESI \rightarrow \end{array}$$

If $K_i = K_{is}$ but ESI cannot break down to products, both K_m and V_{max} are changed by the presence of the inhibitor. This type of inhibition is known as *mixed inhibition*. Most examples of *(pure) non-competitive inhibition* reported are probably in fact mixed inhibition in which the change in K_m is so small that the accuracy of the data did not permit detection.

Pure non-competitive inhibition

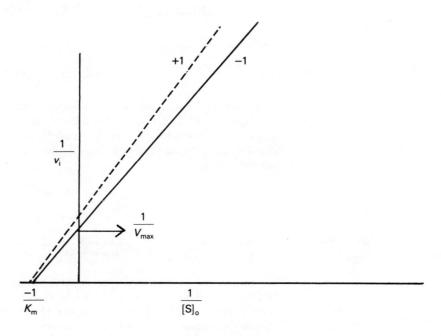

Appendix 13a (*continued*)

A type of mixed inhibition

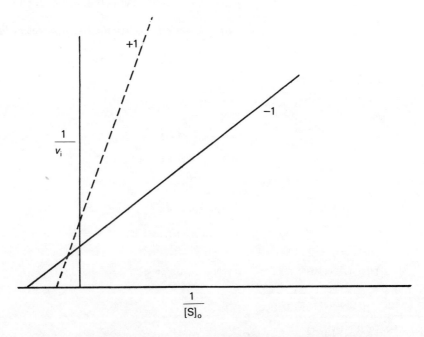

Another rather trivial situation in which K_m is unchanged is when a mixture containing residual active enzyme and some inactive enzyme formed by reactions with an irreversible inhibition is subscribed to v_i–$[S]_o$ analysis. Then clearly the residual enzyme must have unchanged characteristics and K_m will be unaltered; since $[E]_T$ has been decreased, however, V_{max} will be decreased.

Appendix 13b
Example of the mechanism of action of an enzyme

Hydrolysis of acetylcholine catalysed by acetylcholinesterase

Deacylation (again addition followed by elimination)

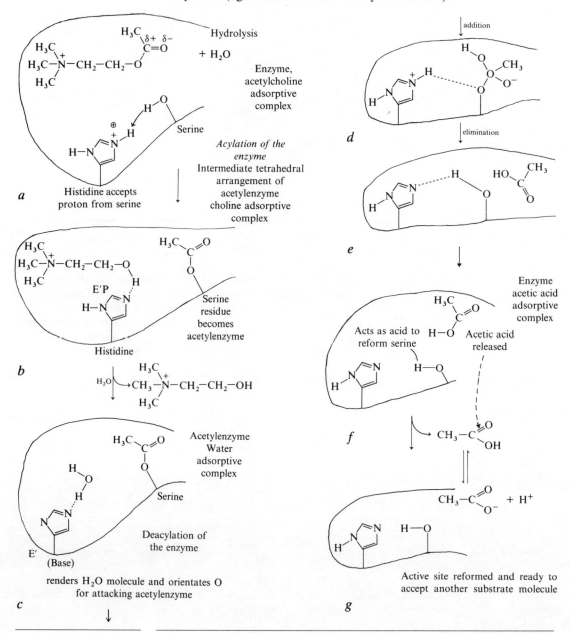

Appendix 14

Glycolysis—outline of stages

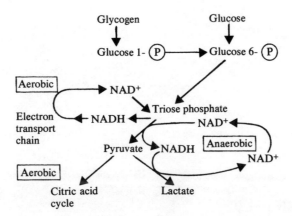

Appendix 15
Glycolysis

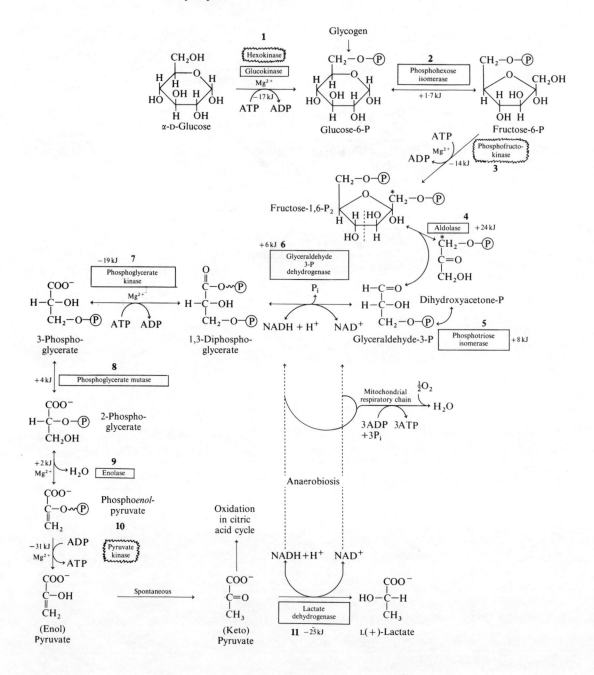

Appendix 16

Pentose phosphate pathway—outline of stages

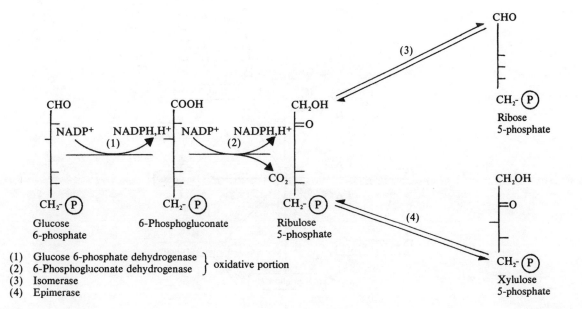

(1) Glucose 6-phosphate dehydrogenase }
(2) 6-Phosphogluconate dehydrogenase } oxidative portion
(3) Isomerase
(4) Epimerase

Appendix 17

Citric acid cycle—ATP formation

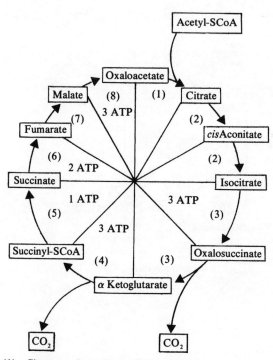

(1) Citrate synthase: irreversible reaction.
(2) Aconitase   ⎫
(3) Isocitrate dehydrogenase   ⎬ reversible reactions
(4) α-Ketoglutarate dehydrogenase: irreversible reaction
(5) Succinyl-ScoA thiokinase
(6) Succinate dehydrogenase   ⎬ reversible reactions
(7) Fumarase
(8) Malic dehydrogenase

Appendix 18
Citric acid cycle—structures of intermediates

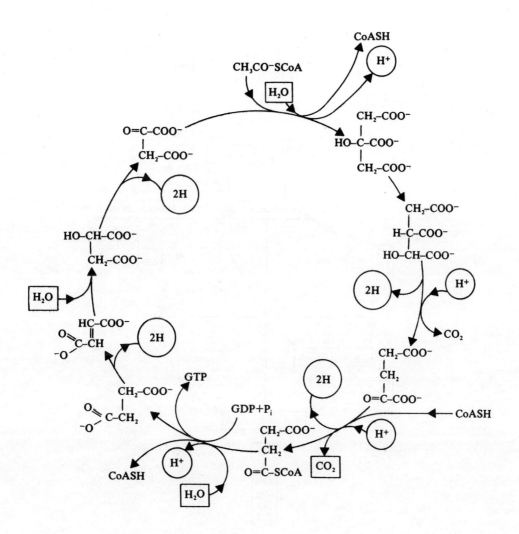

Appendix 19

Citric acid cycle—links with amino acid metabolism

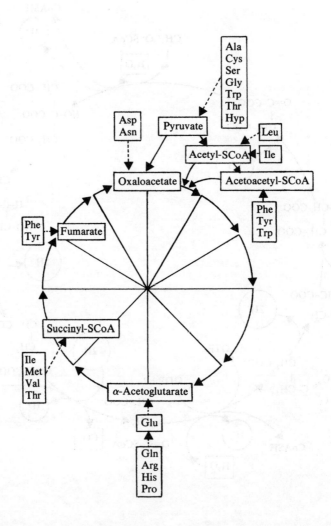

Appendix 20

β-Oxidation of fatty acids

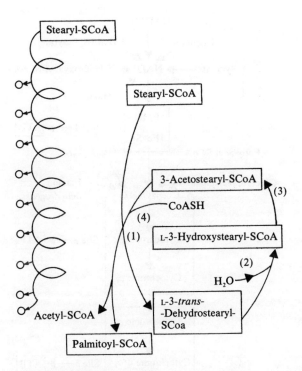

Stearyl-SCoA

Stearyl-SCoA

3-Acetostearyl-SCoA

(3)

CoASH

(4)

(1)

L-3-Hydroxystearyl-SCoA

(2)

H₂O

L-3-*trans*-
-Dehydrostearyl-
SCoa

Acetyl-SCoA

Palmitoyl-SCoA

O = Acetyl-SCoA

(1) Acyl-SCoA dehydrogenase (FAD → FADH₂)
     three different forms
(2) Crotonase
(3) L-3-Hydroxyacyl-SCoA dehydrogenase
     NAD⁺ → NADH,H⁺
(4) β-Ketothiolase

Appendix 21
The electron transport chain

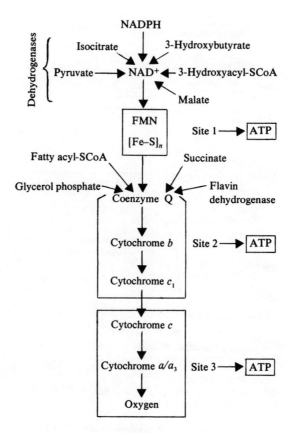

FMN – Flavin mononucleotide

Fe–S – Iron sulphur proteins

Appendix 22

Steroid synthesis: biosynthesis of cholesterol

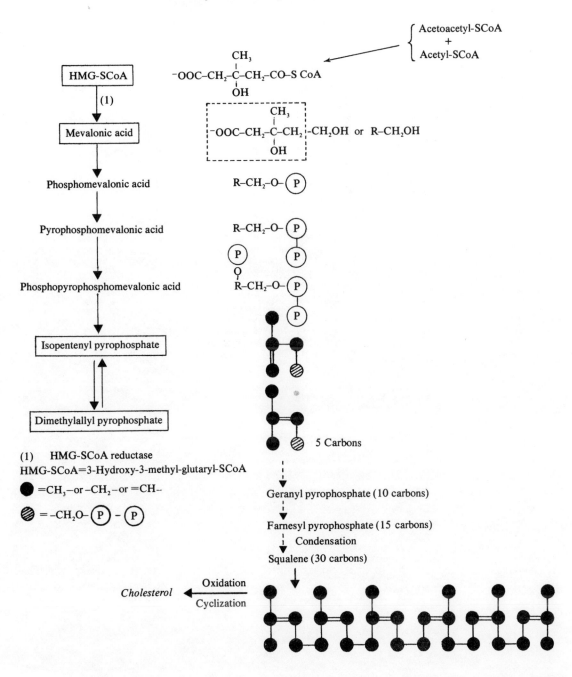

Acetoacetyl-SCoA
+
Acetyl-SCoA

HMG-SCoA

$$^-OOC-CH_2-\underset{\underset{OH}{|}}{\overset{\overset{CH_3}{|}}{C}}-CH_2-CO-S\,CoA$$

(1)

Mevalonic acid

$$^-OOC-CH_2-\underset{\underset{OH}{|}}{\overset{\overset{CH_3}{|}}{C}}-CH_2-CH_2OH \text{ or } R-CH_2OH$$

Phosphomevalonic acid

$$R-CH_2-O-\text{(P)}$$

Pyrophosphomevalonic acid

$$R-CH_2-O-\text{(P)}$$

Phosphopyrophosphomevalonic acid

$$R-CH_2-O-\text{(P)}$$

Isopentenyl pyrophosphate

Dimethylallyl pyrophosphate

5 Carbons

(1)   HMG-SCoA reductase

HMG-SCoA=3-Hydroxy-3-methyl-glutaryl-SCoA

● $=CH_3-$ or $-CH_2-$ or $=CH-$

◉ $=-CH_2O-$ (P) $-$ (P)

Geranyl pyrophosphate (10 carbons)

Farnesyl pyrophosphate (15 carbons)
Condensation
Squalene (30 carbons)

Cholesterol ← Oxidation / Cyclization

Appendix 23

Purine metabolism

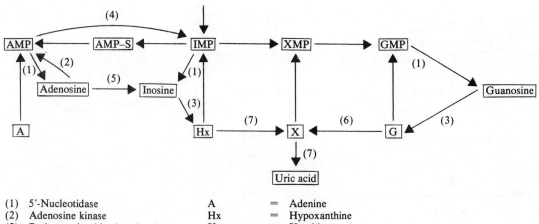

(1) 5'-Nucleotidase   A   =   Adenine
(2) Adenosine kinase   Hx   =   Hypoxanthine
(3) Purine nucleoside phosphorylase   X   =   Xanthine
(4) AMP deaminase   G   =   Guanine
(5) Adenosine deaminase   AMP–S   =   Succinylo-AMP
(6) Guanine deaminase
(7) Xanthine oxidase

Appendix 24

Summary of amino acid metabolism

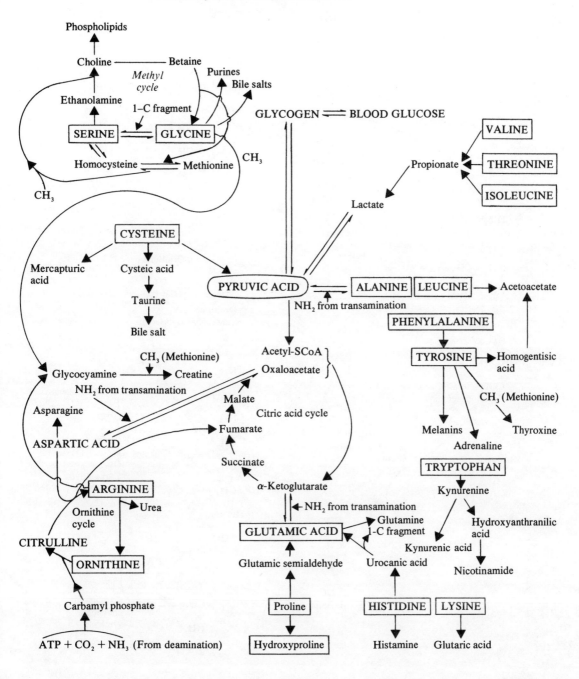

Appendix 25
Ornithine cycle—synthesis of urea

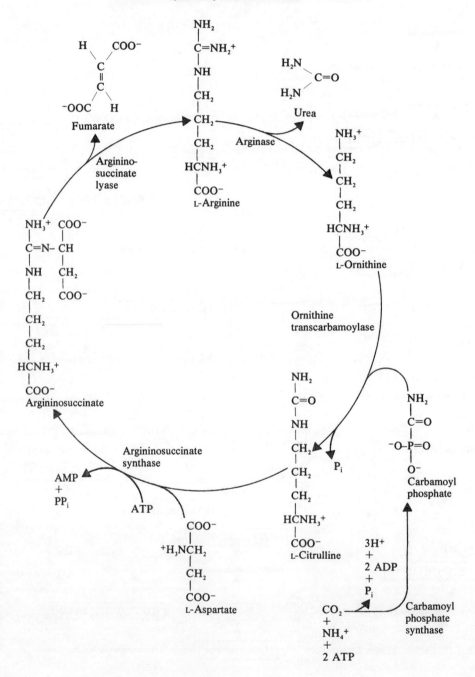

Appendix 26
Synthesis of DNA

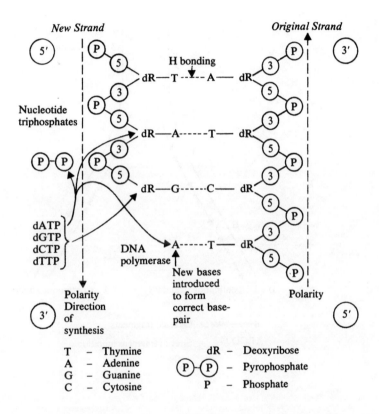

T – Thymine   dR – Deoxyribose
A – Adenine   (P)–(P) – Pyrophosphate
G – Guanine   P – Phosphate
C – Cytosine

Appendix 27
Replication of DNA: formation of Okazaki fragments

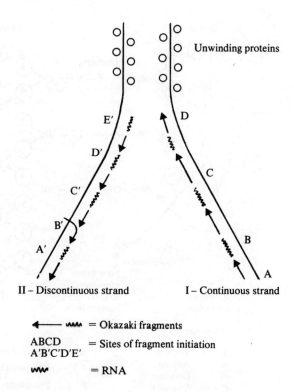

Unwinding proteins

II – Discontinuous strand   I – Continuous strand

← ⌇⌇⌇ = Okazaki fragments

ABCD
A'B'C'D'E'   = Sites of fragment initiation

⌇⌇⌇   = RNA

Appendix 28
Metabolic compartmentation of protein synthesis

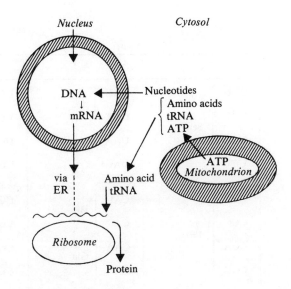

Compartments

1. Cytosol: synthesis of nucleotides and active amino acids
2. Nucleus: synthesis of DNA and mRNA
3. Ribosome: assembly of protein
4. Endoplasmic reticulum (ER)—transport of mRNA?
5. Mitochondrion: ATP

Appendix 29
The genetic code

	Second letter				
First letter	U	C	A	G	Third letter
U	UUU ⎫ Phe UUC ⎭ UUA ⎫ Leu UUG ⎭	UCU ⎫ UCG ⎪ Ser UCA ⎪ UCG ⎭	UAU ⎫ Tyr UAC ⎭ UAA* UAG*	UGU ⎫ Cys UGC ⎭ UGA* UGG Trp	U C A G
C	CUU ⎫ CUC ⎪ Leu CUA ⎪ CUG ⎭	CCU ⎫ CCC ⎪ Pro CCA ⎪ CCG ⎭	CAU ⎫ His CAC ⎭ CAA ⎫ Gln CAG ⎭	CGU ⎫ CGC ⎪ Arg CGA ⎪ CGG ⎭	U C A G
A	AUU ⎫ AUC ⎪ Ile AUA ⎭ AUG Met	ACU ⎫ ACC ⎪ Thr ACA ⎪ ACG ⎭	AAU ⎫ Asn AAC ⎭ AAA ⎫ Lys AAG ⎭	AGU ⎫ Ser AGC ⎭ AGA ⎫ Arg AGG ⎭	U C A G
G	GUU ⎫ GUC ⎪ Val GUA ⎪ GUG ⎭	GCU ⎫ GCC ⎪ Ala GCA ⎪ GCG ⎭	GAU ⎫ Asp GAC ⎭ GAA ⎫ Glu GAG ⎭	GGU ⎫ GGC ⎪ Gly GGA ⎪ GGG ⎭	U C A G

*Termination code

Appendix 30

The genetic code–relationships of amino acid structure to coding symbol

Second (U)	Third U	C	A	G
U	Phe	Phe	Leu	Leu
C	Leu	Leu	Leu	Leu
A	Ile	Ile	Ile	Met
G	Val	Val	Val	Val

Second (C)	Third U	C	A	G
U	Ser	Ser	Ser	Ser
C	Pro	Pro	Pro	Pro
A	Thr	Thr	Thr	Thr
G	Ala	Ala	Ala	Ala

Second (A)	Third U	C	A	G
U	Tyr	Tyr	X	X
C	His	His	Gln	Gln
A	Asn	Asn	Lys	Lys
G	Asp	Asp	Glu	Glu

Second (G)	Third U	C	A	G
U	Cys	Cys	X	Trp
C	Arg	Arg	Arg	Arg
A	Ser	Ser	Arg	Arg
G	Gly	Gly	Gly	Gly

First

○ Hydrophobic   △ Basic   ● Acidic

Appendix 31

Transfer RNAs

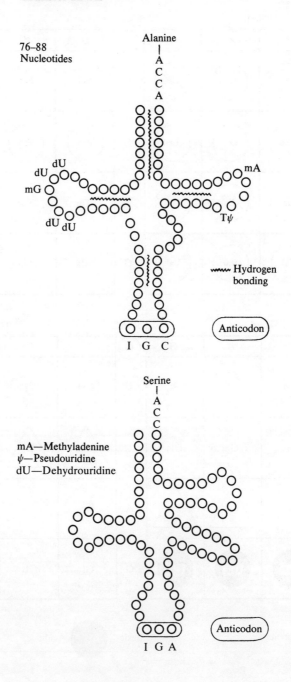

76–88
Nucleotides

Alanine

mA—Methyladenine
ψ—Pseudouridine
dU—Dehydrouridine

Appendix 32

Protein synthesis

Ribosome cycle

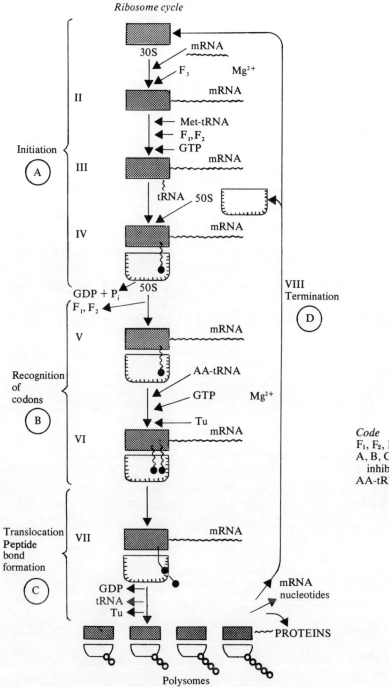

Polysomes

Code
F_1, F_2, F_3, Tu protein factors
A, B, C, D Possible sites of
   inhibition by drugs
AA-tRNA Aminoacyl-tRNA

Further Reading

Part 1 Biochemistry of the cell and its metabolites

Chapter 1 **Ultrastructure of the mammalian cell**

G. D. Byrne (Ed.) (1972) *Subcellular Components—Preparation and Fractionation*. London: Butterworths
K. E. Carr and P. G. Toner (1982) *Cell Structure*, 3rd edn. Edinburgh: Churchill Livingstone
E. Holtzman and A. B. Novikoff (1984) *Cells and Organelles*, 3rd edn. London: Saunders
A. G. E. Pearse (1980) *Histochemistry*, 4th edn. Edinburgh: Churchill Livingstone

Chapter 2 **Role of extracellular and intracellular membranes—membrane structure and membrane transport**

J. A. Lucy (1974) in *The Cell in Medical Science*. Eds F. Beck and J. B. Lloyd. Chap. 2. London: Academic Press
G. Weissman and R. Claiborne (1975) *Cell Membranes*. New York: HP Publishing

Chapter 3 **Role of subcellular organelles: lysosomes**

J. T. Dingle (Ed.) (1969–1976) *Lysosomes in Biology and Pathology*. Vols. 1–5. Amsterdam: North-Holland

Chapter 4 **Role of subcellular organelles: peroxisomes**

N. E. Tolbert (1981) *Annu. Rev. Biochem.* **50**, 133

Chapter 5 **Role of subcellular organelles: metabolism in the cytosol**

A. L. Lehninger (1982) *Principles of Biochemistry*. Chaps. 15, 20. New York: Worth
L. Stryer (1981) *Biochemistry*, 2nd edn. Chaps. 12, 15, 16, 18. San Francisco: W. H. Freeman

Chapter 6   **Role of subcellular organelles: mitochondria and energy conservation**

H. Baum (1974) in *The Cell in Medical Science*. Eds F. Beck and J. B. Lloyd.
    Chap. 6. London: Academic Press
A. L. Lehninger (1964) *The Mitochondrion*. New York: W. A. Benjamin
A. L. Lehninger (1982) *Principles of Biochemistry*. Chap. 17. New York: Worth

Chapter 7   **Role of subcellular organelles: metabolic inter-relationship of mitochondria and the cytosol**

K. F. LaNoue and A. C. Schoolwerth (1979) *Annu. Rev. Biochem.* **48**, 871–922

Chapter 8   **Role of subcellular organelles: the nucleus**

T. Igo-Kemeres, W. Horz and H. G. Zachau (1982) *Annu. Rev. Biochem.* **51**, 89
A. L. Lehninger (1982) *Principles of Biochemistry*. Chaps. 27, 28. New York:
    Worth
A. P. Mathias (1974) in *The Cell in Medical Science*. Eds F. Beck and
    J. B. Lloyd. Chap. 4. London: Academic Press
M. Szekely (1980) *From DNA to Protein*. Basingstoke: Macmillan
J. D. Watson (1977) *Molecular Biology of the Gene*, 3rd edn. New York:
    W. A. Benjamin

Chapter 9   **Role of subcellular organelles: the endoplasmic reticulum**

A. L. Lehninger (1982) *Principles of Biochemistry*. Chap. 29. New York: Worth
H. G. Wittman (1983) *Annu. Rev. Biochem.* **52**, 35

Part 2   Whole body metabolism

Chapter 10   **Nutrition: general aspects**

S. Davidson and R. Passmore, J. F. Brock and A. S. Truswell (1979) *Human
    Nutrition and Dietetics*, 7th edn. Edinburgh: Churchill Livingstone
Methodology
W. N. Pearson (1964) in *Nutrition*. Eds G. H. Beaton and E. W. McHenry. Vol.
    3. Chap. 7. London: Academic Press
Toxicants
B. J. Wilson (1979) *Nutr. Rev.* **37**, 305
Symposium on Nutrition and Toxicology (1981) *Proc. Nutr. Soc.* **40**, 47
Malnutrition
N. S. Scrimshaw and M. Béhar (1964) in *Nutrition*. Eds G. H. Beaton and
    E. W. McHenry. Vol. 2. Chap. 7. London: Academic Press

Chapter 11   **Nutrition: energy requirements and the supply of energy by oxidation of foodstuffs**

R. W. Swift and K. H. Fisher (1964) in *Nutrition*. Eds G. H. Beaton and E. W. McHenry. Vol. 1. Chap. 4. London: Academic Press

Chapter 12   **Nutrition: proteins in the diet**

J. B. Longenecker (1965) in *Newer Methods of Nutritional Biochemistry*. London: Academic Press
R. E. Olson (Ed.) (1975) *Protein–Calorie Malnutrition*. New York: Academic Press
Symposium (1979) *Proc. Nutr. Soc.* **38**, 1
J. C. Waterlow, P. J. Gaslick and D. J. Millward (1977) *Protein Turnover in Mammalian Tissues and in the Whole Body*. Amsterdam: Elsevier/North-Holland

Chapter 13   **Nutrition: dietary fats**

G. J. Brisson (1981) *Lipids in Human Nutrition*. Minneapolis: Burgess
C. J. Glueck (1979) *Am. J. Clin. Nutr.* **32**, 2637, 2703
F. A. Kummerow (1979) *Am. J. Clin. Nutr.* **32**, 58
P. Paul, C. S. Ramesha and J. Gangaly (1980) *Adv. Lipid Res.* **17**, 155
J. J. Rahman and R. T. Holman (1971) in *The Vitamins: Chemistry, Physiology, Pathology, Methods*. Eds W. H. Sebrell and R. S. Harris. Vol. III. New York: Academic Press
J. M. Scott (1976) *Biochem. Soc. Trans.* **4**, 1
H. M. Sinclair (1964) in *Lipid Pharmacology*. Ed. R. Paoletti. Chap. 5. London: Academic Press
D. A. Van Dorp (1975) *Proc. Nutr. Soc.* **34**, 279
A. L. Willis (1983) *Nutr. Rev.* **39**, 289

Chapter 14   **Nutrition: vitamins**

B. M. Barker and D. A. Bender (1980) *Vitamins in Medicine*. Vol. I. London: Heinemann
F. Bicknell and F. Prescott (1953) *The Vitamins in Medicine*. 3rd edn. London: Heinemann
Special articles in *Vitamins and Hormones*, 1943–1984. Vols. 1–40. London: Academic Press

Chapter 15   **Nutrition: inorganic constituents of the diet**

A. S. Rasad (1978) *Trace Elements and Iron in Human Metabolism*. London: Wiley
E. J. Underwood (1977) *Trace Elements in Human and Animal Nutrition*. London: Academic Press

Chapter 16   **Digestion and absorption of foodstuffs**

I. M. McColl and G. E. G. Sladen (Eds) (1975) *Intestinal Absorption in Man*. New York: Academic Press

H. I. Friedman (1980) *Intestinal Fat Digestion and Absorption. Am. J. Clin. Nutr.* **33**, 1108–1139
Fibre
D. P. Burkitt and A. I. Mendeloff (1976) *Dig. Dis. Sci.* **21**, 103
Symposium on Role of Fibre in Health (1978) *Am. J. Clin. Nutr.* **31**, 21
G. V. Vahouny and D. Kritchevsky (Eds) (1982) *Dietary Fibre in Health and Disease.* New York: Plenum

Chapter 17 Hormones

N. P. Christy (Ed.) (1971) *The Human Adrenal Cortex.* New York: Harper & Row
C. H. Gray and V. H. T. James (1983) *Hormones in Blood*, 3rd edn. Vols. 1–5. London: Academic Press
H. L. J. Makin (1975) *Biochemistry of Steroid Hormones.* Oxford: Blackwell
H. V. Rickenberg (1974) *Biochemistry of Hormones.* London: Butterworths

Chapter 18 Plasma glucose and its regulation

P. K. Bondy and L. E. Rosenberg (1980) *Metabolic Control and Disease.* Chap. 6. London: W. B. Saunders

Chapter 19 Plasma lipids and their regulation

M. S. Brown and J. L. Goldstein (1983) *Annu. Rev. Biochem.* **52**, 223
L. C. Smith, H. J. Pownall and A. M. Golto (1978) *Annu. Rev. Biochem.* **47**, 751

Chapter 20 Plasma amino acids and utilization of amino acids by the tissues

T. T. Aoki, R. J. Finley and G. F. Cahill (1978) *Biochem. Soc. Symp.* No. 43, 17
P. Felig (1975) *Annu. Rev. Biochem.* **44**, 933
H. A. Krebs (1972) *Adv. Enzyme Regul.* **10**, 397

Chapter 21 Plasma electrolytes

A. C. Guyton (1981) *Textbook of Medical Physiology.* Chap. 33. London: W. B. Saunders

Chapter 22 Plasma calcium and phosphates: regulation by vitamin D and parathyroid hormone

F. Bronner (ed.) (1976) *Am. J. Clin. Nutr.* **29**, 1257
H. F. deLucan and H. K. Schnoes (1983) *Annu. Rev. Biochem.* **52**, 411
J. T. Irving (1973) *Calcium and Phosphorus Metabolism.* New York: Academic Press
B. E. C. Nordin (1980) *Clin. Endocrinol. Metab.* **9**, 177
L. G. Raisz (1980) *Clin. Endocrinol. Metab.* **9**, 27

Chapter 23   **Starvation**

G. F. Cahill (1970) *N. Engl. J. Med.* **282**, 668–675
G. F. Cahill (1976) *Clin. Endocrinol. Metab.* **5**, 397
C. D. Sandek and P. Felig (1976) *Am. J. Med.* **60**, 117

Part 3   Specialized metabolism of tissues

Chapter 24   **Blood: erythropoiesis—role of folate and vitamin B$_{12}$**

H. A. Barker (1972) *Annu. Rev. Biochem.* **41**, 55
V. Herbert (1970) *Am. J. Med.* **48**, 539
V. Herbert and K. C. Das (1976) *Vitamins and Hormones,* **34**, 2
A. V. Hoffbrand (Ed.) (1976) *Clin. Haematol.* **5**, 3
W. H. Sebrell and R. S. Harris (1968) *The Vitamins.* Vol. II. Chap. 4. London:
   Academic Press

Chapter 25   **Blood: metabolism in the red blood cell**

N. J. Russel (1982) *Blood Chemistry.* London: Croom Helm

Chapter 26   **Blood: blood clotting**

British Medical Bulletin (1976) **33**, No. 3
C. M. Jackson and Y. Nemeson (1980) *Annu. Rev. Biochem.* **49**, 705

Chapter 27   **Blood: haemoglobin catabolism**

P. K. Bondy and L. E. Rosenberg (Eds) (1980) *Diseases of Metabolism*, 8th edn.
   Chap. 14, pp. 1009–1088. Philadelphia: Saunders
C. H. Gray (1961) Bile Pigments in Health and Disease. Springfield, Ill.:
   Thomas

Chapter 28   **Blood: iron metabolism**

A. Jacobs and M. Wormwood (Eds) (1974) *Iron in Biochemistry and Medicine.*
   London: Academic Press
P. Saltman and J. Hegenauer (Eds) (1982) *Biochemistry and Physiology of Iron.*
   New York: Academic Press

Chapter 29   **Functions of the liver**

C. H. Rouiller (1963) *The Liver.* Vols 1 and 2. London: Academic Press

Chapter 30   **The kidney**

R. F. Pitts (1974) *Physiology of the Kidney and Body Fluids*, 3rd edn. Chicago: Year Book Medical Publishing

Chapter 31   **Muscle**

J. M. Murray and A. Weber (1974) *Sci. Am.* **230**, 58–71
D. R. Wilkie (1976) *Muscle*, 2nd edn. London: Arnold

Chapter 32   **Bone and collagen calcification**

Vitamin C
M. J. Barnes and E. Kodiak (1972) *Vit. Horm.* **30**, 1
D. J. Prockop *et al.* (1979) *N. Engl. J. Med.* **301**, 13, 77
Collagen
D. R. Eyre, M. A. Paz and P. M. Gallop (1983) *Annu. Rev. Biochem.* **53**, 717

Chapter 33   **Brain and the central nervous system**

G. J. Siegel, R. W. Albers, B. W. Agranoff and R. Katzman (1981) *Basic Neurochemistry*, 3rd edn. Boston: Little, Brown

Part 4   Environmental hazards—detoxication

Chapter 34   **Toxicology: general aspects**

J. Doull, C. D. Klaasen and M. O. Amdur (Eds) (1980) *Toxicology*, 2nd edn. Basingstoke: Macmillan
D. E. Hathway (1984) *Molecular Aspects of Toxicology*. London: Royal Society of Chemistry
E. Hodgson and F. E. Guthrie (Eds) (1980) *Introduction to Biochemical Toxicology*. Oxford: Blackwell

Chapter 35   **Toxic metals**

F. W. Oehme (1979) *Toxicity of Heavy Metals in the Environment*. Vols. 1 and 2. New York: Dekker

Chapter 36   **Metabolism of xenobiotics—xenobiochemistry**

B. N. LaDu, H. G. Mandell and E. L. Way (1971) *Fundamentals of Drug Metabolism and Drug Disposition*. Baltimore: Williams & Wilkins
R. T. Williams (1959) *Detoxication Mechanisms*. London: Chapman and Hall

Chapter 37   **Alcohol: effects on metabolism**

L. L. Madison (1968) *Adv. Metab. Disord.* **3**, 85–109

Chapter 38   **Chemical carcinogenesis**

E. C. Miller (1978) *Cancer Res.* **36**, 1479–1496

Part 5   Biochemical basis of diagnosis—disease and its treatment

Chapter 39   **Biochemical diagnosis**

N. W. Tietz (Ed.) (1976) *Fundamentals of Clinical Chemistry*, 2nd edn.
    Chap. 12. Philadelphia: Saunders

Chapter 40   **An example of metabolic disturbance: obesity**

M. J. Albrink (Ed.) *Clin. Endocrinol. Metab.* **5**, No. 2. *Obesity*. Philadelphia:
    Saunders
J. S. Garrow (1978) *Energy Balance and Obesity in Man*, 2nd edn. Amsterdam:
    Elsevier/North-Holland
Symposium on Obesity in Man (1973) *Proc. Nutr. Soc.* **32**, 169

Chapter 41   **Biochemical genetics: inborn errors of metabolism**

J. B. Stanbury, J. B. Wyngaarden, D. S. Fredrickson, J. L. Goldstein and
    M. S. Brown (1983) *The Metabolic Basis of Inherited Disease*, 5th edn. New
    York: McGraw-Hill
D. Wellner and A. Meister (1981) *Annu. Rev. Biochem.* **50**, 911

Chapter 42   **Immunology**

O. Bier, W. D. daSilva, D. Golze and I. Mota (1981) *Fundamentals of
    Immunology*. Berlin: Springer Verlag
I. Roitt (1980) *Essential Immunology*, 4th edn. Oxford: Blackwell
Complement
K. B. M. Reid and R. R. Porter (1981) *Annu. Rev. Biochem.* **50**, 433

Chapter 43   **Principles of chemotherapy**

A. A. Albert (1979) *Selective Toxicity*, 6th edn. London: Methuen
E. F. Gale, E. Cundliffe, P. E. Reynolds, M. H. Richmond and M. J. Waring
    (1981) *The Molecular Basis of Antibiotic Action*. Chichester: Wiley
R. J. Schnitzer and F. Hawking (1963) *Experimental Chemotherapy*. Vols. I
    and II. London: Academic Press

Index

N

O

Q

R

S

T